Carol Schneppen

Thermodynamics and Its Applications

Thermodynamics and Its Applications

second edition

MICHAEL MODELL
Modar Inc.
Natick, Massachusetts

ROBERT C. REID
The Chevron Professor
Massachusetts Institute of Technology

PRENTICE-HALL, INC.
Englewood Cliffs, N. J. 07632

Library of Congress Cataloging in Publication Data

Modell, Michael.
Thermodynamics and its applications.

(Prentice-Hall international series in the physical and chemical engineering sciences)
Bibliography: p.
Includes index.
1. Thermodynamics. I. Reid, Robert C. II. Title. III. Series.
QD504.M63 1983 541.3'69 82-15044
ISBN 0-13-915017-x

Editorial/production supervision by Lori Opre
Manufacturing buyer: Anthony Caruso

Printed in the United States of America

10 9 8 7 6 5 4 3 2 1

ISBN 0-13-915017-X

Prentice-Hall International, Inc., *London*
Prentice-Hall of Australia Pty. Limited, *Sydney*
Editora Prentice-Hall do Brasil, Ltda., *Rio de Janeiro*
Prentice-Hall Canada Inc., *Toronto*
Prentice-Hall of India Private Limited, *New Delhi*
Prentice-Hall of Japan, Inc., *Tokyo*
Prentice-Hall of Southeast Asia Pte. Ltd., *Singapore*
Whitehall Books Limited, *Wellington, New Zealand*

Contents

Preface

As long as we can remember, our department has offered a one-semester, graduate-level subject in classical thermodynamics. Traditionally, it has been applications oriented; one of its primary objectives has been to develop competence and self-confidence in handling challenging applications in new and sometimes unusual situations. One-half to two-thirds of the contact hours are usually devoted to problem solving. Over the years, there accumulated many interesting, challenging problems, most of which originated from our consulting work.

We have used a number of texts in conjunction with our graduate subject. None were completely satisfactory. We are convinced that a firm foundation in theory is essential for students who will be asked to fulfill the needs of tomorrow with an increasing demand for talents that are flexible and adaptable. On the other hand, the theory is useless unless the student can effectively bridge the gap between theory and application. Thus, we have attempted to develop a text with a rigorous theoretical and conceptual basis, interspersed with a relatively large number of examples and solutions. We have stressed to our students the desirability of working these examples before reviewing the solutions.

This text is intended to be a *learning text* rather than a teaching text. We have attempted to be thorough; but as a consequence of limited space and the short time a student spends in formal education, it is unreasonable to expect the student to appreciate all of the subtleties that will be apparent to the experienced reader. It is our hope that students will attain a basic level of understanding of theory and rationale of applications in their formal use of this text such that deeper insights can be gained in a self-instructional mode throughout their professional careers, as the need arises.

Following this philosophy, the text contains more material than one could hope to cover in one term—nor do we recommend a two-term sequence at the expense of student flexibility to shape the graduate curriculum to meet individual needs. In three contact hours per week a term, we have covered at a fairly rapid pace all the chapters except 12 and 13.

The theoretical basis of classical thermodynamics is developed in the first six chapters, which can be covered in one-third to one-half of a term. The developments up to the introduction of the Fundamental Equation parallels the historical evolution of the classical body of knowledge (see Chapter 1).

The introduction of the formalism of the Fundamental Equation and Legendre transforms is a departure from traditional chemical engineering texts. The Fundamental Equation is introduced because we believe it is of significant conceptual value in treating one of the central problems in engineering applications: What are the minimum data required to reach a given objective, and how does one manipulate available data to forms that are more appropriate to the problem at hand?

The Fundamental Equation in the energy representation—that is, $\underline{U} = f(\underline{S}, \underline{V}, N_1, \ldots, N_n)$—contains all thermodynamic information for a given single-phase, simple system. All other thermodynamic properties can be derived from it. Although we do not have available the Fundamental Equation for many materials, we can determine what other data sets have equivalent information content. Using Legendre transformations to preserve the information content, it is shown that, for example, $\underline{H} = f(\underline{S}, P, N_1, \ldots, N_n)$ is also a Fundamental Equation, and thus a Mollier diagram contains all thermodynamic information. Similarly, the Fundamental Equation of a pure material can be reconstructed from the equation of state and the heat capacity. Thus, any problem can be solved using P–V–T and C_p data; if these data are available, we need not search for any other data.

The last half of the text covers areas of increasing complexity. Following a discussion of single-phase systems of pure materials (Chapter 7) and mixtures (Chapter 8), the criteria of stability and critical-point phenomena are developed in some depth (Chapter 9).

Phase equilibrium and chemical equilibrium are treated as progressively more complex applications of the building blocks covered previously. In these areas especially, it is stressed that thermodynamics is of little practical utility without sound engineering judgment. A phase diagram can only be constructed when there is prior knowledge of what phases do in fact exist and what properties (e.g., information equivalent to the Fundamental Equation) each phase exhibits. Similarly, the concept of chemical equilibria is of little use in the complex systems that engineers generally face until there are data or insight into the kinetically feasible routes.

The last two chapters of the book deal with the thermodynamics of systems in electric, magnetic, stress, and other potential fields (Chapter 12) and the thermodynamics of surfaces and nucleation (Chapter 13). The approach used is parallel to that developed earlier; that is, the applicable Fundamental

Equation is found and Legendre transforms employed to relate the variables of interest in any real application.

It is impossible to acknowledge everyone who made this book a reality. We have been influenced by authors of previous articles and texts in thermodynamics and by our teachers. Professors J. M. Smith and H. P. Meissner, in particular, excited our interest in this field and illustrated its power to attack and solve real and significant problems.

The revision necessary to prepare this second edition was done by one of us (R.C.R.) while working at the University of Wisconsin, Madison, as the 1980–81 Olaf A. Hougen Professor. Discussions with the Wisconsin faculty were invaluable, and the ChE 710 graduate students were extremely helpful in providing constructive criticism on rough-draft revisions. In particular, Chao-Chang Pai and Randall De Ruiter should be acknowledged, the former for making very valuable suggestions for efficient ways to manipulate second-order derivatives of Legendre transforms and the latter for his aid in working and modifying many of the examples and problems found in the text.

At M.I.T., Wayne Erickson, Pablo Debenedetti, Horacio Valerias, and Ken Allen were extremely helpful in critiquing the text and working all the chapter problems.

Pat O'Callaghan and Donna Gabl were faultless in their typing of the manuscript. The drawings and sketches found throughout the book are the product of Tetsuo Maejima's skill and his obvious mastery of the subject material.

MICHAEL MODELL
Natick, Mass.

ROBERT C. REID
Cambridge, Mass.

The patience and encouragement of my wife, Nancy, were invaluable to me in the preparation of this revision. In the necessary chore of proofreading, she surely is to be distinguished as the only individual who will ever read this book aloud in its entirety, including each and every punctuation mark and all mathematical symbols including those confusing subscripts, superscripts, carats, overbars, underbars, etc.!

ROBERT C. REID

Thermodynamics and Its Applications

Introduction 1

1.1 The Scope of Classical Thermodynamics

To the scientist, classical thermodynamics is one of a few mature fields epitomized by a rather well-defined, self-consistent body of knowledge. The essence of the theoretical structure of classical thermodynamics is a set of natural laws governing the behavior of macroscopic systems. The laws are derived from generalizations of observations and are largely independent of any hypothesis concerning the microscopic nature of matter. From these laws, a large number of corollaries and axioms are derivable by proofs based entirely on logic.

The scientist is sometimes at a loss to understand why the engineer has so much difficulty applying thermodynamics; after all, the theoretical development is rather straightforward. From the engineer's point of view, understanding the theory as developed by the chemist or physicist is not particularly difficult; however, the neat, self-contained presentation of the subject by the scientist is not necessarily amenable to practical application. Real-world processes are usually far from reversible, adiabatic, or well mixed; very rarely are they isothermal or at equilibrium; few mixtures of industrial importance are ideal. Thus, the engineer must take a pragmatic approach to the application of thermodynamics to real systems. One major concern is to redefine the real problem in terms of idealizations to which thermodynamics can be applied.

In the engineering context, almost all problems of thermodynamic importance can be classified into one of three types:

1. For a given process with prescribed (or idealized) internal constraints and boundary conditions, how do the properties of the system vary?

2. To effect given changes in system properties, what external interactions must be imposed? (This is the inverse of type 1.)
3. Of the many alternative processes to effect a given change in a system, what are the efficiencies of each with respect to the resources at our disposal?

Problems of the first two classes require application of the First Law, which is developed in Chapter 3:

$$\Delta E = Q - W \tag{1-1}$$

where E is energy and Q and W are the heat and work interactions, respectively. The First Law may also be viewed as:

$$\text{internal changes} = \sum \text{interactions occurring at boundaries}$$

The change in energy can be related to variations of other internal properties of interest (e.g., T, P, V, etc.).

The third class of problems requires application of the Second Law, for which an idealization—the reversible process—is introduced as a standard for comparison.

There are basically only three steps required to develop a solution to any thermodynamic problem:

1. *Definition of the problem.* The real-world situation must be modelled by specifying the internal constraints and boundary conditions. For example, is a boundary permeable, semipermeable, or impermeable? Is heat transfer fast or slow relative to the time span of interest? Which chemical reactions are known to occur under the conditions of interest?
2. *Application of thermodynamic laws.* As described above, these either relate effects internal to the system with external interactions (the First Law) or they set limits on the extent of internal variations (the Second Law). The combined laws prescribe in part the relationships between property variations, but they do not uniquely specify the magnitude of the change in properties. For example, for a simple system undergoing a process in which the temperature and pressure are observed to change from T_1, P_1 to T_2, P_2, we might wish to calculate the energy change in order to specify the necessary heat and work interactions. We might employ the following analysis:
   (a) From *thermodynamic reasoning*, ΔE is a unique function of T_1, P_1 and T_2, P_2 because E is a state function. Therefore, ΔE can be evaluated over any path between these end states.
   (b) From *mathematical reasoning*, over any path for which E is defined, dE may be expressed as an exact differential, such as

$$\Delta E = \int dE = \int_{T_1}^{T_2} \left(\frac{\partial E}{\partial T}\right)_{P_1} dT + \int_{P_1}^{P_2} \left(\frac{\partial E}{\partial P}\right)_{T_2} dP \tag{1-2}$$

   Note that $(\partial E/\partial T)_P$ and $(\partial E/\partial P)_T$ must be expressed as functions of T and P before Eq. (1-2) can be integrated.

(c) Applying *thermodynamic reasoning*, E is defined as a function of T and P over a reversible path, and thus $(\partial E/\partial T)_P$ and $(\partial E/\partial P)_T$ can be reduced to other variable sets that are more readily quantified:

$$\Delta E = \int_{T_1}^{T_2} \left[C_p - P\left(\frac{\partial V}{\partial T}\right)_P \right] dT - \int_{P_1}^{P_2} \left[T\left(\frac{\partial V}{\partial T}\right)_P + P\left(\frac{\partial V}{\partial P}\right)_T \right] dP \tag{1-3}$$

where C_p is the constant-pressure heat capacity. Note that Eq. (1-3) is a *general result;* it must be satisfied by any material undergoing a change from T_1, P_1 to T_2, P_2. However, the value of ΔE is not unique; it differs from one material to the next, which leads us to the third and final step.

3. *Evaluation of property data.* There are property relationships that are unique characteristics of matter. For example, in Eq. (1-3), thermodynamics does not dictate the functions

$$C_p = f_1(T, P), \qquad V = f_2(T, P) \tag{1-4}$$

required for the integration. Evaluation of these property data lies outside the scope of classical thermodynamics. However, they are essential to the solution of real problems and hence are within the scope of this text. The engineer must make recourse to a variety of methods (e.g., literature, experiments, correlations, or microscopic theories as developed with statistical mechanics) in order to determine or approximate these property relationships.

Before discussing the approach to classical thermodynamics (Section 1.3) used herein, it is instructive to review the historical evolution of this body of knowledge.

1.2 Preclassical Thermodynamics

The origin of classical thermodynamics can be traced back to the early 1600s. The laws, as we know them today, were not formalized until the late 1800s. The interim 250 to 300 years are called the *preclassical period,* during which many of our current concepts were developed.

The chronological development is a fascinating example of the application of scientific methodology. Experimentation (e.g., thermometry) led to the development of hypotheses and concepts (e.g., the adiabatic wall) which, in turn, suggested other experiments (e.g., calorimetry) followed by new concepts (the caloric theory), etc. The historical events also illustrate some of the potential pitfalls in scientific analysis, such as overemphasis on intuitive images (e.g., the nature of heat) which go far beyond the existing body of observations and facts. Consequently, the preclassical period was marked with pedagogical controversy and much confusion.

The beginning of the preclassical period is usually associated with Galileo's attempts to quantify thermometry (ca. 1600). It is interesting to note that seventeenth-century scientists were motivated primarily by a desire to understand phenomena perceived by their senses. In contrast, scientists today require very sensitive and elaborate instrumentation to detect phenomena that are far beyond the reaches of their senses. One of Galileo's principal objectives was to quantify the subjective experiences of hot and cold. The expansion of air upon heating was appreciated in the Hellenistic era, but it was never applied. Galileo used this phenomenon in his bulb-and-stem device with the stem submerged in water. The measurements would change with time, but in the early 1600s there was no reason to assume that they should not vary in such a manner. It was not until 1643, when a student of Galileo, Torricelli, developed the barometer, that it was appreciated that Galileo's device was more of a "barothermoscope" than a thermometer. As glass-blowing technology advanced, the availability of narrow capillaries led to the development of liquid thermometers in the 1630s. As might be expected, water was the first liquid used. Although difficulties that should be obvious to today's students were experienced, 10 years elapsed before the sealed alcohol thermometer gained acceptance. Gas thermometers did not reappear until the 1700s, when gas properties were better understood.

When any new experimental tool is developed, invariably there is the desire to quantify it so that results among different investigators can be compared. The quantification of thermometry required the introduction of at least two fiducial or fixed points. Stop for a moment and reflect on what fixed points you might have chosen had you been a scientist in 1640. The boiling point of water? It varies from day to day. The freezing point of water? It is difficult to find ice during most of the year. Furthermore, there was no reason to believe that materials like water had unique properties that would be reproducible. Thus, it is understandable that our ancestors turned to phenomenological references, such as the "warmest water the hand could stand" or the "most severe winter cold" or the "temperature of the human body." Later, selection of the melting point of butter or the freezing point of aniseed oil were steps toward objectivity, although these transition points were not very sharp. It was not until 1694 that the freezing and boiling points of water gained acceptance.

With the advance of quantitative thermometric experimentation it soon became apparent that different types of containers had different thermal properties. Hot liquids would cool less rapidly in mica or wood vessels than in metals. These observations led to the idealized concept of the adiabatic wall, which could be approached in practice; thus, the science of calorimetry was born.

If two portions of the same fluid were mixed in a calorimeter, the final temperature could be expressed as a weighted mean of the two initial temperatures:

$$t_f = \frac{a_1 t_1 + a_2 t_2}{a_1 + a_2} \tag{1-5}$$

The weighting factor could be mass or volume.

Although Eq. (1-5) is of the form of a conservation law, it is not at all clear *what* is conserved. A simplistic interpretation would have temperature as the conserved quantity. The conservation law was given more structure in the mid-1700s, when Eq. (1-5), with mass or volume as weighting factors, was shown to be invalid when different liquids were mixed. In the 1760s, Joseph Black suggested a modification that was consistent with mixing data for different fluids. The constants a_i in Eq. (1-5) were subdivided into a mass-related component and an intensity parameter, the specific heat, which was a unique property of the liquid. This was the first time that it was proposed that matter had distinctive properties in the thermodynamic sense.

Black's modification indicated that something other than temperature was conserved in the mixing process. This quantity was called *heat*, or *caloric*. The interpretation went far beyond the physical observations into the realm of metaphysics. The caloric theory attempted to define the microscopic nature of the conserved quantity.

Black's ingenious hypothesis led to a flurry of experimentation, during which specific heats were measured and reported in the then flourishing royal societies. The conservation law was repeatedly challenged, but by more exacting experimentation, the theory was enlarged to account for the variation of specific heat with temperature and, later, latent heats were introduced to account for phase transitions.

In the 1780s, no more than 20 years after Black's work, Count Rumford conducted his exhaustive experiments to show that mechanical work was an inexhaustible source of caloric. Hence, caloric could not be conserved and could not be of a material nature. Rumford suggested a revival of the mechanical concept of heat that had been abandoned 50 years earlier. Although we now know that Rumford's suggestion was closer to the truth, it is understandable why very few of his peers followed his lead. The statistical concepts necessary to relate micromechanical energy to the macroscopic energy of calorimetry were not to be introduced by Maxwell, Boltzmann, and Gibbs until a century later. Rumford's hypothesis failed to produce tangible phenomenological results.

Although the caloric theory remained in use for over 50 years after Rumford's work, he emphasized the dilemma between conservation and creation (or conversion) which was to perplex the best minds of the nineteenth century. The conversion crisis was firmly established by the work of Mayer and Joule in the 1840s.

In 1824, Carnot offered a partial reconciliation of the conversion and conservation phenomena with an argument based on the caloric theory (the results of which actually prolonged the life of the theory). Carnot introduced a step-change in the level of complexity and sophistication: he put forth a number of new concepts that were essential to the eventual clarification of preclassical ideas and that later led to the replacement of the caloric theory by the First and Second Laws. These firsts include the concepts of heat reservoirs and reversibility, and the requirement of a temperature difference to generate work from a heat interaction.

Carnot proposed that cyclic operation of an engine working between two heat reservoirs was analogous to water flowing over a dam. Some quantity, in being transferred from a high to a low potential, produced work but the quantity being transferred was conserved in the process. We know today that Carnot's hypothesis is incorrect because he assumed that the conserved quantity was caloric. He carried this reasoning further to prove that there must be a limiting efficiency of heat engines. If such a limitation did not exist, then two engines could be suitably operated in a cyclic process in order to bring the heat reservoirs back to their initial states, the net effect being the production of work. This, Carnot declared, was an impossibility.

We might ask on what basis Carnot ruled out the possibility of what we today call a "perpetual motion machine of the first kind." It was not until 1847 that Helmholtz advanced the hypothesis of the conservation of energy. With few notable exceptions, there appeared to be general agreement among scientists of the period that the basic powers of nature are uncreatable and indestructible. Thus, over a period of many years, we see the gradual acceptance of what we refer to today as a basic postulate.

Carnot's engine reconciled the conversion and conservation phenomena—at least in the reversible limit. But how is this consistent with conservation in the highly irreversible mixing process of calorimetry, or conversion in the equally irreversible process of generating heat through friction? Carnot was cognizant of these difficulties and called for further experimentation and also for reconsideration of the foundations of the theory.

The preclassical period drew to a close with the quantitative work of Joule who established the equivalence of mechanical, electrical, and chemical energy to heat. We can see that at that time there were a number of concepts yet to be clarified. Caloric had to be split into heat quantity, energy, and entropy. It had to be shown that heat and work were forms of energy transfer and that the interconvertibility was asymmetric. It is energy that is conserved in the calorimeter, in Carnot's cycle, and in frictional processes. Entropy is conserved only in the limit of the reversible process.

These developments occurred in relatively rapid succession beginning approximately in 1850 with the genius of Clausius, with important contributions from Kelvin, Maxwell, Planck, Duhem, and Poincaré, and terminating with the brilliance of Gibbs near the end of the century. Although we cannot cover these events here, we should not underestimate their historical importance. We will attempt to reconstruct these developments, applying the insight that we have gained over the last 80 years.

1.3 The Postulatory Approach

Almost all approaches to classical thermodynamics follow one of two extremes: the *historical* approach, which parallels closely the chronological development of concepts and misconceptions, and the *postulatory* approach, in which axioms that cannot be proved from first principles are stated. There are merits and drawbacks in each extreme.

Advocates of the historical approach contend that if we are to expect our students to evolve new concepts and theories, we must expose them to the historical development of existing theories. Existing postulatory approaches make no reference to historical developments. The basis for the laws of thermodynamics is impersonally stated in a small number of postulates that cannot be proved and can only be disproved by showing that consequences derived from them are in conflict with experimental facts. The postulates tend to be mathematical and abstract, but the laws of thermodynamics are derivable from them. Many students are unimpressed because little insight is provided for the necessity to define new concepts or properties.

The approach we follow parallels the historical development in many respects. We begin by assuming the state of mind and body of knowledge available to the seventeenth-century scientist and by proceeding from that point to a logical development of a self-consistent set of rules that applies to the behavior of macroscopic bodies. In the process, we make use of many of the arguments put forth by our ingenious predecessors over the last 300 years, but we use hindsight to avoid the incorrect conclusions that prevailed at many points in the preclassical period and that resulted in much confusion (some of which is usually transferred to the student who studies thermodynamics by the historical approach).

At several junctures in our development we will face obstacles that cannot be obviated by invoking first principles. Our predecessors overcame these obstacles by trial-and-error experimentation until they amassed a large body of knowledge. In this manner, generalities were stated and rules established. We clearly identify those principles that our ancestors learned to accept without proof; these are stated as postulates, but in a form that could be understood by Black, Lavoisier, Kelvin, or Carnot. The ultimate verification of these postulates lies in the success of the formalism derived from them. In this vein, Schrödinger's equation is a basic postulate of quantum mechanics and Newton's laws of motion *were* basic postulates of classical mechanics. It is conceivable that, at some later date, new experimental information will be obtained that will necessitate revision or reformulation of the thermodynamic postulates, just as Newton's laws were found inapplicable to the motion of elementary particles of the atom.

Although the set of postulates presented in this text has the same information content as those developed in other texts, the phrasing may appear significantly different. These are obviously many different sets of postulates[1] that are equally valid bases for the theoretical development. In developing a set of postulates, we have attempted to keep them as real as possible (as opposed to

[1]See, for example, H. B. Callen, *Thermodynamics* (New York: Wiley, 1960), and G. W. Hatsopoulos and J. H. Keenan, *Principles of General Thermodynamics* (New York: Wiley, 1964).

abstract) while retaining a form that will be readily acceptable to the chemical engineer.

SUGGESTED READINGS

ANDRADE, E. N. DA C. (1935). "Two Historical Notes—Humphry Davy's Experiments on the Frictional Development of Heat; Newton's Early Notebook," *Nature*, **135**, 359.

BOMPASS, C. (1817). *An Essay on the Nature of Heat, Light, and Electricity*. London: T & C Underwood.

LARDNER, D. (1833). *A Treatise on Heat*. London: Longmans, Rees, Orme, Brown, Green, and Longmans.

LESLIE, J. (1804). *An Experimental Inquiry into the Nature and Propagation of Heat*. London: J. Mawman.

METCALFE, S. L. (1843). *Caloric* (2 vols.). London: William Pickering.

ROLLER, D. (1950). *The Early Development of the Concepts of Temperature and Heat*. Cambridge, Mass.: Harvard University Press.

RUMFORD, B. T. (1798). "An Experimental Inquiry Concerning the Source of the Heat Which Is Supplied by Friction," *Philos. Trans.*, **88**, 80.

THOMSON, W. (1840). *An Outline of the Sciences of Heat and Electricity*. London: H. Baillière.

TISZA, L. (1966). *Generalized Thermodynamics*, Paper 1. Cambridge, Mass.: MIT Press.

TYNDALL, J. (1880). *Heat, A Mode of Motion*. London: Longmans, Green, and Company.

URE, A. (1818). *New Experimental Researches on Some of the Leading Doctrines of Caloric*. London: William Bulmer and Company.

Basic Concepts and Definitions 2

2.1 The System and Its Environment

If we are to develop a set of fundamental laws of nature without any preconceived notions, we must first develop the facility to perform some experiments. The subject of the experiment will be called the *system*, which will refer to a region that is clearly defined in terms of spatial coordinates. The surface enclosing this region will be referred to as the *boundary*. It may be an actual wall or it may be an imaginary surface whose position is defined during an experiment. The region of space external to the system and sharing a common boundary with the system is referred to as the *environment* or *surroundings*.

As will be seen shortly, work and heat effects are defined in terms of events at system boundaries; thus, the choice of a boundary is usually dictated by the kind of information desired. In any given situation, there are often many different system boundaries one can choose, each having some advantages and disadvantages. Developing a facility for choosing those system boundaries that will result in the shortest path to the desired information is essential to the engineer. Insight into the selection process can be gained by working a given problem using several different system boundaries.

2.2 Primitive Properties

To record events that occur within a system, we must devise experimental tools that are sensitive to changes in the system. *Primitive properties* will refer to characteristics of the system that can be determined or measured by performing a standardized experiment on the system. To ensure that the measurement is a characteristic of only the system, we require that the experiment not disturb the system. The primitive property is of value because it is directly associated with the system at a particular time, and the observer need not know the history of the system to ascertain the value of the property.

A primitive property which is easily measured and is particularly useful is the *thermometric temperature* (denoted by θ). The thermometric temperature can be measured, for example, by noting the volume of a known mass of liquid in a sealed tube when the tube is brought into contact with the system. The mass of the thermometer should be small in relation to that of the system so that the measurement does not alter the system. (The effect of the thermometer on the system can be determined by inserting a second thermometer and noting any change in the reading of the first thermometer.)

It should be emphasized that the value of θ is completely arbitrary and depends on the type of fluid, the materials used in construction of the tube, how the tube is notched, and the labels associated with the notches. Once the notches and labels have been made, however, the device can be used to observe *changes* occurring within a system. It could also be used to rank the θ-property of different systems. We return to a more detailed description of the thermometric temperature in Section 3.4.

Innumerable primitive properties could be defined by outlining suitable experiments. In this sense, volume, mass, pressure, index of refraction, color (given a standardized method of observation), etc., could be called primitive properties and scales for each property could be devised.

Primitive properties are useful in defining a system and in recording the occurrence of events in a system. In an *event*, at least one primitive property changes. An *interaction* is defined as events occurring simultaneously in the system and surroundings, at least one of which would not have occurred if the system were removed from the surroundings and placed in any other arbitrary environment.

2.3 Classification of Boundaries

The interactions between a system and its surroundings are governed by the nature of their common boundary. If the boundary is impermeable to mass flow, the system is called a *closed system*. An *open system* has a boundary that permits a mass flux of at least one component of the system through at least one point. In either case, the boundary may be *rigid* or *movable*.

There is one other set of conjugates needed to complete the classification of boundaries: the *adiabatic* and *diathermal* walls. These boundaries govern the extent of heat interaction between system and surroundings, but since we have not yet defined a heat interaction, this definition is inadmissible. The adiabatic boundary or wall is one of the key concepts in thermodynamics, and we use it to define work and heat interactions. To avoid a circular system of definitions, we treat the adiabatic wall in detail in the next section.

In the absence of external force fields, specification of one of each of the three conjugate sets of boundaries (i.e., permeable and impermeable, rigid and movable, adiabatic and diathermal) is necessary to describe completely the external constraints placed on the system. Of the eight combinations[1] there is one that is of particular importance. This is the system enclosed by impermeable, rigid, and adiabatic walls, called an *isolated system*. It will become evident (Section 3.4) that this system can have no interactions. Any events occurring within this system are independent of events in the environment.

2.4 The Adiabatic Wall

The concept of the adiabatic wall evolves from our experience; it can be illustrated by conducting a series of simple experiments. Consider a closed, rigid system in which a device to measure the thermometric temperature has been placed. Surround this system with a system having a higher thermometric temperature. If the initial system were constructed from aluminum, the variation of the thermometric temperature with time would look like curve 1 of Figure 2.1. Curves 2, 3, and 4 would result if the initial system were constructed from steel, glass, and asbestos, respectively. Finally, if the container were made of a double-walled Dewar flask, the variation of temperature over the time of the experiment would be quite small (curve 5). The adiabatic wall is an idealized concept representing the limiting case of curve A in Figure 2.1. In practice,

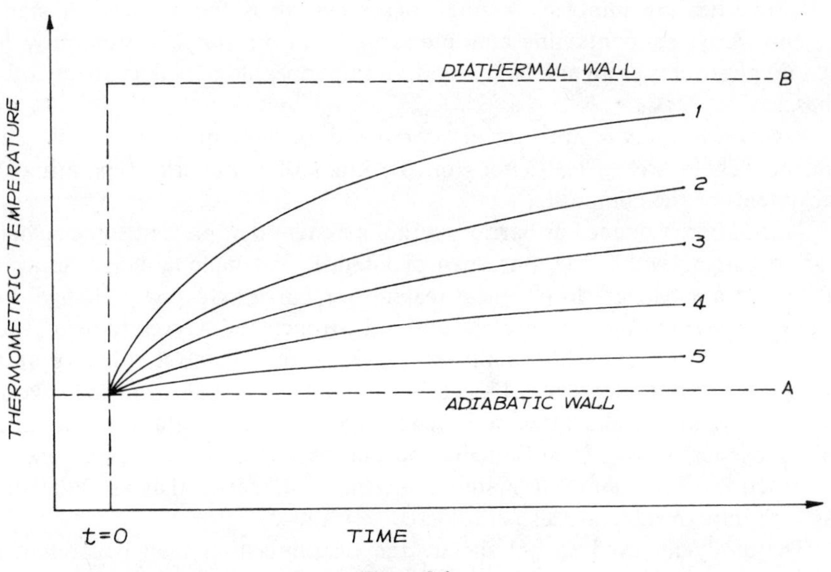

Figure 2.1

[1]It will become evident in discussing the criteria of equilibrium that only six of the eight combinations are meaningful. Of the four combinations involving permeable boundaries, there is no distinction between adiabatic and diathermal walls (see Section 6.4).

adiabatic boundaries are approached in many situations, especially those in which events occur rapidly in relation to the time scale of the experiment.

The case diametric to that of adiabatic is the diathermal wall (curve *B*), in which the change in thermometric temperature is rapid in relation to the time scale of the experiment (i.e., the thermometric temperature of the two systems is always identical).

In our operational definition of an adiabatic wall, we chose the thermometric temperature to relate changes in the environment to those in the system. This selection was made for ease of visualization. Other primitive properties could also have been chosen. Some primitive properties would not be of value to define adiabatic walls. Pressure, for example, could have been varied in the surroundings with no effect on our system with rigid walls. We will specify how to select an appropriate property to define an adiabatic wall after Postulate I has been introduced.

2.5 Simple and Composite Systems

There is a special class of systems that plays a central role in the developments to follow. These systems, referred to as *simple systems*, are devoid of any internal adiabatic, rigid, and impermeable boundaries and are not acted upon by external force fields or inertial forces.

A *phase* is defined as a region within a simple system throughout which all of the properties are uniform. A single-phase system is the simplest of simple systems. A system containing multiple phases is also a simple system provided that no phase acts as an adiabatic, rigid, or impermeable boundary to any other phase.

Composite systems are systems composed of two or more simple subsystems. There are no restrictions on the kinds of boundaries separating the subsystems of the composite.

Restraints are defined as barriers within a system that prevent some changes from occurring within the time span of interest. In simple systems, restraints of interest are barriers to chemical reaction or barriers to phase change. For example, the room-temperature reaction of hydrogen and oxygen to form water can be made to occur within milliseconds if a proper catalyst is incorporated in the system. In the absence of a catalyst, no noticeable reaction occurs within months or years. In the latter case, there is an internal restraint (the activation energy barrier) which, for all intents and purposes, prevents the occurrence of the reaction. For composite systems, internal boundaries that are adiabatic, rigid, or impermeable are also considered restraints.

Thermodynamics does not dictate the restraints that may be present in a given system. In any given situation, one must decide which restraints are present. Two factors that must be considered in making this decision are the laws of matter and the rates of the various conceivable processes: The former are (1) the law of continuity of matter (matter cannot move from one position to another without appearing at some time in the intervening space), (2) the law of conservation of electrical charge (net electrical charge must be conserved in all processes), and (3) the law of conservation of chemical elements (in the absence of nuclear transformations and relativistic effects, mass must be conserved). The application of some of these principles is illustrated in the following example.

Example 2.1

A closed vessel contains water, oil, and air at room temperature. If the system is synthesized by first adding the water and then layering the oil on the water, is this system a simple or composite system? If composite, define each simple subsystem and the internal restraints. If the vessel is shaken vigorously, is the system then a simple or composite system?

Solution

There are clearly no adiabatic or rigid walls within the vessel. Thus, it is only necessary to decide if there are impermeable walls. At room temperature, a system containing water and air should have appreciable water vapor present in the air. If the system is formed by layering the oil on the water without shaking, the law of continuity of matter requires that the water pass through the oil layer before it evaporates into the air. Thus, the water must diffuse through the oil layer. This process is slow and will not occur to any appreciable extent even after several hours. If our interest in this system did not extend beyond this time scale, the oil layer would have to be considered as an impermeable barrier to the water. In this case, the vessel would be considered a composite system of two simple subsystems: air + oil (assuming oil evaporates into the air within the time scale of the experiment), and water. If the system is shaken vigorously, water droplets will contact the air directly. Thus, the entire contents of the vessel would be considered a simple system.

2.6 States of a System

Now that we have the means for conducting experiments on a system, we would like to have a formal way of characterizing a system so that others could reproduce the experiments. For this purpose, we will identify the condition or *state* of the system by the values of those properties that are required to reproduce the system. Although this definition is functional, it is not very practical because we do not always know the number of properties that are required to specify the state of the system. We are fortunate, however, to have available a large body of experimental data, accumulated over several hundred years, which indicates that there are particular types of states that can be specified by delineating only a certain number of properties. These states, which are called *stable equilibrium states*, are defined in the next section. In general, nonequilibrium states can also be specified (for the purpose of reproducing them) from a finite number of properties; the number of such properties, however, is not given by the principles of classical thermodynamics.

2.7 Stable Equilibrium States

The aforementioned body of experimental data indicating the existence of these states is summarized in the following postulate.[2]

> ***Postulate I.*** *For closed simple systems with given internal restraints, there exist stable equilibrium states which can be characterized completely by two independently variable properties in addition to the masses of the particular chemical species initially charged.*

By "two independently variable properties"[3] it is meant that each property could be varied (by at least a small amount) in at least one experiment during which the other property is held constant. For example, consider a closed vessel containing the simple system of a pure-component liquid and its vapor in a stable equilibrium state. Given the amount of material initially charged, the thermometric temperature, and the total volume of the vessel, the system could be reproduced at will because the volume and temperature are independently variable and therefore completely specify the system. If, however, the pressure were specified instead of the total volume, the system could not be reproduced because, as we shall see later, pressure and temperature are not independently variable in this case.

We are now at an impasse. We do not know which two properties to choose to specify the state of the system because we do not yet know which properties are independently variable. For a given system with a given set of restraints, we find that the laws of thermodynamics (which we shall develop from our postulates) will result in relationships among certain variables. The relationships and the variables included therein will depend on the system and the restraints. Only from these relationships will we be able to determine which sets of properties are not independently variable. By a process of elimination, we shall then be able to determine which sets of properties are independently variable. For an example of a closed system of liquid and vapor in a stable equilibrium state, we can show that as a result of the requirement of phase equilibrium (Chapter 10), the vapor pressure is a unique function of temperature (the Clausius–Clapeyron equation). Thus, pressure and temperature cannot be independently variable in this case.

Postulate I can also be used to aid in selecting an appropriate property to *measure* whether a wall is or is not adiabatic (see Section 2.4). Consider the following *thought* experiment. We have a simple, closed, and rigid system in a

[2]This postulate is similar to a conclusion drawn by Duhem in 1899 and is sometimes referred to as Duhem's Theorem. See I. Prigogine and R. Defay as translated by D. H. Everett, *Chemical Thermodynamics* (London: Longmans, Green, 1954), p. 188.

[3]There is a class of thermodynamically trivial properties called *neutral* properties (e.g., the shape of a system of given volume) that cannot be used to determine the stable equilibrium state. See G. N. Hatsopoulos and J. H. Keenan, *Principles of General Thermodynamics* (New York: Wiley, 1965), pp. 30–32.

stable equilibrium state. We can measure the properties of this system. Suppose that we have values for four properties which we designate black, blue, red, and yellow. In our initial state, these properties have the values black-1, blue-1, red-1, and yellow-1. Assume, next, that the system changes to a different stable equilibrium state (by some interaction with the environment) and, in this change, the value of the property yellow changes from yellow-1 to yellow-2. The values of the other three properties remain unchanged. We immediately conclude that property yellow is independent of properties black, blue, or red. We can go further and state that on the basis of this experiment, the properties black, blue, and red *cannot* be independently variable. This statement is easily proved by visualizing a subsequent experiment in which property yellow remains at 2, property red remains at 1, but properties black and/or blue change. But this is not allowed since yellow-2 and red-1 completely define the state of the system and if, in one state, black and blue had values of 1, they cannot subsequently vary.

When this line of reasoning is applied to the study of an adiabatic wall, suppose that, instead of the thermometric temperature, we had selected some other property—yellow in this case. We vary the value of yellow in the surroundings and measure the property yellow in the system. If property yellow varies in the system as property yellow changes in the surroundings, we have the same situation as described in Section 2.4; that is, *yellow* replaces thermometric temperature as the test property. However, if we find *no* change in property yellow in the system as it is varied in the surroundings, either (1) the wall is adiabatic or (2) yellow is not a suitable property for the experiment. To test condition (2), we can repeat the test with *any* other property that is *independent of yellow*. For example, choose property red. This property *must* be a suitable variable to measure the adiabaticity of a wall. To carry out the test, we would vary red in the surroundings (keeping yellow constant) and then measure whether red or yellow (or both) vary in the system. If either red or yellow changed, the wall would not be adiabatic.

Thus, two independently variable primitive properties may be selected to test whether a rigid wall is adiabatic. If the wall were not adiabatic, we may be fortunate to select an appropriate primitive property for the first experiment (e.g., thermometric temperature as used in Section 2.4). However, if the choice had been, for example, pressure, then variations in this variable in such a way as to keep the thermometric temperature constant (and thus independently variable) would have produced no change in pressure inside the rigid, closed system. Then, as shown above, if we keep pressure constant and vary any other independently variable property, by measuring both this new property *and* pressure inside the system we can make an unequivocal statement concerning the adiabaticity of the wall. That is, if neither change, the wall is adiabatic. If either or both change, the wall is not adiabatic.

Knowing that stable equilibrium states exist is not nearly as informative as knowing when they exist. The second postulate is directed toward establishing this fact.

Postulate II. *In processes for which there is no net effect on the environment, all systems (simple and composite) with given internal restraints will change in such a way as to approach one and only one stable equilibrium state for each simple subsystem. In the limiting condition, the entire system is said to be at equilibrium.*

Postulate II is specific to systems which are in effect isolated; also, for processes that consist of a series of steps, the system may interact with the environment in two or more steps, but the net effect of these steps must leave the environment unaltered.

Since the stable equilibrium state is defined as a limiting condition toward which a simple system tends to change, it follows that no property of this state varies with time. Then, from Postulate I, once a simple system has reached a stable equilibrium state, only two independently variable properties and the masses initially charged need to be specified to determine this state completely. Since all other properties are fixed in the stable equilibrium state, it follows that all other properties of the simple system are dependent variables that are determined by the two independently variable properties and the masses of the initial chemical species. Note that this conclusion is valid for each *simple* system at equilibrium, even if the simple systems are part of a composite system. The conclusion, however, does not apply to the composite system at equilibrium: that is, the state of a composite system at equilibrium cannot be specified by two independently variable properties plus the masses initially charged. The difficulty arises from the fact that all properties may vary from one subsystem to another within a composite system at equilibrium. For example, the thermometric temperature of a composite system has little significance if the subsystems of the composite are separated by adiabatic, impermeable walls. Postulate I has been restricted to simple systems in order to avoid such difficulties.

In Postulate II, it is stated that for an isolated system, there exists one and only one set of stable equilibrium states (toward which the subsystems tend) *for a given set of internal restraints.* There will be different sets of stable equilibrium states for different sets of internal restraints. For example, with reference to Example 2.1, there will be a unique equilibrium state if we assume that there is an impermeable barrier preventing water from reaching the air space; there will also be a unique equilibrium state if we assume that no such barrier exists. Although each of these states is unique, the properties of each will be different.[4] Thus, before the equilibrium state can be completely defined, the internal restraints must be specified. Specification of internal restraints is clearly an important part of specifying the system.

[4]In some texts the former state would be referred to as metastable in the sense *that given sufficient time*, the latter state (for which the term "stable equilibrium state" would be reserved) would be reached. The problem with this set of definitions is that almost all systems of interest to chemical engineers would then have to be classified as metastable even though the final stable equilibrium state may not be obtained in any time span of interest.

2.8 Thermodynamic Processes

A change of state[5] of a system is identified by a change in the value of at least one property. For systems initially in stable equilibrium states, changes of state will occur only when the system has an interaction with the environment *or* when internal restraints are altered. "Change of state" is usually applied to systems that are initially in one stable equilibrium state and are found after some event to be in another equilibrium state. The change of state is then fully described by the values of the properties in the two end states.

The *path* refers to the description of all the states that the system traverses during a change of state. Thus, the path is described in terms of the primitive properties that define the intermediate states. Paths for which all the intermediate states are equilibrium states are termed *quasi-static* paths.[6] From Postulate I, quasi-static paths of closed simple systems can be completely described in terms of the values of only two independent properties.

It also follows from Postulate II that if a system progressing along a quasi-static path is isolated at some point (e.g., by temporarily altering a boundary condition), the values of all the properties will remain constant at the values observed just prior to isolation. It may, however, take more than two properties to describe a non-quasi-static path. If the system is isolated during such a path, some primitive properties will change after isolation as the system approaches a stable equilibrium state. For example, consider the system of a gas initially at two atmospheres which is contained in a cylinder fitted with a piston and stops. The stop holding the piston is removed, and the gas expands until the piston reaches a second stop. If the piston is lubricated, the expansion process will be rapid. At any instant during the process, there will be a pressure gradient within the gas phase. To describe such an intermediate state, it will be necessary to determine the pressure (in addition to other properties) at all points within the cylinder. The intermediate states are not stable equilibrium states; if such a state were isolated (by stopping the piston at an intermediate point), the pressure gradient would be damped out as the system approached a stable equilibrium state. Clearly, this frictionless process is not quasi-static. Alternatively, if there were external forces acting against the piston so that the expansion was very slow, no appreciable pressure gradient would be found. If the system were then isolated at an intermediate point, no properties would change because the system was, at all times, in some stable equilibrium state. Thus, the latter process was quasi-static.

The thermodynamic *process* involved in a change of state usually refers to a description of the end states, the phenomena occurring at the system boundaries (i.e., heat and work interactions, which are discussed in Chapter 3) during the process, and the path (which is usually described only for quasi-static

[5]In common usage, a change of state often is synonymous with "change of phase." This is not the meaning we use in this book.

[6]Quasi-static paths are closely related to (and sometimes confused with) reversible processes. The distinction between the two is considered in Section 4.6.

processes). In many instances, however, the term "process" is loosely applied to describe the path without explicitly specifying the boundary conditions. Thus, an isothermal, isobaric, or isochoric process is one in which the temperature, pressure, or volume is constant. In such cases the boundary conditions are usually implied by the nature of the process or are immaterial to the problem in question.

2.9 Derived Properties

Primitive properties were defined in terms of an experiment or measurement made on the system at some point in time. Only experiments that do not disturb the system are allowed. By definition, primitive properties are not restricted to stable equilibrium states.

We have now established two basic postulates dealing with a particular class of states, the stable equilibrium states. These states have innumerable primitive properties associated with them. Each of these properties is measurable by definition. We might also ask ourselves if there are other *properties* of these stable equilibrium states that are not measurable by any method but could be used to characterize the system. No definite answer can be given at this point, but it will become obvious in Chapter 3 and those following that such properties do exist. We will find that we can define such properties only in terms of changes in the system between initial and final stable equilibrium states. (Note that the path need not be quasi-static.) To distinguish these properties from the *primitive* type, we will call them *derived* properties. As defined, derived properties exist only for stable equilibrium states and, as such, may be used as variables to define a system as required in Postulate I.

Since derived properties are functions of state, and since any stable equilibrium state of simple systems can be characterized by the values of two independently variable properties plus the masses, any derived property can be expressed mathematically as a function of two other independently variable properties, derived or primitive.

It will be instructive at this point for the reader to review the mathematical relations of functions of state. A review of this topic is presented in Appendix B.

Energy: Concepts and Consequences 3

In this chapter we adopt the definition of work from the mechanics of rigid bodies and extend this concept to thermodynamic systems. We introduce energy as a unique measure of the work required to reach one stable equilibrium state from another in certain processes. Since we cannot show a priori that this quantity is a function only of the end states, it will be necessary to postulate that energy is a derived property. The conservation law for energy will be a direct consequence of the new postulate. After defining work and postulating the existence of the energy property, we present an operational definition of heat and discuss the directionality of heat interactions.

3.1 Work Interactions

The mechanical work associated with the movement of a rigid body is defined as

$$W = \int_{x_1}^{x_2} (\sum \mathbf{F}_s) \cdot d\mathbf{x} \tag{3-1}$$

or, in differential form,[1]

$$đW = (\sum \mathbf{F}_s) \cdot d\mathbf{x} \tag{3-2}$$

where $\sum \mathbf{F}_s$ is the resultant force acting on the surface or boundary of the rigid body at a point where there is a differential displacement of the boundary, $d\mathbf{x}$. In keeping with the mechanical definition of work, *boundary forces* ($\mathbf{F}_s$) are distinguished from *body forces* or forces associated with external fields

[1] As discussed in Appendix B, the symbol $đ$ is used throughout the text to denote differentials of functions that are not state variables.

($\mathbf{F}_b$)(i.e., centrifugal, gravitational, inertial, coulombic, etc.); in the absence of body forces, only boundary forces are used to calculate work.

For a rigid body acted upon by both boundary and body forces, Newton's Second Law of Motion states that

$$\sum \mathbf{F}_s + \sum \mathbf{F}_b = 0 \tag{3-3}$$

In Eq. (3-3), the inertial force, $-M\mathbf{a}$ is considered a body force and is included in the second summation. For example, consider a weight suspended by a string, as in Figure 3.1. If the body is initially at rest, and if $\mathbf{F}_s$ and $\mathbf{g}$ are col-

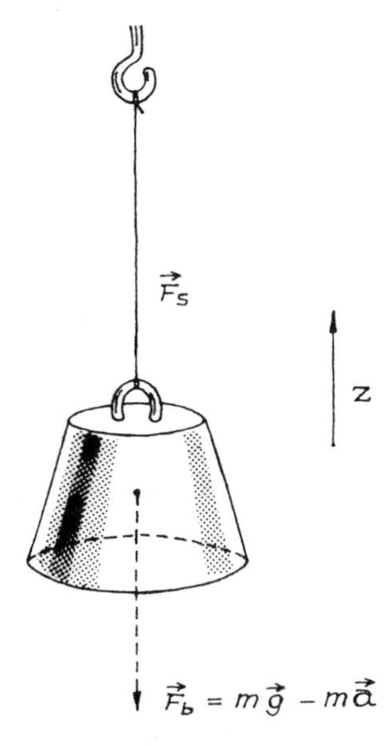

Figure 3.1

linear, then Eq. (3-3) becomes

$$F_s - Mg = 0 \tag{3-4}$$

If F_s is then increased so that the weight rises, at any instant during the motion, Eq. (3-3) is

$$F_s - Mg - M\frac{dv}{dt} = 0 \tag{3-5}$$

The differential work done on the weight at any instant of time is due only to the boundary force, F_s. Thus, using Eqs. (3-2) and (3-5),

$$đW = \mathbf{F}_s \cdot d\mathbf{x} = \left(Mg + M\frac{dv}{dt}\right) dz = Mg\,dz + Mv\,dv \tag{3-6}$$

The total work done on the weight between z_1 and z_2 is

$$W = Mg(z_2 - z_1) + \frac{M}{2}(v_2^2 - v_1^2) \tag{3-7}$$

On the right-hand side of Eq. (3-7), we call the first and second terms the difference in potential and kinetic energy, respectively. (Here we accept the terms of potential and kinetic energies as Mgz and $Mv^2/2$, but in no way have we defined the concept of energy.)

We are now in a position to define the work associated with systems of interest in thermodynamics; in particular, we are concerned with systems having nonrigid as well as rigid boundaries.

Consider the expansion of a gas contained in a cylinder fitted with a piston and surrounded by the atmosphere (Figure 3.2). If we choose the system as

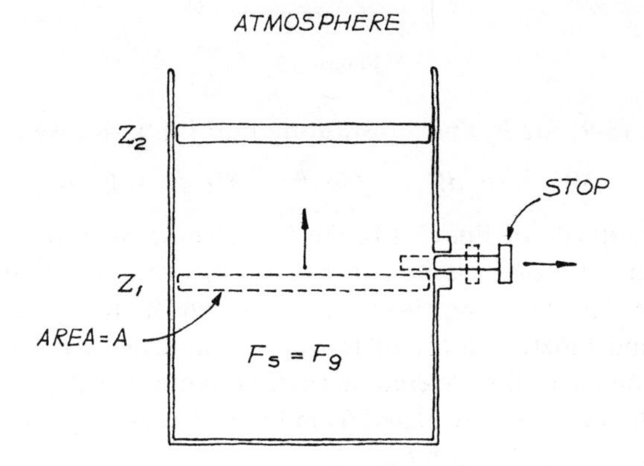

Figure 3.2

the gas, the work done by the system on the environment is

$$đW_g = \mathbf{F}_g \cdot d\mathbf{z} = P_g \, d\underline{V}_g \tag{3-8}$$

where $\mathbf{F}_g$ is the force exerted by the gas on the boundary which is displaced and A is the area of the displaced boundary. (Note that this force may not be the same as that exerted on the other boundaries of the gas system.) Thus, before the work can be calculated, $\mathbf{F}_g$ must be known as a function of $\mathbf{z}$. For the gas, the term F_g is, of course, equal to $P_g A$, where P_g is the pressure on the boundary that is displaced.

If neither F_g nor P_g is known, the work can still be determined by measuring the effects on the environment. For example, consider a differential element of time during the expansion. To illustrate, let us suppose that there is friction between the piston and the cylinder walls. As is shown in Figure 3.3, the unknown force F_g can be evaluated by making a force balance on the object which is acted upon by this force. In this case, the object is the piston. Thus,

$$F_g - P_a A - F_f - Mg - Mv\frac{dv}{dz} = 0 \tag{3-9}$$

where P_a is the pressure of the atmosphere, F_f is the frictional force, and M is the mass of the piston. Note that the inertial force acts in the direction opposite to that of the motion during acceleration.

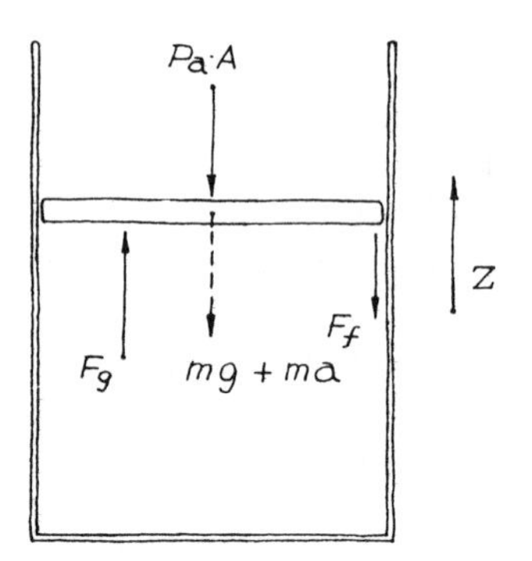

Figure 3.3

Solving Eq. (3-9) for F_g and substituting into Eq. (3-8), we obtain

$$đW_g = P_a \, d\underline{V}_g + Mg \, dz + Mv \, dv + F_f \, dz \tag{3-10}$$

On the right-hand side of Eq. (3-10), the first term is work done by the gas in pushing back the atmosphere and increasing the volume of the gas by $d\underline{V}_g$. The second and third terms represent the work done by the system in increasing the potential and kinetic energy of the piston, and the last term is the work done on the cylinder wall to overcome friction. Note that if the frictional force were known, $đW_g$ could be calculated from Eq. (3-10), and P_g could be calculated directly by solving Eq. (3-9) for F_g.

A different point of view is found if the piston were chosen as the system. The work done by the piston on its surroundings would, then, be

$$đW_p = \mathbf{F}_p \cdot d\mathbf{z} \tag{3-11}$$

where $\mathbf{F}_p$ is the net boundary force exerted by the piston on the environment. According to Newton's third law of motion, the net force is *equal and opposite* to the net boundary force exerted by the environment on the piston. Therefore,

$$\mathbf{F}_p = \frac{(-P_g A + P_a A + F_f)\mathbf{z}}{|z|} \tag{3-12}$$

and

$$đW_p = -P_g \, d\underline{V}_g + P_a \, d\underline{V}_g + F_f \, dz \tag{3-13}$$

Finally, if the atmosphere were chosen as the system, the work done by the atmosphere on its surroundings would be

$$đW_a = \mathbf{F}_a \cdot d\mathbf{z} = P_a \, d\underline{V}_a = -P_a \, d\underline{V}_g \tag{3-14}$$

If we define $-F_f \, dz$ as the "work" done by the walls, $đW_w$, substitution of Eqs. (3-8) and (3-14) into Eq. (3-13) yields

$$đW_p = -đW_g - đW_a - đW_w \tag{3-15}$$

Thus, choosing any object in Figure 3.3 as the system, it is clear from Eq. (3-15) that the work done by the system on the environment is equal and opposite to the work done by the environment on the system. This result, which is valid for work interactions in general, is a consequence of the requirement that the sum of all body and boundary forces on a system is zero.

Returning to the problem of calculating the total work done by the gas in the expansion depicted in Figure 3.2, we see that it is clear from the discussion above that the work is equal to the total work done on the piston, atmosphere, and walls. If we neglect the frictional work for the time being (we shall return to this subject in Section 3.7), then

$$W_g = (P_a A + Mg)(z_2 - z_1) + \frac{M}{2}(v_2^2 - v_1^2) \tag{3-16}$$

There is one other method of measuring the work done by a system which, although not very practical, will be of help in visualizing a "thought" experiment which will be described shortly. We could remove the system of interest from its given environment, such as that shown in Figure 3.2, and place it in a cylinder covered with a weightless piston surrounded by a vacuum. The piston is balanced by weights placed on the top. If we then continually remove the weights so that the pressure–volume history of the gas during this process is identical to that which occurred in the original expansion, the work done by the gas during the hypothetical expansion is equal to the change in level (or potential energy) of the weights. Furthermore, this work is equal to the work done by the gas in the original expansion. Thus, the work done by a system can always be found by measuring the rise or fall of weights in the environment.

3.2 Adiabatic Work Interactions

Consider two closed systems that undergo an interaction through a common boundary. We shall call the interaction an *adiabatic work interaction* if the events occurring in each system could be repeated in such a way that the *sole* effect external to one system could be duplicated by the rise (or fall) of weights in a standard gravitational field and the *sole* effect external to the other system could be duplicated by an equivalent fall (or rise) of weights of equal magnitude. Some examples will help to illustrate the use of the definition.

Consider the situation illustrated in Figure 3.4(a). A vessel containing water initially at 0°C is bounded by adiabatic walls. A rough-surfaced disk is immersed in the water and attached to a drum by a shaft. The drum is rotated by allowing a weight to fall from the first to the second level. The first step in determining if this is an adiabatic work interaction is to designate the boundaries of the systems we wish to study. Let us choose the dashed line in Figure 3.4(a) as the boundary separating systems A and B. If we consider the events that occur external to system A, the sole effect is the fall of a weight. If the interaction is an adiabatic work interaction, we must somehow show that the events that occur external to B could be repeated solely by the raising of an identical weight by an identical amount. If, as in Figure 3.4(b), we replace system A with a drum that has a weight attached to it, and then repeat the event in system

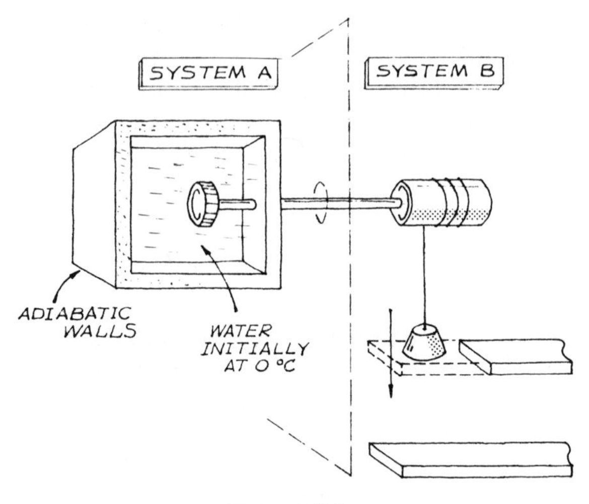

Figure 3.4(a)

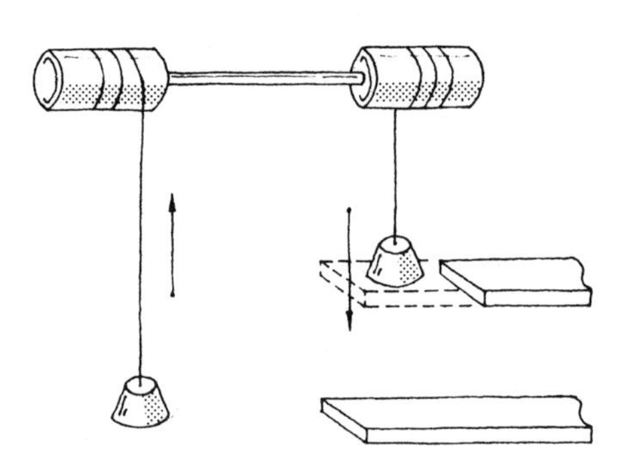

Figure 3.4(b)

B, it is clear that the sole effect external to *B* is the required rise of a weight. Thus, the interaction is an adiabatic work interaction.

Now consider the slightly more complex situation illustrated in Figure 3.5(a). If we consider the interaction of system *C* with the composite system $A + B$, the situation is analogous to that of the previous example and we have an adiabatic work interaction. If, however, we consider the interaction of system *A* with the composite system $B + C$, the analysis is quite different. External to system *A*, the sole effect is the lowering of a weight since the final temperature of the water in system *B* will be 0°C as long as some ice remains in system *A*. (Note that we need only consider the initial and final states; whether or not the water remains at 0°C throughout the process is immaterial.)

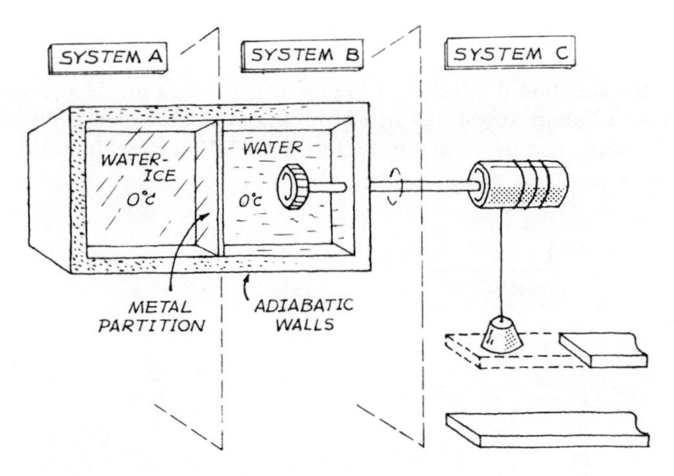

Figure 3.5(a)

Although we can devise processes external to $B + C$ which would result in the rise of a weight, it does not appear possible to devise an experiment in which the *sole* effect external to $B + C$ is the rise of a weight. (At least no such experiment is known.) For example, we could replace system A with a cylinder filled with a gas at 0°C and fitted with a piston which in turn is attached to a flywheel [system A' in Figure 3.5(b)]. As the weight in system C is lowered, the piston is freed to move to the left by removing a stop. As the piston moves out, the weight on the flywheel is raised. By judiciously choosing the components of the system in Figure 3.5(b), it would be possible to have the flywheel weight rise by the same amount as the weight in system C falls. However, the net effect external to system $B + C$ is the expansion of the gas in the cylinder in addition to the rise of the flywheel weight. Therefore, the interaction between A and $B + C$ is not an adiabatic work interaction.

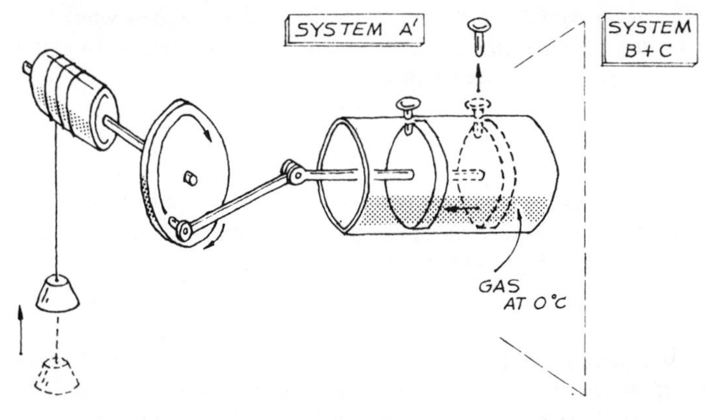

Figure 3.5(b)

Example 3.1

Consider the situation illustrated in Figure 3.6, in which an electric generator is operated by a falling weight and in which the power generated is dissipated in a resistor. Neglect any dissipative processes such as I^2R line drop, friction in bearings, etc. Is this an adiabatic work interaction?

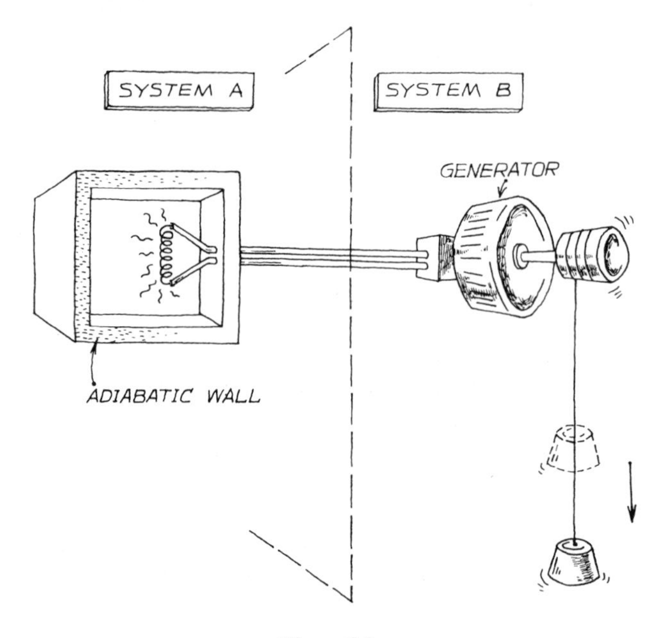

Figure 3.6

Solution

The sole effect external to system A is the fall of a weight. By replacing system A with a motor that has a weight attached to its shaft by a rope (i.e., the inverse of system B), system B could be made to execute the same process while the sole effect external to B would be an equivalent rise in the level of the weight. Hence, an electric current flowing between two systems is an adiabatic work interaction.

Example 3.2

Return to the example shown in Figure 3.4(a). Suppose that the dashed boundary was an adiabatic wall and a bearing was placed in the wall to hold the rotating shaft. Is this process an adiabatic work interaction?

Solution

If the bearing were frictionless, the case is no different from that described in text. However, if there were friction in the bearing, it would not be possible to have *equivalent* fall and rise of weights when each system was considered separately. Thus, this interaction is *not* an adiabatic work interaction.

In summary, it should be clear that the description of a work interaction is meaningful only when systems and boundaries are carefully delineated. An adiabatic work interaction requires that all common boundaries be adiabatic walls and, if any moving shafts or electrical wires penetrate these adiabatic walls, there can be no dissipative processes occurring at the wall.

3.3 Energy

We have gone to the trouble to define an adiabatic work interaction because the magnitude of such interactions allows us to rank stable equilibrium states. The fact that adiabatic work interactions are always possible between stable equilibrium states cannot be developed from first principles, but the truth of such a statement has been borne out by a large body of experimental evidence. Thus, it is presented in the form of a postulate.

> ***Postulate III.*** *For any states* (1) *and* (2), *in which a closed system is at equilibrium, the change of state represented by* (1) $\longrightarrow$ (2) *and/or the reverse change* (2) $\longrightarrow$ (1) *can occur by at least one adiabatic process and the adiabatic work interaction between this system and its surroundings is determined uniquely by specifying the end states* (1) *and* (2).

As implied in the postulate, it is not always possible to go from state (1) to state (2) by an adiabatic process, but when this route is impossible, it must always be possible to find an adiabatic process from state (2) to state (1). For example, consider a system containing 1 mole of oxygen gas at a thermometric temperature θ_1 and pressure P_1. Through interactions with the environment, the temperature is increased to θ_2 but the pressure is decreased to P_2. This change of state can be effected by an adiabatic process in many ways. Suppose that the system was fitted with a movable, frictionless, and adiabatic piston. This piston moves in such a way to expand the oxygen gas and reduce the pressure from P_1 to P_2. This is an adiabatic process since the boundary walls are adiabatic and no dissipative processes occur at the (moving) boundary. During the pressure reduction, θ will also change. When we have attained the desired pressure level, we will hold this value with the use of an external pressure reservoir at P_2. We will then change the value of θ in the system with a scheme as shown in Figure 3.6 to attain the value of θ_2. This scheme is possible only if we have a thermometric scale that increases as electrical power is dissipated by an I^2R drop in the system (i.e., physiologically when the system becomes warmer). Should we have a thermometric scale that behaves in an inverse manner, we could not proceed by an adiabatic process from $1 \longrightarrow 2$, but it should then be obvious that we could then go from $2 \longrightarrow 1$ by similar adiabatic processes where the gas is compressed from P_2 to P_1, etc.

Since the postulate was stated for *any* states of a system at equilibrium, it follows that all stable states can be bridged by adiabatic processes originating from a given initial stable state. Thus, if state A is chosen as a reference state, any change to different states represented by $B_1, B_2, \ldots B_j, \ldots$ can be characterized by measuring experimentally the adiabatic work required for the change in state of A to B_j (or B_j to A, if the former change in state is not possible). Since the adiabatic work is only a function of the end states, the adiabatic work is a derived property of the system. We shall call this derived property the *energy*, $\underline{E}$, of the system and we shall follow the convention that the energy decreases when work is done *by* the system on the surroundings. That is,

$$\underline{E}_{B_j} - \underline{E}_A = -W^a_{A \to B_j} \tag{3-17}$$

where W^a is the adiabatic work and is always calculated as the work done by the system on the surroundings. Although by convention we could associate a value of $\underline{E}$ with each stable state, it is clear that only differences of energy have significance.

Equation (3-17) appears as an abridged form of the First Law of Thermodynamics and leads directly to the *Conservation Law for Energy*: Since the adiabatic work measured in the surroundings must be equal and opposite to the adiabatic work measured in the system (see Section 3.1), the energy change of the system must be equal and opposite to the energy change of the surroundings.

It should be noted that the definitions of adiabatic work and energy are not restricted to simple systems. They are valid for composite systems which may be in the fields of external forces. Throughout the text the symbol $\underline{E}$ is used to denote the energy of such systems. When a development is limited to systems that are not acted upon by external force fields or inertial forces, the symbol $\underline{U}$ is used to denote the energy. The symbol $\underline{U}$ is used for composite and simple systems for which such forces are absent.

For a simple system, the energy can be defined with the aid of Postulate I. If each of the end states, A and B_j, of a simple system is uniquely specified by two independently variable properties plus the masses of the n components, the energy associated with each stable state must also be a unique function of these $n + 2$ independent properties. For example, if θ and P are independently variable for a given system, then

$$\underline{U}_i = f(\theta_i, P_i, M_1, M_2, \ldots M_n) \tag{3-18}$$

It can also be shown that as a consequence of Postulate III, $\underline{U}$ must be first order in the total mass of the system. That is, the function of Eq. (3-18) is such that

$$a\underline{U}_i(\theta_i, P_i, M_1, M_2, \ldots M_n) = \underline{U}_i(\theta_i, P_i, aM_1, aM_2, \ldots aM_n) \tag{3-19}$$

where a is a constant. The proof can be developed by comparing the following two processes, which have the same net effects: (1) a process in which two identical systems are acted upon simultaneously and (2) a process in which each system is acted upon separately.

Because energy is first order in mass, it can be shown that the energy of a composite of simple subsystems is equal to the sum of the energies of the subsystems of the composite. This conclusion is used extensively in the following discussions.

Energy, as introduced above, is a mathematical abstraction with no physical connotation. Readers familiar with statistical mechanics will tend to associate energy with the translational, rotational, vibrational, etc., motions of molecules as well as with intramolecular concepts (electronic, nuclear) and intermolecular effects. This division is very useful in developing theories of molecular behavior, but in classical thermodynamics, we can (and should) dissociate ourselves from physical interpretations and treat $\underline{U}$ (or $\underline{E}$) as the unfathomable, but useful, $\sqrt{-1}$ of thermodynamics.

3.4 Heat Interactions

Heat is an elusive entity that is recognizable only by its effect on material substances. For our discussion of work we were fortunately able to adopt definitions and procedures from mechanics. For a discussion of heat we have no precedent to follow since the onus of developing this concept lies within the realm of thermodynamics. Thus, we must endeavor to define heat by using the definitions and concepts already presented.

The key concept needed is that the energy difference between two states *can always* be determined by measuring the work in an adiabatic process connecting the two states (Postulate III). Now, with the same initial and final states, visualize *any* process (adiabatic or nonadiabatic), as going between these states. The energy difference is the same as that found for the adiabatic process because energy is a function of state only (i.e., it is independent of the path connecting the two states). If the process is not adiabatic, the work interaction will be different from that of the adiabatic process; however, the work can always be determined by one of the methods outlined in Section 3.1. *We then define heat as the sum of the energy change and the actual work performed.* That is,

$$Q = (\underline{E}_{\text{final}} - \underline{E}_{\text{initial}}) + W \tag{3-20}$$

where, by convention, W is the work done *by* the system on the surroundings and Q is the heat "added" to the system.

The definition for heat above, like that given previously for energy, is devoid of any microscopic significance. Nevertheless, it is of great practical utility. We will deduce shortly under what circumstances a heat interaction is to be expected and develop a method for ranking systems with respect to the direction of heat interactions.

In Eq. (3-20), heat is defined as the difference between the actual work in the process and the adiabatic work that would be required to effect the same change in state. Since any system completely enclosed by adiabatic walls can only undergo adiabatic work interactions, it follows that for such systems

$Q = 0$. Alternatively, a system must have at least one diathermal wall if it is to undergo a heat interaction. The converse, however, is not necessarily true; namely, systems connected by a diathermal wall will not necessarily have a heat interaction.

In Section 2.3, an isolated system was defined as one having adiabatic, rigid,[2] and impermeable walls. Since the walls are adiabatic, the system cannot have any heat interactions. Since the walls are rigid, the system cannot have any work interactions. Hence, an isolated system can have no interactions with the environment, and therefore the energy of an isolated system is invariant.

We will define a *pure heat interaction* as one for which the actual work is zero, and therefore $\Delta \underline{E} = Q$, where Δ is defined as final minus initial. For a system to undergo a pure heat interaction, the system must be surrounded by rigid walls, at least one of which is diathermal. As discussed below, the pure heat interaction is helpful in defining more specifically a thermometric temperature.

Consider two systems, A and B, which are closed, have rigid walls, and have adiabatic walls except for their common boundary, as indicated in Figure 3.7. When these two systems are first brought together, they may or may not

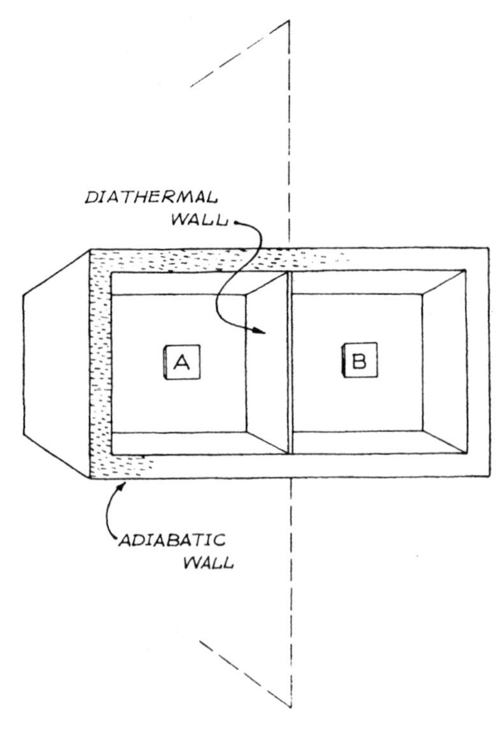

Figure 3.7

[2]The term "rigid" is used in the generic sense throughout. That is, not only are the walls immovable, but no shafts or electrical conductors pass through them.

have a heat interaction. In this case, any heat interaction will be a pure heat interaction. We can tell if any interaction occurs by observing whether or not the primitive properties of A and B change after they are brought together.

If no heat interaction occurs when the systems are brought together, it follows from Postulate II that the composite is at equilibrium. Subsystems that are at equilibrium across a diathermal wall are said to be in *thermal equilibrium.* But if there is a heat interaction, it follows from Postulate II that the interaction must eventually cease since the subsystems of the isolated composite system will approach thermal equilibrium. If a heat interaction occurs, Q_A must be equal to $-Q_B$ because the total energy change of the isolated composite system, which is the sum of the energy changes of the subsystems, must be zero.

Let us now turn to the problem of determining the direction of a heat interaction. Let us redefine the procedure for measuring the thermometric temperature. We construct a sealed liquid-in-glass thermometer in such a way that the walls are rigid and adiabatic except for one diathermal surface which is placed in contact with the system under study. Thus, the thermometer–system interaction is equivalent to the A–B interaction of Figure 3.7.[3] By heating or cooling the thermometer on a hot plate or in an ice bath, we can eventually find a liquid level for which no heat interaction occurs when the thermometer is placed in contact with system B, and hence the measurement of the temperature of system B can be made without disturbing system B. (Alternatively, we can make the thermometer small in relation to the system and let the two come into thermal equilibrium with only a minor change in the properties of system B.) In this manner the liquid level can be used as a measure of the stable equilibrium state of the simple system, B. In order to give additional significance to the θ-property, an additional postulate is required. If we put the thermometer in contact with system B and then with system C, and if we find the same liquid level, there is still no way to prove that systems B and C are in thermal equilibrium with each other. Thus, we shall postulate such a requirement.

> ***Postulate IV.*** *If the sets of systems A–B and A–C each have no heat interactions when connected across nonadiabatic walls, there will be no heat interaction if systems B and C are also so connected.*

This postulate is sometimes referred to as the *zeroth law of thermodynamics.* It requires that systems with the same thermometric temperature be in thermal equilibrium and that no heat interaction occur between these systems.

It can be shown that the converse to Postulate IV is also true. If A–B are in thermal equilibrium, and A–C result in a heat interaction, then B–C must also result in a heat interaction. The proof involves assuming that B–C does not result in a heat interaction and then showing that this is contrary to the

[3]Other primitive properties may be equally suitable for studying a heat interaction having no work or mass exchange. We chose the thermometric temperature as a convenient device that is easy to visualize (see Section 2.4).

initial statement. Thus, composite systems attain thermal equilibrium only when the temperatures are uniform throughout the subsystems.

When systems A and B undergo a pure heat interaction such that $\Delta \underline{E}_A = -\Delta \underline{E}_B < 0$, we shall use the imprecise statement that "heat is transferred from A to B" instead of the longer, more precise statement that energy is transferred from A to B as a result of a pure heat interaction. In shorthand notation, we shall say $Q_{A \to B}$ is positive, which implies that $Q_A = -Q_B < 0$.

We are now in a position to show that the thermometric temperature can be used to rank systems with respect to the direction of heat interactions. We shall prove that for the three systems, A, B, and C, if $Q_{A \to B}$ is positive (i.e., if heat is transferred from A to B) and if $Q_{B \to C}$ is positive, then θ_B must lie between θ_A and θ_C. Let us choose a thermometer scale so that $\theta_A > \theta_B$ and prove that θ_C cannot be equal to or greater than θ_B.

If $\theta_C = \theta_B$, then $Q_{B \to C} = 0$, which is in contradiction to the initial statement.

If $\theta_C > \theta_B$, we could have chosen C so that $\theta_A = \theta_C$. If we then allow the three systems to interact in such a way that $Q_{A \to B} = Q_{B \to C}$, so that θ_B does not change in the process (i.e., $\Delta \underline{E}_B = 0$), the net result is a heat interaction from A to C. This, however, is not allowed because $\theta_A = \theta_C$ and, therefore, θ_C cannot be greater than θ_B. Thus, the thermometric temperature can be used to rank different systems according to the directions in which heat interactions will occur.

It should be emphasized that the thermometric-temperature scale is still arbitrary. We could define another scale, called the ξ-scale, for which the freezing point of water is 100°ξ and the boiling point is 0°ξ. All that we require is that $\Delta\xi \longrightarrow 0$ when systems of different ξ undergo heat interactions. Of course, once we choose a temperature scale by some convention, we must be consistent throughout.

We shall adopt the *convention* that when $\theta_A > \theta_B$, the heat interaction is such that the energy of A decreases and the energy of B increases, or

$$\frac{d\underline{E}}{d\theta} > 0 \tag{3-21}$$

3.5 The Ideal Gas

Before exploring the applications of the First Law, it is convenient to point out here another temperature scale that will be very useful in later developments.

It has been found that for a gas, such as helium, a very simple experiment may be carried out to define a particular temperature scale. Suppose that we confine a given quantity of gas in a container fitted with a piston and keep the pressure on this gas constant. We will immerse this gas system in boiling water at 1 atm pressure and measure the volume; we then repeat the experiment in an ice-water bath (the pressure above the bath should be that corresponding to water vapor, but for this discussion 1 atm pressure will not significantly affect the results). We denote the ideal-gas temperature by T instead of θ. We call the first temperature 373.16 and the second temperature 273.16 and then plot $\underline{V}$ against T, drawing a straight line between the two $\underline{V}$–T points and extrapolate to zero volume; it will be found that at $\underline{V} = 0$, our temperature scale also is zero. This $\underline{V}$–T relationship gives us an experimental method to measure temperature. Repeating the experiments at other (but low) pressures will give different lines but all will intersect the $\underline{V} = 0$, $T = 0$ singular point. Any of the various $\underline{V}$–T lines (each at a different pressure) can be used to define a temperature scale. This particular scale is called the "ideal gas" scale. If we were clever enough, we could bring all the $\underline{V}$–T lines (at different pressures) together by noting that if the same number of moles of gas were used in each experiment, then the product $P\underline{V}$ (rather than just $\underline{V}$) lines, which were drawn between the same two temperature points, fall on the same line and pass through the point $(P\underline{V}) = 0$, $T = 0$. This linearity is given analytically as $P\underline{V}/T =$ constant. Further, if we vary the moles of gas in the system, we can obtain by similar experiments the expression $P\underline{V}/NT =$ constant. This latter constant is called the *gas constant*, R, and the relation when written as

$$P\underline{V} = NRT \tag{3-22}$$

is the *ideal gas law*. If the pressure is given in pascals (N/m^2), the volume in m^3/mol, the temperature in kelvins, and N in gram-moles, then R has the value 8.314 J/mol K (N-m/mol K).

Now review carefully the development of this particular temperature scale. We carried out two experiments to measure the volume (or $P\underline{V}$ or $P\underline{V}/N$ products) and then we defined temperatures at these two points as some numbers so that a straight line drawn between these two points intersects the origin.

If the ideal gas law is applied to any two experimental conditions, the expression $T_2 = T_1(P_2\underline{V}_2/P_1\underline{V}_1)$ results. In particular, if the two conditions are the freezing and normal boiling points[4] of water, the ratio of $P_2\underline{V}_2/P_1\underline{V}_1$ is equal to 1.366. The ideal gas temperature scale is called the *Kelvin scale* (K) if the number 273.16 is assigned to the freezing point of water. If desired, other scales for which $T = 0$ at $\underline{V} = 0$ could be devised by assigning a different value to the freezing point of water and by assigning the normal boiling point of water a value of 1.366 times the freezing point.

An *ideal gas* is defined as one that obeys the ideal-gas law, Eq. (3-22), and as one whose total energy is a function only of N and T, that is,

$$\underline{U} = NU_0 + N \int \frac{dU}{dT}\, dT \tag{3-23}$$

The integrand is expressed as a total derivative since, if U is a function only of T, $dU/dT = (\partial U/\partial T)_v = (\partial U/\partial T)_p$, etc. U_0 is a reference energy per mole (or

[4]The normal boiling point is referenced to one standard atmosphere or 1.01325 bars.

mass) of material. It is obvious that in dealing with ideal gases, there is a great simplification since only the variables N and T need be considered. For non-ideal gases (and liquids or solids), $\underline{U}$ would have to be expressed as a function of N plus *two* other independently variable properties, in accordance with Postulate I.

3.6 The First Law for Closed Systems

Many authors refer to Eq. (3-24) as the *First Law.*

$$\Delta \underline{E} = Q - W \tag{3-24}$$

Since Q and W are defined only in terms of interactions at boundaries, Eq. (3-24) has significance only when applied to a specific system.

For a closed system interacting with the surroundings, the composite of system and surroundings can always be considered as a system of constant volume surrounded by adiabatic walls. Since such an isolated system can have no heat or work interactions with its environment, the energy of the composite system is invariant. Therefore,

$$\Delta \underline{E}_{\text{system}} = -\Delta \underline{E}_{\text{surroundings}} \tag{3-25}$$

where Δ is defined as final minus initial. Since we have already shown (Section 3.1) that

$$W_{\text{system}} = -W_{\text{surroundings}} \tag{3-26}$$

it follows that

$$Q_{\text{system}} = -Q_{\text{surroundings}} \tag{3-27}$$

For a differential change in state, the First Law can be written as

$$d\underline{E} = đQ - đW \tag{3-28}$$

where $đQ$ signifies that heat, like work, is a path function.

It should be noted that there may be many processes for going from one stable state to another; for each process, $\Delta \underline{E}$ is the same, but Q and W may be quite different.

Equations (3-24) and (3-28) apply to all processes of closed systems. The restriction of closed systems relates to the manner in which Postulate III was stated. Extension of the First Law to systems with permeable boundaries is presented in Section 3.8.

To the engineer the significance of the First Law may not be readily apparent from Eqs. (3-24) and (3-28). The fact that energy is conserved when two or more systems (forming an isolated composite) interact may have some aesthetic value, but since we do not usually measure energy directly, the statements of the First Law by themselves do not have real utility. We usually measure work effects and properties such as pressure, temperature, volume, and mass because these measurements are relatively simple. It is only when we relate energy (and other derived properties which we introduce later) to other measurable properties by using Postulate I that we obtain the full utility of the First Law. These relationships between properties (both derived and primitive) are not fixed by any principle or law of classical thermodynamics (although they place some restrictions on the form of the functionality, e.g., $\underline{U}$ must be first order in mass). In general, these relationships must be determined empirically, as discussed in Chapters 7 and 8. Equation (3-22) is an example of such a relationship.

Almost all engineering applications of the First Law fall within two categories: (1) for given or measured interactions at the boundaries of a system, what are the corresponding changes in the properties of the system; and (2) for given changes in the properties, what interactions may occur at the boundaries?

3.7 Applications of the First Law for Closed Systems

Applications of the First Law are illustrated by two examples. In both, ideal gases have been chosen. Nonideal gases (and liquids) are treated in Chapters 7 and 8.

In presenting numerical examples where only ideal gases are considered, Eq. (3-22) is used to relate the pressure–temperature–volume variables. To employ Eq. (3-23) we define a term, C_v, the heat capacity at constant volume, to relate energy to temperature:

$$C_v = \left(\frac{\partial U}{\partial T}\right)_v \tag{3-29}$$

(This definition is *not* limited to ideal gases, but applies in all cases.)

Thus, in Eq. (3-23), for an ideal gas,

$$\underline{U} = NU_0 + N \int C_v \, dT \tag{3-30a}$$

or, per mole (or unit mass),

$$dU = C_v \, dT \tag{3-30b}$$

The convenience of limiting the change of U with T to constant-volume cases is evident when simple systems ($E = U$) are treated by Eq. (3-28). Then, if the system has rigid walls ($đW = 0$),

$$dU = C_v \, dT = đQ \qquad (N, V = \text{constant}) \tag{3-31}$$

and $C_v = dQ/dT$ for a case involving constant mass and volume. Again, it is emphasized that Eqs. (3-29) and (3-31) are general and not limited to ideal gases as are Eqs. (3-30a) and (3-30b).

Example 3.3

Two well-insulated cylinders are placed as shown in Figure 3.8. The pistons in both cylinders are of identical construction. The clearances between piston and wall are also made identical in both cylinders. The pistons and the connecting rod are metallic.

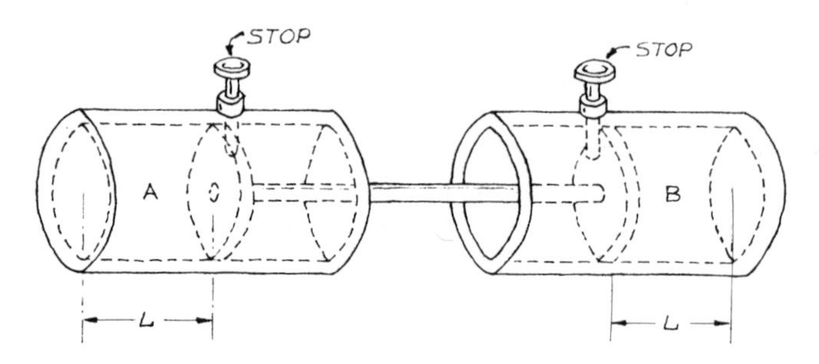

Figure 3.8

Cylinder A is filled with gaseous helium at 2 bar and cylinder B is filled with gaseous helium at 1 bar. The temperature is 300 K and the length L is 10 cm. Both pistons are only slightly lubricated.

The stops are removed. After all oscillations have ceased and the system is at rest, the pressures in both cylinders are, for all practical purposes, identical.

Assuming that the gases are ideal with a constant C_v and, for simplicity, assuming that the masses of cylinders and pistons are negligible (i.e., any energy changes of pistons and cylinders can be neglected), what are the final temperatures?

Solution

Let us choose the gases in compartments A and B as systems A and B, respectively. Since these are simple systems, we denote the energies by $\underline{U}$ instead of $\underline{E}$. Since the mass and initial temperature are known for the gas in each compartment, the final temperature can be calculated for each compartment, if the final energy—or energy change—of each compartment is first determined. Thus, from Eqs. (3-22) and (3-30a) as applied for each system,

$$N_A = \frac{P_{A_i}\underline{V}_{A_i}}{RT_{A_i}} \tag{3-32a}$$

$$N_B = \frac{P_{B_i}\underline{V}_{B_i}}{RT_{B_i}} \tag{3-32b}$$

$$T_{A_f} = T_{A_i} + \frac{\underline{U}_{A_f} - \underline{U}_{A_i}}{N_A C_v} \tag{3-32c}$$

and

$$T_{B_f} = T_{B_i} + \frac{\underline{U}_{B_f} - \underline{U}_{B_i}}{N_B C_v} \tag{3-32d}$$

where $\underline{U}_{A_f}$ and $\underline{U}_{B_f}$ are the only unknowns [$\underline{U}_{A_i}$ and $\underline{U}_{B_i}$ can be chosen at will because U_0 in Eq. (3-30a) is an arbitrary constant].

Since the pistons and shaft have been assumed to be good conductors of heat, the final temperatures of compartments A and B will be equal:

$$T_{A_f} = T_{B_f} \tag{3-33}$$

To determine the energy change, let us apply the First Law, Eq. (3-24), to system A:

$$\underline{U}_{A_f} - \underline{U}_{A_i} = Q_A - W_A \tag{3-34}$$

The work done by this system is equal to the work done on the gas in B and the frictional work done on the walls of A and B, or

$$W_A = -W_B - W_{W_A} - W_{W_B} \tag{3-35}$$

where W_B is the work done *by* the gas in B and W_{W_A} is the work done *by* the wall of A, etc. Now consider the wall of A as the system and apply the First Law to this system:

$$\Delta\underline{U}_{W_A} = Q_{W_A} - W_{W_A} \tag{3-36}$$

If we neglect the change in energy of the wall, the frictional work done by the gas on the wall will be transmitted back to the gas in the form of a heat interaction; that is,

$$Q_{W_A} = W_{W_A} \tag{3-37}$$

and furthermore,

$$Q_A = -Q_{W_A} - Q_{AB} \tag{3-38}$$

where Q_{AB} is the heat conducted from A to B through the pistons. Substituting Eqs. (3-35), (3-37), and (3-38) into Eq. (3-34),

$$\Delta\underline{U}_A = W_B + W_{W_B} - Q_{AB} \tag{3-39}$$

By analogy to Eqs. (3-37) and (3-38) as applied to the walls of B,

$$W_{W_B} = Q_{W_B} = -Q_B + Q_{AB} \tag{3-40}$$

and by analogy to Eq. (3-36),

$$W_B - Q_B = -\Delta\underline{U}_B \tag{3-41}$$

Equation (3-39) becomes

$$\Delta\underline{U}_A = -\Delta\underline{U}_B \tag{3-42}$$

Of course, Eq. (3-42) could have been stated at the outset because the composite system $A + B$ is equivalent to an isolated system; this result, however, may not have been immediately obvious. Combining Eq. (3-42) with Eqs. (3-32c) and (3-32d), and making use of Eq. (3-33), we obtain

$$T_f = \frac{N_A T_{A_i} + N_B T_{B_i}}{N_A + N_B} \tag{3-43}$$

or, for this special case wherein $T_{A_i} = T_{B_i} = T_i$,

$$T_f = T_i = 300 \text{ K}$$

Let us now use hindsight to reevaluate the problem. We had a qualitative feeling for the path in that we knew frictional work was involved, but we could not describe the path quantitatively because we did not know the coefficient of friction of either piston-cylinder. Nevertheless, we were able to determine the final conditions and, therefore, we did not have to describe the path to find the solution. Such a situation would be expected if the end state was independent of the path. This is obviously the case in the present example: Since the composite system $A + B$ is an isolated simple system, there is only one state to which it can go, and that is the one for which $\underline{U} = \underline{U}_{A_i} + \underline{U}_{B_i}$ and $T = T_{A_f} = T_{B_f}$. For an ideal gas, $\underline{U}$ is a unique function of T and N and thus the final temperature can be determined.[5]

[5]If the gas were not ideal, $\underline{U} = f(T, \underline{V}, N)$. Thus, since the final energy, volume, and mass are known, the final temperature could still be determined.

Example 3.4

Consider the situation described in Example 3.3, but with well-insulated pistons and connecting rods of low thermal conductivity. What are the final temperatures after the oscillations have ceased and the pressures have equalized?

Solution

The composite system of $A + B$ is no longer a simple system because it contains an internal adiabatic wall. Therefore, the final composite cannot be described by a single equilibrium state; instead, the final conditions will depend on the path of the process.

With the exception of Eq. (3-33), Eqs. (3-32) through (3-42) are still valid. Combining Eqs. (3-32c) and (3-32d) with Eq. (3-42) results in

$$N_A C_v(T_{A_f} - T_{A_i}) + N_B C_v(T_{B_f} - T_{B_i}) = 0 \tag{3-44}$$

which now gives us one equation in two unknowns, T_{A_f} and T_{B_f}. We could, of course, try to juggle the other equations to find another relationship between T_{A_f} and T_{B_f}, but until we make some assumptions regarding the path, our efforts will be in vain.

If we have no information on the coefficient of friction, we are forced to use our engineering judgment to simplify the situation while obtaining a close approximation to the actual conditions.

Let us assume that there is friction only in compartment B. This will give us a lower bound for T_{A_f} and an upper bound for T_{B_f}. We can then treat the case of friction only in compartment A, which will give us an upper bound for T_{A_f}. In this manner, we can bracket the true solution.

If there is no friction in compartment A and if we assume that the process is quasi-static (i.e., no pressure gradients within the compartment), we can write Eq. (3-28) for the simple system of the gas in A as

$$d\underline{U}_A = -đW_A = -P_A d\underline{V}_A \tag{3-45}$$

since

$$đW_{W_A} = đQ_A = đQ_{AB} = 0$$

Substituting for $\underline{U}_A$ in Eq. (3-45) from Eq. (3-32c), and for $\underline{V}_A$ from Eq. (3-22), Eq. (3-45) becomes

$$N_A C_v dT_A = -N_A R\left(dT_A - \frac{T_A}{P_A} dP_A\right) \tag{3-46}$$

or

$$\left(\frac{C_v + R}{R}\right)\frac{dT_A}{T_A} = \frac{dP_A}{P_A} \tag{3-47}$$

Integrating between initial and final conditions, we obtain

$$\frac{T_{A_f}}{T_{A_i}} = \left(\frac{P_{A_f}}{P_{A_i}}\right)^{R/(C_v+R)} \tag{3-48}$$

Equations (3-48) and (3-44) give us two equations in three unknowns, T_{A_f}, T_{B_f}, and $P_{A_f} = P_f$. The final pressure can be eliminated in the following manner.

Since

$$\underline{V}_{A_f} + \underline{V}_{B_f} = \underline{V}_{A_i} + \underline{V}_{B_i} = \underline{V}_T \tag{3-49}$$

and $\underline{V}_T$ is known, substitution of Eq. (3-22) into Eq. (3-49) gives

$$\underline{V}_T = (N_A T_{A_f} + N_B T_{B_f})\frac{R}{P_f} \tag{3-50}$$

Equation (3-50) together with Eqs. (3-48) and (3-44) give us three equations in these unknowns. From Eqs. (3-50) and (3-44),

$$P_f = \frac{P_{A_i}\underline{V}_{A_i} + P_{B_i}\underline{V}_{B_i}}{\underline{V}_{A_i} + \underline{V}_{B_i}} = 1.5 \text{ bar} \tag{3-51}$$

From Eq. (3-48), with $C_v = 12.6$ J/mol K

$$T_{A_f} = \left(\frac{1.5}{2}\right)^{0.4} T_{A_i} = 267 \text{ K}$$

From Eq. (3-44),

$$T_{B_f} = 366 \text{ K}$$

If it is assumed that there is friction only in compartment A, then we would have found $T_{B_f} = 353$ K. Since in the actual case the friction is distributed between A and B, a better approximation might be $T_{B_f} = (366 + 353)/2 = 360$ K and from Eq. (3-44), $T_{A_f} = 270$ K.

The fact that the adiabatic wall in Example 3.4 prevents a direct solution to the problem in the absence of a complete description of the path is sometimes referred to as the "adiabatic dilemma." In fact, it is no dilemma at all, but results from the difference between heat and work interactions.

3.8 The First Law for Open Systems

A system was defined as being open if it has at least one region of its boundary which is permeable to mass. The First Law, as stated in Section 3.6, was restricted to closed systems. However, since an open system can always be considered as a closed system by redefining the boundaries, the extension of the First Law to open systems requires no additional postulates.

Consider an open system bounded by the σ-surface as illustrated in Figure 3.9.[6] Part of the σ-surface is diathermal and part is movable, so that there may be heat and work interactions, Q_σ and W_σ, with the surroundings. The σ-surface also contains a region through which mass can enter or leave. The boundary at that region may consist of an opening or a permeable membrane.

Consider a time, δt, during which a small quantity of mass, δn_e, enters the system bounded by the σ-surface.[7] The properties of the entering material are pressure, P_e, specific volume, v (volume per mole), and specific energy, e (energy per mole). Although the region bounded by the σ-surface is an open system, the composite system of $\sigma + \delta n_e$ is a closed system. Defining $\underline{E}$ as the total

[6]Throughout the text, underlined capitals are used for *total* properties of the *system*, capitals without underline for *specific* properties of the *system*, and lowercase letters for *specific* properties of entering and leaving streams of open systems.

[7]Rather than employ moles, one could use mass.

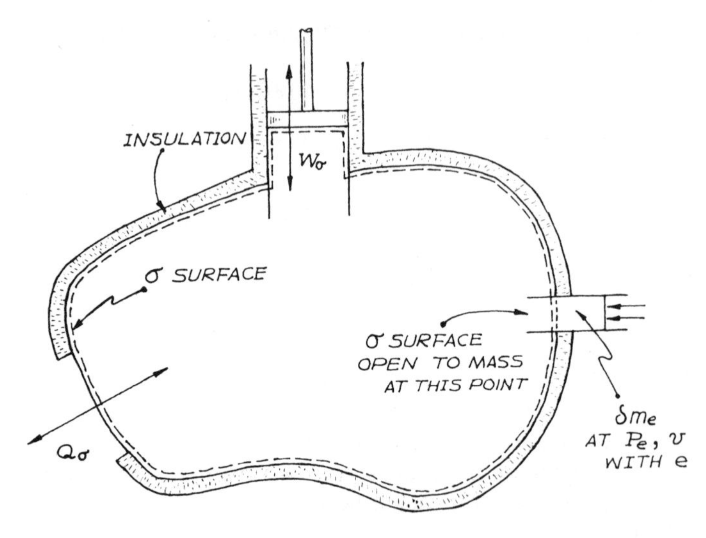

Figure 3.9

energy of the system bounded by the σ-surface, let us apply the First Law to the closed composite system:

$$\underline{E}_2 - (\underline{E}_1 + e\,\delta n_e) = Q_\sigma - (W_\sigma - P_e v\,\delta n_e) \tag{3-52}$$

where subscripts 1 and 2 refer to initial and final conditions, respectively, and $P_e v\,\delta n_e$ is the P–$\underline{V}$ work required to push δn_e into the region bounded by the σ-surface.

Equation (3-52) can be rearranged to a form similar to the First Law for a closed system:

$$\Delta\underline{E} = Q_\sigma - W_\sigma + (e + P_e v)\,\delta n_e \tag{3-53}$$

where $\Delta\underline{E}$, Q_σ, and W_σ apply only to the region bounded by the σ-surface (i.e., the open system), and the last term applies to the mass flux entering the system.

The differential form of Eq. (3-53) is

$$d\underline{E} = đQ_\sigma - đW_\sigma + (e + P_e v)\,dn_e \tag{3-54}$$

and

$$d\underline{E} = d(NE) = N\,dE + E\,dN \tag{3-55}$$

For the general case wherein multiple streams enter and leave the system, the First Law for the open system is

$$d\underline{E} = đQ_\sigma - đW_\sigma + \sum_e (e_e + Pv_e)\,dn_e - \sum_\ell (e_\ell + Pv_\ell)\,dn_\ell \tag{3-56}$$

or, in the integrated form,

$$\Delta\underline{E} = Q_\sigma - W_\sigma + \sum_e \int (e_e + Pv_e)\,dn_e - \sum_\ell \int (e_\ell + Pv_\ell)\,dn_\ell \tag{3-57}$$

where the summations are taken over all entering and leaving streams, and where dn_e and dn_ℓ are both taken as positive quantities. That is,

$$dN = \sum_e dn_e - \sum_\ell dn_\ell \tag{3-58}$$

When the system defined by the σ-surface is a simple system, E may be replaced by U. Similarly, if the entering and leaving masses are simple systems, e may be replaced by u. (Note that the composite closed system is not a simple system if the boundaries at the points of mass flux are semipermeable.) Defining a term called the specific enthalpy (in units of energy per unit mass or mole) as

$$h = u + Pv \tag{3-59}$$

the general form of the First Law for such open systems is, in differential form,

$$d\underline{U} = đQ_\sigma - đW_\sigma + \sum_e h_e\, dn_e - \sum_\ell h_\ell\, dn_\ell \tag{3-60}$$

3.9 Application of the First Law for Open Systems

The variable, enthalpy, introduced in Eq. (3-59) is treated in detail in Chapter 5. However, it has considerable utility in working problems with open systems. Enthalpy is a property as is clear from its definition, written for the system as

$$\underline{H} = \underline{U} + P\underline{V} \tag{3-61}$$

or per mole (or unit mass)

$$H = U + PV \tag{3-62}$$

Also, since U is a function only of temperature [see Eq. (3-23)] and $PV = RT$ for an *ideal gas*, then H also depends only on temperature for this special case.

Analogous to the introduction of C_v [Eq. (3-29)], we define a heat capacity at constant pressure as

$$C_p = \left(\frac{\partial H}{\partial T}\right)_p \tag{3-63}$$

For an *ideal gas*, the restriction to constant-pressure cases is not necessary since, as noted above, H is a function only of temperature.

Finally, again *only* for an *ideal gas*, if the differential of Eq. (3-62) is used with Eqs. (3-29) and (3-63), then

$$C_p = C_v + R \tag{3-64}$$

Example 3.5

A 4-m^3 storage tank containing 2 m^3 of liquid is to be pressurized with air from a large, high-pressure reservoir through a valve at the top of the tank to permit rapid ejection of the liquid (see Figure 3.10). The air in the reservoir is maintained at 100 bar and 300 K. The gas space above the liquid contains initially air at 1 bar and 280 K. When the pressure in the tank reaches 5 bar, the liquid transfer valve is opened and the liquid is ejected at the rate of 0.2 m^3/min while the tank pressure is maintained at 5 bar.

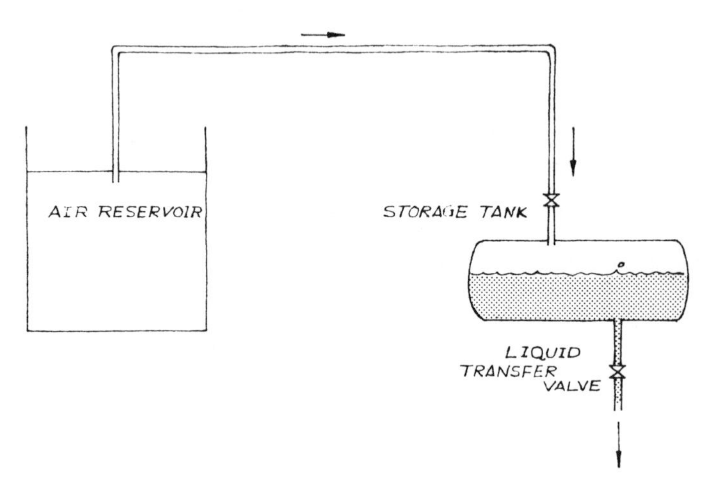

Figure 3.10

What is the air temperature when the pressure reaches 5 bar and when the liquid has been drained completely?

Neglect heat interactions at the gas–liquid and gas–tank boundaries. It may be assumed that the gas above the liquid is well mixed and that air is an ideal gas with a constant value of $C_v = 20.9$ J/mol K.

Solution

Let us treat the process in two steps: (a) the period during which the pressure rises from 1 bar to 5 bar and the volume of the gas in the tank is constant, and (b) the period during which liquid is drained.

Step (a). The most convenient system is the gas in the tank at any time; this is then an open, simple system at constant total volume. With Eq. (3-60) where

$$d\underline{U} = U\,dN + N\,dU$$

$$đQ_\sigma = đW_\sigma = 0$$

$$h_e = \text{constant} \qquad \text{(reservoir large relative to system)}$$

$$dn_e = dN$$

$$\int \frac{dU}{h_e - U} = \int \frac{dN}{N}$$

or

$$\frac{h_e - U_i}{h_e - U} = \frac{N}{N_i} = \frac{P}{P_i}\frac{T_i}{T} \tag{3-65}$$

where the subscript i denotes the initial state. Since the heat capacity of air is assumed constant with temperature, and the gas behaves ideally,

$$h_e = C_p(T_e - T_0) + h_0$$

$$U = C_v(T - T_0) + U_0$$

and

$$h_0 = U_0 + RT_0$$

$$C_p - C_v = R$$

Then Eq. (3-65) becomes

$$\frac{C_pT_e - C_vT_i}{C_pT_e - C_vT} = \frac{P}{P_i}\frac{T_i}{T}$$

or

$$T = \frac{\kappa T_e}{1 + (P_i/P)[\kappa(T_e/T_i) - 1]} \tag{3-66}$$

where

$$\kappa = \frac{C_p}{C_v} = \frac{C_v + R}{C_v} = \frac{20.9 + 8.314}{20.9} = 1.4$$

With $T_e = 300$ K, $T_i = 280$ K, $P_i = 1$ bar, and $P = 5$ bar,

$$T = \frac{(1.4)(300)}{1 + (1/5)[1.4(300/280) - 1]} = 382 \text{ K}$$

At this time

$$N = \frac{P\underline{V}}{RT} = \frac{(5 \times 10^5)(2)}{(8.314)(382)} = 315 \text{ mol}$$

It is interesting to note that, contrary to what might have been anticipated, the final temperature of the gas is higher than that of either the initial temperature or the temperature of the incoming gas. In the limit where $P \gg P_i$, the temperature approaches κT_e independent of the initial conditions in the tank.

Step (*b*). We choose the same system as in step (a). Equation (3-60) is still applicable, but in this case, the total volume varies while the pressure remains constant. Then with

$$d\underline{U} = U\,dN + N\,dU$$

$$đQ_\sigma = 0$$

$$đW_\sigma = P\,d\underline{V} = PN\,dV + PV\,dN$$

$$P = \text{constant}$$

$$dn_e = dN$$

$$h_e = \text{constant}$$

$$U\,dN + N\,dU = -PN\,dV - PV\,dN + h_e\,dN$$

$$N(dU + P\,dV) = N\,d(U + PV) = N\,dH = (h_e - U - PV)\,dN$$

or

$$\int \frac{dH}{h_e - H} = \int \frac{dN}{N}$$

$$\frac{h_e - H_i}{h_e - H} = \frac{N}{N_i} = \frac{V}{V_i}\frac{T_i}{T} = \frac{T_e - T_i}{T_e - T} \tag{3-67}$$

The subscript i in this case represents the conditions existing at the end of step (a). Solving for T, we obtain

$$T = \frac{T_e}{(V_i/V)[(T_e/T_i) - 1] + 1}$$

When the tank has been completely drained, $\underline{V} = 4$ m³ $= 2V_i$. With $T_e = 300$ K and T_i [from (a)] $= 382$ K,

$$T = \frac{300}{(1/2)[(300/382) - 1] + 1} = 336 \text{ K}$$

PROBLEMS

3.1. A small, well-insulated cylinder and piston assembly (Figure P3.1) contains an ideal gas at 10.13 bar and 294.3 K. A mechanical lock prevents the piston from moving. The length of the cylinder containing the gas is 0.305 m and the piston

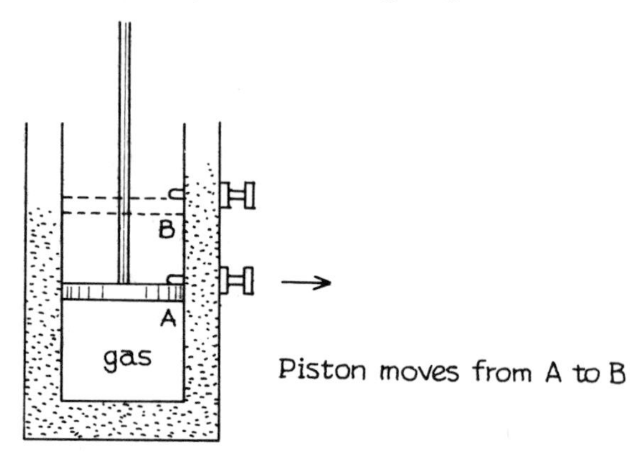

Figure P3.1

cross-sectional area 1.858×10^{-2} m^2. The piston, which weighs 226 kg, is tightly fitted and when allowed to move, there are indications that considerable friction is present. When the mechanical lock is released, the piston moves in the cylinder until it impacts and is engaged by another mechanical stop; at this point, the gas volume has just doubled. The heat capacity of the ideal gas is 20.93 J/mol K, independent of temperature and pressure. Consider the heat capacity of the piston and cylinder walls to be negligible.

(a) As an engineer, can you estimate the temperature and pressure of the gas after such an expansion? Clearly state any assumptions.

(b) Repeat the calculations if the cylinder were rotated both 90° and 180° before tripping the mechanical lock.

3.2. Many very large liquefied natural gas (LNG) storage tanks have been built or are under construction. The LNG is predominantly liquid methane with a boiling point near 111 K at 1 bar. To avoid excessive loss, the tanks are very well insulated. In the construction of such tanks, to encourage the contractor to do the best job, there is normally a clause written into the contract which awards the builder a bonus if the heat leak into the tank is below some agreed value—but there is also a penalty clause if this specified heat leak is exceeded.

Since such penalty (or bonus) values are large, it is crucial to specify a detailed testing procedure to "prove" the heat leak after the tank has been built and filled with LNG. Normally in such a proof test, the filled tank is allowed to attain an equilibrium state with the internal pressure held constant. Ambient conditions should not vary greatly during this period. At the end of this pretest period, the tank and liquid are assumed to be in thermal equilibrium at the existing tank pressure. The actual proof test then consists of measuring the boil-off vapor over a period of several days while keeping the tank pressure equal to the pretest value.

Consider a real test. The LNG tank contains 40,000 m^3 of pure liquid methane. The tank pressure is 1.044 bar. Over the test period, the measured boil-off rate was 4.267×10^4 mol/h. Calculate as accurately as you can the heat leak into the tank (J/h). In this particular case, the contract specified a $500,000 penalty clause if the heat leak exceeded 0.35 GJ/h (0.35×10^9 J/h). A bonus of the same amount was to be awarded to the contractor if the heat leak were less than 0.35 GJ/h.

Data (from the National Bureau of Standards Report NBSIR 73-342): For methane at 112 K,

$$P = 1.044 \text{ bar}$$
$$V^v = 8.6036 \times 10^{-3} \text{ m}^3/\text{mol}$$
$$V^L = 0.0380 \times 10^{-3} \text{ m}^3/\text{mol}$$
$$\frac{dV^v}{dT} = -6.5015 \times 10^{-4} \text{ m}^3/\text{mol K}$$
$$\frac{dV^L}{dT} = 1.309 \times 10^{-7} \text{ m}^3/\text{mol K}$$
$$H^v = 1.27982 \times 10^4 \text{ J/mol}$$
$$H^L = 4.6052 \times 10^3 \text{ J/mol}$$
$$dH^v/dT = 26.8 \text{ J/mol K}$$
$$dH^L/dT = 55.96 \text{ J/mol K}$$
$$dP/dT = 0.0854 \text{ bar/K}$$

Assume that the 40,000-m^3 tank is filled with 99% liquid by volume.

(a) Would the contractor gain or lose the $500,000?

(b) The true facts for the example above were only slightly different: over the 24-h period where the boil-off vapor averaged 4.267×10^4 mole/h, despite the best intentions of the operators, the tank pressure fell from 1.044 to 1.043 bar. Would this fact change the award?

3.3. (Refer to Figure P3.3 for notation.) A piston (*A*) and piston rod (*B*) are fitted inside a cylinder of length 0.508 m and area 6.45×10^{-3} m^2. Although the piston

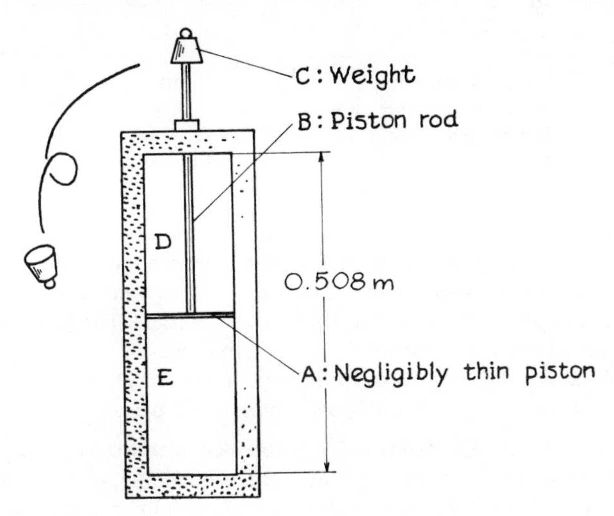

Figure P3.3

is quite thin, it weighs 9.07 kg; the piston rod is 1.29×10^{-3} m^2 in area and weighs 4.53 kg. On top of the rod, but outside the cylinder, an 18.14-kg weight (C) is placed. Originally, gas in D is at atmospheric pressure while the piston is positioned in the middle of the cylinder. Gases D and E are helium and under these conditions may be considered ideal with a constant $C_v = 12.6$ J/mol K. The initial temperature everywhere is 311 K.

Assuming the cylinder, piston, and piston rod to be nonconducting and having a negligible heat capacity, discuss any heat or work interactions if weight C should fall off. What is the final state of the system when the piston has stopped and there is a balance of forces across the piston? Do not neglect the fact that during motion there may be some friction between moving parts. Consider cases in which the piston is (a) diathermal and (b) adiabatic.

3.4. Two cylinders are attached as shown in Figure P3.4. Both cylinders and pistons are adiabatic and have walls of negligible heat capacity. The connecting rod is nonconducting.

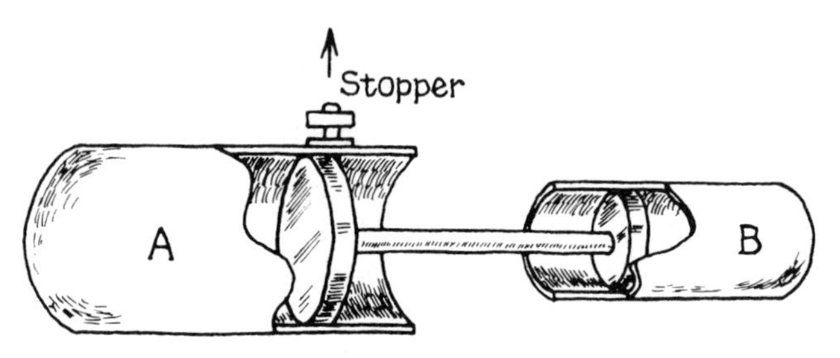

Figure P3.4

The initial conditions and pertinent dimensions are as follows:

	Cylinder	
	A	B
Initial pressure (bar)	10	1
Initial temperature (K)	300	300
Initial volume (m^3)	6.28×10^{-3}	1.96×10^{-4}
Piston area (m^2)	3.14×10^{-2}	1.96×10^{-3}

The pistons are, initially, prevented from moving by a stop on the outer face of piston A. When the stop is removed, the pistons move and finally reach an end state characterized by a balance of forces on the connecting rod. There is some friction in both piston-cylinders during this process. The gases A and B are ideal and have constant values of $C_v = 20.9$ J/mol K.

What are the final pressures in both A and B? Do two cases, one where the ambient pressure is 0 bar and one where it is 1 bar.

3.5. A horizontal cylinder 0.457 m long is divided into two parts, A and B (see Figure P3.5) by a latched piston. As illustrated, the volume of A is twice that of B and contains helium at 311 K and 10.13 bar. B contains hydrogen at 311 K and 1.01 bar. Both of these gases may be considered ideal with constant heat capacities as

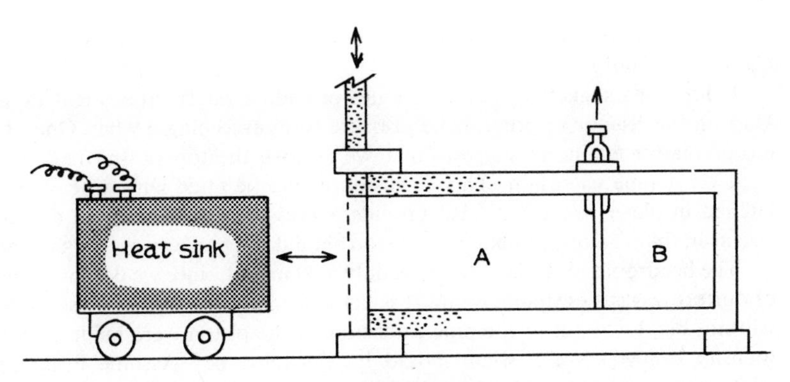

Figure P3.5

follows: C_v(helium) = 12.56 J/mol K, C_v(hydrogen) = 20.9 J/mol K. Provision has also been made to connect volume A to a constant temperature reservoir at 311 K. When the latch is removed, the piston is allowed to seek an equilibrium position so that there are equal pressures on each side.

Without neglecting friction but assuming no heat transfer to the cylinder walls, compute the final position of the piston and the temperature and pressure in both A and B for the following four cases:

No contact of volume A with the 311 K reservoir:
- Piston is diathermal
- Piston is adiabatic

Volume A is in diathermal contact with the 311 K reservoir:
- Piston is diathermal
- Piston is adiabatic

3.6. Our next research experiment is to be carried out in a vertical, cylindrical reactor 0.0929 m^2 in cross-sectional area and 0.605 m long (see Figure P3.6). A gas mix-

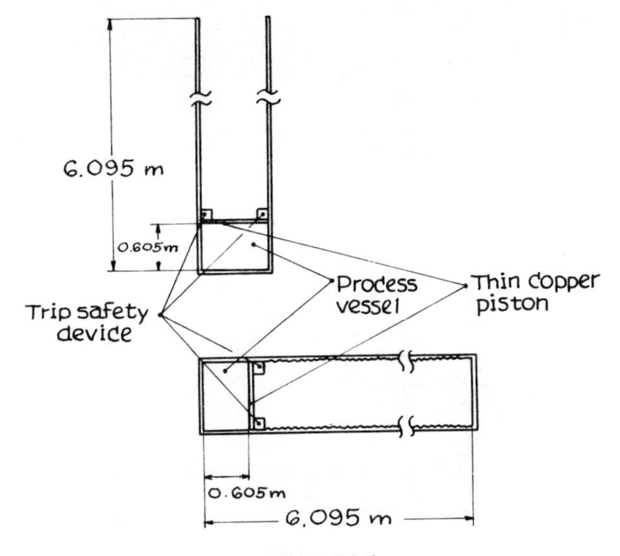

Figure P3.6

ture in the reactor is at 288.8 K and may be considered to be ideal with a constant $C_p = 29.3$ J/mol K.

It has been suggested that we should provide a safety attachment of some kind on the reactor to prevent the pressure from exceeding 6.9 bar. One of our more creative engineers suggests that we remove the top of the reactor and weld on a pipe extension. This extension would be fitted with a heavy piston latched in place. We would also provide a pressure transducer and activation circuit in the reactor to unlatch the piston should the pressure exceed 6.9 bar.

The headroom in the laboratory is only 5.49 m and, since we do not want any of our process gas escaping, we must select a piston of the correct mass so that it will not be blown out of the pipe extension. If the piston were made of copper (density = 8660 kg/m^3), what should its thickness be? Assume insignificant friction and neglect any heat transfer from the gas to the cylinder walls or piston.

Another engineer, however, has been advocating an alternative technique. He, too, wishes to remove the reactor top and weld to it a pipe extension with a latched piston. But he also wishes to put a cap on the top of this extension and rotate the cylinder on to its side. In this case, should the reactor pressure exceed 6.9 bar, the piston would move horizontally and compress the gas in the pipe extension. He also wishes to roughen the walls in the pipe so that there is friction between the moving piston and pipe walls but no gas leakage.

(a) Again assuming negligible heat transfer, what is your best estimate of the final gas temperatures on both sides of the piston after pressures on both sides are equal? Assume initially that the gas in the closed pipe extension is air at 1 bar and 288.8 K.

(b) Which of these two techniques would you select? Why?

3.7. Advertised is a small toy that will send up a signal flare and the operation "is so simple that it is amazing" (see Figure P3.7). Our examination of this device

Figure P3.7

indicates that it is a sheet metal tube 2.13 m long and 645 mm^2 in area. A plug shaped into the form of a piston fits into the tube and a mechanical trigger holds it in place 0.61 m above the bottom. The piston weighs 1.57 kg and contains the necessary parachute and pyrotechnics to make the show exciting. To operate the device, the volume below the piston is pumped up to a pressure of about 4.05 bar with a small hand pump, and then the trigger is depressed, allowing the piston to fly out the top. The pyrotechnic and parachute devices are actuated by the acceleration force during ejection.

When we operated this toy last summer, the ambient temperature was 305 K.

(a) Assuming no friction in the piston and no heat transfer or other irreversibilities in the operation, how high would you expect the piston to go? What would be the time required from the start to attain this height?

(b) Since you are an engineer who is never satisfied with a commercial object, suggest improvements to make the piston go even higher. What is the maximum height that could be obtained if it were limited to 4.05 bar pressure?

(c) Comment on the way you might analyze the expected performance if the restrictions in part (a) are removed.

3.8. Bottles of compressed gases are commonly found in chemistry and chemical engineering laboratories. They present a serious safety hazard unless they are properly handled and stored. Oxygen cylinders are particularly dangerous. Pressure regulators for oxygen must be kept scrupulously clean, and no oil or grease should ever be applied to any threads or on moving parts within the regulator. The rationale for this rule comes from the fact that if oil *were* present—and if it *were* to ignite in the oxygen atmosphere—this "hot" spot could lead to ignition of the metal tubing and regulator and cause a disastrous fire and failure of the pressure container. Yet it is hard to see how a trace of heavy oil or grease could become ignited even in pure, compressed oxygen since ignition points probably are over 800 K if "nonflammable" synthetic greases are employed.

Let us model the simple act of opening an oxygen cylinder that is connected to a closed regulator (see Figure P3.8). Assume that the sum of the volumes of the connecting line and the interior of the regulator is V_R. V_R is negligible compared to the bottle volume. Opening valve A pressurizes V_R from some initial

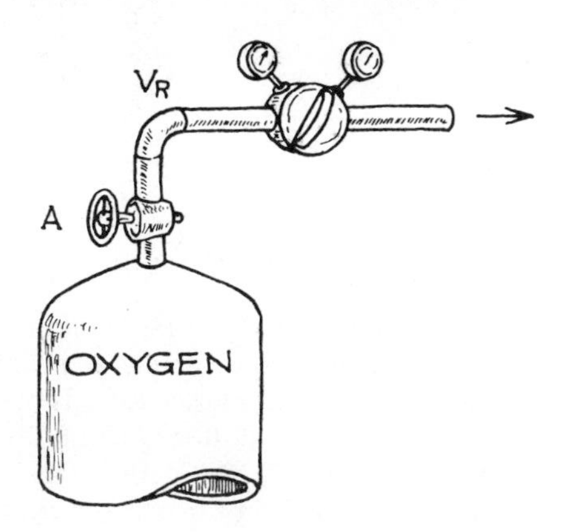

Figure P3.8

pressure to full bottle pressure. Presumably, the temperature in V_R also changes. The question we would like to raise is: Can the temperature in V_R ever rise to a sufficiently high value to ignite any traces of oil or grease in the line or regulator?

Data: The oxygen cylinder is at 15.17 MN/m² and 311.0 K. The connecting line to the regulator and the regulator interior (V_R) are initially at 0.101 MN/m², 311 K, and contain pure oxygen.

Assume no heat transfer to the metal tubing or regulator during the operation. Oxygen is essentially an ideal gas and $C_p = 29.3$ J/mol K, $C_v = 20.9$ J/mol K, independent of pressure or temperature.

(a) If the gas entering V_R mixes completely with the initial gas, what is the final temperature in V_R?

(b) An alternative model assumes that there is *no* mixing between the gas originally in V_R and that which enters from the bottle. In this case, after the pressures are equalized, we would have two identifiable gas slugs which presumably are at different temperatures. Assuming no axial heat transfer between the gas slugs, what is the final temperature of each?

(c) Comment on your assessment of the hazard of this simple operation of bottle opening. Do you think the models in (a) and (b) are realistic? Can you suggest other models?

3.9. During an emergency launch operation to fill a missile with RP-4 (a kerosene-based fuel), the ullage volume of the fuel storage tank is first pressurized with air from atmospheric pressure (1.01 bar) to a pressure of 10.34 bar (see Figure P3.9).

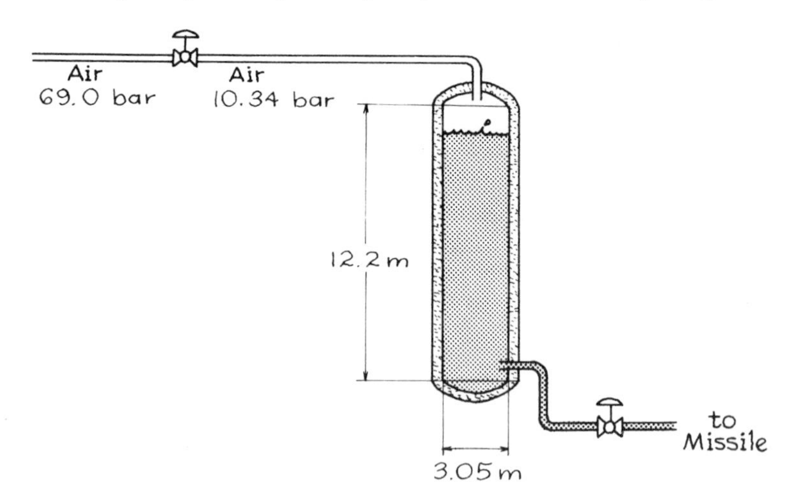

Figure P3.9

The air is available from large external storage tanks at high pressure (69.0 bar). This operation is to be completed as rapidly as possible. After the 10.34-bar pressure level is reached, the main transfer valve is opened and fuel flows at a steady rate until the missile is loaded. It is necessary to maintain a constant gas pressure of 10.34 bar inside the fuel tank during transfer.

The fuel storage tank can be approximated as a right circular cylinder 12.2 m tall and 3.05 m in diameter and is originally filled to 90% of capacity. Transfer of fuel to a residual volume of 10% must be completed in 18 min. Assume ideal gases and that the operation is adiabatic and all hardware has negligible heat capacity. Initial temperatures may range from 242 K (arctic sites) to 333 K (equatorial sites), but for the purposes of a first estimate, use 294 K as an initial temperature.

(a) Comment on any safety hazards that might be encountered.

(b) What problems would you anticipate if the inlet gas control valve were to malfunction and the gas space above the fuel were to reach full storage tank pressure (69.0 bar)? (The fuel tank has been hydrostatically tested to 276 bar.)

(c) What is your estimate of the time–temperature history of the gas above the fuel during the entire operation? These data are needed to size the inlet air lines.

3.10. In many installations in the chemical industry, occasions arise when compressed gas bottles are rapidly blown down. These bottles are constructed of carbon steel that becomes dangerously brittle at low temperatures. Certainly, for rapid-blowdown situations the gas temperature could drop to such a low value that if rapid heat transfer with the cylinder were to occur, a hazardous operation would result. It is believed that the only place where very high heat transfer rates are possible is in the cylinder neck, because velocities are highest in this region. To calculate neck wall temperatures, however, the time variation of the bulk gas temperature of the bottle must be available.

Demonstrate your ability to estimate the bulk gas temperature of the bottle gas as a function of time for the first 2 min in the following case:

Inside wall area = 9.290 m^2

Volume = 0.7788 m^3

Thickness of wall = 8.38 mm

Specific heat of wall = 0.419 J/g K

Density of wall = 8.46×10^3 kg/m^3

Initial bottle pressure = 137.9 bar

Rate of pressure decay: bottle pressure is reduced by factor of 2 every 1.6 min

Bottle gas: nitrogen (assume ideal gas behavior)

Initial bottle gas and wall temperatures = 241.5 K

It may assumed that heat transfer from the cylinder wall to the gas occurs by a natural convection process. For purposes of computation, assume that the heat transfer coefficient is 45.4 W/m K for all bottle pressures in excess of 69 bar and 34.1 W/m K for pressures less than 69 bar.

3.11. A rigid laboratory gas cylinder of 5.66×10^{-2} m^3 volume is charged with air at 138 bar and 277.8 K. An experiment is to be carried out whereby the cylinder is rapidly vented by opening the valve on the cylinder top. The pressure in the cylinder is always so high that the flow is choked (i.e., sonic) in the valve throat. If one assumes that the gas is ideal and that there is negligible heat transfer between the gas and walls:

(a) Derive a relation to calculate the bottle pressure as a function of time.

(b) Calculate the time when the bottle pressure drops to 69 bar.

For sonic flow through a round, sharp-edged orifice (assumed to apply to the valve throat),

$$\text{mass flow} = C_a A P \left[\frac{\kappa m}{RT} \left(\frac{2}{\kappa + 1} \right)^{(\kappa+1)/(\kappa-1)} \right]^{1/2}$$

where C_a = discharge coefficient (assume = 0.6)
A = orifice area = 9.29×10^{-4} m²
P = cylinder pressure, N/m²
R = gas constant, 8.314 J/mol K
$\kappa = C_p/C_v = 1.4$ for air (assume constant)
m = molecular weight = 29×10^{-3} kg/mol
mass flow in kg/s

3.12. A well-insulated pipe of 2.54 cm inside diameter carries air at 2 bar pressure and 366.5 K. It is connected to a 0.0283-m³ insulated "bulge," as shown in Figure P3.12.

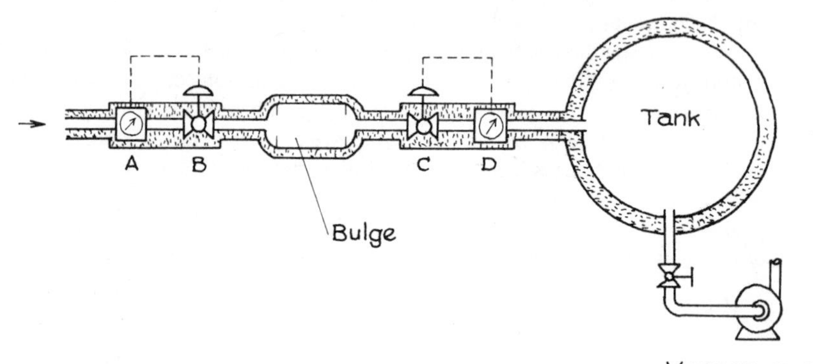

Figure P3.12

The air in the bulge is initially at 1 bar pressure and 311 K. A and D are flow meters which accurately measure the mass rate of airflow. Valves B and C control the airflow into and out of the bulge. Connected to the bulge is a 0.283-m³ rigid, adiabatic tank which is initially evacuated to a very low pressure.

At the start of the operation, valve B is opened to allow 4.54 g/s of air to flow into the bulge; simultaneously, valve C is operated to transfer exactly 4.54 g/s from the bulge into the tank. These flows are maintained constant as measured by the mass flow meters.

Air may be assumed an ideal gas with a constant C_p of 29.3 J/mol K. Assume also that the gases, both in the bulge and large tank, are completely mixed so that there are no temperature or pressure gradients present.

(a) What is the temperature and pressure of the gas in the bulge after 6 s?
(b) What is the temperature and pressure of the air in the large tank after 3 s?

3.13. A vessel containing a reactive compound is about 0.06 m³ in volume (gas space). There is an inert atmosphere of helium maintained at 1 bar and 311 K. If, however, the compound shows signs of decomposition, it is desired to increase very rapidly the helium overpressure to 10 bar. This higher pressure will then be used to dump the reactive compound to a water-soak tank. To accomplish this rapid pressurization, the vessel is connected by a short transfer line and valve to another vessel filled with high-pressure helium. This vessel is 0.18 m^3 and contains helium originally at 20 bar and 311 K.

(a) What is the pressure of the helium supply vessel after a pressurization of the reactor? Assume ideal gases and adiabatic operation.

(b) How many reactors could the supply vessel serve simultaneously? (Each reactor is 0.06 m^3 and is pressurized from 1 to 10 bar.)

3.14. To reduce gas storage costs, two companies, *A* and *B*, have built a common storage tank in the shape of a horizontal right circular cylinder 0.3 m in diameter and 30 m long (see Figure P3.14). To decide how much gas each company uses

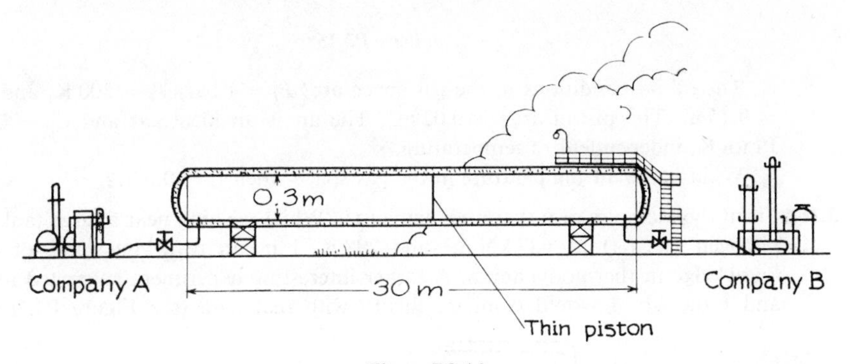

Figure P3.14

between refills, a thin piston was placed in the tank. The piston moves freely, that is, there is essentially no friction present, and the pressure is the same on both sides. Thus, as company *A* uses gas, the piston moves left and as company *B* uses gas, the piston moves right.

When the gas company refills the tank, it must decide how much gas has been used by each company. It can easily measure the position of the piston and can, if necessary, install other instrumentation such as thermometers or pressure gauges in either or both ends of the tank.

Assume that (1) the gas is ideal; (2) the piston is adiabatic; (3) the walls are well insulated and have a low heat capacity; and (4) at the start of each month after filling the tank, the gas company positions the piston in the center of the tank, equalizes the temperature on both ends, and carefully meters the total amount of gas added.

List the minimum instrumentation that you would recommend, and show from this list how the amount of gas consumed by both companies could be determined at the time of refilling.

3.15. A thermodynamicist is attempting to model the process of balloon inflation by assuming that the elastic casing behaves like a spring opposing the expansion (see Figure P3.15). As air is admitted, the spring is compressed. The pressure in the gas space is given by

$$P - P_i = k(L - L_i), \qquad k = \text{constant} = 5 \text{ bar/m}$$

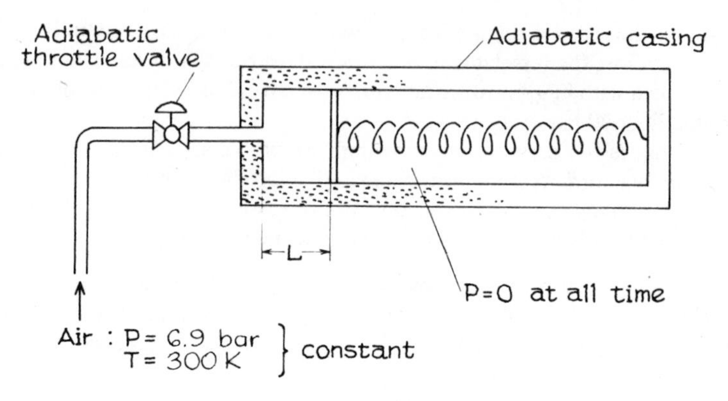

Figure P3.15

The initial conditions in the gas space are: $P_i = 1$ bar, $T_i = 300$ K, and $L_i = 0.15$ m. The piston area is 0.02 m^2. The air is an ideal gas and $C_v = 20.9$ J/mol K, independent of temperature.

What is the air temperature in the gas space when $L = 0.6$ m?

3.16. From the memoirs of a thermodynamicist: "While relaxing near a large tank of nitrogen gas (A) at 687 kN/m^2 and 298 K, I began reviewing some of my knowledge in thermodynamics. A rather interesting experiment suggested itself and I thought I would compare theory with real data (see Figure P3.16). I

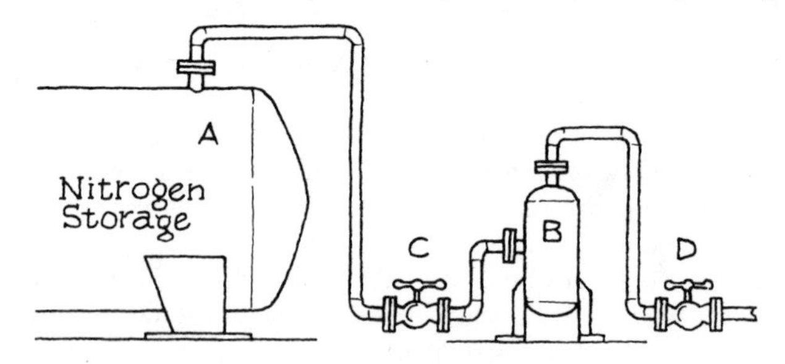

Figure P3.16

obtained a small high-pressure vessel (B) and two valves (C) and (D). I first filled B with nitrogen gas at 101 kN/m^2 and 298 K and connected it as shown. Then working quickly, I opened valve C (with D closed) and allowed the pressures in B and A to equalize. Then, I quickly closed C and opened D to blow down vessel B to its original pressure. I repeated this sequence a number of times. Tank A was so large that I did not cause any significant drop in pressure in it by my experiments. Second, I pressurized and blew down B so rapidly that little heat transfer probably occurred during this time." Nitrogen has a value of $C_p = 29.33$ J/mol K and is an ideal gas, so $C_p - C_v = R$.

(a) Guess the temperature of the gas in B after the second pressurization and after the second blowdown.

(b) What do you think these temperatures were after a very large number of cycles?

3.17. An all-quartz Dewar flask is filled initially with liquid hydrogen at 1 bar (see Figure P3.17). The inner walls cool to the normal boiling point of hydrogen 20.6 K. The liquid is then quickly poured out and the flask evacuated to a very low pressure. Assume that at the end of the evacuation the walls are still at 20.6 K.

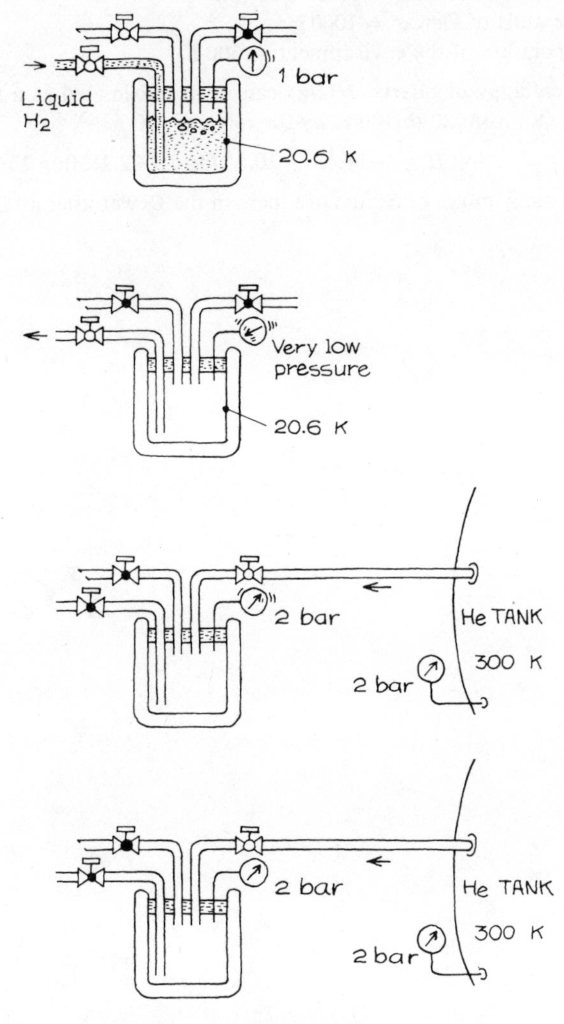

Figure P3.17

The Dewar is then connected to a large tank of helium at 2 bar and 300 K and pressurized very rapidly to 2 bar. After pressurization, the connecting line is left open to allow additional flow to occur in order to maintain a pressure of 2 bar. There is heat transfer between the helium gas and inner Dewar walls, but assume no heat transfer by radiation, convection, or conduction across the walls of the Dewar.

Data:

C_p(helium) = 20.9 J/mol K
C_v(helium) = 12.6 J/mol K
Helium is an ideal gas
Dewar flask volume = 8.206×10^{-3} m^3
Inner walls of Dewar = 1000 g
Temperature of the environment = 300 K

The enthalpy of quartz, H(J/g), can be approximated as a function of temperature (K) from 20 to 100 K by the relation

$$\log H = -11.43 + 10.64 \log T - 2.215(\log T)^2$$

How many moles of helium are there in the Dewar after all flow has ceased?

Reversibility: Concepts and Consequences 4

In Postulate III we implied that only certain processes between stable equilibrium states may be possible. We have in fact already stated all the postulates necessary for determining which processes are and are not possible. In this chapter we explore in greater depth the consequences of these postulates.

One of the underlying concepts in the developments to follow is the reversible process, which is discussed in Section 4.2. This process has not been found to occur in reality; it is in fact a limiting condition that cannot be attained, but it can be closely approached. Consequently, this concept cannot be developed fully by examining real systems in the laboratory; instead, we must utilize thought experiments. In these thought experiments, we shall use a device called a heat engine, which is described in the next section. Although many kinds of heat engines (also referred to as power cycles) have been in use for centuries, we will employ almost exclusively the fictitious reversible heat engine.

The approach we follow is in many ways similar to the historical development of thermodynamics in that it was the introduction of a reversible heat engine by Carnot in 1824 that formed the basis of the introduction of the concept of entropy by Clausius in 1850.

4.1 Heat Engines

A *heat engine* is a closed device that undergoes heat interactions with one or more systems and work interactions with a *work reservoir*. The work reservoir is a device that operates adiabatically and quasi-statically and is used for storing energy. For example, a system of weights at different levels in a gravitational field can be used as a work reservoir. The heat engine always undergoes a *cyclic* process such that any net effects appear only in the external systems and in the work reservoir. The definition above excludes various open-system power cycles, such as the internal combustion engine.

There is great diversity in kinds of heat engines. Although we need not specify any particular one for the developments to follow, it may be instructive to illustrate the operation with a common Rankine cycle used in most stationary power plants. The internal working fluid, usually water, is vaporized at a high temperature and high pressure in a boiler. Useful work is obtained by expanding the vapor to a low pressure in a turbine. The low-pressure vapor is liquefied in a condenser, and the liquid is pressurized and returned to the boiler.

The heat engine is a convenient device for evaluating processes (real and imaginary). We shall use heat engines to determine whether or not a process is consistent with the postulates and concepts that have been developed previously. A process will be considered allowable if we know of a real case in which the process occurs. (If such a process violates one of our postulates, we would have to revise the postulates.) If we have no prior experience for a given process, the process will be considered *impossible* if we can prove that the process leads to a violation of one or more of our postulates. If however, we cannot prove that the process violates a postulate, the process may be possible or impossible since we may not have been clever enough to show that a violation exists.

In the processes shown in Figure 4.1, we will choose systems A and B that are in stable equilibrium states prior to their participation in the heat engine

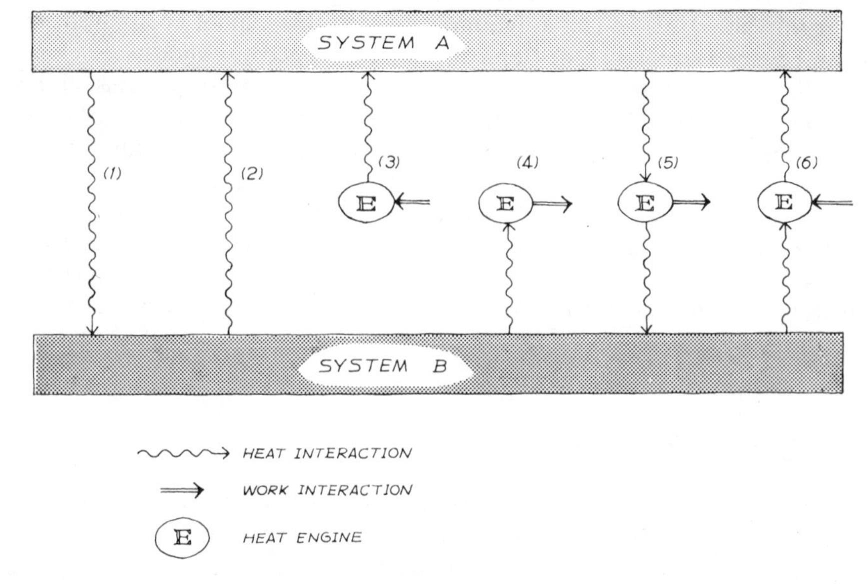

Figure 4.1

interaction. We will also choose them so that $\theta_A > \theta_B$ (recall that we have chosen by convention that a rise in θ corresponds to an increase in energy). Now let us consider a number of conceivable processes.

Case 1. A heat interaction occurs from A to B (i.e., the energy of A decreases and the energy of B increases) without any work being performed. Since we know of real cases in which such interactions exist, case (1) is clearly allowable.

Case 2. A heat interaction occurs from B to A without any work being performed. Since we have notched our thermometer so that $dE/d\theta > 0$, and since this process leads to an increase in energy of A and a decrease in energy of B, the net effect of this process is to increase $\Delta\theta = \theta_A - \theta_B$. This process is in violation of Postulate II since the composite system $A + B$ does not tend to a state of equilibrium (i.e., $\Delta\theta$ does not tend to zero). Since we have no prior knowledge of real cases in which such processes occur, this process is impossible. Thus, it is concluded that any process in which the net effect is the transfer of heat from a cooler to a hotter system is impossible. A similar conclusion was drawn by Clausius over a hundred years ago, and some authors refer to this result as Clausius's statement of the Second Law.

Case 3. A work interaction occurs whereby work from the reservoir passes to the engine and results in a heat interaction with A. This process is allowable since we know of real cases in which this process occurs [e.g., in Figure 3.5(a) consider B as the engine and C as the work reservoir]. Note that the heat interaction could also have been to system B instead of A.

Case 4. A heat interaction involving only B occurs to decrease the energy of B and all of this energy appears as work in the work reservoir. This process can be shown to lead to a violation of Postulate II. If process (4) could occur, we could use process (3) to extract the work produced in (4) and convert this to a heat interaction with A. The net result of the combined processes is equivalent to process (2), which is impossible. Since process (3) is possible, process (4) must be impossible. Thus, it is concluded that any cyclic process for which the net result is the conversion of energy of a single system to work is impossible. Many authors refer to this conclusion as the Kelvin–Planck statement of the Second Law. Impossible processes of this kind are sometimes referred to as perpetual-motion machines of the second kind (PMM2).[1]

Case 5. A heat interaction occurs between A and the engine. Some work is produced in the reservoir and there is also a simultaneous heat interaction between the engine and system B to increase the energy of B. There is nothing in the postulates to prevent such an occurrence, and processes of this kind

[1]Note that these processes do not necessarily violate the First Law. Perpetual-motion machines of the first kind (PMM1) refer to processes that lead to a net change in the energy of the universe.

are well known; the Rankine power cycle is an example. Note that although the direction of the work interaction can be reversed, altering either of the heat interaction vectors leads to an impossible process.

Case 6. In case 5, all of the arrows are reversed, the net effects being extraction of work from the reservoir, decrease in energy of B and increase in energy of A. Again, there is no violation of the postulates and real cases of process 6 are known as refrigerators.

Of all the cases discussed, 5 and 6 are of the most immediate interest. We note that 5 and 6 are opposites, and any enterprising person could immediately conjure up some interesting combined processes. As yet we have not placed any quantitative value on the heat and work interactions except to ensure that we have not violated the First Law, which necessitates the conservation of energy. Why couldn't we use case 5 and take 100 units of energy from A, put 90 of them in the work reservoir, and reject 10 to B? Then, following this, we could use case 6 to take 50 units from B, 50 units from the work reservoir, and reject 100 units to A. In this sequence, system A has undergone a cycle—but the work reservoir has gained a net 40 units and system B has lost 40 units. This particular combination of cases 5 and 6, carried out as prescribed, is in reality case 4, which we found to be impossible. Thus, some combinations of cases 5 and 6 lead to impossible processes, whereas other combinations are allowable. To avoid combinations that lead to a violation of our postulates, we must specify some limitations to the way in which the heat and work effects are split. We will find out later that the split depends on the temperatures of systems A and B, but as of now we simply recognize that there is a limitation.

A convenient way to delineate the split is to specify an efficiency, η, for case 5 as

$$\eta_5 \equiv \frac{\text{work done by engine}}{\text{heat transferred from hot system}} \tag{4-1}$$

or

$$\eta_5 = -\frac{W_E}{Q_A} \qquad \text{(4-2)}^2$$

That is, the more efficient we are, the more work we can get out of a given heat interaction between system A and the engine.

We found from case 4 that the efficiency cannot equal unity; what, then, does limit the efficiency? In a practical sense, we realize that factors such as friction and other resistances will decrease engine efficiency, but we have made no mention of such factors in deciding that there exists a limiting value of the efficiency. In fact, we assume in the thought experiments to follow that we can

[2]Following the conventions described in Chapter 3, W_E is the work for the system comprising the engine and is positive when work is done *by* the engine. Q_A is the heat interaction for system A and is positive when the interaction *increases* the energy of A. In all allowable processes, only one of the two terms is negative, and therefore η is positive.

construct an engine that is not plagued by friction and other resistances; such engines will be referred to as *reversible* heat engines.

The efficiency of a heat engine must be less than 1:

$$\eta_5 < 1 \tag{4-3}$$

If we define the efficiency for case 6 by Eq. (4-1),[3] then to avoid a combination of processes 5 and 6 which violates our postulates (as discussed above), we require that

$$\eta_5 \leq \eta_6 < 1 \tag{4-4}$$

Note that Eq. (4-4) must be satisfied regardless of the kind of engine.

4.2 Reversible Processes

It can be seen by inspection of Eq. (4-4) that the most efficient engines that could be conceived would correspond to $\eta_5 = \eta_6$. For these engines we could operate process 5 to obtain the maximum work and then use process 6 to restore systems *A* and *B and* the work reservoir to their original states. The combination is an example of a *reversible cycle*, and the engines involved are referred to as reversible engines.

In the general case, *a process will be called reversible if a second process could be performed in* <u>*at least one way*</u> *so that the system and all elements of its environment can be restored to their respective initial states, except for differential changes of second order.*[4]

It can readily be shown that each step in the path of a reversible process must be reversible. It can also be shown that in a reversible process, all systems must be in states of equilibrium at all times (i.e., all subsystems must traverse quasi-static paths). The proof follows from Postulate II: If a system in a nonequilibrium state is isolated, it will tend toward a state of equilibrium. Since there is no way to transform a system from a state of equilibrium to the nonequilibrium state without removing it from isolation, any process involving an intermediate nonequilibrium state is irreversible. Many useful corollaries follow directly from this last conclusion. For example, it can be shown that simple systems involved in reversible processes can have no internal pressure or temperature gradients.

Finally, it can be shown that friction and similar resistances must not be present if a process is to be reversible. The proof follows from the fact that

[3]The efficiency of refrigeration cycles such as case 6 are more commonly measured by the coefficient of performance, ω, where

$$\omega \equiv -\frac{\text{heat transferred from cold system}}{\text{work done by engine}}$$

[4]It can be proved that the maximum work obtainable from, for example, an expansion process corresponds to the hypothetical case in which the boundary is moved at an infinitesimal rate for an infinite time. Thus, the difference in the maximum work obtained in an expansion and the minimum work required for the reverse compression is of the order of $(dP)(d\underline{V})$.

the work required in such processes exceeds the minimum (or the work obtained is less than the maximum) because a finite unbalance of boundary forces is required to effect the changes involved.

Example 4.1

A rigid, closed vessel with adiabatic walls is divided internally by a strong diaphragm. On one side, air is present at 300 K and 1 bar; the other side is evacuated. The diaphragm is broken and air fills the entire vessel. Is this a reversible process?

Solution

It is not easy to employ the definition of a reversible process to answer the question posed above. An alternative technique to use in such cases is to assume that the given process *was* indeed reversible and then to examine whether the consequences of the assumption lead to a violation of any of the postulates.

In this case, if the expansion were reversible, then, since the heat and work interaction between the system and the environment are zero, it would be possible to propose the opposite process. That is, given a vessel filled with air at some P, T, there would be a way to have a change of state of the gas, with $Q = W = 0$, so that at the end all the gas is located at one end while there would be a vacuum in the rest of the vessel. Then, visualize a cyclic process wherein we begin with the system as initially described. But we have a piston rather than a diaphragm. We allow the piston to move into the evacuated portion and obtain work to be stored in an external work reservoir. Since the gas will cool, we will allow a heat interaction to occur so that at the end of the expansion we have the same state of the air as in the diaphragm-breaking case, but we have had a heat interaction with the environment and obtained work. Then we invoke the assumed reversibility to go back to the initial composite case with no further Q or W interactions. It is clear that this cyclic process is identical to case 4 of Figure 4.1, which we showed violated Postulate II.

Thus, the original assertion that the expansion was reversible is untrue and the process is irreversible.

4.3 Thermodynamic Temperature

One of the major results of the previous section can be summarized as follows: The efficiency of all cycles involving reversible heat engines that operate between two given systems with different thermometric temperatures is a constant. It is a simple matter to extend this reasoning to show that the efficiency of any reversible engine is dependent on the thermometric temperatures of both systems with which it interacts. For example, we can prove that the efficiency must depend on the temperature of the cold system in the following manner. Let us assume that η is only a function of the temperature of the hot system and then show that this assumption cannot be valid. If we were to operate a reversible engine as in case 5 between two systems at θ_A and θ_B, and also operate a reversible refrigeration cycle as in case 6 between two systems at θ_A and θ_C, the efficiencies of the two processes will be the same (if the initial assumption is correct). If we had chosen systems such that $\theta_A > \theta_B > \theta_C$, the net effect of the combined process would have been a transfer of heat from system C to system B. But since θ_B was chosen to be greater than θ_C, the net effect, which is equivalent to case 2, is in violation of our postulates. Therefore, our assumption was incorrect; the efficiency of the reversible engines must involve the temperature of the cold system. In this manner, it can be shown that the efficiency is a function of the temperatures of both systems. Thus,

$$\eta = -\frac{W_E}{Q_A} = f_1(\theta_A, \theta_B) \tag{4-5}$$

Since

$$W_E = -(Q_A + Q_B) \tag{4-6}$$

where Q_B is positive for heat transferred *to* system B, Eq. (4-5) could be expressed in the equivalent form

$$\frac{Q_B}{Q_A} = f_2(\theta_A, \theta_B) = \eta - 1 \tag{4-7}$$

As a result of the foregoing deductions, the primitive property, thermometric temperature, has assumed a role of prime significance. Of course, we could have used any number of clever techniques to define properties that have the same information content as the thermometric temperature (e.g., electrical resistance, thermocouples, thermal electron emission detectors, infrared emission analyzers, etc.),[5] and we would have arrived at the same result expressed by Eqs. (4-5) and (4-7). In fact, until we specify the form of the function f_1 or f_2, any primitive temperature measurement would be equally acceptable.

Let us now look at the problem of determining the form of the function f_1 or f_2. We imagine our systems A and B (with θ_A greater than θ_B) operating as in case 5 to produce work with a reversible heat engine, E_1 [see Figure 4.2(a)]. Imagine that these systems are so large that they do not change in state by a

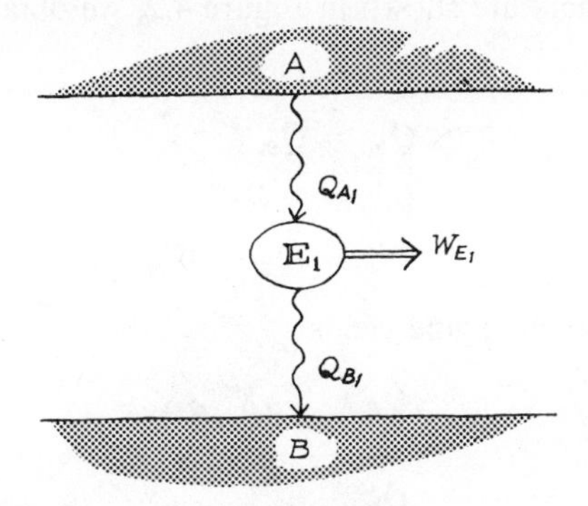

Figure 4.2(a)

[5] Recall the definition of a "temperature"-measuring device. It is a system, closed and rigid, which shows a variation in at least one primitive property when allowed to undergo a finite heat interaction with another system whose temperature is being measured.

significant amount during the process. Now connect to system A *another* heat engine, E_2, which in this case rejects heat to a new system C. System C has a temperature intermediate between A and B. Also, any heat transferred to C is immediately transferred to a third reversible engine, E_3, which rejects heat to system B [see Figure 4.2(b)].

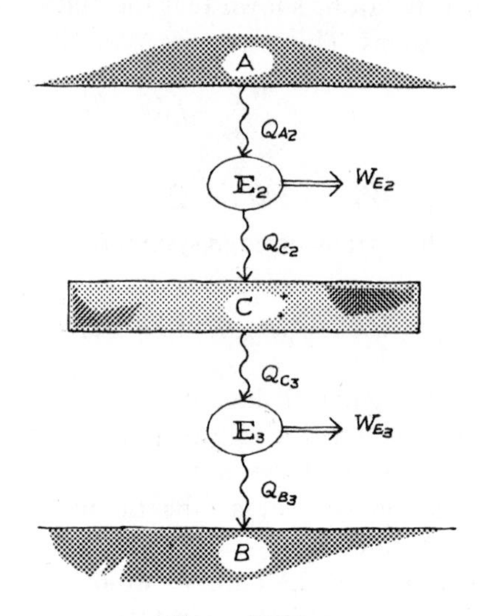

Figure 4.2(b)

If each of the processes shown in Figure 4.2(a) and (b) reduces the energy of system A by an equal amount, and if there is no accumulation of energy in system C, it is obvious that the work obtained from engine E_1 is equal to the sum of the work from engines E_2 and E_3. If this were not so, the more efficient of the two procedures could be reversed, resulting in a violation of Postulate II.

Using the nomenclature shown in Figure 4.2, we obtain

$$Q_{A_1} = Q_{A_2} \tag{4-8}$$

$$Q_{B_1} = Q_{B_3} \tag{4-9}$$

$$Q_{C_2} = -Q_{C_3} \tag{4-10}$$

$$W_1 = W_2 + W_3 \tag{4-11}$$

Using Eq. (4-7) for each engine yields

$$\frac{Q_{B_1}}{Q_{A_1}} = f_2(\theta_A, \theta_B) \tag{4-12}$$

$$\frac{Q_{C_2}}{Q_{A_2}} = f_2(\theta_A, \theta_C) \tag{4-13}$$

$$\frac{Q_{B_3}}{Q_{C_3}} = f_2(\theta_C, \theta_B) \tag{4-14}$$

Multiplying Eq. (4-13) by Eq. (4-14) and equating the result to Eq. (4-12), we get

$$[f_2(\theta_A, \theta_C)][f_2(\theta_C, \theta_B)] = -f_2(\theta_A, \theta_B) \tag{4-15}$$

This result places a definite restriction on the form of the function f_2; that is,

$$f_2(\theta_j, \theta_i) = -\frac{f_3(\theta_j)}{f_3(\theta_i)} \quad \text{or} \quad -\frac{f_4(\theta_i)}{f_4(\theta_j)} \tag{4-16}$$

If we adopt the former form, Eq. (4-12) reduces to

$$\frac{Q_{B_1}}{Q_{A_1}} = -\frac{f_3(\theta_A)}{f_3(\theta_B)} = \eta - 1 \tag{4-17}$$

and

$$\eta = \frac{f_3(\theta_B) - f_3(\theta_A)}{f_3(\theta_B)} \tag{4-18}$$

If we adopt the latter form,

$$\frac{Q_{B_1}}{Q_{A_1}} = -\frac{f_4(\theta_B)}{f_4(\theta_A)} = \eta - 1 \tag{4-19}$$

and

$$\eta = \frac{f_4(\theta_A) - f_4(\theta_B)}{f_4(\theta_A)} \tag{4-20}$$

This is a very profound result. The efficiency of a reversible, but otherwise arbitrary, heat engine has been related to some function of the temperatures of the systems *A* and *B*. According to our preceding arguments, since the efficiency of a reversible engine cannot be multivalued, it is apparent that f_3 and f_4 must be single-valued functions of θ.

Let us recapitulate the progress we have made thus far. We have concluded that any heat engine or refrigerator operating reversibly must have a single-valued efficiency that is a function only of the primitive property, temperature (thermometric or any other empirical temperature scale which has the same information content). The relation between efficiency and temperature must be of the form given by Eq. (4-18) or (4-20). It is clear that for given systems, *A* and *B*, the experimental technique used to define temperature is arbitrary; the efficiency, however, is not arbitrary. Systems *A* and *B* having been chosen, the efficiency is set by nature. If we could build a reversible engine, we could measure experimentally the value of the efficiency. Since the efficiency is not arbitrary and the temperature scale is arbitrary, the form of the functions in Eqs. (4-18) and (4-20) cannot be arbitrary. Once a temperature scale has been decided upon, there will be one and only one valid form of each of the functions f_3 and f_4.

We have said that we could measure empirically the efficiency if we could construct a reversible engine and, in this manner, we could empirically determine the functions f_3 and f_4. Nevertheless, we still do not have access to any real reversible engines and we still do not have any reason to believe that one can be constructed. We can, however, conduct a thought experiment for a hypothetical reversible engine in much the same manner as Carnot did over a century ago. It is only through such analysis that we are able to define unequivocally the efficiency of reversible engines and a consistent temperature scale.

Thus, let us construct our reversible heat engine cycle using an ideal gas as a working fluid. Such a cycle, involving heat input from system A at T_A and rejection of heat to system B at T_B, is called a *Carnot cycle*. There are four steps to this cycle:

1. Heat flows to an ideal gas contained in a piston-cylinder device at temperature T_A.
2. The ideal gas system is then isolated from system A and allowed to expand adiabatically and reversibly to a lower pressure so that the temperature is T_B.
3. At this lower pressure, the ideal gas system is connected to system B and the gas compressed, heat being rejected to system B at a constant temperature, T_B.
4. At a particular point, the ideal gas system is isolated from system B and compressed adiabatically and reversibly to the original pressure. The point of initiation of this step is determined so that the final temperature after compression is T_A.

The heat engine (i.e., the ideal gas system) has undergone a cycle: Work has been produced in the work reservoir, the energy of system A has decreased, and the energy of system B has increased. By calculating the actual work and heat flows, and assuming that the ideal gas has the properties such that $P\underline{V} = NRT$ and the energy of the gas is not a function of pressure but only of temperature, it is possible to show that

$$\eta = \frac{T_A - T_B}{T_A} \tag{4-21}$$

The proof is left as an exercise.

From Eq. (4-21) it is clear that the functions f_3 and f_4 are of a very simple form if the temperature is measured by the ideal gas scale.[6] The functions would, of course, have been much more complicated if we had used a conventional sealed mercury thermometer. Fortunately, the ideal gas temperature scale has been adopted as *the* thermodynamic temperature scale.

We can measure approximately an ideal gas temperature since empirically we have found that simple gases at low pressures will obey the two relations specified above to a high degree of approximation. From this point on we will use the T-scale instead of the θ-scale and assume that we can accurately measure T.

[6]We developed Eq. (4-21) without specifying a priori any functional form for f_3 or f_4. We shall have no further need for these functions since we now have a practical temperature scale and a consistent expression for the efficiency.

4.4 The Theorem of Clausius

We are now in a position to extend our analysis of heat engines to the development of an additional derived property, the entropy. There are, perhaps, as many ways to proceed logically to infer this property as there are textbooks in thermodynamics. In whichever way we choose, however, we must limit ourselves to using the stated postulates or the conclusions obtained from them. We first derive a quantitative definition of entropy and then proceed to show that it has a very significant bearing on the concept of equilibrium.

Let us first rephrase the results of the last section. Combining Eqs. (4-7) and (4-21) for a reversible heat engine cycle, we have

$$\frac{Q_B}{Q_A} = -\frac{T_B}{T_A} \tag{4-22}$$

or

$$\frac{Q_A}{T_A} + \frac{Q_B}{T_B} = 0 \tag{4-23}$$

The temperatures of A and B are assumed not to change in the heat engine cycle; if they were to change, one could still write the equations but in a differential form:

$$\frac{đQ_A}{T_A} + \frac{đQ_B}{T_B} = 0 \tag{4-24}$$

Let us now shift our attention to the interior of the closed, reversible engine. We imagine that it is charged with some material which we call the *working substance*. The only restriction we have placed thus far on the internals of the engine is that it operate reversibly. The engine may be a composite or simple system.

We assume, for simplicity, that there are no external body force fields and that only P–$\underline{V}$ work need be considered; the results, however, are valid for systems in external body force fields. Since we are considering *reversible* heat engine cycles, and since all steps within a reversible process must be reversible, it follows that we cannot have any pressure gradients across moving boundaries or any temperature gradients across diathermal boundaries. Therefore, in any $P\,d\underline{V}$ work terms, P is both the external pressure and the internal pressure at the region of the moving boundary. Similarly, T_A is the temperature of external system A and also the temperature of that portion of the working substance at the diathermal boundary with system A. A similar conclusion holds for T_B. We may have many different events occurring in our system (i.e., adiabatic compressions and expansions, isothermal compressions and expansions, etc.). In Figure 4.3(a) we show a path representing an arbitrary reversible change in state of our system from i to f (i.e., a change that might occur during some part of the heat engine cycle). In Figure 4.3(b) the same path is shown and we have drawn through points i and f curves that represent those paths which would occur if adiabatic expansions or compressions were to take place starting from

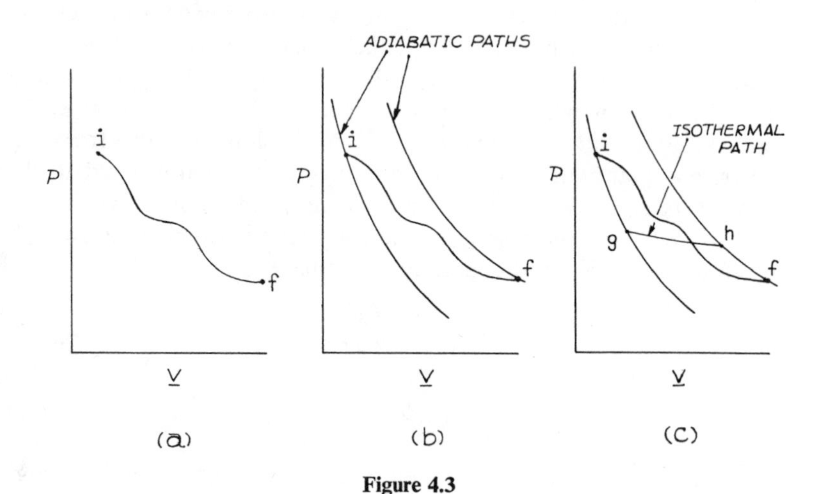

Figure 4.3

either i or f. Next, in Figure 4.3(c) we repeat (a) and (b) but add another path curve to represent the behavior of our system if it were at point g and were expanded isothermally (and reversibly) to h. This path, g–h, of course, represents a particular temperature level. The exact position of point g is established as shown below.

The work proceeding along the real path i–f is the integral under the path curve. We choose point g so that the area under the path curve $ighf$ is equal to the actual work.

For these two alternative paths, $\Delta \underline{E}_{if} = \Delta \underline{E}_{ighf}$ and $W_{if} = W_{ighf}$. Therefore, $Q_{if} = Q_{ighf}$. Since it was specified that $Q_{ig} = Q_{hf} = 0$, then $Q_{if} = Q_{gh}$. This simple little scheme is known as the *Theorem of Clausius*: Given any reversible process in which the temperature changes in any prescribed manner, it is always possible to find a reversible zigzag process consisting of adiabatic–isothermal–adiabatic steps such that the heat interaction in the isothermal step is equal to the heat interaction in the original process.

4.5 Entropy

The Theorem of Clausius is useful in analyzing the entire cycle carried out by a reversible heat engine or, for that matter, any system undergoing a reversible, cyclic process. This is shown in Figure 4.4 by the curve *jifkdc*. The unusual curve is drawn to emphasize that an actual P-$\underline{V}$ path is not necessarily simple. The heat interactions in various portions of the cycle may result from contacts with heat reservoirs at different temperatures. Choose for particular examination a portion of the cycle represented by the terminal points i and f. Draw path

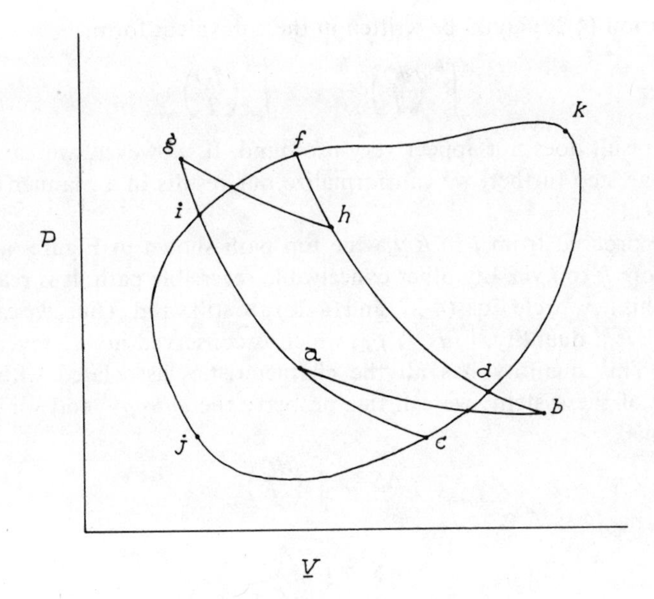

Figure 4.4

lines through these points representing adiabatic processes starting from i and f. These adiabatic path lines cut the cycle at points c and d, respectively. From Clausius's theorem we know that the path curves i–f and c–d can be broken down into a series of adiabatic and isothermal paths with the same work and heat interactions as occur in the original path. Let the heat interaction in path i–f be from a reservoir at T_A and in path c–d from a reservoir at T_B. We note that this net inner cycle *ighfdbac* is, in reality, a Carnot cycle. Thus,

$$\frac{Q_{if}}{T_{if}} + \frac{Q_{cd}}{T_{cd}} = 0 \tag{4-25}$$

If the paths i–f and c–d are made infinitesimal,

$$\frac{đQ_{if}}{T_{if}} + \frac{đQ_{cd}}{T_{cd}} = 0 \tag{4-26}$$

where T_{if} and T_{cd} become equal to the temperature of that part of the system in contact with the diathermal boundary at the respective points in the *actual* path. (The heat interactions in the differential cycle must still be equal to the heat interactions in the actual path.)

An infinite number of infinitesimal cycles, the first beginning at j and the last ending at k (see Figure 4.4), cover the entire cycle; summing, we obtain

$$\oint \left(\frac{đQ}{T}\right)_{\text{rev}} = 0 \tag{4-27}$$

where T and $đQ$ are the actual values associated with the original path, and the subscript "rev" reminds us that this result is only valid if the original path is reversible.

Equation (4-27) could be written in the equivalent form,

$$\int_j^k \left(\frac{dQ}{T}\right)_{\text{rev}} = -\int_k^j \left(\frac{đQ}{T}\right)_{\text{rev}} \tag{4-28}$$

This result does not appear very profound. If, however, we carry our reasoning one step further, we can formalize our results in a manner that will be most useful.

If we proceed from j to k via the top path shown in Figure 4.4 and then return from k to j via any other conceivable reversible path, it is readily shown that for this new cycle Eqs. (4-27) and (4-28) are still valid. Thus, we can conclude that there is a quantity, $\int (đQ/T)_{\text{rev}}$, which is conserved in *any* reversible cyclic process. This quantity has all the characteristics associated with a derived property of the system; we call this property the *entropy*, and we represent it by $\underset{\sim}{S}$. Thus,

$$\Delta \underset{\sim}{S} \equiv \int \left(\frac{đQ}{T}\right)_{\text{rev}} \tag{4-29}$$

or

$$d\underset{\sim}{S} \equiv \left(\frac{đQ}{T}\right)_{\text{rev}} \tag{4-30}$$

Equations (4-29) and (4-30) were derived for systems in which only force-displacement work occurs. The derivation can be extended to include systems that are acted on by body forces as well. The result is not changed by including other forces. Clausius's theorem is still valid because we can always find a reversible adiabatic–isothermal–adiabatic path having the same work interaction as the actual path. Furthermore, Eqs. (4-29) and (4-30) follow directly from Clausius's theorem. Thus, the defining equations for entropy, Eqs. (4-29) and (4-30), are valid for any system (simple or composite) undergoing a reversible process. In the case of a composite system consisting of two or more subsystems at different temperatures (which are separated from one another by internal adiabatic boundaries), the change in entropy of the composite is

$$\sum_i \frac{đQ_i}{T_i}$$

where $đQ_i$ is the heat transferred to subsystem i at T_i.

Note that in the derivation given above, only differences in entropy take on any significance, and these differences apply only between states at equilibrium.

If we follow the same line of reasoning used when energy was introduced as a property (see Section 3.3), it can be shown that entropy must also be first order in mass. Unlike the energy, there is nothing in our previous developments or in our postulates to require that the entropy of the universe should always be conserved. On the contrary, we shall soon prove that the entropy of the universe must increase in any irreversible process and that entropy is conserved only in reversible processes. In fact, the entropy change is of great use as a means for determining how closely real processes approach reversible processes.

Finally, we present an operational method for determining the entropy difference between two stable equilibrium states of a closed, simple system. There are no entropy meters available, and there are no meters to measure directly the quantity of heat transferred. Thus, in order to determine an entropy change between two stable equilibrium states, we must visualize any convenient *reversible* path and calculate the reversible work that would be done. Then, using Eqs. (3-28) and (4-30), we have

$$T\,d\underline{S} = d\underline{E} + đW_{\text{rev}} \tag{4-31}$$

which, for a closed simple system, reduces to

$$T\,d\underline{S} = d\underline{U} + P\,d\underline{V} \tag{4-32}$$

or, over the entire reversible path,

$$\Delta\underline{S} = \int \frac{1}{T}\,d\underline{U} + \int \frac{P}{T}\,d\underline{V} \tag{4-33}$$

The units of either entropy or temperature are completely arbitrary; all that we require is that the product of the entropy and temperature units be equivalent to the units of energy. Hence, we are free to choose one or the other at will, but, once we choose one, the units of the other are fixed. The accepted convention is to specify temperature in kelvins and entropy in J/K.

Example 4.2

A bar of aluminum is placed in a large bath of ice and water (Figure 4.5). Current is passed through the bar until, at steady state, there is a power dissipation of 1000 W. A thermocouple on the surface of the aluminum reads 640 K. Film boiling is occurring at the interface with a subsequent, noisy collapse of the bubbles. What is the entropy change of the bar, water, and universe during 2 min of operation for this highly irreversible operation? Assume that there is ice remaining at the end of the 2 min.

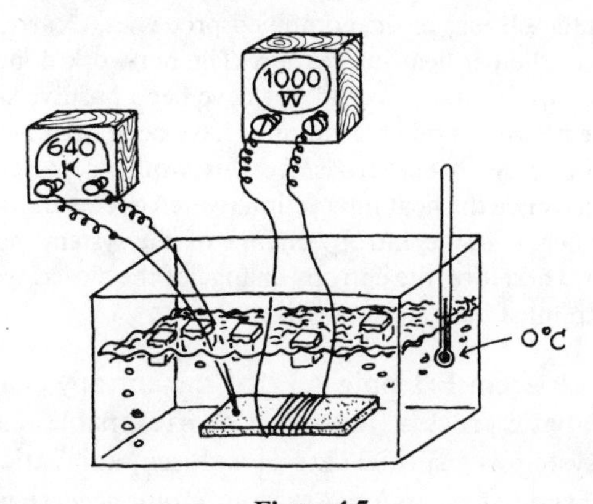

Figure 4.5

Solution

The aluminum bar undergoes no change in properties during the 2 min of operation. Since entropy is a property, $\Delta\underline{S}$(aluminum) = 0.

For the ice–water bath, the net result is that ice is melted but the temperature of the water does not change. (We neglect the small volume change due to melting.) To calculate the entropy change of this ice–water system, we need to devise a *reversible* process between the same initial and end states. For example, we could visualize that we contacted the system with some external heat reservoir only differentially above 0°C and allow a heat interaction to occur to melt the requisite amount of ice. This is now a reversible process so, with Eq. (4-33), assuming that $d\underline{V} \approx 0$, and $T = 273.2$ K,

$$\Delta\underline{S}(\text{water–ice}) = \frac{\Delta\underline{U}}{T} = \frac{Q}{T}$$

But Q = power × time = $10^3 \times 2 \times 60 = 1.2 \times 10^5$ J, and $\Delta\underline{S}$(water–ice) = $(1.2 \times 10^5)/273.2 = 439$ J/K. The entropy change of the universe is then 439 J/K.

In effect, we simply calculated $\Delta\underline{S}$(universe) as the power dissipation rate divided by the (constant) bath temperature. The reason why this is allowed is discussed further in Section 4.6.

Example 4.3

Prove that for an *adiabatic* process between defined states I and II of a closed system, the entropy change of the system must be equal to or greater than zero.

Solution

If the process were reversible, Eq. (4-29) would show that $\Delta\underline{S}_{\text{system}}$ must be zero. If the process were irreversible, we could follow this process by a second process to bring the system back to state I. Let us make this second process reversible and let us make it such that all heat transfer between system and surroundings occurs during an isothermal step. Such a reversible process can always be found (see Clausius's theorem, Section 4.4). In the combination of the two processes, the system has undergone a complete cycle so that the overall energy and entropy changes must be zero. In the reversible portion of the cycle, there may have been work and heat interactions; in the combined processes, clearly, the net work must have been equal to the net heat interaction. The net work done by the system and the net heat added to the system could not have been positive since this would be a PMM2. Thus, the net work and net heat must have been negative. (The net work and net heat could not have been zero since this would correspond to a reversible adiabatic process.) Since the heat interaction (which occurs during the return reversible step only) is negative, the entropy change of the system must also be negative in the return step. Therefore, the entropy change of the closed system in the irreversible adiabatic step must have been positive.

It should be clear from Example 4.3 that the entropy change of any closed system in any adiabatic process (reversible or irreversible) cannot be negative. Since an isolated system is a special case of a closed, adiabatic system, it follows that *the entropy change of an isolated system in any process must be equal to or greater than zero.* Furthermore, since the universe can be considered an isolated system, we are led to the same conclusion first stated by Clausius over a century ago: "*Die Entropie der Welt strebt einem Maximum zu.*"

4.6 Internal Reversibility

In Section 4.2 it was shown that in a reversible process all interacting systems must be in states of equilibrium at all times. Hence, a process involving a temperature gradient across a diathermal boundary separating two systems is an irreversible process because the composite system is not in thermal equilibrium. Thus, case (1) illustrated in Figure 4.1 is an irreversible process. Let us focus, however, our attention on system A in Figure 4.1. By altering the environment external to system A from that of case (1) to that of case (5), we could make the process reversible if we were to use a reversible heat engine. Since we could have effected the same changes *in system A* through either of these processes, the irreversibility in case (1) must be external to system A. If, however, during the heat interaction a temperature gradient had been established within system A, the process would have been irreversible and the irreversibility would have occurred within system A.

In many engineering applications it may be necessary to minimize irreversibilities in a process in order to, for example, reduce power costs. Here it would be helpful to know where the irreversibilities occur in a process. In order to identify subsystems that proceed along reversible paths within a process that is irreversible *in toto, we will define a process of a system as being internally reversible if the process can be performed in at least one way with another environment selected such that the system and all elements of this environment can be restored to their respective initial conditions, except for differential changes of second order in external work reservoirs.*

As discussed in Section 4.2, it then follows that a system undergoing an internally reversible process must traverse a quasi-static path. Nevertheless, an entire process is not necessarily reversible even if all subsystems traverse quasi-static paths because there may be irreversibilities occurring at the *boundaries* between the subsystems.

It should be obvious that for internally reversible systems there is no need to devise artificial processes to calculate entropy changes for the system in question; the actual interactions will suffice for this purpose.

Example 4.4

Consider a heat engine operating between two systems, 1 and 2, at T_1 and T_2, respectively, with $T_1 > T_2$. If the two systems are internally reversible, and if the overall process is irreversible, show that the entropy change of the universe must be greater than zero.

Solution

If the overall process were reversible, then for a complete cycle of the engine,

$$\Delta \underline{S}_{\text{universe}} = \Delta \underline{S}_1 + \Delta \underline{S}_2 = \frac{Q_1}{T_1} + \frac{Q_2}{T_2} = 0 \tag{4-34}$$

If we denote the work produced by a reversible and an irreversible engine as W_R and W_I, it can be shown that W_I must be less than W_R in order to avoid the possibility of a PMM2. If we run the reversible and irreversible engines in a manner such that the heat transferred from system 1 was the same in both cases, it is clear that Q_2 must be greater for the irreversible case than for the reversible case. Therefore,

$$\Delta \underline{S}_{2I} > \Delta \underline{S}_{2R}$$

while

$$\Delta \underline{S}_{1I} = \Delta \underline{S}_{1R}$$

so that for the irreversible case,

$$\Delta \underline{S}_{\text{universe}} = \Delta \underline{S}_{1I} + \Delta \underline{S}_{2I} > 0 \tag{4-35}$$

4.7 The Combined Law for Single-Phase Systems

We begin by considering only closed, simple systems. In this case, if the system is undergoing a reversible process, it was shown in Section 4.5 [Eq. (4-32)] that

$$d\underline{S} = \frac{1}{T}\, d\underline{U} + \frac{P}{T}\, d\underline{V} \tag{4-36}$$

It should be apparent from the discussion in Section 4.6 that Eq. (4-36) is equally valid for all internally reversible or quasi-static processes of closed, simple systems. Rearranging Eq. (4-36), we obtain

$$d\underline{U} = T\, d\underline{S} - P\, d\underline{V} \tag{4-37}$$

Either Eq. (4-36) or (4-37) is commonly referred to as the combined *First and Second Law of Thermodynamics* for a *closed, simple system.* They represent the special case of the First Law, Eq. (3-28), applied to an infinitesimal quasi-static path of a closed, simple system.

However, from Postulate I, for a single-phase system in a stable equilibrium state, we know that any property can be expressed as a function of any two independently variable properties in addition to the masses or mole numbers. Thus, we could express the energy as

$$\underline{U} = f(\underline{S}, \underline{V}, N_1, \ldots N_n) \tag{4-38}$$

Over an infinitesimal quasi-static path for this system, Eq. (4-38) can be written in differential form:

$$d\underline{U} = \left(\frac{\partial \underline{U}}{\partial \underline{S}}\right)_{\underline{V},N} d\underline{S} + \left(\frac{\partial \underline{U}}{\partial \underline{V}}\right)_{\underline{S},N} d\underline{V} + \sum_{i=1}^{n} \left(\frac{\partial \underline{U}}{\partial N_i}\right)_{\underline{S},\underline{V},N_j[i]} dN_i \tag{4-39}$$

By comparison with Eq. (4-37), we see that the summation term must be zero since $\underline{S}$ and $\underline{V}$ are independently variable properties. It must also be zero for a single-phase closed system undergoing a quasi-static process, which is simply a traverse of stable equilibrium states. Hence, Eq. (4-39) reduces to

$$d\underline{U} = \left(\frac{\partial \underline{U}}{\partial \underline{S}}\right)_{\underline{V},N} d\underline{S} + \left(\frac{\partial \underline{U}}{\partial \underline{V}}\right)_{\underline{S},N} d\underline{V} \tag{4-40}$$

It then follows that since Eqs. (4-37) and (4-40) are valid for the infinitesimal process under consideration, the coefficients of the differentials must be equal. Therefore,

$$T = \left(\frac{\partial \underline{U}}{\partial \underline{S}}\right)_{\underline{V},N} \tag{4-41}$$

and

$$-P = \left(\frac{\partial \underline{U}}{\partial \underline{V}}\right)_{\underline{S},N} \tag{4-42}$$

A similar treatment beginning with Eq. (4-36) and expressing

$$\underline{S} = f(\underline{U}, \underline{V}, N_1, \ldots, N_m) \tag{4-43}$$

would lead to

$$d\underline{S} = \left(\frac{\partial \underline{S}}{\partial \underline{U}}\right)_{\underline{V},N} d\underline{U} + \left(\frac{\partial \underline{S}}{\partial \underline{V}}\right)_{\underline{U},N} d\underline{V} \tag{4-44}$$

and

$$T^{-1} = \left(\frac{\partial \underline{S}}{\partial \underline{U}}\right)_{\underline{V},N} \tag{4-45}$$

$$\frac{P}{T} = \left(\frac{\partial \underline{S}}{\partial \underline{V}}\right)_{\underline{U},N} \tag{4-46}$$

Equations (4-36) through (4-46), excepting Eqs. (4-38), (4-39) and (4-43), apply only to closed systems. In this case, the mole numbers may change as a result of chemical reactions, but the masses of all constituent atoms are constant.

To extend our treatment to simple, yet open, systems, we could return to Eqs. (4-38) and (4-39). Here we may have material flow across the boundary of the system (in addition to chemical reactions internal to the system). Equation (4-39) is then the general relationship for the combined First and Second Law of Thermodynamics for a simple, open system. The definitions of temperature and pressure as partial derivatives are still valid [see Eqs. (4-41) and (4-42)] and, in addition, we define a new property, the *chemical potential*, as

$$\mu_i = \left(\frac{\partial \underline{U}}{\partial N_i}\right)_{\underline{S},\underline{V},N_j[i]} \tag{4-47}$$

where the subscripting indicates the total entropy, total volume, and all mole numbers (except i) are held constant. Then, Eq. (4-39) becomes

$$d\underline{U} = T\,d\underline{S} - P\,d\underline{V} + \sum_i \mu_i\,dN_i \tag{4-48}$$

This relationship applies only to systems undergoing quasi-static processes, and no irreversibility may occur at the boundaries where mass enters or leaves. In Chapter 5 we identify Eq. (4-48) with the Fundamental Equation of Thermodynamics and show that it has great utility in deriving relationships between properties of a material.

Had Eq. (4-43) been employed as a starting equation, the analog to Eq. (4-48) would be

$$d\underline{S} = \frac{1}{T}\, d\underline{U} + \frac{P}{T}\, d\underline{V} - \sum_i \frac{\mu_i}{T}\, dN_i \tag{4-49}$$

This relation could also have been obtained from Eq. (4-48) by resolving for entropy.

4.8 Reversible Work of Expansion or Compression in Flow Systems

The minimum work required to pressurize a flowing fluid or the maximum work obtainable in expanding a flowing fluid is often required in process design calculations. In such an analysis, reversible engines must be employed. Since all such engines have the same efficiency, we could choose any specific, reversible engine to study. Alternatively, we could simply consider the engine as a device into which a fluid flows at some defined state, and, out of which, fluid exists at some other state. The device has heat and work interactions with the environment. We will also assume that the device and its contents do not change with time, or, if they do, the variation in the total energy of the device is small relative to the enthalpy fluxes in and out and the work and heat interactions.

Let us focus on some finite element of the device and write the open-system energy balance [Eq. (3-60)]:

$$\delta\underline{U} = 0 = Q_\sigma - W_\sigma + h_e\,\delta n_e - h_\ell\,\delta n_\ell \tag{4-50}$$

Since we have neglected any accumulation of mass (or energy) within the device,

$$\delta n_e = \delta n_\ell = \delta n \tag{4-51}$$

For our finite section, dividing by δn, we get

$$h_\ell - h_e = \frac{Q_\sigma}{\delta n} - \frac{W_\sigma}{\delta n} \tag{4-52}$$

$$= Q' - W' \tag{4-53}$$

Q' is the heat interaction in our section per mole of fluid passing through the engine; W' is the work interaction on the same basis. Now let us shrink the finite section to one of differential size; then Eq. (4-53) becomes

$$dh = dQ' - dW' \tag{4-54}$$

Equation (4-54) would be applicable whether the engine were or were not reversible. However, if we consider the fluid to traverse a quasi-static path and assume all heat interactions at the boundaries to be reversible, then

$$dh = du + P\,dv + v\,dP = T\,ds + v\,dP \tag{4-55}$$

and

$$đQ' = T\,ds \tag{4-56}$$

Substituting Eqs. (4-55) and (4-56) into Eq. (4-54) yields

$$đW = -v\,dP \tag{4-57}$$

$$W' = -\int_{P_{\text{in}}}^{P_{\text{out}}} v\,dP = \frac{đW_\sigma}{dn} \tag{4-58}$$

or

$$W_\sigma = -\int_n \int_{P_{\text{in}}}^{P_{\text{out}}} v\,dP\,dn \tag{4-59}$$

The work term in Eq. (4-59) refers to the interaction of the chosen system (i.e., the engine) with the surroundings. If positive, the engine does work on the surroundings; if negative, the converse is true.

We note, in Eq. (4-59), that the limits of integration with respect to pressure may vary with the quantity of fluid flowing through the engine. This is illustrated in Example 4.5.

Example 4.5

A 1-m³ tank that initially contains air at 1 bar and 300 K is to be evacuated by pumping out the contents, as illustrated in Figure 4.6. The tank contents are maintained at 300 K throughout the operation by heat transfer through the walls. The compressor discharges the air at 1 bar and is operated isothermally at 300 K.

What is the total work done by the compressor? Assume that the compressor operates reversibly and that air is an ideal gas.

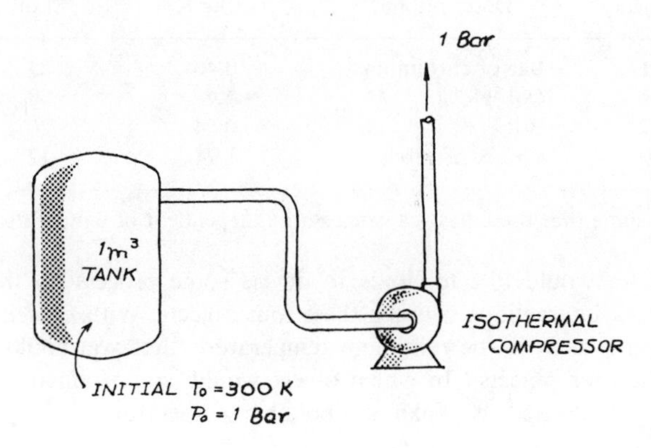

Figure 4.6

Solution

If it is assumed that the amount of air processed during one cycle of the compressor is small in relation to the total quantity of air expelled, the work of any one cycle can be treated as a differential and the properties within the tank can be assumed to vary smoothly with the amount of air expelled. Thus, for a differential amount processed, the work done by the gas on the compressor is given by Eq. (4-59):

$$đW = -\,dn \int_{P_t}^{P_a} v\,dP \tag{4-60}$$

where P_a is the discharge pressure (1 bar) and P_t is the tank pressure during the cycle under consideration.

Introducing the ideal gas law and integrating, we obtain

$$đW = RT \ln \frac{P_t}{P_a} dn \tag{4-61}$$

Since the amount of gas processed, dn, is equal to the decrease in gas in the tank, $-dN$, and since

$$dN = \frac{\underline{V}_t}{RT} dP_t \tag{4-62}$$

Eq. (4-61) becomes

$$đW = -\underline{V}_t \ln \frac{P_t}{P_a} dP_t \tag{4-63}$$

Integrating between $P_t = 1$ bar and $P_t = 0$,

$$W = -\underline{V}_t[P_t \ln P_t - P_t]_1^0 = -\underline{V}_t \cdot (1 \text{ bar}) \tag{4-64}$$

Thus, the work done by the compressor is 10^5 J.

PROBLEMS

4.1. We have four objects in our possession; the masses, heat capacities, and initial temperatures are as follows:

Object	Description	Heat capacity[a] (kJ/kg K)	Mass (kg)	T_i (K)
A	Bar of chromium	0.46	25	500
B	Oak block	2.4	8	300
C	Brick	0.84	20	400
D	Chunk of rubber	1.74	12	200

[a]Assume that these heat capacities are independent of temperature.

What we would like to do is to devise some process(es) that allow heat and/or work interactions between these four objects. With *no net* change in the environment, what is the minimum temperature that you could attain in any one of the four objects? In which object would you attain this temperature? Repeat if you desired the maximum possible temperature.

4.2. Most of us have seen, in novelty stores, small glass birds that appear to enjoy taking an endless series of drinks of water from a glass as illustrated in Figure P4.2. If we looked closely, we could see that these toys are simply two hollow glass bulbs separated by a tube and mounted on a swivel joint. The lower bulb is partially filled with a volatile liquid such as ethyl ether.

Ether boils at room temperature and some of the vapor condenses in the upper bulb, which is kept cold by evaporation of water on a wick placed over the bulb. In the drinking step the wick is moistened. The ether condensed in the upper bulb is prevented from returning to the lower bulb by the upward flow of vapor. Also, some liquid is pumped to the upper bulb by a "coffee percolator action." When the upper bulb is nearly full, the bird's center of gravity shifts,

Figure P4.2

the bird swings to a horizontal position (and "drinks"), thereby allowing the ether to flow back down the tube to the lower bulb. The bird then becomes upright and the cycle is repeated.

We would like to connect this bird to some mechanism and allow it to pump water from a lower level. If we assume that *all* the water pumped is eventually evaporated from the bird's head, what is the greatest height from which it can pump water for steady-state operation? Assume normal ambient conditions and neglect any heat losses or other irreversibilities.

4.3. A Hilsh vortex tube for sale commercially is fed with air at 300 K and 5 atm into a tangential slot near the center (point A in Figure P4.3). Stream B leaves

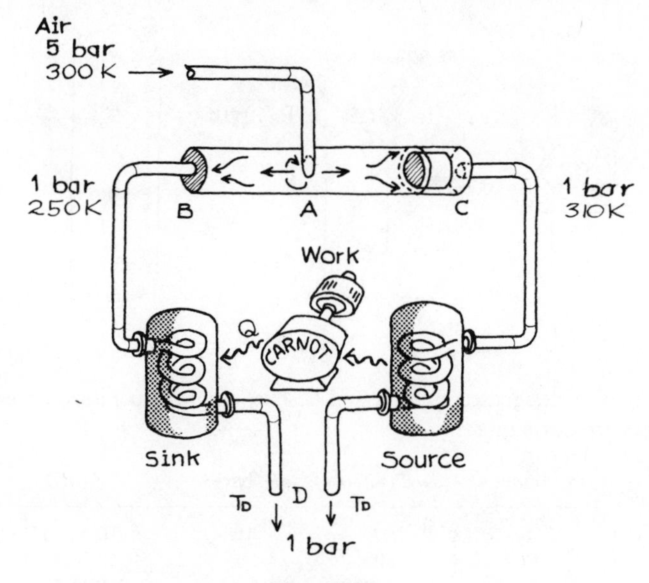

Figure P4.3

from the left end at 1 atm and 250 K; stream C leaves at the right end at 1 atm and 310 K. These two streams then act as a sink and source for a Carnot engine and both streams leave the engine at 1 atm and T_D. Assume ideal gases that have a constant heat capacity $C_p = 29.3$ J/mol K.

(a) If stream A is 1 mole/s, what are the flow rates of streams B and C?
(b) What is T_D?
(c) What is the Carnot power output per mole of stream A?
(d) What is the entropy change of the overall process per mole of A?
(e) What is the entropy change in the Hilsh tube per mole of A?
(f) What is the maximum power that one could obtain by any process per mole of A if all heat were rejected or absorbed from an isothermal reservoir at T_D?

4.4. We have at our disposal three cylinders of pure helium gas (see Figure P4.4).

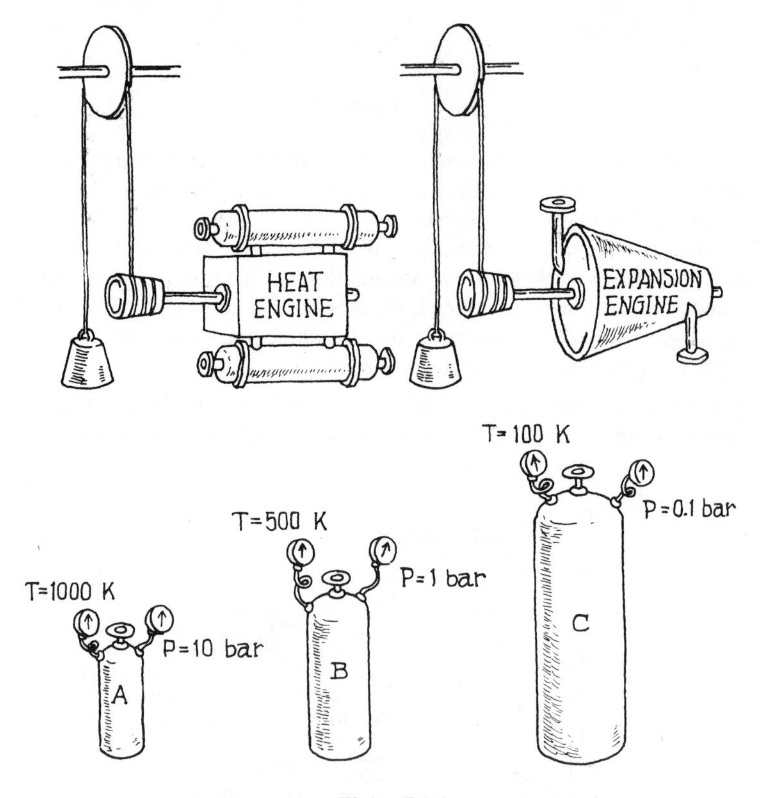

Figure P4.4

These may be designated as A, B, and C. The initial temperatures, pressures, and volumes are as follows:

Cylinder	T(K)	P (bar)	$\underline{V}$ (m³)
A	1000	10	10×10^{-3}
B	500	1	50×10^{-3}
C	100	0.1	100×10^{-3}

We will allow interactions among these three cylinders. C_p for helium is 20.9 J/mol K. Helium may also be considered to behave as an ideal gas.

What is the *maximum* amount of work that can be produced? Heat engines, expansion engines, etc., are allowed; the only restraint is that there must be no net effect in the environment other than the rise of weights in a gravitational field (i.e., in a work reservoir). (This would then imply that the environment not be compressed or expanded; that is, the total volume of the three cylinders initially must equal their combined volumes at the end.)

4.5. As you are no doubt aware, we in the academic world are most anxious to prepare you to solve technical problems that may arise in the future. Our Advanced Planning Section has been examining some unusual projected problems and requests your assistance on the particular one described below.

Decades from now, the present method of supplying energy to households (i.e., with electricity, gas, or oil) may not be possible. Instead, housepersons will shop for their energy in supermarkets (Figure P4.5). Cylinders of gas (let us

Figure P4.5

assume that the cylinders contain air) may be purchased and connected to any number of Carnot engines or other such efficient devices to be stocked in the home. Work is then obtained which may be utilized by the houseperson's family. When purchased, the cylinders are packed in well-insulated bags which may be removed (if desired) when connecting to a work-producing device.

The problem we face is to devise a convenient method to allow the houseperson to compare prices for the various gas cylinders available at the supermarket. The usual size for most cylinders is 1 m³, but the initial air pressure and temperature vary widely. One produced by R. Jones, Jr. is widely advertised to be quite economical, but our analysis indicated that the cylinder contained no air at all!

R. Nader III is expected to object to this deplorable situation and to require that we provide a simple equation to allow housepersons to calculate quickly (on their HP-1001 or equivalent) the unit cost of work energy in joules per dollar knowing only the initial temperature (K) and pressure (N/m²) of the air in the cylinder as well as the selling price.

Data: Assume that the ambient temperature and pressure are 300 K and

1×10^5 N/m^2. Air has a heat capacity at constant volume of 20.7 J/mol K, and, at constant pressure, of 29.0 J/mol K. The gas constant is 8.314 J/mol K and the gravitational acceleration is 9.81 m/s^2.

Derive an equation for the unit cost of work energy and demonstrate its application for a cylinder 1 m^3 in volume initially at 8×10^5 N/m^2 and 400 K which sells for $0.32.

4.6. Under ordinary operation a steady flow of helium gas equal to 50 g/s passes from a large storage manifold through an expansion engine as shown in Figure P4.6. There are, however, certain times when the engine must be shut down for

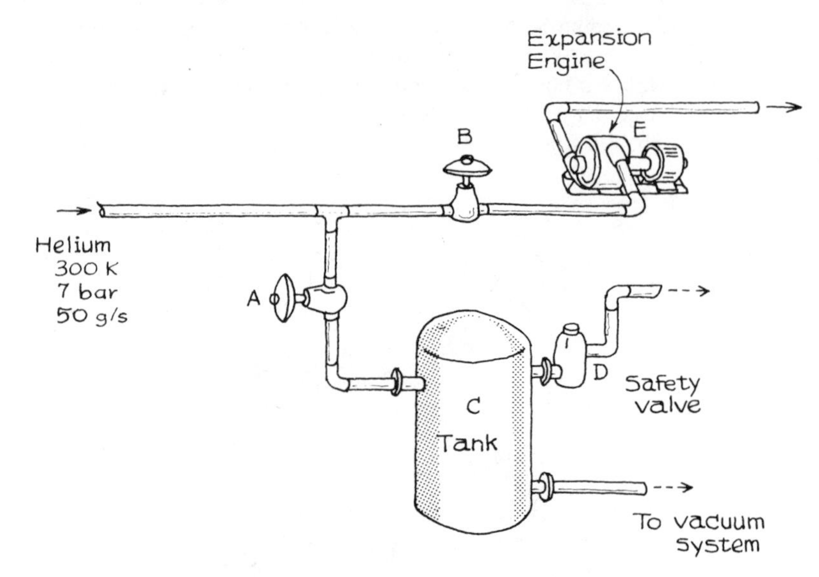

Figure P4.6

short intervals; the inlet flow cannot, however, be decreased at such times and thus the helium stream must be diverted. It is proposed to employ an adjacent system which at present is not being used. The latter system consists of a large (4 m^3) insulated tank (C in Figure P4.6) with a safety valve venting to the atmosphere through the plant ducting system.

The emergency diverting system is operated in the following manner. If the expansion engine must be shut down (or operates improperly), valve B is shut and A is opened, letting helium into tank C. When the pressure in C rises to 2.8 bar, safety valve D operates, venting gas.

Initially, tank C is evacuated to a very low pressure. We will assume that when gas enters this tank, it is well mixed but has negligible heat transfer with the tank walls. Also assume that helium is an ideal gas with a constant $C_v =$ 12.6 J/mol K.

(a) How long will flow enter tank C before the pressure increases to 2.8 bar and the safety valve opens?

(b) If the safety valve on tank C should operate, we would like to maintain a constant mass of gas in this tank equal to the mass at the time the safety tripped. The flow into the tank is, as noted above, 50 g/s; the flow out of the safety valve may be expressed as

$$\text{flow (g/s)} = \frac{KAP}{T}$$

where K is a constant, A the valve throat area, and P and T are the pressure and temperature in tank C. To keep the tank mass constant, the valve flow area A will be varied. What are the pressure and temperature in tank C 10 s after the safety valve opens? What is the variation of A with time necessary to keep the mass of gas in tank C constant?

(c) What is the total entropy change of the gas, the surroundings, and the universe during the time between the opening of valve A and just prior to the opening of relief valve D?

(d) What is the total entropy change of the gas, the surroundings, and the universe during the time between the opening of the relief valve D and a time 10 s later? The vented gas mixes with an infinite amount of air exterior to the tank and cools to 300 K. Leave the entropy change of mixing as an undetermined constant.

(e) Ten seconds after venting begins, the original expander comes back into operation so that valves A and D are shut and B is opened. It is desired to restore the tank C to its original evacuated state with the least possible work. You are free to select any technique that you deem feasible; all heat is to be rejected to surroundings at 300 K, and the final state of the gas should be 1 bar, 300 K. What is the minimum work required? (Do not consider any work obtainable from mixing helium with air.)

4.7.

TRANS-GALAXY-SPACELINES

— A DIVISION OF MITY INC.—

Welcome New Summer Employee:

As your first job with our famous old spaceline, we want you to answer a small technical problem that has arisen in our new line of boosters known to you as the Super Dodos.

We have a number of small jets in this series that will be used in attitude control in space. These jets will be powered by low-pressure nitrogen gas heated by an arc at the jet nozzle. Your problem deals with the nitrogen storage system.

The nitrogen is stored in a large well-insulated, 0.4-m^3 sphere at 1 bar pressure. At takeoff the temperature is 280 K. The mass rate of flow of N_2 will be constant and be equal to 7 g/s. Since the pressure inside the sphere must always be kept at 1 bar, a heater will be used inside the sphere (see the drawing).

(a) Under these conditions, what will be the temperature of the N_2 in the sphere, the instantaneous rate of heat flow to the heater, and the total heat required after 10 s of operation?

(b) The energy requirement of the heater may be difficult to meet. Rocky Jones, our local genius, has made a suggestion that we would like you to evaluate. Rocky wishes to take the nitrogen from the sphere that leaves at a pressure

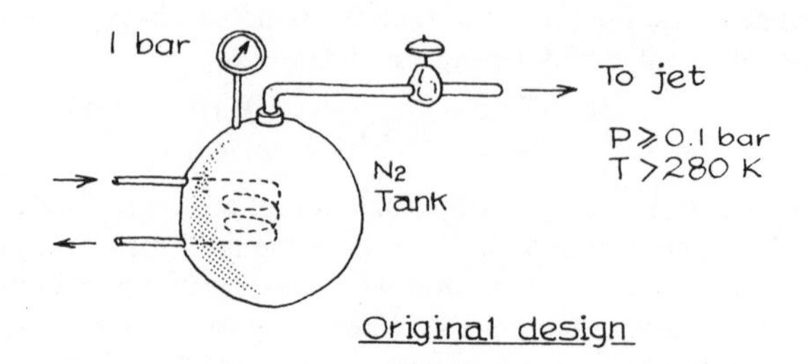

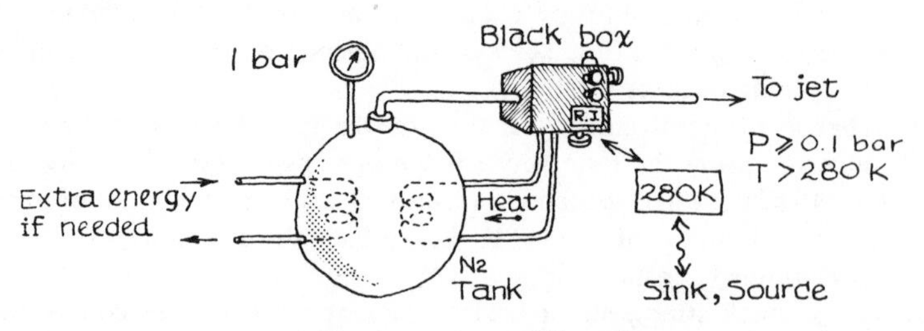

of 1 bar and sphere-gas temperature and put it through a "black box" to extract the maximum work possible before sending it to the arc jets. The interior metal of the booster may act as a heat source or sink at 280 K, and the final nitrogen pressure must be no less than 0.1 bar as fed to the jets. The work from the black box will be converted to heat and fed back to supply the sphere-heat energy. Nitrogen is an ideal gas with a $C_p = 29.3$ J/mol K, independent of temperature. Work may be stored in the black box if more power is generated than is needed during some periods of the process. What is the instantaneous additional heat required (W) at the start of operation? Will the black box supply all the sphere heat required, and, if so, for how long? What is the temperature in the sphere at this time?

4.8. MITY, Inc. has developed another unusual device; and as usual, you are requested to aid in the evaluation.

This time, Rocky Jones has proposed that on occasion, one might desire a low-capacity high (or low)-temperature source that could be stored for indefinite times before usage. For example, long-range space flight experiments may require a short-term, high (or low)-temperature source years after launch. Obviously, one cannot hope, in such cases, to carry high-temperature (or cryogenic) fluids and expect no heat transfer to occur. Rocky says that his back-of-the-envelope calculations indicate that the way to solve this problem is simply to carry small pressurized gas tanks and, by appropriate means, convert these rapidly to low-pressure, high- or low-temperature vessels.

Our engineering department has just finished the construction of a small vessel to hold 1 mole of air at 300 K and 100 bar. This is to be the prototype unit. Upon demand, we will insert this into our conversion unit and convert the high-pressure, low-temperature gas into a low-pressure, high (or low)-temperature gas. The gas volume may vary but no gas is lost to the environment. Air is an ideal gas with a constant isobaric heat capacity of 29.3 J/mol K.

However, the exact design of this conversion unit is not yet firm and we would like your help.

(a) Assume that we want the highest possible temperature from the unit described above. Suggest a sequence of processes within the conversion unit that will attain this value. You are allowed to have heat transfer to a large sink at 300 K and with the atmosphere (air) around the unit constant at 1 bar. What is the *maximum* temperature that can be obtained?

(b) Repeat part (a), but assume now that we want the lowest possible temperature. In this case, what is the minimum temperature that can be obtained?

(c) One of our younger engineers states that we should use some gas in our device other than air. Do you perceive any advantages in doing so? If you do, please elaborate.

Note: You do not have to be practical in your experimental plans. Yet it would be appreciated if you would not violate any of the laws of nature or even of thermodynamics.

4.9. Horace J., one of our less responsible students, neglected to close tightly a valve on his compressed air bottle last Friday. Over the weekend, the bottle pressure dropped slowly to 1 bar. Sally Z., his laboratory partner, berated him for his thoughtlessness. She pointed out that his action had resulted in a serious loss of available energy.

Her concern was real as it was −10°C outside. She insisted that if the bottle had had to be blown down, *she* could have devised a more efficient method to provide considerable energy to heat the laboratory. As she was but a Freshwoman, she had not had the opportunity to benefit from our thorough thermodynamics program and thus was unable to express her feelings in a quantitative manner. Could you come to her assistance and indicate the maximum number of joules that could have been made available for heating if the bottle had been blown down in a less thoughtless manner? (Assume ideal gases.)

Data:

Laboratory temperature	20°C = 293.2 K = constant
Outside temperature	−10°C = 263.2 K
Bottle volume	0.06 m^3
Initial bottle pressure	150 bar
Final bottle pressure	1 bar
C_p, air	29.3 J/mol K

(*Note:* The department is well stocked with various sizes of Carnot engines, if needed.) Be certain that you use your most perceptive imagination to work this problem. The bottle initially is at 20°C. The final pressure is 1 bar. You may carry out the blowdown process inside or outside the laboratory, and you may employ any type of compressor, expander, and heat exchanger you desire. Neglect, however, the heat capacity of the bottle itself.

4.10. Mr. Rocky Jones, the well-known inventor and entrepreneur, visited me yesterday with a most interesting idea that we should examine carefully in this age of ever-dwindling energy sources. It appears that Mr. Jones is concerned about the inefficient use of hot water in our sinks. As he described it, one normally turns on both the hot water and the cold water and, with soap, rapidly moves the hands to-and-fro to obtain the overall sensation of warm water. The time the hands are actually in any stream of water is quite short and most of the water used is mixed in the sink and flows down the drain.

What Mr. Jones is proposing is a new kind of sink, as shown in Figure P4.10.

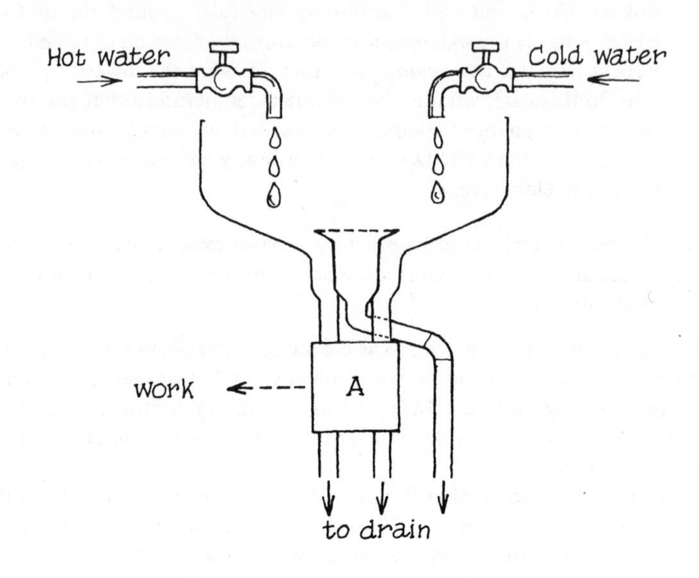

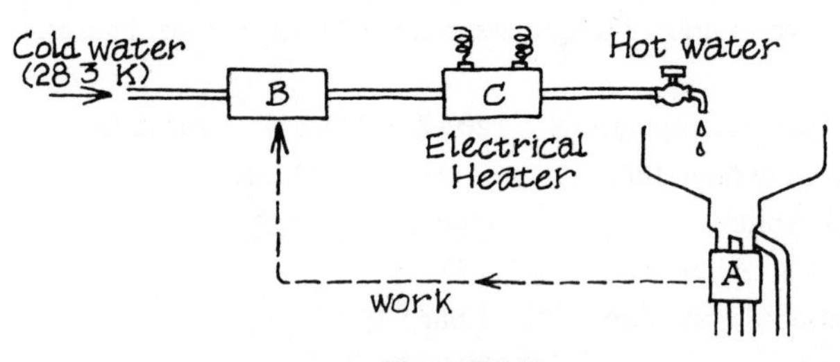

Figure P4.10

Except in the brief intervals when the actual washing is done, the hot and cold water from the sink exit through separate drains into a device *A*. Mr. Jones claims he can obtain work from this device to reheat more hot water in *B*. The work from *A* may not be quite sufficient to heat all the water, so an auxiliary electric heater in *C* is used to maintain the desired inlet hot-water temperature. We want to supply about 0.2 gal/min (757 cm^3/min) of hot water to our sinks at about 70°C (343.2 K). The average cold-water flow is twice the hot-water flow and is normally at 10°C (283.2 K).

(a) If our sinks operate at steady state, what do you estimate is the necessary extra energy flow input required in unit *C*? Express your results in watts of electrical power. Assume no interaction with the environment in device *A*. You may, however, extract energy from the environment, if desirable, when considering the operation of *B*; the environment remains at 10°C.

(b) Can you accept the challenge to improve on Mr. Jones's device so as to yield our hot water with even less electrical energy?

4.11. In several parts of the world, there exist ocean currents of differing temperatures that come into contact. An example of this takes place off the coast of southern Africa, where the warm Agulhas and the cold Benguela currents meet at Cape Point, near Cape Town (see Figure P4.11). It has been proposed that work may

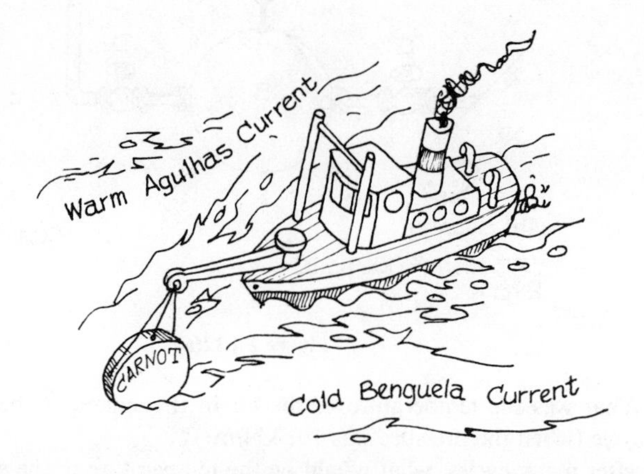

Figure P4.11

be obtained from these ocean currents if a heat engine could operate between the warm current as a source and the cold current as a sink.

Assume that the system may be simplified into two channels of water in contact and flowing cocurrently. Furthermore, assume that no mixing occurs between the streams and that the effect of heat conduction between the streams is negligible.

(a) Derive a general expression for the maximum amount of power that could be obtained from the system. Express your result in terms of temperatures, flow rates, and physical properties of the streams.

(b) Repeat part (a) if the two streams flow countercurrent rather than cocurrent. What is the pinch temperature, and where does it occur?

(c) It has been estimated that the Benguela current is 16×10^6 m^3/s and its initial temperature is 278 K. The Agulhas current is 20×10^6 m^3/s and its initial temperature is 300 K. Calculate and compare the power obtainable from these two currents assuming that they flow (1) cocurrently and (2) countercurrently.

4.12. Recently, we carried out an interesting experiment in our laboratory. We connected a *constant*-pressure air supply line to a small spherical vessel (*B*) as shown in Figure P4.12. With valves *C* and *D* we could either pressurize the sphere with high-pressure air or vent it to the atmosphere. In the test, with vent valve *D* closed, we pressurized the sphere very rapidly to the air-supply pressure. Then, quickly, we closed valve C and opened valve D to vent the sphere to atmospheric pressure. As soon as this pressure level was achieved, we closed valve D and repeated the cycle. This sequence was followed for many cycles. Assume that all steps are carried out so rapidly that heat transfer from the air to any piping or vessel walls is negligible. Air may be assumed to be an ideal gas with $C_p = 29.3$ J/mol K and $C_v = 21.0$ J/mol K. Pressures are shown in Figure P4.12(a).

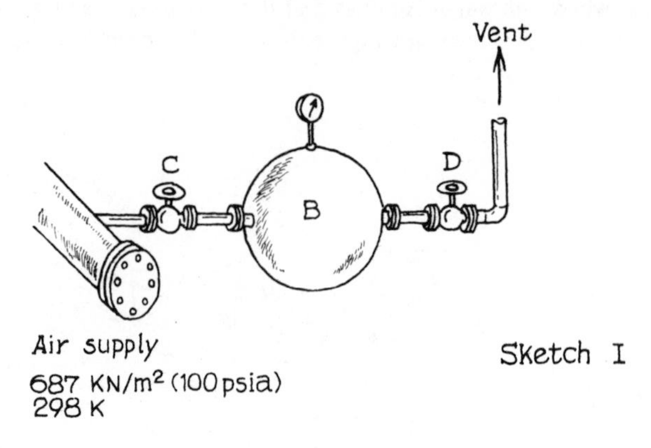

Figure P4.12(a)

(a) What was the temperature of the air in the sphere at the end of the first cycle (when the pressure was 101 kN/m^2)?

(b) After many cycles, what would be the temperature in the sphere at the end of the pressurizing step? At the end of the venting step?

As we were conducting these experiments, Rochelle Jones wandered into our laboratory. After she had watched the operation for many cycles so that the conditions at the end of the pressurization and venting steps were time invariant [see part (b)], she became excited. During the venting, she saw that the gas outlet temperature was initially high but decreased as venting proceeded. "Why not get some 'free' work from the device?" she asked. She began to explain her ideas, but I am afraid I could not understand everything she said.

She apparently wants to replace valve D with a Y tube attached to valves D_1 and D_2 [see Figure P4.12(b)]. She says she can build a sequencer to allow flow through valve D_1 for the first part of the venting step and through valve D_2 for the rest of the venting. Both flow into what she calls her Hilsch box. Both streams exit at the same temperature and at 101 kN/m^2.

No interaction with the environment is evident except for the transfer of work. We are not certain, but we believe that Rochelle's idea is to include in her Hilsch box some expansion engines and Carnot engines to transfer heat between the two streams since the average inlet temperature of the air through valve D_1 exceeds that through valve D_2.

Rochelle is returning on October 7 to discuss her idea in more detail. Now, we don't wish to appear too ignorant, so by 9 A.M. of that day, carry out the following calculations:

(c) What is the *maximum* work Rochelle can obtain in her Hilsch box, J/kg air?

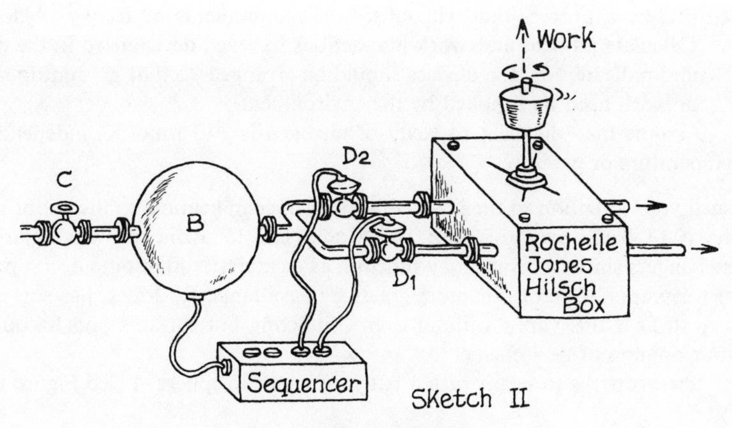

Figure P4.12(b)

(d) At what pressure in the sphere B should the sequencer operate to direct flow from D_1 to D_2?

(e) If you were going to obtain work from the venting of sphere B, could you suggest improvements to Rochelle's scheme? For example, if you were to allow heat transfer to the environment in her Hilsch box, what is the maximum work you could obtain per kilogram of air?

4.13. Our highly efficient plant has a nitrogen stream at 2.53 bar and 305.5 K which is presently vented to the atmosphere. The management would like to use this stream to satisfy some of the heating or cooling requirements of other processes. Our old friend, Rocky Jones, boy genius, has devised a black box that will produce equal amounts of a hot stream at 500 K and a cold stream at 111 K and thus satisfy simultaneously some heating and cooling requirements (see Figure P4.13). Furthermore, Rocky claims that his device will be self-sustaining because no additional heat or work need be supplied to the device.

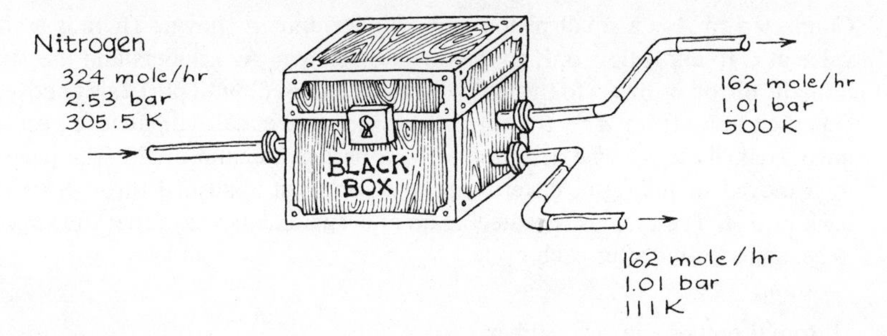

Figure P4.13

One of our new engineers, Barry Goldfinder, claims that he has a black-box device that will produce equal amounts of a hot stream of 533 K and a cold stream of 77.7 K. His device will also be self-sustaining.

(a) Are either (or both) of these devices possible? Explain.

(b) Present a process that will satisfy the requirements of Rocky's black box. Calculate all heat and work interactions for each device used in the process and indicate how the devices should be arranged so that no additional heat or work need be supplied by the environment.

Assume that the heat capacity of nitrogen is 29.3 J/mol K, independent of temperature or pressure.

4.14. I call your attention to the problem we now face in heating up the vapor feed to the X-13 batch synthesis unit. We would like to avoid any customary heat exchangers since hot walls may function as a catalyst and initiate decomposition of this vapor. One of our more creative consultants, R. Jones, has suggested a way to heat this vapor without even contacting hot surfaces and I would like your opinion of her scheme.

She proposes to begin with a batch of vapor in sphere A (see Figure P4.14).

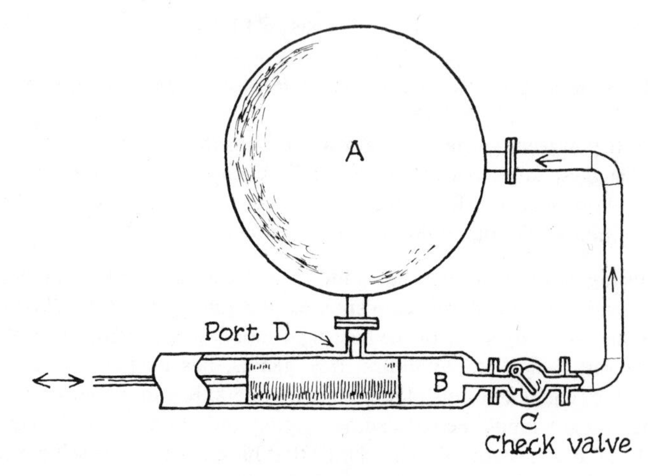

Figure P4.14

Connected to A is a small piston and cylinder unit as shown. There is a check valve at C to allow flow only in the direction shown. As I understand the operation, the piston is drawn to the left (with valve C closed) until port D is uncovered. Gas then flows from A to B until the pressures are equalized. (Before port D is uncovered, there may be assumed to be a perfect vacuum in B.) The piston is then moved to the right, covering port D, and gas is pushed through valve C back into A. The cycle is repeated again and again. Jones says that the vapor in A becomes hotter after each cycle.

Data:

Initial pressure in A: 10 bar

Volume of A: 1 m^3

Volume of cylinder B when the piston uncovers port D: 0.01 m^3

There is no significant heat transfer between the unit and the environment. The vapor being processed may be considered to be an ideal gas with a heat capacity at constant volume of 33.256 J/mol K independent of temperature.

Also, when the vapor is pushed from B to A, assume that there is perfect mixing of the gas in A. The piston–cylinder operation is adiabatic and reversible.

Can you estimate how many cycles it would take to heat the vapor from 300 to 600 K?

4.15. We have been requested to design a small system that will allow a manufacturer to test helium vacuum pumps both under transient conditions and also for long steady-state periods. The system we now propose is shown in Figure P4.15.

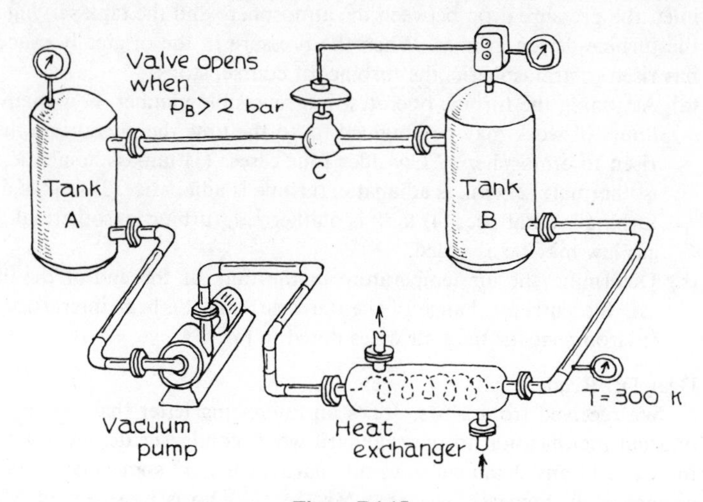

Figure P4.15

Tanks A and B are thin-walled and well insulated, each with a volume of 0.1 m^3. The vacuum pump will take suction from A and discharge into B. Between the pump and B there is a heat exchanger to cool the gases to a constant temperature of 300 K.

Initially, both tanks A and B are charged with helium at 1 bar and 300 K. The vacuum pump is started and gas flows through the heat exchanger into B until the pressure in B reaches 2 bar. At this time valve C automatically opens and vents gas from B to A, always keeping the pressure in B at 2 bar.

Data:

Helium is an ideal gas with a constant $C_p = 20.9$ J/mol K.

The volumes of the connecting lines and pump are negligible compared to the volume of spheres A and B.

The vacuum pumps to be tested are adiabatic and reversible.

(a) Just at the time the pressure in B reaches 2 bar, what is the temperature in both A and B, and what is the pressure in A?

(b) After the system has been running for a long time, with P_B = constant = 2 bar, what is the temperature of B and what is the pressure in A?

(c) What is the change in entropy of all the helium gas in the system from the start until the pressure in B just reaches 2 bar?

(d) Can you suggest any improvements in the system to simplify it and still test the vacuum pumps over transient and long-term periods?

(e) What is the work supplied to the vacuum pump from the start of the test until the pressure in B just reaches 2 bar?

(f) What is the heat load in the heat exchanger during the same period as described in part (e)?

4.16. It is proposed that a device be constructed to operate as follows: An evacuated tank of 1 m^3 is attached to the exhaust of an air-turbine-driven grinding wheel. Air at atmospheric pressure and 300 K will be allowed to enter the turbine inlet, the pressure drop between the atmosphere and the tank serving to operate the turbine-driven grinder. When the pressure in the originally evacuated tank has risen to atmospheric, the turbine, of course, stops.

(a) Assuming the turbine operates in a reversible manner, how many kilowatt-hours of work may be obtained up to the time the pressure in the tank has risen to atmospheric? Consider four cases: (1) tank is adiabatic, turbine is isothermal; (2) tank is adiabatic, turbine is adiabatic; (3) tank is diathermal, turbine is adiabatic; (4) tank is diathermal, turbine is isothermal. The ideal-gas law may be assumed.

(b) Determine the air temperature in the tank at the end of the filling process, the entropy change of the universe, and the heat interaction with the environment for the four cases noted in (a).

4.17. Dear Dr. Reader:

We received from a Ms. Jones an interesting letter that poses a somewhat unusual modification to our standard water condenser design. I won't go into the letter in any detail since, to me, parts of it were somewhat perplexing. The essence of it, however, was that Ms. Jones (who is president of MITY, Inc.) indicates that her company is diversifying and has recently bought out Carnoco, a manufacturer of diminutive reversible heat engines. Ms. Jones suggests that in all our new plants, wherein we cool and condense stream S-13 (see Figure A) from 422 K to 300 K, we eliminate the heat exchangers entirely and, instead, substitute a set of Carnoco engines. If I understand Ms. Jones correctly, her process diagram would appear as in Figure B. Ms. Jones claims that we could reduce the cooling-water requirement (and still maintain the outlet cooling water temperature at 289 K), obtain a great deal of "free" work, and eliminate completely the cost of our heat exchanger.

Using the following data, would you please evaluate Ms. Jones's proposal and indicate the maximum amount we could afford to pay for the purchase and installation of the Carnoco reversible engines?

Cooling water cost	\$1/100 m^3
Value of any work output	2.0 cents/kWh
Present cost of heat exchanger, installed	\$60,000
Operational time per year	7200 h
Investment charges	15%

Your attention to this request should be given priority.

Sincerely yours,

Godfrey Cross

Godfrey Cross
Boss

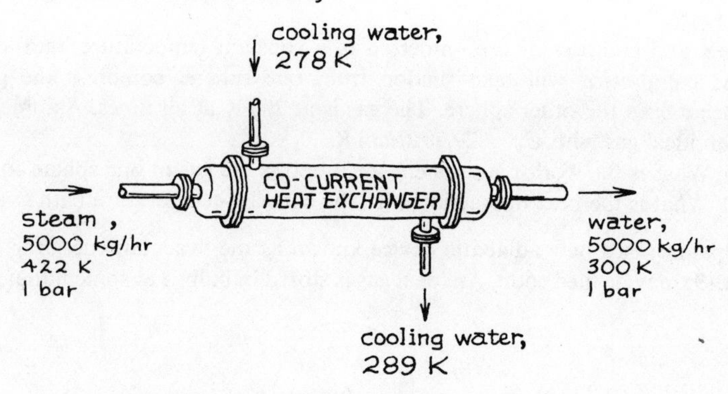

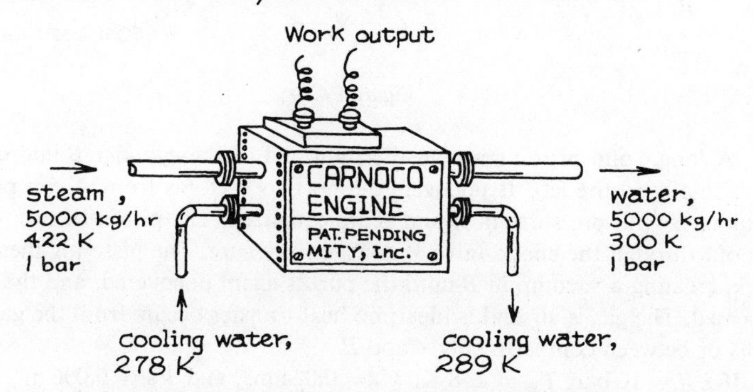

4.18. Two large gas storage spheres (0.1 m^3) each contain air at 2 bar (Figure P4.18). They are connected across a small reversible compressor. The tanks, connecting

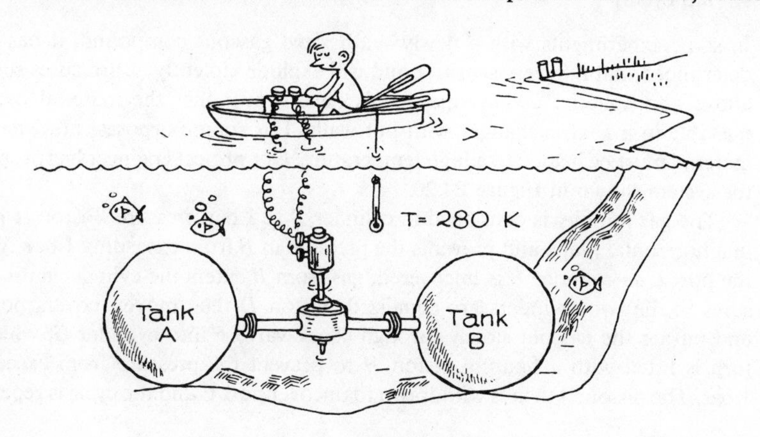

Figure P4.18

lines, and compressor are immersed in a constant temperature bath at 280 K. The compressor will take suction from one sphere, compress the gas, and discharge to the other sphere. The gas is at 280 K at all times. Assume that air is an ideal gas with $C_p = 29.30$ J/mol K.

(a) What is the work requirement to compress the gas in one sphere to 3 bar?
(b) What is the heat interaction with the constant-temperature bath?

4.19. A patent on a new adiabatic device known as the "vacuum energizer" (Figure P4.19) may be filed soon. An ideal gas is stored in bulb A at some initial pressure

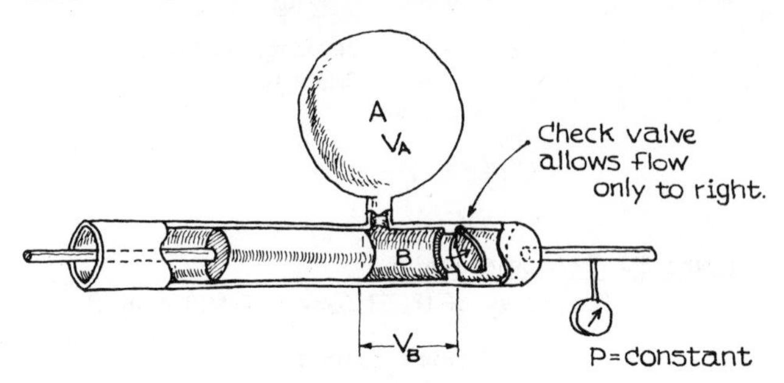

Figure P4.19

P_{A_0}. A long, solid piston starts at the right end of the cylinder B and is pulled horizontally to the left. It uncovers the port so that gas from A can pass into cylinder B until pressures in A and B are equalized. The piston then pushes the gas out through the check valve at constant pressure. The piston is then drawn back, creating a vacuum in B until the port is again uncovered, and the cycle is repeated. The gas is air and is ideal; no heat transfer occurs from the gas to the walls or between compartments A and B.

(a) If $P_{A_0} = 10$ bar, $T_{A_0} = 278$ K, $\underline{V}_A = 0.024$ m^3, and $\underline{V}_B = 0.006$ m^3, what is the temperature and pressure of the gas leaving B during the first cycle and during the nth cycle?
(b) What is the temperature and pressure in A after the first cycle and after the nth cycle?

4.20. In some experiments with a newly synthesized gaseous compound, it has been determined that it is very sensitive and will explode violently at pressures slightly above atmospheric. Other experiments have shown that the material may be unstable in a heat exchanger with hot walls. For some purposes, however, the material must be heated to a high temperature. Our project engineer has proposed the system shown in Figure P4.20.

The gas initially is contained in cylinder B at 1 bar pressure. Piston A floats in a horizontal plane and prevents the pressure in B from exceeding 1 bar. When the port C in cylinder E is uncovered, gas from B enters the cylinder until pressures in the two cylinders are equalized. Piston D then moves, covers port C, and pushes the gas out slowly through check valve F into cylinder G, which in turn is fitted with a floating piston H to prevent the pressure from exceeding 1 bar. The piston D then is withdrawn to uncover port C and the cycle is repeated.

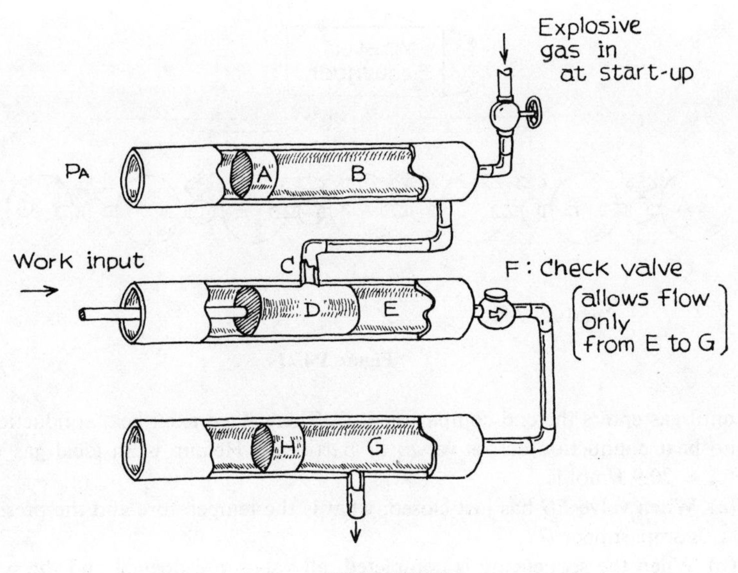

Figure P4.20

Valve F only allows flow from E to G, but not the reverse. The pistons have essentially no friction.

It is claimed that the gas in G is hotter than in B. Any desired temperature may be attained by cascading such devices in series (i.e., feeding gas out from G through J into another tube, similar to the outlet of B).

Data:

Cylinders are 7 cm inside diameter.

Assume no clearance in cylinder E.

Initial mass of gas in B is 1 mole and has a temperature of 311 K.

Assume ideal gases with a constant $C_p = 41.86$ J/mol K.

What is the temperature of the gas after *two* cascading steps?

4.21. Our old friends, Rochelle and Rocky Jones, have been tinkering in the laboratory and have produced an interesting device that they call an "integral pulsed shock tube." As yet, we do not see a large commercial market. In fact, we are not sure what it really does! Can you analyze the device and answer the brief questions given later?

A long, insulated tube (Figure P4.21) is divided into chambers of equal volume by rigid, adiabatic partitions. Each partition has a fast-operating valve to allow flow of gas between compartments when the valve is open. The operation of this device is as follows. Compartment A is initially filled with helium gas at 300 K and 64 bar. The remaining compartments are evacuated to zero pressure. All valves are closed. At time zero, valve AB is quickly opened and gas flows from A to B. Just when the pressures are equalized in A and B, valve AB is shut and valve BC opened. As before, just at the time the pressures in B and C are equal, valve BC is shut and valve CD opened. This sequence is continued

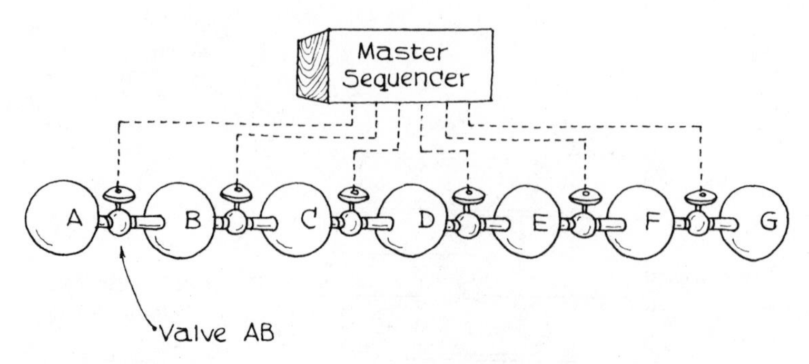

Figure P4.21

until gas enters the end compartment G. There is no axial heat conduction and no heat conduction across valves or partitions. Helium is an ideal gas with a $C_p = 20.9$ J/mol K.

(a) When valve FG has just closed, what is the temperature and the pressure in compartment G?

(b) When the sequencing is completed, all valves are opened and the pressure allowed to equalize in all compartments. Slow axial conduction also equalizes the temperature in all compartments. What is the equilibrium temperature and pressure?

(c) The Joneses are uncertain how much work it requires to prepare their integral pulse shock tube for firing. Estimate the minimum work per mole of helium initially in compartment A. In this calculation, assume that all compartments of the shock tube initially contain helium at 1 bar, 300 K. Any helium that is removed must be pumped into a large pipeline containing helium at a constant pressure of 2 bar, 300 K. Also, any helium used to charge compartment A to 64 bar is to be taken from the same pipeline. If, in your calculations, any heat transfer occurs, assume that you have a large heat sink or source at 300 K.

(d) Can you suggest any use for Rochelle and Rocky's new device?

4.22. In a laboratory experiment you must evacuate as rapidly as possible some process equipment. Initially, the equipment contains nitrogen gas at 280 K and 1 bar pressure. The volume is 0.14 m^3. You have at your disposal a typical rotary vacuum pump. It is rated to pump 0.03 m^3/min of actual volume.[7] There is one complication. The equipment is adiabatic except for a small heater that must be left on at all times and liberates 2 kJ/min at a constant rate. For N_2, $C_p =$ 29.3 J/mol K, independent of temperature.

(a) What is the pressure inside the equipment after 2 min of pumping?

(b) What is the temperature of the nitrogen remaining in the process equipment after 2 min of pumping?

(c) One of your friends kindly offers to lend you his vacuum pump (which is exactly like yours) and he says that you can reach the same pressure in half the time if you connect it up in parallel with yours. Do you agree?

[7]This pump may be considered to be a reciprocating one, with no clearance, and with an intake stroke of $\delta \underline{V}$ m^3, where $(\delta \underline{V})$ (rpm) $= 0.03$ m^3/min.

(d) What is the lowest pressure you can pump on your system using one pump? Explain.

(e) If you wish only to pump down to 0.3 bar, what minimum horsepower motor do you recommend? The vacuum pump is neither adiabatic nor isothermal, but some tests indicate that the relationship

$$PV^{1.3} = \text{constant}$$

is applicable in the compressor. Base your minimum estimate on the assumption that the pump is reversible.

Thermodynamic Relations for Simple Systems 5

In this chapter we treat simple systems and develop general property relationships that are consequences of the postulates and laws discussed in previous chapters. These relationships, which must be obeyed by all simple systems, should be distinguished from the *physical* property relationships. Physical properties refer to relationships that are specific for each substance and are not dictated by the laws of classical thermodynamics: They must be obtained by experimentation, extrapolation of existing data for similar materials, or development from molecular models using statistical mechanics.

We restrict our attention to simple systems. For such systems we have seen in Section 4.7 that the properties $\underline{S}, \underline{V}, N_1, \ldots, N_n$ form an independently variable set for the energy, $\underline{U}$. We explore the nature of this relationship and the many equivalent forms that this relationship assumes.

5.1 The Fundamental Equation

As will be seen presently, the relationship

$$\underline{U} = f_U(\underline{S}, \underline{V}, N_1, \ldots, N_n) \tag{5-1}$$

completely describes all of the stable equilibrium states of a simple system. By solving Eq. (5-1) explicitly for $\underline{S}$, an alternative form is obtained:

$$\underline{S} = f_S(\underline{U}, \underline{V}, N_1, \ldots, N_n) \tag{5-2}$$

Either relationship is called the *Fundamental Equation:* Eq (5-1) is termed the *energy representation* and Eq. (5-2) the *entropy representation.*

The Fundamental Equation represents a surface in $(n + 3)$-dimensional space. The points on this surface represent stable equilibrium states of the simple system. Quasi-static processes can be represented by a curve on this surface. Processes that are not quasi-static are not identified with points on this surface. (Recall that derived properties such as $\underline{U}$ and $\underline{S}$ are not defined for nonequilibrium states.)

The $n + 2$ first-order partial derivatives of the Fundamental Equation are the traces of the tangent planes taken parallel to one of the coordinate axes. The significance of these tangent plane traces can be seen by expressing the Fundamental Equation in differential form. For the energy representation, Eq. (5-1),

$$d\underline{U} = \left(\frac{\partial f_U}{\partial \underline{S}}\right)_{\underline{V},N} d\underline{S} + \left(\frac{\partial f_U}{\partial \underline{V}}\right)_{\underline{S},N} d\underline{V} + \sum_{i=1}^{n} \left(\frac{\partial f_U}{\partial N_i}\right)_{\underline{S},\underline{V},N_j[i]} dN_i \tag{5-3}$$

where subscript $N_j[i]$ indicates that all N_j except $j = i$ are held constant in the differentiation. If we compare Eq. (5-3) to the combined law for a simple system (see Section 4.7), namely

$$d\underline{U} = T\,d\underline{S} - P\,d\underline{V} + \sum_{i=1}^{n} \mu_i\,dN_i \tag{5-4}$$

it is clear that

$$\left(\frac{\partial f_U}{\partial \underline{S}}\right)_{\underline{V},N} = T = g_T(\underline{S}, \underline{V}, N_1, \ldots, N_n) \tag{5-5}$$

$$-\left(\frac{\partial f_U}{\partial \underline{V}}\right)_{\underline{S},N} = P = g_P(\underline{S}, \underline{V}, N_1, \ldots, N_n) \tag{5-6}$$

and

$$\left(\frac{\partial f_U}{\partial N_i}\right)_{\underline{S},\underline{V},N_j[i]} = \mu_i = g_i(\underline{S}, \underline{V}, N_1, \ldots, N_n) \tag{5-7}$$

where the functions g_T, g_P, and g_i could be obtained directly from the Fundamental Equation if it were available. Equations (5-5) through (5-7) are called the *equations of state*. As shown in Sections 5.2 and 5.3, only two of the three equations of state of a pure substance are independent.

The second-order partial derivatives of the Fundamental Equation are also related to quantities that can be measured experimentally. For example, for a *pure material*, there are four second-order partial derivatives at constant mass or moles:

$$\frac{\partial^2 \underline{U}}{\partial \underline{S}^2} = \frac{\partial}{\partial \underline{S}}\left[\left(\frac{\partial \underline{U}}{\partial \underline{S}}\right)_{\underline{V},N}\right]_{\underline{V},N} = \left(\frac{\partial T}{\partial \underline{S}}\right)_{\underline{V},N} \tag{5-8}$$

$$\frac{\partial^2 \underline{U}}{\partial \underline{V}^2} = \frac{\partial}{\partial \underline{V}}\left[\left(\frac{\partial \underline{U}}{\partial \underline{V}}\right)_{\underline{S},N}\right]_{\underline{S},N} = -\left(\frac{\partial P}{\partial \underline{V}}\right)_{\underline{S},N} \tag{5-9}$$

$$\frac{\partial^2 \underline{U}}{\partial \underline{S}\,\partial \underline{V}} = \frac{\partial}{\partial \underline{S}}\left[\left(\frac{\partial \underline{U}}{\partial \underline{V}}\right)_{\underline{S},N}\right]_{\underline{V},N} = -\left(\frac{\partial P}{\partial \underline{S}}\right)_{\underline{V},N} \tag{5-10}$$

$$\frac{\partial^2 \underline{U}}{\partial \underline{V}\,\partial \underline{S}} = \frac{\partial}{\partial \underline{V}}\left[\left(\frac{\partial \underline{U}}{\partial \underline{S}}\right)_{\underline{V},N}\right]_{\underline{S},N} = \left(\frac{\partial T}{\partial \underline{V}}\right)_{\underline{S},N} \tag{5-11}$$

Of these four derivatives, only three are independent because the last two are related by the reciprocity theorem of Maxwell. That is, for any smoothly varying function for which $X = f(Y, Z)$, the order of differentiation is immaterial and, therefore,

$$\frac{\partial}{\partial Z}\left[\left(\frac{\partial X}{\partial Y}\right)_Z\right]_Y = \frac{\partial}{\partial Y}\left[\left(\frac{\partial X}{\partial Z}\right)_Y\right]_Z \tag{5-12}$$

Thus,

$$-\left(\frac{\partial P}{\partial \underline{S}}\right)_{\underline{V},N} = \left(\frac{\partial T}{\partial \underline{V}}\right)_{\underline{S},N} \tag{5-13}$$

The Fundamental Equation for an n-component mixture is

$$\underline{U} = f_U(\underline{S}, \underline{V}, N_1, \ldots, N_n) \tag{5-14}$$

For an *n-component mixture* there are $n + 1$ first-order and $(n + 2)(n + 1)/2$ second-order *independent* partial derivatives of the Fundamental Equation. These derivatives are particularly important because they form a basis for all other partial derivatives involving thermodynamic properties. That is, any partial derivative can be expressed in terms of an independent set of first-order and second-order derivatives of any form of the Fundamental Equation. Proof of this statement for a pure material is given in Section 7.2.

It should be clear that Fundamental Equations would be of great use if they were available. The problem is that the complete form of the Fundamental Equation is not specified by classical thermodynamics; each substance has its own peculiarities that are reflected in different functionalities of the Fundamental Equation. Thus, there is no single Fundamental Equation governing the properties of all materials.

The postulates of classical thermodynamics place some restrictions on the form of the Fundamental Equation. Let us examine the differential form of the Fundamental Equation in the energy representation, Eq. (5-3). Since $\underline{U}$ must be first order in mass or mole number, we can apply Euler's theorem (see Appendix C) to obtain the integrated form of Eq. (5-3):

$$\underline{U} = \left(\frac{\partial f_U}{\partial \underline{S}}\right)_{\underline{V},N}\underline{S} + \left(\frac{\partial f_U}{\partial \underline{V}}\right)_{\underline{S},N}\underline{V} + \sum_{i=1}^{n}\left(\frac{\partial f_U}{\partial N_i}\right)_{\underline{S},\underline{V},N_j[i]} N_i \tag{5-15}$$

Equation (5-15) is a linear partial differential equation of the first order. Therefore, the solution must be of the form

$$\underline{U} = x\left[g\left(\frac{y}{x}, \frac{z}{x}, \ldots\right)\right] \tag{5-16}$$

where $x, y, z, \ldots$ can be $\underline{S}, \underline{V}, N_1, \ldots, N_n$ or any permutation of these variables. For a one-component system, it is most convenient to choose $x = N$, $y = \underline{S}$, and $z = \underline{V}$; we then obtain

$$\underline{U} = N\left[g\left(\frac{\underline{S}}{N}, \frac{\underline{V}}{N}\right)\right] \tag{5-17}$$

or, since $\underline{U} = NU$,

$$U = g(S, V) \tag{5-18}$$

The only other requirements our prior developments place on the form of the Fundamental Equation are that $\underline{U}$ should be a single-valued function of $\underline{S}$, $\underline{V}$, and N (see Postulate III), and that $(\partial f_U/\partial \underline{S})_{\underline{V},N} = T$ should be nonnegative.

5.2 Intensive and Extensive Properties

At this point, let us digress to a special case of Postulate I. In Eq. (5-18), we see that for a one-component system, only two properties, S and V, are required to obtain the specific energy of a system. This is in no way a violation of Postulate I; by delineating the *specific* properties (expressed in terms of unit mass or mole number), we have determined the "intensity" of the system but not the "extent" of the system. To specify the system completely (e.g., so that it can be reproduced by others), we must specify the mass of the system in addition to S and V.

The variables that express intensity of the system are zero order in mass and are called *intensive variables.* Variables that relate to extent of the system are first order in mass and are called *extensive variables.*

We now prove that for a single-phase simple system of n components, any intensive property can be defined by the values of $n + 1$ other intensive properties.[1] Let us call $b, c_1, c_2, \ldots, c_{n+1}$ *intensive* properties of a single-phase simple system containing n components. In general, we can express b as a function of $n + 2$ other properties according to Postulate I. Let us choose these $n + 2$ as $c_1, c_2, \ldots, c_{n+1}$, and the total moles (or mass) N. Thus,

$$db = \left(\frac{\partial b}{\partial c_1}\right)_{c_j[1],N} dc_1 + \cdots \left(\frac{\partial b}{\partial c_{n+1}}\right)_{c_j[n+1],N} dc_{n+1} + \left(\frac{\partial b}{\partial N}\right)_{c_1,\ldots,c_{n+1}} dN \tag{5-19}$$

Integrating Eq. (5-19) by using Euler's theorem (see Appendix C), we have

$$\left(\frac{\partial b}{\partial N}\right)_{c_1,\ldots,c_{n+1}} N = 0 \tag{5-20}$$

Since N can be nonzero, $(\partial b/\partial N)_{c_1,\ldots,c_{n+1}}$ must be zero. Therefore, Eq. (5-19) reduces to a function of $n + 1$ intensive variables. Of course, these $n + 1$ intensive variables must be independent, so that we clearly cannot use all of the n mole fractions $x_1, \ldots, x_n$. We could, however, use $n - 1$ mole fractions in addition to two other intensive variables to obtain the required $n + 1$.

Note that this result is valid because we limited the original set of $n + 2$ variables to include only one extensive variable; if we had included two extensive variables in the original $n + 2$ set, no partial derivative would have to be zero. Thus, we could state as a corollary to Postulate I: *For a single-phase simple system, the change of any intensive variable can be expressed as a function*

[1]The proof is restricted to a single-phase system to allow us to choose any $n + 2$ variables as an independent set. The proof can be extended to specific cases of composite simple systems provided that the $n + 2$ variables chosen form an independently variable set.

of any $n+1$ other independent intensive variables. We shall use this corollary frequently in Chapter 8 in dealing with the properties of mixtures.

At this point, a word of caution is in order when dealing with intensive and extensive variables in partial derivatives. For example, if for a pure material we express U as a function of S and V, then using Eqs. (5-19) and (5-20), we find $(\partial U/\partial N)_{S,V} = 0$. But $(\partial \underline{U}/\partial N)_{\underline{S},\underline{V}}$ is not zero; from Eq. (5-7) applied to a pure material,

$$\left(\frac{\partial \underline{U}}{\partial N}\right)_{\underline{S},\underline{V}} = \mu \tag{5-21}$$

Since $\underline{U} = NU$, we also have

$$\left(\frac{\partial \underline{U}}{\partial N}\right)_{\underline{S},\underline{V}} = U + N\left(\frac{\partial U}{\partial N}\right)_{\underline{S},\underline{V}} = \mu \tag{5-22}$$

Since U is not equal to μ, $(\partial U/\partial N)_{\underline{S},\underline{V}}$ is not equal to zero. Thus, each of the three derivatives, $(\partial U/\partial N)_{S,V}$, $(\partial \underline{U}/\partial N)_{\underline{S},\underline{V}}$, and $(\partial U/\partial N)_{\underline{S},\underline{V}}$ have different connotations. The first represents the change in the specific energy as we add more material while maintaining constant specific entropy and specific volume. Since we are holding two intensive variables constant during the process, all other intensive variables (e.g., T, P, etc.) for the pure material must remain unchanged. The only way to conduct the process is to enlarge the system in direct proportion to the added mass. The second and third cases, however, represent changes in the total and specific energy during a process in which we maintain constant total entropy and total volume. Since we are adding mass to the system, the only way to keep total entropy and total volume constant is to change the specific entropy and specific volume (e.g., by varying T and P during the addition of mass). Thus, the specific energy changes as the state of the system is varied. The total energy changes because both the specific energy and mass vary.

Example 5.1

In the entropy representation, the Fundamental Equation for a monatomic ideal gas is

$$\underline{S} = N\left[S^0 + R\ln\left\{\left(\frac{U}{U^0}\right)^{3/2}\frac{V}{V^0}\right\}\right] \tag{5-22a}$$

where S^0, U^0, and V^0 are constants representing values in a reference or base state. From the energy representation in the form of Eq. (5-1), determine the three equations of state in the form of Eqs. (5-5) through (5-7).

Solution

Solving Eq. (5-22a) explicitly for U yields

$$U = U^0\left(\frac{V^0}{V}\right)^{2/3} e^{(2/3)(S-S^0)/R} \tag{5-22b}$$

and, thus,

$$\underline{U} = NU^0 \left(\frac{V^0}{V}\right)^{2/3} e^{(2/3)(S-S^0)/R} \tag{5-22c}$$

The equations of state can be found directly by partial differentiation of Eq. (5-22b) or (5-22c):

$$T = \left(\frac{\partial \underline{U}}{\partial \underline{S}}\right)_{\underline{V},N} = \left(\frac{\partial U}{\partial S}\right)_V = \frac{2}{3}\left(\frac{U^0}{R}\right)\left(\frac{V^0}{V}\right)^{2/3} e^{(2/3)(S-S^0)/R} \tag{5-22d}$$

$$-P = \left(\frac{\partial \underline{U}}{\partial \underline{V}}\right)_{\underline{S},N} = \left(\frac{\partial U}{\partial V}\right)_S = -\frac{2}{3}\frac{U^0(V^0)^{2/3}}{V^{5/3}} e^{(2/3)(S-S^0)/R} \tag{5-22e}$$

$$\begin{aligned} \mu &= \left(\frac{\partial \underline{U}}{\partial N}\right)_{\underline{S},\underline{V}} = \frac{\partial}{\partial N}\left[NU^0\left(\frac{V^0}{V}\right)^{2/3} e^{(2/3)(S-S^0)/R}\right]_{\underline{S},\underline{V}} \\ &= U^0\left(\frac{V^0}{V}\right)^{2/3} e^{(2/3)(S-S^0)/R}\left(\frac{5}{3} - \frac{2}{3}\frac{S}{R}\right) \end{aligned} \tag{5-22f}$$

The results given in Eqs. (5-22d) through (5-22f) may be simplified if Eq. (5-22a) is used to evaluate the exponential terms. If this is done, then

$$\begin{aligned} T &= \frac{2U}{3R} \\ P &= \frac{2U}{3V} \\ \mu &= U\left(\frac{5}{3} - \frac{2}{3}\frac{S}{R}\right) \end{aligned} \tag{5-22g}$$

In fact, the assumptions behind the development of Eq. (5-22a) are that $U = \frac{3}{2}RT$ and $PV = RT$.

5.3 Reconstruction of the Fundamental Equation

We now consider the minimum information necessary for reconstruction of a Fundamental Equation for a given material. Although we are not necessarily interested in knowing the Fundamental Equation (in general, they are much too cumbersome to use), we are interested in knowing what information content is equivalent to the Fundamental Equation. Once we have this, we need not look for any more because all other information of thermodynamic interest can be obtained therefrom.

As shown in Section 5.1, if the Fundamental Equation were known, the properties T, P, μ_i, could be determined by partial differentiation as expressed in the equations of state in Eqs. (5-5) through (5-7). Alternatively, the Fundamental Equation can be recovered, if all the equations of state were known, by substituting these equations into Eq. (5-15) or the more common form

$$\underline{U} = T\underline{S} - P\underline{V} + \sum_{i=1}^{n} \mu_i N_i \tag{5-23}$$

As shown in Section 5.2, the $n + 2$ intensive variables, T, P, μ_i, which are expressed explicitly by the equations of state, are not all independently variable. Any one of these variables can be expressed in differential form in terms of the other $n + 1$ variables and, upon integration, an expression between the $n + 2$ variables can be determined to within an arbitrary constant. It thus follows that only $n + 1$ equations of state are necessary to determine the Fundamental Equation to within an arbitrary constant.[2] In Example 5.1, if Eq. (5-22g) is combined with Eq. (5-22a) to elimimate S, μ can be expressed in terms of U^0, V^0; T, V, and an arbitrary constant S^0.

5.4 Legendre Transformations

In the energy representation of the Fundamental Equation, Eq. (5-1), the properties $\underline{S}, \underline{V}, N_1, \ldots, N_n$ are treated as independent variables. This is not always an appropriate set of independent parameters. For example, since temperature can be measured much more conveniently than entropy, we might like to use $T, \underline{V}, N_1, \ldots, N_n$ as the independent variables. For a single-phase simple system, we can always express a property such as $\underline{U}$ in terms of $n + 2$ other properties such as $T, \underline{V}, N_1, \ldots, N_n$. Thus, we know that a function f exists such that

$$\underline{U} = f(T, \underline{V}, N_1, \ldots, N_n) \tag{5-24}$$

and

$$d\underline{U} = \left(\frac{\partial f}{\partial T}\right)_{\underline{V},N} dT + \left(\frac{\partial f}{\partial \underline{V}}\right)_{T,N} d\underline{V} + \sum_{i=1}^{n} \left(\frac{\partial f}{\partial N_i}\right)_{T,\underline{V},N_j[i]} dN_i \tag{5-25}$$

Given the Fundamental Equation, the function of Eq. (5-24) can be found by differentiating Eq. (5-1) to obtain Eq. (5-5),

$$T = g_T(\underline{S}, \underline{V}, N_1, \ldots, N_n)$$

and then solving Eqs. (5-1) and (5-5) simultaneously in order to eliminate $\underline{S}$. The result is an equation of the form

$$\underline{U} = f(T, \underline{V}, N_1, \ldots, N_n) = f\left[\left(\frac{\partial f_U}{\partial \underline{S}}\right)_{\underline{V},N}, \underline{V}, N_1, \ldots, N_n\right] \tag{5-26}$$

Although Eq. (5-26) is of the form desired [i.e., Eq. (5-24)], the information content of Eq. (5-26) is less than that of the Fundamental Equation. Equation (5-26) is a partial differential equation that can be integrated to yield the Fundamental Equation only to within an arbitrary function of integration.

We must now ask whether or not there are other functions with the same information content as that of the Fundamental Equation but with independent

[2] Since only differences in $\underline{U}$ have physical significance, we need not specify the energy representation more definitively than to within an arbitrary constant.

variables other than $\underline{S}$, $\underline{V}$, $N_1, \ldots, N_n$. The answer is that there are such functions if we are willing to restrict ourselves to a set of independent variables in which we choose only one from each of the following pairs: $\underline{S}, T$; $\underline{V}, P$; N_i, μ_i. These pairs of variables are usually referred to as *conjugate coordinates.*

To formulate these functions a *Legendre transform* is employed. Such a transformation stems from a basic theorem in line geometry,[3] and although the rigorous proof is no simple task, the results are easy to apply. The basic principle is that a curve consisting of a locus of points can be described completely by the tangent lines that form the envelope of the curve. Since no information is lost in shifting from one description of the curve to another, the process can be reversed. For example, the function

$$y^{(0)} = f(x_1, \ldots, x_m) \tag{5-27}$$

represents a locus of points in $m + 1$ space. There are m first-order partial derivatives of $y^{(0)}$ with respect to each of the m independent variables, $x_1, \ldots, x_m$. Defining these derivatives as ξ_i,

$$\xi_i \equiv \left(\frac{\partial y^{(0)}}{\partial x_i}\right)_{x_1,\ldots,[x_i],\ldots,x_m} \qquad \text{(5-28)[4]}$$

or

$$\xi_i = f(x_1, \ldots, x_m) \tag{5-29}$$

it follows that the variation of $y^{(0)}$ with x_1 could be described by the envelope of tangents in the $y^{(0)}$–x_1 plane. If $y^{(1)}$ is the intercept of the tangent corresponding to ξ_1,

$$y^{(1)}(\xi_1, x_2, \ldots, x_m) = y^{(0)} - \xi_1 x_1 \tag{5-30}$$

The function $y^{(1)}(\xi_1, x_2, \ldots, x_m)$ is called the *first* Legendre transform of $y^{(0)}$ with respect to x_1. In other words, a Legendre transform results in a new function in which one or more independent variables is replaced by its slope. There are obviously m different first transforms, depending on the ordering of the variables $x_1, \ldots, x_m$.

For the simple case where there is only one independent variable [i.e., $y^{(0)} = f(x)$], the envelope of tangents is shown in Figure 5.1. The slope is ξ and the intercept of each tangent is $y^{(1)}$. Similarly, if $y^{(0)} = f(x_1, x_2)$, the transforms $y^{(1)}$ and $y^{(2)}$ are shown in Figure 5.2. In this case, the transforms apply to point A on the surface.

[3]H. B. Callen, *Thermodynamics* (New York: Wiley, 1960), Chaps. 5, 6, 8; L. Tisza, *Generalized Thermodynamics* (Cambridge, Mass : MIT Press, 1966), pp. 61, 136, 236; R. Aris and N. R. Amundson, *Mathematical Methods in Chemical Engineering*, Vol. 2 (Englewood Cliffs, N.J.: Prentice-Hall, 1973), pp. 197–201.

[4]The symbol $[x_i]$ in the subscript of the partial derivative indicates that x_i is not held constant.

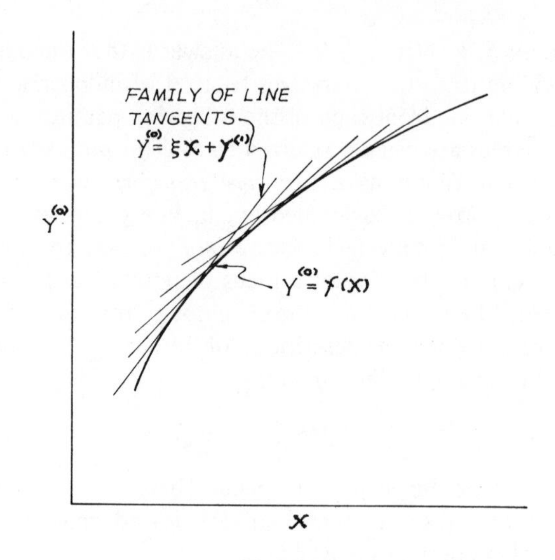

Figure 5.1

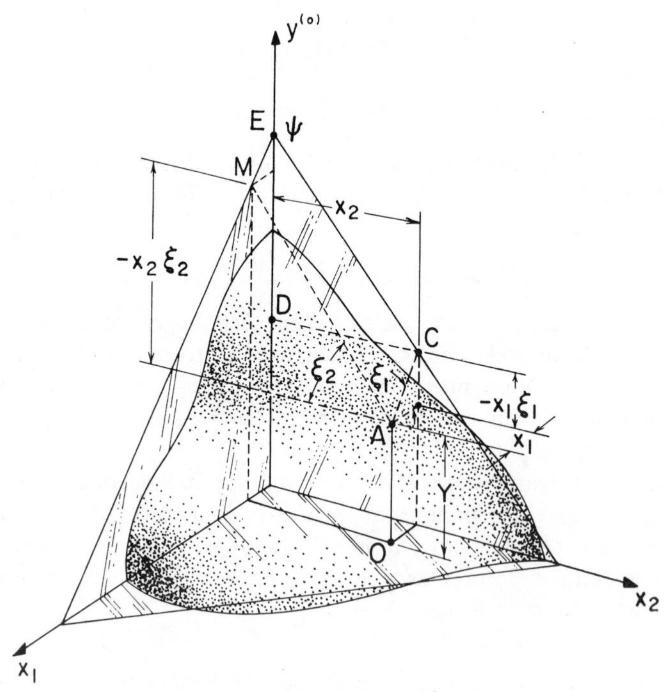

Legendre Transform for $y^{(0)} = f(x_1, x_2)$

IF: $y^{(1)} = f(\xi_1, x_2)$, $y^{(1)} =$ Ⓒ

$y^{(1)} = f(x_1, \xi_2)$, $y^{(1)} =$ Ⓜ

$y^{(2)} = f(\xi_1, \xi_2)$, $y^{(2)} =$ Ⓔ

Figure 5.2

Higher-order transforms are defined in a similar manner; thus, $y^{(k)}(\xi_1, \ldots, \xi_k, x_{k+1}, \ldots, x_m)$ is the kth Legendre transform:

$$y^{(k)} = y^{(0)} - \sum_{i=1}^{k} \xi_i x_i \tag{5-31}$$

The differential of $y^{(0)}$ can be expressed as

$$dy^{(0)} = \sum_{i=1}^{m} \xi_i \, dx_i \tag{5-32}$$

and the differential of the kth transform can be obtained by differentiating Eq. (5-31) and substituting Eq. (5-32) into the result.

$$dy^{(k)} = -\sum_{i=1}^{k} x_i \, d\xi_i + \sum_{i=k+1}^{m} \xi_i \, dx_i \tag{5-33}$$

Since $y^{(k)}$ is a function of $\xi_1, \ldots, \xi_k, x_{k+1}, \ldots, x_m$, it follows from Eq. (5-33) that for the transformed variables $(1 \leq i \leq k)$,

$$\left(\frac{\partial y^{(k)}}{\partial \xi_i}\right)_{\xi_1, \ldots, [\xi_i], \ldots, \xi_k, x_{k+1}, \ldots, x_m} = -x_i \tag{5-34}$$

whereas for the untransformed variables $(k < i \leq m)$,

$$\left(\frac{\partial y^{(k)}}{\partial x_i}\right)_{\xi_1, \ldots, \xi_k, x_{k+1}, \ldots, [x_i], \ldots, x_m} = \xi_i \tag{5-35}$$

Equation (5-34) is sometimes called the *inverse* Legendre transform. Equation (5-35) is applicable for all cases where $i > k$, and thus one may generalize the result as

$$\frac{\partial y^{(i-1)}}{\partial x_i} = \frac{\partial y^{(i-2)}}{\partial x_i} = \cdots \frac{\partial y^{(0)}}{\partial x_i} = \xi_i \tag{5-36}$$

The partial derivatives in Eq. (5-36) were expressed without indicating the set of variables to be held constant. However, it is clear from the discussion above that the degree of the transform determines the set, and the only exception would be that variable used in the actual differentiation. For example, the restraints on the term $\partial y^{(i-1)}/\partial x_i$ would be that $\xi_1, \ldots, \xi_{i-1}, x_{i+1}, \ldots, x_m$ would be held constant.

To illustrate the application of these relations, let $y^{(0)}$ be the total internal energy of a system $\underline{U}$; then the Fundamental Equation would be given in Eq. (5-1). Suppose that we desired the transform $y^{(2)}$, where

$$y^{(2)} = f(\xi_S, \xi_V, N_1, \ldots, N_n) \tag{5-37}$$

with

$$\xi_S \equiv \left(\frac{\partial \underline{U}}{\partial \underline{S}}\right)_{\underline{V}, N} = T \tag{5-38}$$

$$\xi_V \equiv \left(\frac{\partial \underline{U}}{\partial \underline{V}}\right)_{\underline{S}, N} = -P \tag{5-39}$$

Then, with Eq. (5-31),

$$y^{(2)} = \underline{U} - T\underline{S} - (-P\underline{V}) \equiv \underline{G} \tag{5-40}$$

where $\underline{G}$ is the total Gibbs energy. The analogs of Eqs. (5-32) and (5-33) are Eqs. (5-4) and (5-41).

$$dy^{(2)} = d\underline{G} = -\underline{S}\,dT + \underline{V}\,dP + \sum_{i=1}^{n} \mu_i\,dN_i \tag{5-41}$$

where the chemical potential μ_i is defined as

$$\mu_i \equiv \left(\frac{\partial \underline{U}}{\partial N_i}\right)_{\underline{S},\underline{V},N_1,\ldots,[N_i],\ldots,N_n} \tag{5-42}$$

$$= \left(\frac{\partial \underline{G}}{\partial N_i}\right)_{T,P,N_1,\ldots,[N_i],\ldots,N_n} \tag{5-43}$$

The significance of the Legendre transform is thus evident. The important thermodynamic property $\underline{G}$ is simply a partial Legendre transform of the energy $\underline{U}$ from $(\underline{S}, \underline{V}, N_1, \ldots, N_n)$ space to $(T, P, N_1, \ldots, N_n)$ space. Equation (5-41) is also a Fundamental Equation and no loss in information content has resulted in the transform from the $\underline{U}$ to the $\underline{G}$ representation.

The *total* Legendre transform of Eq. (5-1) is

$$\begin{aligned} y^{(n+2)} &= \underline{U} - T\underline{S} + P\underline{V} - \sum_{i=1}^{n} \mu_i N_i \\ &= 0 \qquad \text{by Eq. (5-15)} \end{aligned} \tag{5-44}$$

Thus, Eq. (5-33) becomes

$$dy^{(n+2)} = 0 = -\underline{S}\,dT + \underline{V}\,dP - \sum_{i=1}^{n} N_i\,d\mu_i \tag{5-45}$$

Equation (5-45) is the *Gibbs–Duhem equation.*

It is important to note certain generalities in the use of Legendre transformations. We began with a general functional equation [Eq. (5-27)] with an arbitrary ordering of $x_1, \ldots, x_m$. We noted that for each x_i there was a conjugate coordinate variable, ξ_i [Eq. (5-28)]. We then illustrated that one could readily derive a functional relation with the same information content wherein we replaced independent variables $x_1, \ldots, x_k$ by $\xi_1, \ldots, \xi_k$. This was the kth Legendre transform defined in Eq. (5-31) and shown in differential form in Eq. (5-33).

One may also redefine this kth Legendre transform as the $y^{(0)}$ basic function if care is given in defining the correct independent variables and conjugate coordinates. For example, we have shown that beginning with Eq. (5-1), we obtained the Gibbs energy potential function by a Legendre transform of $\underline{U}$ into $T, P, N_1, \ldots, N_n$ space. We could now use the Gibbs energy function as our basic function $y^{(0)}$, but the independent variable set $(x_1, \ldots, x_m)$ would be $(T, P, N_1, \ldots, N_n)$ with arbitrary ordering. The conjugate coordinate variables $\xi_1, \ldots, \xi_m$ would still be defined by Eq. (5-28); for example, for the variable T, $\xi_T = (\partial \underline{G}/\partial T)_{P,N} = -\underline{S}$; for P, $\xi_P = \underline{V}$, and for N_j, $\xi_j = \mu_j$.

In fact, we may select any Legendre transform as the basic $y^{(0)}$ function by redefining the independent variable set.

5.5 Relationships between Partial Derivatives of Legendre Transforms

Single variable transforms

Starting with Eq. (5-27), we wish to investigate the relations between derivatives of $y^{(0)}$ and the first transform $y^{(1)}$, where

$$y^{(1)} = f(\xi_1, x_2, \ldots, x_m) \tag{5-46}$$

$$\xi_1 = \left(\frac{\partial y^{(0)}}{\partial x_1}\right)_{x_2,\ldots,x_m} \tag{5-47}$$

There are several ways to obtain the desired results, but the most expeditious involves the use of the derivative operators[5]

$$\left(\frac{\partial}{\partial x_i}\right)_{\xi_1, x_2,\ldots,[x_i],\ldots,x_m} \quad (i > 1) \quad \text{and} \quad \left(\frac{\partial}{\partial \xi_1}\right)_{x_2,\ldots,x_m}$$

$$\left(\frac{\partial}{\partial x_i}\right)_{\xi_1, x_2,\ldots,[x_i],\ldots,x_m} = \left(\frac{\partial}{\partial x_i}\right)_{x_1,\ldots,[x_i],\ldots,x_m} - \left[\frac{y_{1i}^{(0)}}{y_{11}^{(0)}}\right]\left(\frac{\partial}{\partial x_1}\right)_{x_2,\ldots,x_m} \quad (i \neq 1) \tag{5-48}$$

and

$$\left(\frac{\partial}{\partial \xi_1}\right)_{x_2,\ldots,x_m} = \frac{1}{y_{11}^{(0)}}\left(\frac{\partial}{\partial x_1}\right)_{x_2,\ldots,x_m} \tag{5-49}$$

The terms $y_{1i}^{(0)}$ and $y_{11}^{(0)}$ are second-order derivatives:

$$y_{1i}^{(0)} \equiv \frac{\partial^2 y^{(0)}}{\partial x_1\, \partial x_i} \tag{5-50}$$

$$y_{11}^{(0)} = \frac{\partial^2 y^{(0)}}{\partial x_1^2} \tag{5-51}$$

Equations (5-48) and (5-49) are of little value to obtain first derivatives since, in view of Eqs. (5-34) and (5-35),

$$\left(\frac{\partial y^{(1)}}{\partial \xi_1}\right)_{x_2,\ldots,x_m} \equiv y_1^{(1)} = -x_1$$

$$\left(\frac{\partial y^{(1)}}{\partial x_i}\right)_{\xi_1, x_2,\ldots,[x_i],\ldots,x_m} = y_i^{(1)} = y_i^{(0)} = \xi_i \quad (i > 1)$$

However, to relate, for example, $y_{1i}^{(1)}$, to second-order derivatives of $y^{(0)}$, one can employ the operator equation [Eq. (5-49)] on $y_i^{(1)}$.

$$\begin{aligned}\left(\frac{\partial y_i^{(1)}}{\partial \xi_1}\right)_{x_2,\ldots,x_m} &\equiv y_{1i}^{(1)} = \frac{1}{y_{11}^{(0)}}\left(\frac{\partial y_i^{(1)}}{\partial x_1}\right)_{x_2,\ldots,x_m} \\ &= \frac{1}{y_{11}^{(0)}}\left(\frac{\partial y_i^{(0)}}{\partial x_1}\right)_{x_2,\ldots,x_m} = \frac{y_{1i}^{(0)}}{y_{11}^{(0)}} \quad (i \neq 1)\end{aligned} \tag{5-52}$$

[5]The derivation of these operators is given in an appendix to the paper "Legendre Transforms and Their Application in Thermodynamics," *AIChE J.*, **20**, 1194–1200 (1974).

where Eq. (5-36) was used to simplify the third step. In a similar manner, other second- and third-order derivatives may be readily transformed. A list of these is presented in Table 5.1.

TABLE 5.1
SECOND- AND THIRD-ORDER DERIVATIVES OF $y^{(1)}$ IN TERMS OF $y^{(0)}$

Derivative in $y^{(1)}$ system	Derivative in $y^{(0)}$ system	Quantity operated upon:	Equation used
$y_{11}^{(1)}$	$-\dfrac{1}{y_{11}^{(0)}}$	$y_1^{(1)}$	(5-49)
$y_{1i}^{(1)}$	$\dfrac{y_{1i}^{(0)}}{y_{11}^{(0)}} \quad (i \neq 1)$	$y_i^{(1)}$	(5-49)
$y_{ij}^{(1)}$	$y_{ij}^{(0)} - \dfrac{y_{1i}^{(0)} y_{1j}^{(0)}}{(y_{11}^{(0)})} \quad (i, j \neq 1)$	$y_j^{(1)}$	(5-48)
$y_{111}^{(1)}$	$\dfrac{y_{111}^{(0)}}{(y_{11}^{(0)})^3}$	$y_{11}^{(1)}$	(5-49)
$y_{11i}^{(1)}$	$\dfrac{y_{11i}^{(0)}}{(y_{11}^{(0)})^2} - \dfrac{y_{111}^{(0)} y_{1i}^{(0)}}{(y_{11}^{(0)})^3} \quad (i > 1)$	$y_{1i}^{(1)}$	(5-49)
$y_{1ij}^{(1)}$	$\dfrac{y_{1ij}^{(0)}}{y_{11}^{(0)}} - \left\{\dfrac{y_{11i}^{(0)} y_{1j}^{(0)} + y_{11j}^{(0)} y_{1i}^{(0)}}{(y_{11}^{(0)})^2}\right\} + \dfrac{y_{111}^{(0)} y_{1i}^{(0)} y_{1j}^{(0)}}{(y_{11}^{(0)})^3} \quad (i, j \neq 1)$	$y_{ij}^{(1)}$	(5-49)
$y_{ijk}^{(1)}$	$y_{ijk}^{(0)} - \dfrac{(y_{1i}^{(0)} y_{1jk}^{(0)} + y_{1j}^{(0)} y_{1ik}^{(0)} + y_{1k}^{(0)} y_{1ij}^{(0)})}{y_{11}^{(0)}} + \dfrac{(y_{1i}^{(0)} y_{1j}^{(0)} y_{11k}^{(0)} + y_{1i}^{(0)} y_{1k}^{(0)} y_{11j}^{(0)} + y_{1j}^{(0)} y_{1k}^{(0)} y_{11i}^{(0)})}{(y_{11}^{(0)})^2} - \dfrac{y_{1i}^{(0)} y_{1j}^{(0)} y_{1k}^{(0)} y_{111}^{(0)}}{(y_{11}^{(0)})^3} \quad (i, j, k \neq 1)$	$y_{jk}^{(1)}$	(5-48)

Example 5.2

Assume that the basic function is $\underline{U} = f(\underline{S}, \underline{V}, N_1, \ldots, N_n)$. Determine $y_{112}^{(1)}$ in terms of derivatives of this basic function.

Solution

With $y^{(1)} = f(T, \underline{V}, N_1, \ldots, N_n)$ and $y^{(0)} = f(\underline{S}, \underline{V}, N_1, \ldots, N_n)$,

$$y^{(1)} = \underline{A} = \underline{U} - T\underline{S} \tag{5-53}$$

$\underline{A}$ is the Helmholtz energy. From Table 5.1,

$$y_{112}^{(1)} = A_{TTV} = \frac{\partial^3 \underline{A}}{\partial T^2 \, \partial \underline{V}}$$

$$= \frac{y_{112}^{(0)}}{(y_{11}^{(0)})^2} - \frac{y_{111}^{(0)} y_{12}^{(0)}}{(y_{11}^{(0)})^3}$$

$$= \frac{U_{SSV}}{U_{SS}^2} - \frac{U_{SSS} U_{SV}}{U_{SS}^3}$$

or

$$-\frac{\partial^2 \underline{S}}{\partial T\,\partial \underline{V}} = -\left(\frac{\partial^2 P}{\partial T^2}\right)_{\underline{V},N} = \frac{\partial^2 T/\partial \underline{S}\,\partial \underline{V}}{(\partial T/\partial \underline{S})^2_{\underline{V},N}} - \frac{(\partial^2 T/\partial \underline{S}^2)_{\underline{V},N}(\partial T/\partial \underline{V})_{\underline{S},N}}{(\partial T/\partial \underline{S})^3_{\underline{V},N}}$$

Multiple variable transforms

The development shown above was limited to the case where only a single variable was transformed. Should one wish to transform more than a single variable, it is always possible to proceed a step at a time and transform each separately as was shown above. It is also possible to develop a more general technique to allow one to express the partial derivatives of a Legendre transform $y^{(j)}(\xi_1, \ldots, \xi_j, x_{j+1}, \ldots, x_m)$ in terms of the basis function $y^{(0)}$ or, in general, to some other Legendre transform $y^{(j-q)}$, where $q = 1, 2, \ldots, j$. The equations to obtain these second-order derivatives are shown in Table 5.2 for the basis function and in Table 5.3 for other functions. In Table 5.2 there are three cases for $y^{(j)}_{ik}$: that is, $j > i, k$; $k \geq j > i$; and $j \leq i, k$. Each case is illustrated in Example 5.3.

TABLE 5.2

RELATIONS BETWEEN SECOND-ORDER DERIVATIVES OF THE jth LEGENDRE TRANSFORM AND THE BASIS FUNCTION

$$y^{(j)}_{ik} = \frac{\begin{vmatrix} \mathfrak{D}^{(0)}_j & \mathrm{B} \\ \mathrm{A} & 0 \end{vmatrix}}{\mathfrak{D}^{(0)}_j} \qquad j > i, k$$

$$y^{(j)}_{ik} = \frac{\begin{vmatrix} \mathfrak{D}^{(0)}_j & \begin{matrix}\mathrm{D}\\ \mathrm{F}\end{matrix} \\ \mathrm{A} & 0 \end{vmatrix}}{\mathfrak{D}^{(0)}_j} \qquad \begin{matrix} j > i \\ j \leq k \end{matrix}$$

$$y^{(j)}_{ik} = \frac{\begin{vmatrix} \mathfrak{D}^{(0)}_j & \begin{matrix}\mathrm{D}\\ \mathrm{F}\end{matrix} \\ \mathrm{C} \;|\; \mathrm{E} & \mathrm{G} \end{vmatrix}}{\mathfrak{D}^{(0)}_j} \qquad \begin{matrix} j \leq i \\ j \leq k \end{matrix}$$

δ_{ij}: Kronecker delta, $= 1$ if $i = j$, $= 0$ if $i \neq j$.
A: j terms; each with a value of $(-\delta_{mi})$, where $m = 1, 2, \ldots, j$.
B: j terms; each with a value of $(-\delta_{mk})$, where $m = 1, 2, \ldots, j$.
C: $(j-1)$ terms; each with a value of $(1 - \delta_{ji})y^{(0)}_{mi}$, where $m = 1, 2, \ldots, (j-1)$.
D: $(j-1)$ terms; each with a value of $(1 - \delta_{jk})y^{(0)}_{mk}$, where $m = 1, 2, \ldots, (j-1)$.
E: One term; $(1 - \delta_{ji})y^{(0)}_{ji} - \delta_{ji}$.
F: One term; $(1 - \delta_{jk})y^{(0)}_{jk} - \delta_{jk}$.
G: $(1 - \delta_{ji})(1 - \delta_{jk})y^{(0)}_{ik}$.

and

$$\mathfrak{D}^{(0)}_j = \begin{vmatrix} y^{(0)}_{11} & y^{(0)}_{12} & \cdots & y^{(0)}_{1j} \\ y^{(0)}_{21} & & & \vdots \\ \vdots & & & \vdots \\ y^{(0)}_{j1} & \cdots & \cdots & y^{(0)}_{jj} \end{vmatrix}$$

TABLE 5.3

RELATIONSHIPS BETWEEN SECOND-ORDER DERIVATIVES OF THE jth LEGENDRE TRANSFORM AND THE $(j-q)$ TRANSFORM[a]

$$y_{ik}^{(j)} = \frac{\left|\begin{array}{cc|c} \multicolumn{2}{c|}{\mathfrak{D}_q^{(j-q)}} & \begin{array}{c} B' \\ \hline B'' \end{array} \\ \hline A' & A'' & H' \end{array}\right|}{\mathfrak{D}_q^{(j-q)}} \qquad j > i, k \qquad\qquad y_{ik}^{(j)} = \frac{\left|\begin{array}{cc|c} \multicolumn{2}{c|}{\mathfrak{D}_q^{(j-q)}} & \begin{array}{c} D' \\ \hline F' \end{array} \\ \hline A' & A'' & I' \end{array}\right|}{\mathfrak{D}_q^{(j-q)}} \qquad \begin{array}{l} j > i \\ j \leq k \end{array}$$

$$y_{ik}^{(j)} = \frac{\left|\begin{array}{cc|c} \multicolumn{2}{c|}{\mathfrak{D}_q^{(j-q)}} & \begin{array}{c} D' \\ \hline F' \end{array} \\ \hline J' & K' & L' \end{array}\right|}{\mathfrak{D}_q^{(j-q)}} \qquad \begin{array}{l} j \leq i \\ j \leq k \end{array}$$

and

$$\mathfrak{D}_q^{(j-q)} = \begin{vmatrix} y_{(j-q+1)(j-q+1)}^{(j-q)} & y_{(j-q+1)(j-q+2)}^{(j-q)} & \cdots & y_{(j-q+1)j}^{(j-q)} \\ y_{(j-q+2)(j-q+1)}^{(j-q)} & y_{(j-q+2)(j-q+2)}^{(j-q)} & \cdots & \\ \cdot & & & \\ \cdot & & & \\ \cdot & & & \\ y_{j(j-q+1)}^{(j-q)} & \cdots\cdot & & y_{jj}^{(j-q)} \end{vmatrix}$$

[a] $Z_{rs} = 0 \quad \text{if } r \leq s$

$\quad = 1 \quad \text{if } r > s$

A′: $(q-1)$ terms; each with a value of

$$[Z_{(j-q+1)i}\, y_{(j-p)i}^{(j-q)} - \delta_{(j-p)i}]$$

where $p = (q-1), (q-2), \ldots, 1$

A″: $Z_{(j-q+1)i}\, y_{ji}^{(j-q)}$

B′: $(q-1)$ terms, each with a value of

$$[Z_{(j-q+1)k}\, y_{(j-p)k}^{(j-q)} - \delta_{(j-p)k}]$$

where $p = (q-1), (q-2), \ldots, 1$

B″: $Z_{(j-q+1)k}\, y_{jk}^{(j-q)}$

H′: $Z_{(j-q+1)i} Z_{(j-q+1)k}\, y_{ik}^{(j-q)}$

D′: $(q-1)$ terms, each with a value of

$$(1 - \delta_{jk})\, y_{(j-p)k}^{(j-q)}$$

where $p = (q-1), (q-2), \ldots, 1$

F′: $(1 - \delta_{jk})\, y_{jk}^{(j-q)} - \delta_{jk}$

I′: $(1 - \delta_{jk}) Z_{(j-q+1)i}\, y_{ik}^{(j-q)}$

J′: $(q-1)$ terms each with a value of

$$(1 - \delta_{ji})\, y_{(j-p)i}^{(j-q)}$$

where $p = (q-1), (q-2), \ldots, 1$

K′: $(1 - \delta_{ji})\, y_{ji}^{(j-q)} - \delta_{ji}$

L′: $(1 - \delta_{ji})(1 - \delta_{jk})\, y_{ik}^{(j-q)}$

Example 5.3

Relate the derivative $y_{ik}^{(j)}$ to second-order derivatives of $y^{(0)}$ for four cases:

(a) $j = 3, i = 1, k = 2$
(b) $j = 3, i = 1, k = 3$
(c) $j = 2, i = 3, k = 4$
(d) $j = 3, i = 3, k = 3$

Solution

Table 5.2 is employed. The results are:

(a) $j = 3, i = 1, k = 2$

$$y_{12}^{(3)} = \frac{\left|\begin{array}{ccc|c} & & & -\delta_{12} \\ & \mathfrak{D}_3^{(0)} & & -\delta_{22} \\ & & & -\delta_{32} \\ \hline -\delta_{11} & -\delta_{12} & -\delta_{13} & 0 \end{array}\right|}{\mathfrak{D}_3^{(0)}} = \frac{\begin{vmatrix} y_{12}^{(0)} & y_{13}^{(0)} \\ y_{23}^{(0)} & y_{33}^{(0)} \end{vmatrix}}{\mathfrak{D}_3^{(0)}}$$

(b) $j = 3, i = 1, k = 3$

$$y_{13}^{(3)} = \frac{\left|\begin{array}{ccc|c} & & & (1-\delta_{33})y_{13}^{(0)} \\ & \mathfrak{D}_3^{(0)} & & (1-\delta_{33})y_{23}^{(0)} \\ & & & (1-\delta_{33})y_{33}^{(0)} - \delta_{33} \\ \hline -\delta_{11} & -\delta_{21} & -\delta_{31} & 0 \end{array}\right|}{\mathfrak{D}_3^{(0)}} = -\frac{\begin{vmatrix} y_{12}^{(0)} & y_{13}^{(0)} \\ y_{22}^{(0)} & y_{23}^{(0)} \end{vmatrix}}{\mathfrak{D}_3^{(0)}}$$

(c) $j = 2, i = 3, k = 4$

$$y_{34}^{(2)} = \frac{\left|\begin{array}{cc|c} \multicolumn{2}{c|}{\mathfrak{D}_2^{(0)}} & (1-\delta_{24})y_{14}^{(0)} \\ & & (1-\delta_{24})y_{24}^{(0)} - \delta_{24} \\ \hline (1-\delta_{23})y_{13}^{(0)} & (1-\delta_{23})y_{23}^{(0)} - \delta_{23} & (1-\delta_{23})(1-\delta_{24})y_{34}^{(0)} \end{array}\right|}{\mathfrak{D}_2^{(0)}}$$

$$= \frac{\begin{vmatrix} y_{11}^{(0)} & y_{12}^{(0)} & y_{14}^{(0)} \\ y_{21}^{(0)} & y_{22}^{(0)} & y_{24}^{(0)} \\ y_{13}^{(0)} & y_{23}^{(0)} & y_{34}^{(0)} \end{vmatrix}}{\mathfrak{D}_2^{(0)}}$$

(d) $j = 3, i = 3, k = 3$

$$y_{33}^{(3)} = \frac{-\mathfrak{D}_2^{(0)}}{\mathfrak{D}_3^{(0)}}$$

Example 5.4

Using Table 5.3, express:

(a) $y_{23}^{(4)}$ in terms or derivatives of $y^{(1)}$.
(b) $y_{26}^{(5)}$ in terms of derivatives of $y^{(2)}$.

Solution

(a) In this case $j = 4, i = 2, k = 3, q = 3$. We will use the case where $j > i, k$. Notice that $Z_{(j-q+1)i} = Z_{(4-3+1)i} = Z_{2i}$. Since $2 = i$, then $Z_{2i} = 0$. Also, $Z_{(j-q+1)k} = 0$.

$$\mathfrak{D}_q^{(j-q)} = \mathfrak{D}_3^{(1)}$$

where

$$\mathfrak{D}_3^{(1)} = \begin{vmatrix} y_{22}^{(1)} & y_{23}^{(1)} & y_{24}^{(1)} \\ y_{23}^{(1)} & y_{33}^{(1)} & y_{34}^{(1)} \\ y_{24}^{(1)} & y_{34}^{(1)} & y_{44}^{(1)} \end{vmatrix}$$

Then

$$y_{23}^{(4)} = \frac{\begin{vmatrix} & \mathfrak{D}_3^{(1)} & & \begin{matrix} -\delta_{23} \\ -\delta_{33} \\ 0 \end{matrix} \\ -\delta_{22} & -\delta_{32} & 0 & 0 \end{vmatrix}}{\mathfrak{D}_3^{(1)}} = \frac{\begin{vmatrix} y_{23}^{(1)} & y_{24}^{(1)} \\ y_{34}^{(1)} & y_{44}^{(1)} \end{vmatrix}}{\mathfrak{D}_3^{(1)}}$$

(b) Here

$$j > i$$

$$j < k$$

$$j = 5, i = 2, k = 6, q = 3, \qquad Z_{(j-q+1)i} = Z_{32} = 1$$

$$y_{26}^{(5)} = \frac{\begin{vmatrix} y_{33}^{(2)} & y_{34}^{(2)} & y_{35}^{(2)} & y_{36}^{(2)} \\ y_{34}^{(2)} & y_{44}^{(2)} & y_{45}^{(2)} & y_{46}^{(2)} \\ y_{35}^{(2)} & y_{45}^{(2)} & y_{55}^{(2)} & y_{56}^{(2)} \\ y_{32}^{(2)} & y_{42}^{(2)} & y_{52}^{(2)} & y_{26}^{(2)} \end{vmatrix}}{\mathfrak{D}_3^{(2)}} = -\frac{\begin{vmatrix} y_{23}^{(2)} & y_{24}^{(2)} & y_{25}^{(2)} & y_{26}^{(2)} \\ y_{33}^{(2)} & y_{34}^{(2)} & y_{35}^{(2)} & y_{36}^{(2)} \\ y_{34}^{(2)} & y_{44}^{(2)} & y_{45}^{(2)} & y_{46}^{(2)} \\ y_{35}^{(2)} & y_{45}^{(2)} & y_{55}^{(2)} & y_{56}^{(2)} \end{vmatrix}}{\mathfrak{D}_3^{(2)}}$$

5.6 Restructuring of Thermodynamic Transforms

Beginning with Eq. (5-1), we can now transform one or all of the independent variable set $\underline{S}, \underline{V}, N_1, \ldots, N_n$. Let us choose only the variable $\underline{V}$ and reorder so that $\underline{V}$ represents x_1. Then

$$y^{(1)} = y^{(0)} - \xi_1 x_1 = \underline{U} - (-P)\underline{V} = \underline{U} + P\underline{V} = \underline{H} \tag{5-54}$$

where this particular Legendre transform is called the *enthalpy*. We note that

$$\underline{H} = f(P, \underline{S}, N_1, \ldots, N_n) \tag{5-55}$$

We call Eq. (5-55) a Fundamental Equation in the same way that we refer to Eq. (5-1); then P is x_1, $\underline{S}$ is x_2, etc., and $\xi_1 = (\partial \underline{H}/\partial P) = \underline{V}$. We can recover Eq. (5-1) by carrying out a Legendre transform assuming that Eq. (5-55) is the $y^{(0)}$ function; that is,

$$y^{(1)} = y^{(0)} - \xi_1 x_1 = \underline{H} - (\underline{V})P = \underline{U} \tag{5-56}$$

This transform can be readily shown in Figure 5.3 for a common pressure–

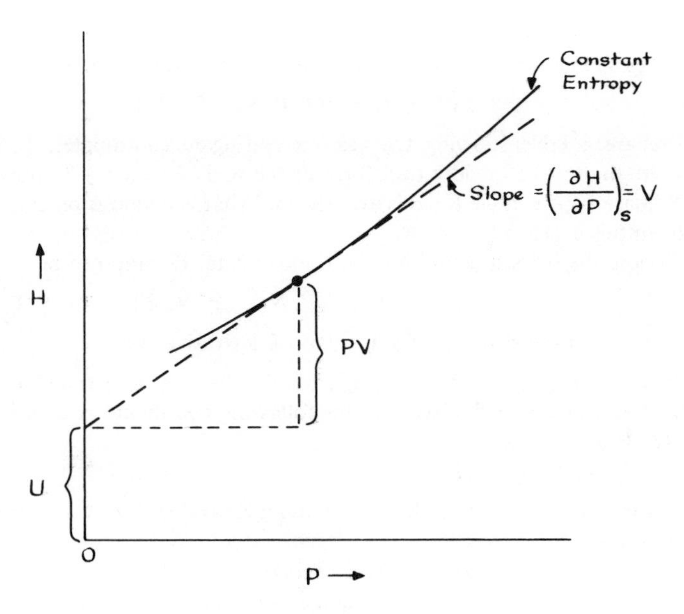

Figure 5.3

enthalpy diagram. If a curve of constant entropy is considered, the slope is $\underline{V}$. The intercept of this tangent on the enthalpy axis is, as shown, equal to the internal energy $\underline{U}$.

With internal energy as the basis function, there are $n + 2$ permutations of first Legendre transforms: the two common functions, $\underline{H}(P, \underline{S}, N_1, \ldots, N_n)$ and $\underline{A}(T, \underline{V}, N_1, \ldots, N_n)$, and n other functions for the independent variable set of $\underline{S}, \underline{V}, N_1, \ldots, N_{i-1}, \mu_i, N_{i+1}, \ldots, N_n$. Since the ordering of components is arbitrary, we shall refer to the n functions as $\underline{U}'(\underline{S}, \underline{V}, \mu_1, N_2, \ldots, N_n)$. In a similar manner, there are $(n + 2)(n + 1)/2$ second Legendre transforms: one is the Gibbs energy $\underline{G}(T, P, N_1, \ldots, N_n)$; there are n of the form $\underline{A}'(T, \mu_1, \underline{V}, N_2, \ldots, N_n)$ and n of the form $\underline{H}'(P, \mu_1, \underline{S}, N_2, \ldots, N_n)$, and $(n)(n - 1)/2$ of the form $\underline{U}''(\mu_1, \mu_2, \underline{S}, \underline{V}, N_3, \ldots, N_n)$. Third Legendre transforms would involve $\underline{G}'$, $\underline{A}''$, $\underline{H}''$, and $\underline{U}'''$ potential functions.

PROBLEMS

5.1. (a) Demonstrate the utility of Eq. (5-36) for a system with six components and where $i = 4$. Order the Fundamental Equation as

$$\underline{U} = f(\underline{S}, \underline{V}, N_1, \ldots, N_6)$$

(b) If one were to write the Fundamental Equation as

$$\underline{S} = f(\underline{U}, \underline{V}, N_1, \ldots, N_n)$$

What would the derivative of the total Legendre transform be?

(c) Suppose that one were to write the Fundamental Equation as

$$y^{(0)} = \underline{G} = f(T, P, N_1, \ldots, N_n)$$

Prepare a table showing the various conjugate coordinates, ξ_i, x_i. Next, write the third Legendre transform of the basis function $y^{(0)}$ shown above. Prepare a ξ_i, x_i table for this transform. What generalization can you infer from this exercise?

(d) Choose the basis function for the Fundamental Equation to be

$$y^{(0)} = \underline{A} = f(\underline{V}, N_1, \ldots, N_n, T)$$

Obtain an expression for $y^{(1)}_{22}$ in terms of derivatives $y^{(0)}$.

5.2. Given $y^{(0)} = f(\underline{S}, \underline{V}, N_1, N_2, \ldots, N_n)$, obtain the Legendre transform and its differential if one wished to work in the following coordinate systems:

(a) $f(T, \underline{V}, \mu_1, \ldots, \mu_n)$

(b) $f[(1/T), \underline{V}, N_1, N_2, \ldots, N_n]$

5.3. Express the following partial derivatives in an equivalent form using the Fundamental Equation

$$\underline{U} = U(\underline{S}, \underline{V}, N_1, N_2, \ldots, N_n)$$

and Maxwell's relations. Indicate, if possible, how they might be measured experimentally.

$$\left(\frac{\partial y^{(2)}}{\partial \underline{V}}\right)_{U_S, N}$$

$$\left(\frac{\partial U_{N_j}}{\partial U_S}\right)_{\underline{V}, N}$$

$$\left(\frac{\partial y^{(1)}}{\partial \underline{V}}\right)_{U_S, N}$$

$$\left(\frac{\partial \underline{S}}{\partial U_S}\right)_{y^{(2)}, N}$$

5.4. Carry out the following transformations.

(a) Express $(\partial \underline{S}/\partial P)_{T, N_1, \ldots, N_n}$ as a function of P, $\underline{V}$, T, and their derivatives.

(b) Express $(\partial \underline{H}/\partial P)_{T, N_1, \ldots, N_n}$ as a function of $\underline{G}$ and its derivatives and show how these may be given in terms of P, $\underline{V}$, T, N, and C_p.

(c) Express $[\partial(\underline{G}/T)/\partial(1/T)]_{P, N_1, \ldots, N_n}$ as a function of $\underline{H}$ and its derivatives.

(d) Express $(\partial \underline{H}/\partial \underline{V})_{T, N_1, \ldots, N_n}$ as a function of $\underline{G}$ and its derivatives and show how these may be given in terms of P, $\underline{V}$, T, N, and C_p.

(e) Express $(\partial T/\partial N_A)_{\underline{V}, \underline{S}, \mu_B, N_c, \ldots}$ as a function of $\underline{U}$ and its *independent* derivatives.

(f) Express $(\partial T/\partial N_A)_{\underline{S}, P, \mu_B, N_c, \ldots}$ as a function of $\underline{U}$ and its *independent* derivatives.

5.5 For a one-component system, show:

(a) $$\left(\frac{\partial \mu}{\partial N}\right)_{T, \underline{V}} = \left(-\frac{\underline{V}}{N}\right)\left(\frac{\partial \mu}{\partial \underline{V}}\right)_{T, N}$$

(b) $$\left(\frac{\partial P}{\partial N}\right)_{T, \underline{V}} = \left(-\frac{\underline{V}}{N}\right)\left(\frac{\partial P}{\partial \underline{V}}\right)_{T, N}$$

5.6. The following discussion is limited to one-dimensional motion along the x coordinate and for a constant-mass system.

Newtonian mechanics relates the force on a particle to the mass and acceleration; that is,

$$F = ma = m\ddot{x} \qquad \left(\ddot{x} = \frac{d^2x}{dt^2}\right)$$

Another way in which to study the dynamics of motion is with Lagrangian mechanics. In this case, a function $\mathcal{L}$ is defined as $\mathcal{L} = \mathcal{T} - U$, where $\mathcal{T}$ = the kinetic energy = $m\dot{x}^2/2$ and U is the potential energy = $f(x)$. Newton's Law in Lagrangian mechanics is given as

$$\left(\frac{\partial \mathcal{L}}{\partial x}\right)_{\dot{x}} = \left(\frac{\partial}{\partial t}\right)\left(\frac{\partial \mathcal{L}}{\partial \dot{x}}\right)_x$$

$$\mathcal{L} = f(x, \dot{x}) \qquad \dot{x} = \frac{\partial x}{\partial t} = \text{velocity}$$

Another branch of mechanics uses a function $(-\mathcal{H})$, which is a function of x and the momentum of a particle, p; that is,

$$-\mathcal{H} = f(p, x)$$

with

$$p = \left(\frac{\partial \mathcal{L}}{\partial \dot{x}}\right)_x = \left(\frac{\partial T}{\partial \dot{x}}\right)_x = m\dot{x}$$

(a) Using the concepts of Legendre transforms, define $-\mathcal{H}$ in terms of $\mathcal{L}$, p, and $\dot{x}$.

(b) In the latter branch of mechanics, $\mathcal{H}$ is called the Hamiltonian; with your definition of $\mathcal{H}$ complete the following equations:

$$\left(\frac{\partial \mathcal{H}}{\partial p}\right)_x =$$

$$\left(\frac{\partial \mathcal{H}}{\partial x}\right)_p =$$

5.7. Express the following in terms of C_p, P, V, T and derivatives of these variables.

(a) $(\partial S/\partial T)_{G,N}$

(b) $(\partial A/\partial G)_{T,N}$

5.8. Prove that for an n-component mixture there are $(n + 2)(n + 1)/2$ independent second-order derivatives of the Fundamental Equation.

5.9. A spherical tank contains 1 mol of helium gas at 10 bar and 300 K (see Figure P5.9). We would like to carry out an experiment in which helium is released to

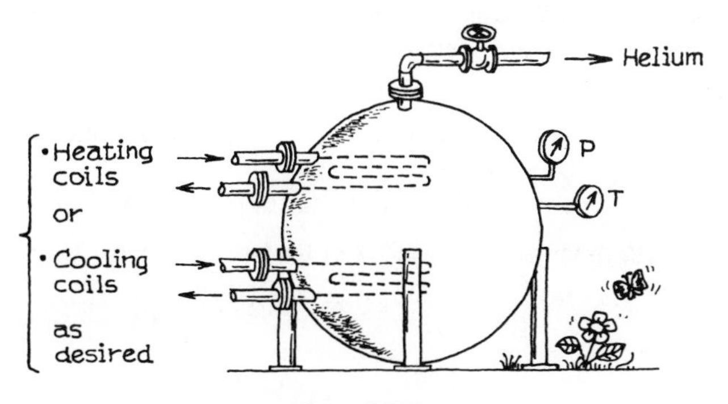

Figure P5.9

the atmosphere, but at the same time, the remaining contents of the sphere maintain a constant total energy, $\underline{U}$. Heating or cooling coils may be used to keep $\underline{U}$ constant during venting. Helium behaves as an ideal gas with a constant value of $C_v = 12.6$ J/mol K. Choose a base state where the specific enthalpy is zero at 300 K.

(a) When one-half of the helium has been vented, what is the temperature and pressure of the residual helium? What is the heat interaction?
(b) Determine $(\partial T/\partial P)_{\underline{U},\underline{V}}$ at the instant when venting begins.
(c) Repeat parts (a) and (b) if the base state were chosen so that the specific internal energy, U, were zero at 300 K.
(d) For part (a), what would be the residual helium temperature when 60% of the gas had been removed?

5.10. Choose as a basis function

$$y^{(0)} = f(\underline{S}, N_1, N_2, N_3, P)$$

and obtain $y_{12}^{(1)}$ in terms of derivatives of the basis function. Discuss how experiments could be designed and conducted to obtain numerical values of the $y^{(0)}$ derivatives.

Using your result, consider the following problem. We have a system containing, initially, N_i moles of a material. We wish to remove $N_i/2$ mol under conditions where the total entropy and the pressure remain constant. If the initial temperature is 400 K, what is the final temperature? Assume that the base state for entropy is such that at 400 K and the system pressure, the specific entropy is 10 J/mol K. Also assume that the heat capacity is 10 J/mol K, independent of temperature.

5.11. If the basis function is chosen as

$$y^{(0)} = \underline{U} = f(\underline{S}, \underline{V}, N_A, N_B, \ldots)$$

show that

$$y_{14}^{(2)} = G_{TN_B} = \frac{\begin{vmatrix} U_{VV} & U_{VN_B} \\ U_{SV} & U_{SN_B} \end{vmatrix}}{\begin{vmatrix} U_{SS} & U_{SV} \\ U_{SV} & U_{VV} \end{vmatrix}}$$

by performing two single-step transforms from $y^{(2)}$ to $y^{(1)}$ and then to $y^{(0)}$.

5.12. Assume that the basis Legendre transform is $y^{(0)} = f(P, T, N_1, N_2, \ldots)$. Express $y_{22}^{(1)}$ in terms of derivatives of $y^{(0)}$ and interpret the results on a physical basis.

5.13. Express $y_{44}^{(3)}$ in terms of derivatives of $y^{(1)}$ for a ternary mixture when the basis function

$$y^{(0)} = \underline{U} = f(\underline{S}, \underline{V}, N_A, N_B, N_C)$$

5.14. Express $y_{24}^{(2)}$ in terms of derivatives of the basis function

$$y^{(0)} = \underline{U} = f(\underline{S}, \underline{V}, N_A, \ldots)$$

5.15. Show that

$$y_{(n+1)(n+1)}^{(n)} = \left(\frac{\partial \xi_{n+1}}{\partial x_{n+1}}\right)_{\xi_1, \ldots, \xi_n, x_{n+2}}$$

5.16. $y^{(m-2)}$ is a Legendre transform of a basis function $y^{(0)}$, where

$$y^{(0)} = f(x_1, x_2, \ldots, x_m)$$

$$y^{(m-2)} = f(\xi_1, \xi_2, \ldots, \xi_{m-2}, x_{m-1}, x_m)$$

Derive a general equation to express the derivative $y^{(m-2)}_{(m-1)(m-1)}$ in terms of derivatives of a Legendre transform $y^{(r)}$, where $0 \leq r < m - 2$. Define any derivatives and show what variables are held constant.

Equilibrium 6

In Postulate I it was stated that stable equilibrium states exist for all simple systems (see Section 2.7). In Postulate II it was stated that all complex systems will approach an equilibrium condition (consistent with any internal restraints) in which each simple system of the composite approaches a stable equilibrium state. It was shown in Example 4.3 that for any permissible process occurring within a closed, adiabatic (as well as isolated) system, the change in total entropy must be equal to or greater than zero. The total entropy is, of course, equal to the sum of the entropies of all subsystems of the isolated system.

In this chapter we examine the criteria that can be applied to determine if a system has reached a stable equilibrium state under various constraints. Of more importance, we investigate the consequences of the fact that a system is in a stable equilibrium state. In this way we can specify the criteria for membrane equilibrium, for phase equilibrium, and for chemical reaction equilibrium.

6.1 Classification of Equilibrium States

Up to this point we have used the term *equilibrium* only in connection with stable equilibrium states. In broader usage, any system that does not undergo a change with time is said to be in an equilibrium state. The adjective *stable* is reserved for those equilibrium states which, following a perturbation, will revert to the original equilibrium state. Alternatively, there are equilibrium states that are not stable (i.e., states may be permanently altered as a result of even a small perturbation).

There are four classes of equilibrium states. For ease of conceptualization,

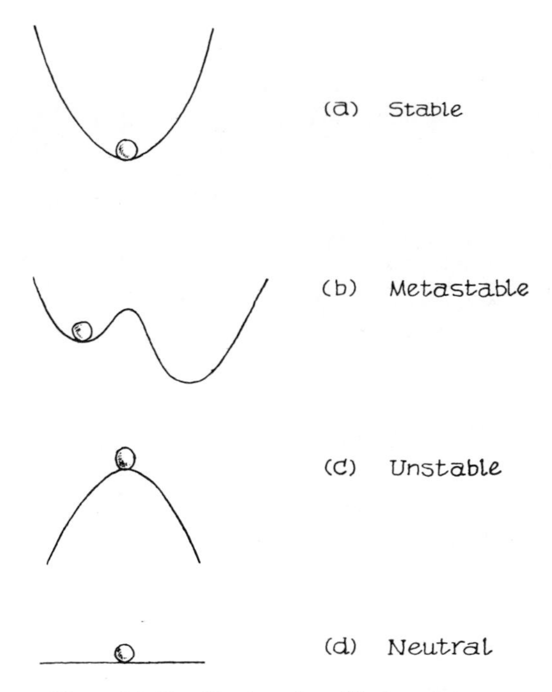

Figure 6.1 Classification of equilibrium states.

they can be described with the mechanical analogy of a ball on a solid surface in a gravitational field (see Figure 6.1). If the ball were pushed to the right or left and if it were to return to its original position, the state is stable [case (a)]. If the original position were metastable [case (b)], the ball would revert to the original position after a small perturbation, but there is the possibility that a large perturbation would displace the ball to a state of lower potential energy. If the original state were unstable [case (c)], then even a minor perturbation would displace it to a position of lower potential energy. A system in a state of neutral equilibrium [case (d)] would be altered by any perturbation, but the potential energy would remain unchanged.

In terms of these mechanical analogies, most real systems would be classified as metastable. All organic materials in the presence of oxygen could attain a more stable state by reacting to form CO_2 and H_2O; similarly, most metals are metastable in relation to their oxides. In many such cases, the barriers to transition to the more stable state may be large enough to prevent the change from occurring within the time span of interest. Thus, for all practical purposes, these states can be considered stable if the barriers are large in relation to the magnitude of the perturbation.

Since almost all real systems are metastable with respect to some perturbation yet stable with respect to minor perturbations, it is of greater utility to approach equilibrium more pragmatically than that suggested by the simple mechanical analogies of Figure 6.1. We shall consider only minor perturba-

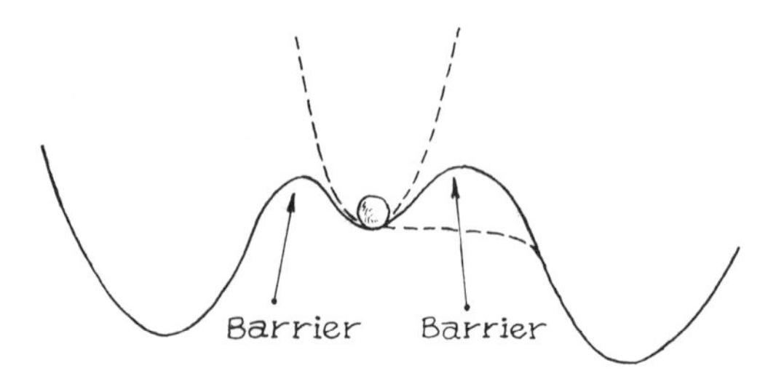

(a) A state stable with respect to minor perturbations.

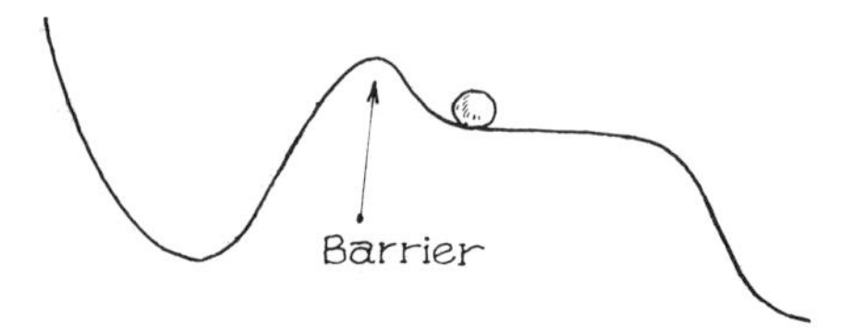

(b) The state in (a) with the right barrier removed. This state is unstable with respect to perturbations which move the ball to the right.

Figure 6.2

tions to a given system. If minor perturbations leave the system unchanged, we define the original state as a stable equilibrium state [see Figure 6.2(a)].

In real systems, the extent to which a barrier exists is dictated not by thermodynamic reasoning but by rate or kinetic considerations. If the rate of a possible transition is too slow to be significant within the time span of interest, we consider the barrier as an impenetrable internal restraint. For simple systems, such barriers may be visualized as activation energies that prevent chemical reactions or as the lack of solid nuclei needed to initiate phase transitions. Complex systems may contain additional barriers such as adiabatic, rigid, or impermeable internal walls.

For practical purposes, metastable systems are treated as stable if the barriers to transition are large in relation to minor perturbations. If one of the barriers is relatively small or nonexistent, the original state is treated as unstable with respect to that particular kind of transition. It should be emphasized that the delineation of internal barriers forms part of the definition of a system. Alteration or elimination of an internal barrier can change a stable system to an unstable system.

An interesting state is depicted by the mechanical analogy in Figure 6.2(b). A minor perturbation to the right leads to an unstable state, whereas a push to the left indicates that the system is stable with respect to this kind of variation. Although the analysis may not reveal what the final state of the system might be, it will indicate that the original system is intrinsically unstable and some transition must result from a minor perturbation. Although it may be difficult experimentally to force systems to their *limit of intrinsic stability*, there is sufficient evidence to suggest that such limits exist.

In general, potential transitions in real systems are more complicated than the horizontal motion of a ball on a surface, which has only two kinds of perturbations (i.e., displacement to either the left or right). Instead of testing a system for all possible transitions, we usually test for only those that we suspect may occur. Hence, we classify the stability of a system with respect to particular kinds of perturbations.

The process by which we test for stability is to perform thought experiments in which we envision a transition occurring as a result of a small, finite change in one or more properties of the system. We then determine if the proposed variation leads to a more favored state. This kind of thought experiment is referred to as a *virtual process*.[1]

6.2 Extrema Principles

The *stable equilibrium state* was introduced in Postulate II as that state which all isolated systems approach and eventually attain. We shall now develop criteria for testing whether or not a given system has attained a stable equilibrium state. The results of Example 4.3 provide the basis for our quantitative treatment. There, it was proved that, for an adiabatic process occurring in a closed system, the total change in entropy must be either zero if the process is reversible or positive if the process is spontaneous or irreversible. This result is also applicable to any isolated system, which is a special case of a closed, adiabatic system. If all real processes in an isolated system occur with a zero or positive entropy change, we can conclude that if an isolated system were at equilibrium, the entropy must have a maximum value with respect to any allowed variations. Thus, to test whether or not a given isolated system is, in fact, in equilibrium and stable, we propose virtual processes to test certain variations. If for such variations we can show that the total entropy decreases, the proposed variation was impossible and the original state was an equilibrium and stable state—at least with respect to the proposed variation.

Almost all systems will be stable with respect to some variations and unstable with respect to others. Thus, the entropy maximum principle is relative and, in the last analysis, we will only be able to propose and test a finite number of variations for a given system. Therefore, we will be able to make only a qualified statement about the stability of an equilibrium state.

[1] Any process that can be conceived—whether or not the process is allowed—is called a *virtual* process. This process, of course, must be consistent with any internal and external restraints placed on the system.

The method we employ to test for an entropy maximum is illustrated in Figure 6.3 for the simple case of a single allowed variation in parameter z_1, where $\underline{S}$ is a function of variables $z_1, z_2, \ldots, z_{n+2}$. As z_1 is varied, there is a value $z_1 = z_1^e$, at which the system entropy is maximized. At this point, $(\partial \underline{S}/\partial z_1) = 0$, $(\partial^2 \underline{S}/\partial z_1^2) < 0$, and the system is in a stable equilibrium state with respect to variations of z_1. If we looked at a system originally at $z_1 = z_1^e$ and proposed a virtual process in which z_1 was varied by $\pm\delta z_1$, the resulting change in $\underline{S}$, or $\Delta\underline{S}$, can be calculated by expanding $\underline{S}$ in a Taylor series about *the conditions of the original state provided that δz_1 is a small perturbation*. Thus,

$$\Delta\underline{S} = \delta\underline{S} + \frac{1}{2!}\delta^2\underline{S} + \frac{1}{3!}\delta^3\underline{S} + \cdots \frac{1}{m!}\delta^m\underline{S} + \cdots \tag{6-1}$$

where $\Delta\underline{S}$ is the resultant change in $\underline{S}$ due to the small perturbation, $\delta\underline{S}$ is the *first-order variation* of $\underline{S}$, and $\delta^m\underline{S}$ is the *mth-order variation* of $\underline{S}$. By definition,

$$\delta\underline{S} \equiv \sum_{i=1}^{n+2} \frac{\partial \underline{S}}{\partial z_i}\delta z_i = \sum_{i=1}^{n+2} S_{z_i}\delta z_i \tag{6-2}$$

$$\delta^2\underline{S} \equiv \sum_{i=1}^{n+2}\sum_{j=1}^{n+2} \frac{\partial^2 \underline{S}}{\partial z_i\,\partial z_j}\delta z_i\,\delta z_j = \sum_{i=1}^{n+2}\sum_{j=1}^{n+2} S_{z_i z_j}\delta z_i\,\delta z_j \tag{6-3}$$

$$\delta^3\underline{S} \equiv \sum_{i=1}^{n+2}\sum_{j=1}^{n+2}\sum_{k=1}^{n+2} \frac{\partial^3 \underline{S}}{\partial z_i\,\partial z_j\,\partial z_k}\delta z_i\,\delta z_j\,\delta z_k = \sum_{i=1}^{n+2}\sum_{j=1}^{n+2}\sum_{k=1}^{n+2} S_{z_i z_j z_k}\delta z_i\,\delta z_j\,\delta z_k \tag{6-4}$$

where $\delta z_i = z_i - z_i^0$ (superscript 0 denotes the value in the original state) and each of the partial derivatives is evaluated at the conditions prevailing in the original state. The shorthand notation for the partial derivatives will be used throughout this chapter.

If $\Delta\underline{S}$ represents the entropy change from the original state to the perturbed

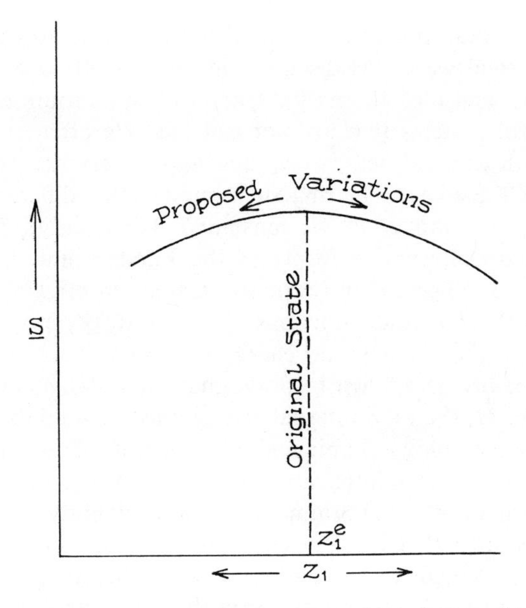

Figure 6.3 Variation in the system entropy with a single variable z_1.

state, and if $\underline{S}$ is a maximum in the former state, the mathematical equivalent of the entropy maximum principle is

$$\Delta \underline{S} < 0 \tag{6-5}$$

If $\underline{S}$ is a smoothly varying function of z_i, it is necessary and sufficient for a maximum in $\underline{S}$ that $\delta \underline{S} = 0$ and $\delta^m \underline{S} < 0$, where $\delta^m \underline{S}$ is the lowest-order, non-vanishing variation of $\underline{S}$. That is,

$$\delta \underline{S} = 0 \qquad \text{and} \tag{6-6}$$

$$\begin{aligned} \delta^2 \underline{S} &\leq 0 \qquad \text{but, if} = 0, \text{ then} \\ \delta^3 \underline{S} &\leq 0 \qquad \text{but, if} = 0, \text{ then—etc.} \end{aligned} \tag{6-7}$$

The appropriate inequality in Eq. (6-7) forms the *criterion of stability* and the equality in Eq. (6-6), $\delta \underline{S} = 0$, the *criterion of equilibrium* in the entropy representation. In this chapter we are concerned primarily with Eq. (6-6); the stability of thermodynamic systems is covered in Chapter 9.

The stability and equilibrium criteria, as stated, apply only to isolated systems. Since the parameters $z_1, z_2, \ldots, z_{n+2}$ are related to $\underline{U}, \underline{V}, N_1, \ldots, N_n$, the isolation requirement places restraints upon the allowed variations of these independent parameters. In particular, if the system in question is a composite of two subsystems, then any proposed virtual process must be consistent with the restraining equations of isolation:

$$\delta \underline{U} = \delta \underline{U}^{(1)} + \delta \underline{U}^{(2)} = 0 \tag{6-8}$$

$$\delta \underline{V} = \delta \underline{V}^{(1)} + \delta \underline{V}^{(2)} = 0 \tag{6-9}$$

$$\delta M = \delta M^{(1)} + \delta M^{(2)} = 0 \tag{6-10}$$

The proposed virtual processes must also be consistent with any internal restraints (e.g., rigid walls), as discussed in more depth in Section 6.4.

We shall now consider alternative criteria of equilibrium and stability that are applicable for systems that are not isolated. We often encounter systems that interact with external heat, work, and mass reservoirs, and we must also be able to treat these cases. During such interactions, different restraints will be imposed (e.g., constant $\underline{S}$, $\underline{V}$, M; constant T, $\underline{V}$, M; etc.). The insight gained in Chapter 5 with alternative forms of the Fundamental Equation should, however, lead us to expect that alternative extremum principles could also be developed with the potential functions, $\underline{U}$, $\underline{A}$, $\underline{H}$, $\underline{G}$, $\underline{G}'$, etc.

The duality of the entropy and energy representation of the Fundamental Equation can readily be applied to prove that for a stable equilibrium system at constant $\underline{S}$, $\underline{V}$, M, the total internal energy must be a minimum. To demonstrate this we shall employ a simple logical proof and follow this by an example to show that the energy minimization and entropy maximization do indeed yield identical criteria of equilibrium and intrinsic stability.

Consider a system that is supposed to be in a state of stable equilibrium. In proving this statement to be true, we isolated the system and showed that *all* allowable variations led to a decrease in the total entropy; that is, for every variation that we considered at constant $\underline{U}$, $\underline{V}$, M, we found $\underline{S}_f$ to be less than $\underline{S}_i$. Now, however, we wish to consider variations at constant $\underline{S}$, $\underline{V}$, M, but to allow $\underline{U}$ to vary. We now take the system at $\underline{S}_f$ and allow it to interact reversibly with an external system to return the value of entropy to $\underline{S}_i$, maintaining $\underline{V}$, M constant. Since $\underline{S}_f - \underline{S}_i < 0$, we must transfer energy *into* the system to return to $\underline{S}_i$. In so doing, we increase $\underline{U}$ in this two-step process. Thus, one concludes that for any variation in a system at constant $\underline{S}$, $\underline{V}$, M, the internal energy will increase—if the system were initially at equilibrium. The converse is also true; that is, if a variation within a system held at constant $\underline{S}$, $\underline{V}$, M leads to a decrease in internal energy, the system was not initially in a state of equilibrium.

In an analogous form to Eqs. (6-5), (6-1), (6-6), and (6-7), at constant $\underline{S}$, $\underline{V}$, M, we obtain

$$\Delta\underline{U} > 0 \qquad \text{(for variations originating from a stable equilibrium state)} \tag{6-11}$$

where

$$\Delta\underline{U} = \delta\underline{U} + \frac{1}{2!}\delta^2\underline{U} + \frac{1}{3!}\delta^3\underline{U} + \cdots \tag{6-12}$$

$$\delta\underline{U} = \sum_{i=1}^{n+2}\left(\frac{\partial\underline{U}}{\partial z_i}\right)\delta z_i\,\text{(equilibrium)} = 0 \tag{6-13}$$

$$\delta^2\underline{U} \geq 0 \quad \text{and} \quad \delta^m\underline{U} > 0 \tag{6-14}$$

where $\delta^m\underline{U}$ is the lowest nonvanishing variation. The restraining equations of isolation are Eqs. (6-9), (6-10), and

$$\delta\underline{S} = \delta\underline{S}^{(1)} + \delta\underline{S}^{(2)} = 0 \tag{6-15}$$

which replaces Eq. (6-8).

Example 6.1

In Figure 6.1(a) the ball is shown at the bottom of the well to represent a case of stable equilibrium. Prove this to be so by considering a *virtual* process wherein the ball moves to a point on the wall, above the bottom. Develop two proofs, one using Eq. (6-6) and the other with Eq. (6-11).

Solution

In this virtual process, the ball has gained energy (potential), but the entropy, volume, and mass have remained constant. Thus, Eq. (6-11) applies and we conclude that the ball was, initially, in a stable equilibrium state with respect to the proposed variation. If we had desired to use the equilibrium criterion given by Eq. (6-6) rather than the equivalent form, Eq. (6-11), we would have to keep the energy, volume, and mass constant during the proposed variation. Since the energy increases due to a gain in potential energy, we would have had to cool the ball sufficiently that the energy in the higher location was the same as when it was at the bottom. Neglecting any change in volume due to the cooling, we have a variation at constant $\underline{E}$, $\underline{V}$, and M. It is obvious, however, that due to the cooling, $\Delta \underline{S} < 0$. Thus, Eq. (6-11) applies and again we conclude that the initial system was in an equilibrium state relative to the proposed variation.

6.3 Use of Other Potential Functions to Define Equilibrium States

In Section 6.2 criteria of equilibrium and stability were developed for systems at constant $\underline{U}$, $\underline{V}$, and M or at constant $\underline{S}$, $\underline{V}$, and M. In the former, the total entropy was maximized and, in the latter, the total energy was minimized. It is desirable to be able to obtain alternative criteria for systems under different constraints. To accomplish this, let us consider a system that may be simple or complex, with no restrictions on the number of components or the number of phases, but with no significant body force fields. We also have available large thermal and work reservoirs (R_T and R_P) to which the system may be connected, if desired, to hold the temperature and/or pressure constant (see Figure 6.4). We still maintain our system at constant total mass during any variation, not because we could not also have used external mass reservoirs, but because the inclusion of mass variations is treated later.

We require that (1) the thermal gate, when operating, be impermeable, rigid, and diathermal and (2) the piston between the system and work reservoir be impermeable, frictionless, adiabatic, and capable of being latched when desired. Thus, the thermal reservoir is held at constant $\underline{V}$, M while the work reservoir is held at constant $\underline{S}$, M. The system restraints may be varied depending on our desire to have interactions with one or both of the adjacent reservoirs. Finally, we assume that the reservoirs are large compared to the system so that small variations in the system as a result of movements of the piston connecting

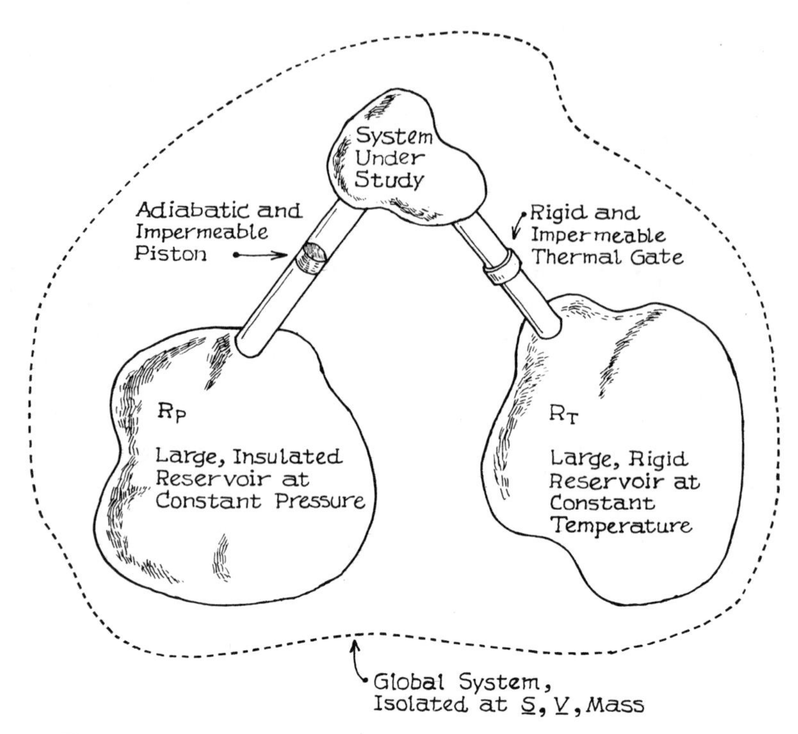

Figure 6.4 System interactions with constant pressure or temperature reservoirs.

the system to R_P, or heat transfer with R_T, will not change the pressure level in R_P or the temperature level in R_T by any significant degree.

Our global system then consists of the original system together with R_P and R_T. These three subsystems are placed in an environment such that the *total* $\underline{S}$, $\underline{V}$, and mass are held constant. We shall investigate some possible variations of this global system and determine the consequences as we apply the energy-minimization criteria.

In general, if the global system is originally in a stable equilibrium state, then for any proposed virtual process,

$$\Delta \underline{U}^{\Sigma} = \Delta(\underline{U} + \underline{U}^{R_T} + \underline{U}^{R_P}) > 0 \tag{6-16}$$

$$\Delta \underline{S}^{\Sigma} = \Delta(\underline{S} + \underline{S}^{R_T}) = \Delta \underline{S}^{R_P} = 0 \tag{6-17}$$

$$\Delta \underline{V}^{\Sigma} = \Delta(\underline{V} + \underline{V}^{R_P}) = \Delta \underline{V}^{R_T} = 0 \tag{6-18}$$

$$\Delta M^{\Sigma} = \Delta M = \Delta M^{R_T} = \Delta M^{R_P} = 0 \tag{6-19}$$

where the superscript Σ refers to the global system, R_T and R_P to the thermal and work reservoirs, and the original system under study is not superscripted.

Case (a). The thermal gate is inoperative. The piston is unlatched in order to allow an interaction between the system and R_P in order to keep the system at constant pressure. Since the movement of the piston is also assumed to be frictionless and thus will operate in a reversible manner, its displacement will vary neither the entropy of the system nor the entropy of R_P. The results of any such variation will then apply to a system at constant $\underline{S}$, P, and M.

For the pressure reservoir, the First Law yields

$$\Delta \underline{U}^{R_P} = -P\,\Delta \underline{V}^{R_P} \tag{6-20}$$

or, using Eq. (6-18),

$$\Delta \underline{U}^{R_P} = P\,\Delta \underline{V} \tag{6-21}$$

Substituting Eq. (6-21) into Eq. (6-16), we obtain

$$\Delta \underline{U}^{\Sigma} = \Delta \underline{U} + P\,\Delta \underline{V} > 0 \tag{6-22}$$

or

$$\Delta \underline{H} > 0 \qquad (\underline{S}, P, M \text{ constant}) \tag{6-23}$$

That is, for a system maintained at constant $\underline{S}$, P, M, the enthalpy is a minimum for a stable equilibrium state. For such a system, the criteria of equilibrium and stability are, respectively,

$$\delta \underline{H} = 0 \qquad (\underline{S}, P, M \text{ constant}) \tag{6-24}$$

$$\delta^m \underline{H} > 0 \qquad (\underline{S}, P, M \text{ constant}) \tag{6-25}$$

where $\delta^m \underline{H}$ is the lowest-order nonvanishing variation of $\underline{H}$.

Case (*b*). Let us lock the piston and allow no interaction of the original system with R_P. Now open the thermal gate. The small system is held at constant $\underline{V}$, T, mass. A possible variation is to allow heat transfer between R_T and the small system. Then, since $\Delta \underline{V} = \Delta \underline{V}^{R_T} = 0$,

$$\Delta \underline{U}^{R_T} = T\,\Delta \underline{S}^{R_T} = -T\,\Delta \underline{S} \tag{6-26}$$

Hence, Eq. (6-16) becomes

$$\Delta \underline{U}^{\Sigma} = \Delta \underline{U} - T\,\Delta \underline{S} > 0 \tag{6-27}$$

or

$$\Delta \underline{A} > 0 \qquad (T, \underline{V}, M \text{ constant}) \tag{6-28}$$

or

$$\delta \underline{A} = 0 \qquad (T, \underline{V}, M \text{ constant}) \tag{6-29}$$

$$\delta^m \underline{A} > 0 \qquad (T, \underline{V}, M \text{ constant}) \tag{6-30}$$

Thus, our extremum principle for systems held at constant total volume, temperature, and mass is that Helmholtz energy should be a minimum with respect to all allowable variations.

Case (*c*). If we allow simultaneous interactions of our small system with both R_P and R_T, by an approach similar to that used in cases (a) and (b),

$$\Delta \underline{G} > 0 \qquad (T, P, M \text{ constant}) \tag{6-31}$$

or

$$\delta \underline{G} = 0 \qquad (T, P, M \text{ constant}) \tag{6-32}$$

$$\delta^m \underline{G} > 0 \qquad (T, P, M \text{ constant}) \tag{6-33}$$

Thus, the Gibbs energy appears as the potential function to be minimized for systems at constant temperature, pressure, and mass.

6.4 Membrane Equilibrium

For an isolated system, it was shown in the preceding sections that a necessary (but not sufficient) condition of a system in a stable equilibrium state is that $\delta \underline{S} = 0$ for any virtual process. In this section and the following ones we examine the consequences of this condition, first for the general case of a complex system and then for multiphase and chemically reacting simple systems.

Consider an isolated, complex system containing two subsystems, each of which contains a nonreacting binary mixture of components A and B (see Figure 6.5). The criterion of equilibrium for the isolated composite, Eq. (6-6), can be

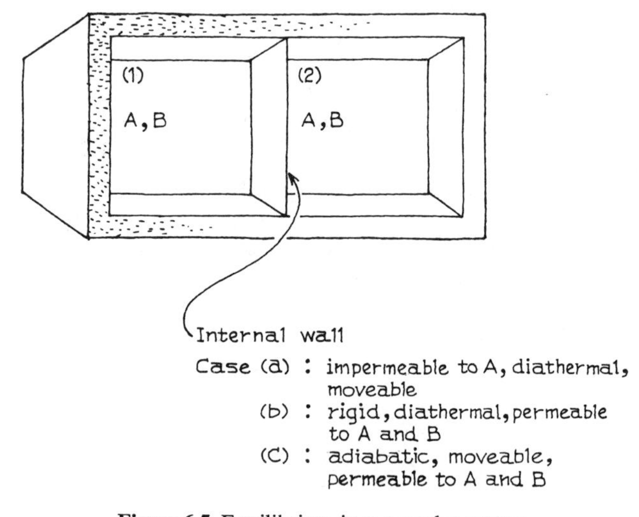

Figure 6.5 Equilibrium in a complex system.

expanded in terms of the properties of the two subsystems:

$$\begin{aligned}\delta \underline{S} &= \delta \underline{S}^{(1)} + \delta \underline{S}^{(2)} \\ &= \frac{1}{T^{(1)}}\,\delta \underline{U}^{(1)} + \frac{1}{T^{(2)}}\,\delta \underline{U}^{(2)} + \frac{P^{(1)}}{T^{(1)}}\,\delta \underline{V}^{(1)} + \frac{P^{(2)}}{T^{(2)}}\,\delta \underline{V}^{(2)} \\ &\quad - \frac{\mu_A^{(1)}}{T^{(1)}}\,\delta N_A^{(1)} - \frac{\mu_A^{(2)}}{T^{(2)}}\,\delta N_A^{(2)} - \frac{\mu_B^{(1)}}{T^{(1)}}\,\delta N_B^{(1)} - \frac{\mu_B^{(2)}}{T^{(2)}}\,\delta N_B^{(2)} = 0\end{aligned} \tag{6-34}$$

We note that the restraining equations of isolation, Eqs. (6-8) through (6-10), place restrictions on the allowable variations in any virtual process that is considered. That is, $\delta \underline{U}^{(1)}$ and $\delta \underline{U}^{(2)}$ are not independently variable. Substituting the restraining equations of isolation into Eq. (6-34) and simplifying, one obtains

$$\delta \underline{S} = \left(\frac{1}{T^{(1)}} - \frac{1}{T^{(2)}}\right) \delta \underline{U}^{(1)} + \left(\frac{P^{(1)}}{T^{(1)}} - \frac{P^{(2)}}{T^{(2)}}\right) \delta \underline{V}^{(1)}$$
$$- \left(\frac{\mu_A^{(1)}}{T^{(1)}} - \frac{\mu_A^{(2)}}{T^{(2)}}\right) \delta N_A^{(1)} - \left(\frac{\mu_B^{(1)}}{T^{(1)}} - \frac{\mu_B^{(2)}}{T^{(2)}}\right) \delta N_B^{(1)} = 0 \tag{6-35}$$

The variations $\delta \underline{U}^{(1)}$, $\delta \underline{V}^{(1)}$, $\delta N_A^{(1)}$, $\delta N_B^{(1)}$ are independently variable only if the composite is a simple system. By definition, a complex system contains some additional internal restraints that must be recognized in any virtual process imposed upon the system. Although there is no universal result applicable to all complex systems, we illustrate three cases that demonstrate the method of obtaining the appropriate criteria of equilibrium (see Figure 6.5).

1. The internal boundary is semipermeable to B, diathermal, and movable. The additional restraining equation is then

$$\delta N_A^{(1)} = \delta N_A^{(2)} = 0 \tag{6-36}$$

and Eq. (6-35) reduces to

$$\delta \underline{S} = \left(\frac{1}{T^{(1)}} - \frac{1}{T^{(2)}}\right) \delta \underline{U}^{(1)} + \left(\frac{P^{(1)}}{T^{(1)}} - \frac{P^{(2)}}{T^{(2)}}\right) \delta \underline{V}^{(1)}$$
$$- \left(\frac{\mu_B^{(1)}}{T^{(1)}} - \frac{\mu_B^{(2)}}{T^{(2)}}\right) \delta N_B^{(1)} = 0 \tag{6-37}$$

For all virtual processes in which variations in $\delta \underline{U}^{(1)}$, $\delta \underline{V}^{(1)}$, and $\delta N_B^{(1)}$ are considered, if $\delta \underline{S}$ is to vanish, the coefficients of each variation must be zero. Therefore, the criteria of equilibrium are $T^{(1)} = T^{(2)}$, $P^{(1)} = P^{(2)}$ and $\mu_B^{(1)} = \mu_B^{(2)}$. Note that there is no apparent restriction on $\mu_A^{(1)}$ or $\mu_A^{(2)}$.

2. The internal wall is rigid, diathermal, and permeable to both A and B. With the same approach, but with

$$\delta \underline{V}^{(1)} = \delta \underline{V}^{(2)} = 0 \tag{6-38}$$

used instead of Eq. (6-36), the temperatures and chemical potentials of both components are found to be equal in both subsystems. Note that there is no apparent restriction on $P^{(1)}$ and $P^{(2)}$.

3. The internal wall is adiabatic, movable, and permeable. By analogy to cases 1 and 2, it might be thought that $\delta \underline{U}^{(1)}$ and $\delta \underline{U}^{(2)}$ were zero. Mass interchange between the subsystems, however, can also vary the energy of each compartment; thus we have, in reality, no additional restraints and, at equilibrium, the temperature, pressure, and component chemical potentials are equal in each subsystem. A similar result would be found even if the boundary were rigid. Thus, the adiabatic–permeable case is very similar to the diathermal–permeable case.

An interesting dilemma may be noted in case 2. At equilibrium, in this binary system μ_A, μ_B, and T were the same on both sides of the internal wall. Since each side is a simple system, the corollary to Postulate I introduced in Section 5.2 then states that only $n + 1$ intensive variables need be specified to define completely all other intensive variables for a single-phase simple system. If each side is indeed a single phase, then $P^{(1)} = P^{(2)}$. If one or both sides contain more than one phase, then, as will be discussed in Section 10.1, fewer than $n + 1$ variables are required to define all other intensive properties, but in any case, $P^{(1)}$ would equal $P^{(2)}$. Similar reasoning may be applied to case 1 to show that $\mu_A^{(1)} = \mu_A^{(2)}$.

It is clear that each complex system with internal membranes or walls must be examined as a special case.

6.5 Phase Equilibria

A system with more than a single phase may be considered to be a composite of simple systems with phase-separating membranes that are movable, diathermal, and permeable to all components. Thus, as will be proved below, the intensive properties T, P, and μ_j are identical in all phases.

In the proof, consider an isolated, simple multicomponent system containing π phases. The equilibrium criterion, Eq. (6-6), then becomes

$$\begin{aligned} \delta \underline{S} &= \sum_{s=1}^{\pi} \delta \underline{S}^{(s)} \\ &= \sum_{s=1}^{\pi} S_U^{(s)}\, \delta \underline{U}^{(s)} + \sum_{s=1}^{\pi} S_V^{(s)}\, \delta \underline{V}^{(s)} + \sum_{s=1}^{\pi} \sum_{j=1}^{n} S_{N_j}^{(s)}\, \delta N_j^{(s)} \\ &= 0 \end{aligned} \tag{6-39}$$

where superscript (s) is a dummy variable denoting the phase. The constraints resulting from isolation, with no chemical reaction, are

$$\delta \underline{U} = \sum_{s=1}^{\pi} \delta \underline{U}^{(s)} = 0 \tag{6-40}$$

$$\delta \underline{V} = \sum_{s=1}^{\pi} \delta \underline{V}^{(s)} = 0 \tag{6-41}$$

$$\delta N_j = \sum_{s=1}^{\pi} \delta N_j^{(s)} = 0 \tag{6-42}$$

In Eq. (6-39), as indicated before, $S_U^{(s)} = (\partial \underline{S}^{(s)}/\partial \underline{U}^{(s)})_{\underline{V},N} = (1/T^{(s)})$, $S_V^{(s)} = P^{(s)}/T^{(s)}$, and $S_{N_j}^{(s)} = -\mu_j^{(s)}/T^{(s)}$.

To include the $n + 2$ restraining equations in order to eliminate $(n + 2)$ dependent variables, it is convenient to use the method of Lagrange undetermined multipliers. Let us define the arbitrary multipliers to Eqs. (6-40), (6-41), and (6-42) as $(1/T^{(1)})$, $(P^{(1)}/T^{(1)})$, and $(-\mu_j^{(1)}/T^{(1)})$, respectively. Then, multiplying each constraint equation by its respective multiplier and subtracting all from Eq. (6-39), there results

$$\begin{aligned} \delta \underline{S} = \sum_{s=2}^{\pi} \left(\frac{1}{T^{(s)}} - \frac{1}{T^{(1)}}\right) \delta \underline{U}^{(s)} + \sum_{s=2}^{\pi} \left(\frac{P^{(s)}}{T^{(s)}} - \frac{P^{(1)}}{T^{(1)}}\right) \delta \underline{V}^{(s)} \\ - \sum_{s=2}^{\pi} \sum_{j=1}^{n} \left(\frac{\mu_j^{(s)}}{T^{(s)}} - \frac{\mu_j^{(1)}}{T^{(1)}}\right) \delta N_j^{(s)} = 0 \end{aligned} \tag{6-43}$$

Since each variation in $\delta \underline{U}^{(s)}$, $\delta \underline{V}^{(s)}$, and $\delta N_j^{(s)}$, is now independent, it is immediately obvious that

$$T^{(1)} = T^{(2)} = \cdots T^{(\pi)} \tag{6-44}$$

$$P^{(1)} = P^{(2)} = \cdots P^{(\pi)} \tag{6-45}$$

$$\mu_j^{(1)} = \mu_j^{(2)} = \cdots \mu_j^{(\pi)} \tag{6-46}$$

These temperature, pressure, and chemical potential equalities between phases hold for all simple multiphase systems, that is, those with no external force field and with no internal barriers to restrict heat, work, or mass interactions.

6.6 Chemical Reaction Equilibria

The treatment for multiphase systems can be extended in the following manner to cases in which chemical reactions occur. The procedure shown in Section 6.5 is applicable except that the constraint on mole numbers, Eq. (6-42), must be modified for those components that react chemically. If a reaction involving components 1 through i occurs, the general form may be expressed as

$$\nu_1 C_1 + \nu_2 C_2 + \cdots + \nu_i C_i = 0 \tag{6-47}$$

or

$$\sum_{j=1}^{i} \nu_j C_j = 0 \tag{6-48}$$

Equations (6-47) and (6-48) are in reality atom balances; C_j could be considered the chemical formula for j and ν_j the molar stoichiometric multiplier or coefficient. *For products, ν_j is always defined as a positive number and, for reactants, a negative number.*

If the reaction is stoichiometrically balanced, there is no net change in mass, and

$$\sum_{j=1}^{i} m_j \nu_j = 0 \tag{6-49}$$

where m_j is the molecular weight of j. The stoichiometric coefficient defines the ratio of mole changes of reacting components. That is, for a single reaction,

$$\frac{\delta N_1}{\nu_1} = \frac{\delta N_2}{\nu_2} = \cdots \frac{\delta N_i}{\nu_i} \tag{6-50}$$

Since the δN_j variations are related, only one may be varied independently. To simplify bookkeeping, a new variable, the *extent of reaction*, ξ, is introduced,

$$\delta \xi = \frac{\delta N_j}{\nu_j}, \qquad j = 1, 2, \ldots, i \tag{6-51}$$

so that all δN_j for reacting species may be expressed in terms of $\delta \xi$. Equation (6-42) is then modified, for reacting components, to

$$\left.\begin{aligned} \delta N_j &= \nu_j\, \delta\xi \qquad (6\text{-}52) \\ \sum_{s=1}^{\pi} \delta N_j^{(s)} &= \nu_j\, \delta\xi \qquad (6\text{-}53) \end{aligned}\right\} \quad j = 1, 2, \ldots, i$$

For inert species, $\delta N_j = 0$ and

$$\sum_{s=1}^{\pi} \delta N_j^{(s)} = 0 \qquad j = i+1, \ldots, n \tag{6-54}$$

Again employing the same Lagrange undetermined multipliers as used in Eq. (6-43), with Eqs. (6-53) and (6-54) in place of (6-42), Eq. (6-43) is again obtained except that there is one additional term,

$$-\frac{1}{T^{(1)}} \sum_{j=1}^{i} (\nu_j \mu_j)\, \delta\xi$$

Since ξ can be varied independently from $\underline{U}^{(s)}$, $\underline{V}^{(s)}$, and $N_j^{(s)}$, we have, at equilibrium, in addition to the equalities of temperature, pressure, and chemical potential between phases, the requirement that

$$\sum_{j=1}^{i} \nu_j \mu_j = 0 \tag{6-55}$$

The superscript on μ_j is deleted since the chemical potential of component j is equal in all phases.

For cases in which multiple reactions occur, it is readily shown by an identical treatment that an equation of the form of Eq. (6-55) applies for each reaction in which there is chemical equilibrium. This general criterion will be developed further in Chapter 11.

Example 6.2

At low temperatures, a mixture of water and excess ferric chloride forms a solid phase of $FeCl_3 \cdot 6H_2O$. If equilibrium existed in an isolated system of water and ferric chloride such that there was a gas phase consisting only of water vapor, a liquid phase with only H_2O and $FeCl_3$, and a solid phase of the hexahydrate, clearly specify all equilibrium criteria that apply to this system. (Neglect any ionization of the $FeCl_3$.)

Solution

Equation (6-39) as written for this three-phase system is

$$\delta \underline{S} = \frac{1}{T^V}\,\delta \underline{U}^V + \frac{P^V}{T^V}\,\delta \underline{V}^V - \frac{\mu_w^V}{T^V}\,\delta N_w^V + \frac{1}{T^L}\,\delta \underline{U}^L + \frac{P^L}{T^L}\,\delta \underline{V}^L$$
$$- \frac{\mu_{FeCl_3}^L}{T^L}\,\delta N_{FeCl_3}^L - \frac{\mu_w^L}{T^L}\,\delta N_w^L + \frac{1}{T^S}\,\delta \underline{U}^S + \frac{P^S}{T^S}\,\delta \underline{V}^S$$
$$- \frac{\mu_{hyd}^S}{T^S}\,\delta N_{hyd}^S = 0$$

The constraints placed on the system are

$$\delta \underline{U}^V + \delta \underline{U}^L + \delta \underline{U}^S = 0 \tag{6-56}$$
$$\delta \underline{V}^V + \delta \underline{V}^L + \delta \underline{V}^S = 0 \tag{6-57}$$

and there is also the reaction

$$6H_2O + FeCl_3 = FeCl_3 \cdot 6H_2O$$

so that

$$\delta\xi = \frac{\delta N_w}{-6} = \frac{\delta N_{\text{FeCl}_3}}{-1} = \frac{\delta N_{\text{hyd}}}{1}$$

and

$$\delta N_w = \delta N_w^L + \delta N_w^V = -6\,\delta\xi \tag{6-58}$$

$$\delta N_{\text{FeCl}_3} = \delta N_{\text{FeCl}_3}^L = -\delta\xi \tag{6-59}$$

$$\delta N_{\text{hyd}} = \delta N_{\text{hyd}}^S = \delta\xi \tag{6-60}$$

Using Lagrange multipliers $(1/T^V)$ to Eq. (6-56), (P^V/T^V) to Eq. (6-57), $(-\mu_w^V/T^V)$ to Eq. (6-58), $(-\mu_{\text{FeCl}_3}^L/T^V)$ to Eq. (6-59), and $(-\mu_{\text{hyd}}^S/T^V)$ to Eq. (6-60), subtracting from the $\delta\underline{S}$ expression, we see immediately, since all variations are now independent, that

$$T = T^V = T^L = T^S$$

$$P = P^V = P^L = P^S$$

$$\mu_w^V = \mu_w^L$$

$$6\mu_w^L + \mu_{\text{FeCl}_3}^L = \mu_{\text{hyd}}^S$$

These are the equilibrium criteria.

Example 6.3

Three reactors in an isolation chamber are initially charged as follows: A has pure methane, B pure hydrogen, and C pure ammonia. They are interconnected by rigid, semipermeable membranes as shown in Figure 6.6. All reactors have suitable

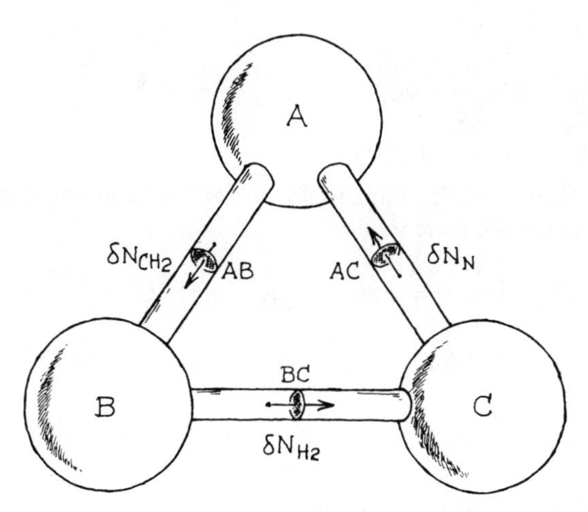

Figure 6.6 Sketch for Example 6.3.

catalysts so that the reactions

$$CH_4 = CH_2 + H_2 \tag{1}$$

$$NH_3 = N + \tfrac{3}{2}H_2 \tag{2}$$

are always in equilibrium. Membrane AB is permeable only to CH_2, BC to H_2, and AC to nitrogen atoms. What are the equilibrium criteria for this system?

Solution

Reactions (1) and (2) can occur in A; call the extent of reactions for these $\delta\xi_1^A$ and $\delta\xi_2^A$, respectively. In B only (1) occurs with $\delta\xi_1^B$ and in C only (2) with $\delta\xi_2^C$. With $\delta\underline{V}^A = \delta\underline{V}^B = \delta\underline{V}^C = 0$, the entropy variation of the total system is

$$\begin{aligned}\delta\underline{S} &= \delta\underline{S}^A + \delta\underline{S}^B + \delta\underline{S}^C\\ &= \frac{1}{T^A}\,\delta\underline{U}^A - \frac{\mu_{CH_4}^A}{T^A}\,\delta N_{CH_4}^A - \frac{\mu_{CH_2}^A}{T^A}\,\delta N_{CH_2}^A - \frac{\mu_{H_2}^A}{T^A}\,\delta N_{H_2}^A\\ &\quad - \frac{\mu_N^A}{T^A}\,\delta N_N^A - \frac{\mu_{NH_3}^A}{T^A}\,\delta N_{NH_3}^A + \frac{1}{T^B}\,\delta\underline{U}^B - \frac{\mu_{CH_4}^B}{T^B}\,\delta N_{CH_4}^B\\ &\quad - \frac{\mu_{CH_2}^B}{T^B}\,\delta N_{CH_2}^B - \frac{\mu_{H_2}^B}{T^B}\,\delta N_{H_2}^B + \frac{1}{T^C}\,\delta\underline{U}^C - \frac{\mu_{H_2}^C}{T^C}\,\delta N_{H_2}^C\\ &\quad - \frac{\mu_N^C}{T^C}\,\delta N_N^C - \frac{\mu_{NH_3}^C}{T^C}\,\delta N_{NH_3}^C = 0\end{aligned}$$

The conditions of restraint are:

(1) $\delta\underline{U}^A + \delta\underline{U}^B + \delta\underline{U}^C = 0$
(2) $\delta N_{CH_2}^A = -\delta N_{CH_2} + \delta\xi_1^A$
(3) $\delta N_{H_2}^A = \delta\xi_1^A + \frac{3}{2}\,\delta\xi_2^A$
(4) $\delta N_{CH_4}^A = -\delta\xi_1^A$
(5) $\delta N_N^A = \delta N_N + \delta\xi_2^A$
(6) $\delta N_{NH_3}^A = -\delta\xi_2^A$
(7) $\delta N_{H_2}^B = -\delta N_{H_2} + \delta\xi_1^B$
(8) $\delta N_{CH_4}^B = -\delta\xi_1^B$
(9) $\delta N_{CH_2}^B = \delta N_{CH_2} + \delta\xi_1^B$
(10) $\delta N_{H_2}^C = \delta N_{H_2} + \frac{3}{2}\,\delta\xi_2^C$
(11) $\delta N_{NH_3}^C = -\delta\xi_2^C$
(12) $\delta N_N^C = -\delta N_N + \delta\xi_2^C$

Multiplying (1) by $-1/T'$, adding to $\delta\underline{S}$, and also substituting (2) through (12) into the expression for $\delta\underline{S}$, there results

$$\begin{aligned}&\left(\frac{1}{T^A} - \frac{1}{T'}\right)\delta\underline{U}^A + \left(\frac{1}{T^B} - \frac{1}{T'}\right)\delta\underline{U}^B + \left(\frac{1}{T^C} - \frac{1}{T'}\right)\delta\underline{U}^C + \left(\frac{\mu_{CH_2}^A}{T^A} - \frac{\mu_{CH_2}^B}{T^B}\right)\delta N_{CH_2}\\ &\quad + \left(\frac{\mu_N^C}{T^C} - \frac{\mu_N^A}{T^A}\right)\delta N_N + \left(\frac{\mu_{H_2}^B}{T^B} - \frac{\mu_{H_2}^C}{T^C}\right)\delta N_{H_2} - \frac{1}{T^A}(\mu_{CH_2}^A + \mu_{H_2}^A - \mu_{CH_4}^A)\,\delta\xi_1^A\\ &\quad - \frac{1}{T^A}(\mu_N^A + \tfrac{3}{2}\mu_{H_2}^A - \mu_{NH_3}^A)\,\delta\xi_2^A - \frac{1}{T^B}(\mu_{CH_2}^B + \mu_{H_2}^B - \mu_{CH_4}^B)\,\delta\xi_1^B\\ &\quad - \frac{1}{T^C}(\mu_N^C + \tfrac{3}{2}\mu_{H_2}^C - \mu_{NH_3}^C)\,\delta\xi_2^C = 0\end{aligned}$$

Therefore, the criteria of equilibrium are

$$T^A = T^B = T^C$$

$$\mu_{CH_2}^A = \mu_{CH_2}^B = \mu_{CH_4}^A - \mu_{H_2}^A = \mu_{CH_4}^B - \mu_{H_2}^B$$

$$\mu_{H_2}^B = \mu_{H_2}^C$$

$$\mu_N^A = \mu_N^C = \mu_{NH_3}^A - \tfrac{3}{2}\mu_{H_2}^A = \mu_{NH_3}^C - \tfrac{3}{2}\mu_{H_2}^C$$

Note that no statement was made on whether the semipermeable membranes were diathermal or adiabatic. Identical equilibrium criteria would result in either case.

PROBLEMS

6.1. A well-insulated steel bomb initially contains liquid and vapor of compound A (see Figure P6.1). The vapor is known to decompose under the conditions of the

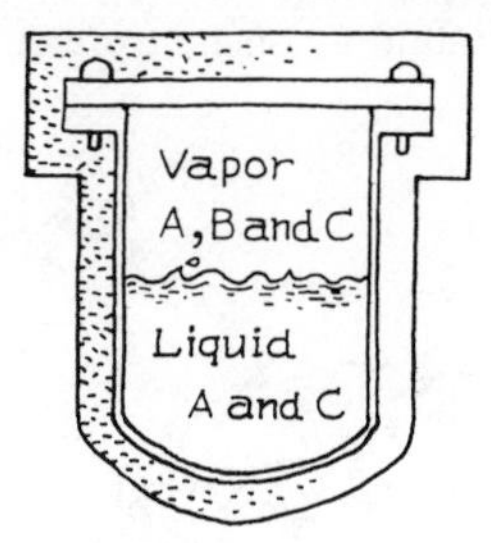

Figure P6.1

experiment to B and C by the reaction

$$A = 2B + C$$

Compound C is soluble in liquid A, but compound B does not dissolve to any measurable extent. Also, the reaction does not occur to any appreciable extent in the liquid. Derive the criteria of equilibrium for the system of liquid plus vapor.

6.2. A closed vessel of volume $\underline{V}$ contains liquid and vapor phases of a pure material. The vessel is immersed in a constant-temperature bath at temperature T^* (see Figure P6.2). Determine the criteria of equilibrium in terms of the intensive properties of the phases.

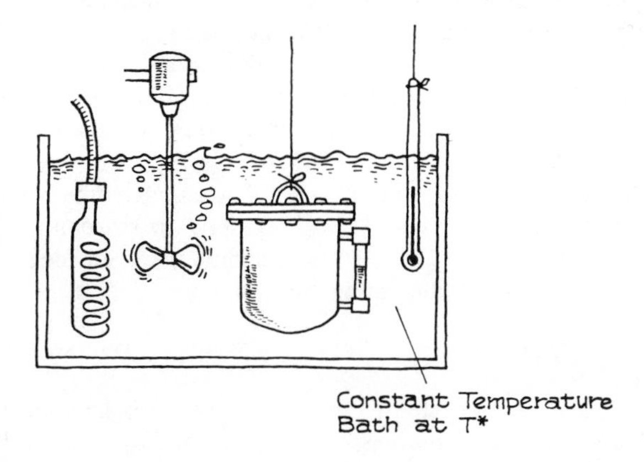

Figure P6.2

6.3. The criteria for chemical equilibrium in a multicomponent, single-phase system,

$$\sum \nu_i \mu_i = 0 \tag{A}$$

is derived from systems at constant ($\underline{S}$, $\underline{V}$), ($\underline{S}$, P), (T, $\underline{V}$), or (T, P), by applying the minimization principle from $\underline{U}$, $\underline{H}$, $\underline{A}$, or $\underline{G}$, respectively. There is some question about the criteria of equilibrium for a system at constant P and $\underline{V}$. For example, a constant-volume bomb may initially contain compound A and an anticatalyst to prevent the reaction

$$A \rightleftharpoons 2B \tag{B}$$

The anticatalyst is removed, and as the reaction proceeds, heat is added or removed in order to maintain a constant pressure (see Figure P6.3).

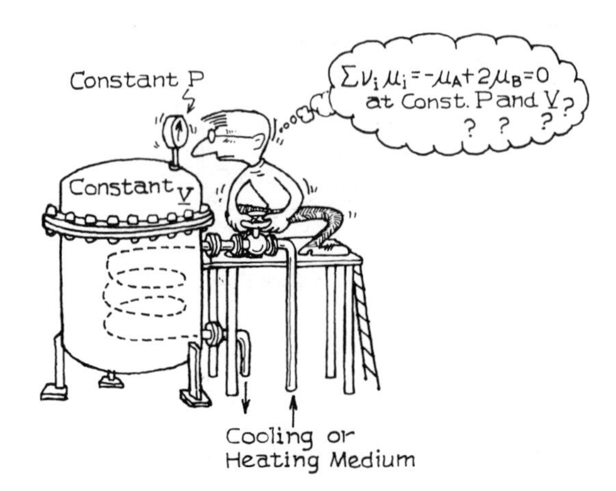

Figure P6.3

One student claims that the system will always reach a final state in which Eq. (A) is satisfied. Another student is not convinced and claims that under certain conditions the system may go from pure reactants to pure products, while Eq. (A) may or may not be satisfied at some intermediate point. He gives as an example the case of an exothermic reaction with the stoichiometric relation (B). When some A reacts to 2B, the system is cooled to maintain constant P, but the reduction in temperature results in a more favorable condition for B formation (Le Châtelier principle), so that more A reacts, and consequently more cooling is required, which in turn drives the reaction to B, etc., until all of A is depleted.

You are asked to resolve this dilemma: Is Eq. (A) a valid criteria of equilibrium for a system at constant P and $\underline{V}$? Are there some conditions for which Eq. (A) need not be satisfied? Clearly explain your reasoning. You may use the gas-phase reaction $A \rightleftharpoons 2B$ with the assumption of an ideal mixture of ideal gases for the purpose of illustrating your point.

6.4. A rigid, diathermal sphere contains a mixture of H_2, N_2, and NH_3. This is subsystem I.

There is a tube connected to the sphere with a palladium membrane at the point of contact. This tube is closed at the other end by a free-floating, diathermal piston. The membrane is permeable only to H_2. This is subsystem II.

The environment (subsystem III) consists of a large isothermal and isobaric reservoir that changes by a negligible amount for variations in subsystems I and II.

(a) What are the equilibrium criteria for the entirety of subsystems I, II, and III considered as a single system?

(b) If only subsystem I were considered, what thermodynamic potential function would be maximized (minimized) in an equilibrium state?

6.5. In Figure P6.5, the tube is well insulated and the initial conditions are:

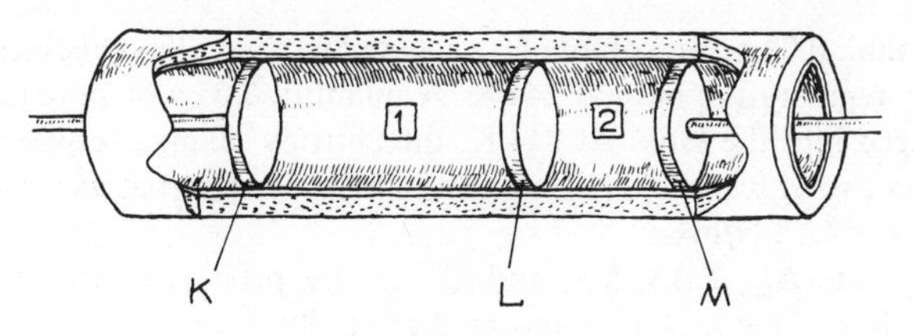

Figure P6.5

Section 1:

$$\underline{V} = 9.086 \times 10^{-2}\ \text{m}^3$$
$$T = 273.2\ \text{K}$$
$$P = 1\ \text{bar}$$
gas, pure A

Section 2:

$$\underline{V} = 2.271 \times 10^{-2}\ \text{m}^3$$
$$T = 273.2\ \text{K}$$
$$P = 1\ \text{bar}$$
gas, pure B

Piston L is rigid, diathermal, and initially impermeable. Suddenly, however, it is made permeable to gas A, but not to B. Piston K is simultaneously unlatched and connected to a large pressure reservoir at 2 bar, and piston M is unlatched and connected to a pressure reservoir at 3 bar. There is no friction in the pistons, but the pressure changes are effected so that oscillations are reduced. Assume ideal gases and $C_v = 20.9$ J/mol K.

(a) After equilibrium is attained, what are the criteria of equilibrium?

(b) Describe quantitatively the states of the gases on either side of the piston.

(c) Repeat parts (a) and (b) assuming that the piston L is permeable to both gases.

6.6. We have a cylinder whose external walls are rigid and adiabatic. There is an internal piston that is, initially, adiabatic, impermeable, and fixed in place. The two sides of the cylinder are charged with a binary mixture of ideal gases as follows:

	Side 1	Side 2
T (K)	300	400
P (bar)	2	1
Total moles	6	5
Mole fraction A	0.4	0.8
Mole fraction B	0.6	0.2

At the start of an experiment, the dividing piston is replaced by one that is movable, diathermal, and permeable *only* to component A. Assume that $C_{v_A} = C_{v_B} = 20.9$ J/mol K.

(a) Derive the criteria of equilibria after all changes in the cylinder cease.

(b) What are the final equilibrium temperatures, pressures, and mole fractions of A in sides 1 and 2?

(c) Repeat part (b) if all the initial conditions were similar except that *no* B was initially present in side 2.

6.7. A rigid tank 0.03 m^3 in volume is evacuated to a very low pressure and then placed in a low-temperature bath at 244 K. A quantity 0.05 g of pure (solid) water ice is introduced into the tank. At 244 K, the entropy change between saturated water vapor and solid water ice is 11.62 J/g K and the reported vapor pressure of solid water ice is 42.8 N/m^2.

(a) Calculate $\Delta \underline{U}$, $T \Delta \underline{S}$, $\Delta \underline{A}$, and Q for the process of ice evaporation. What fraction of the ice has evaporated at equilibrium?

(b) What conclusions can you reach about the general equilibrium criterion in this case?

Single-Phase Simple Systems of Pure Materials 7

In Chapter 5 the theoretical basis for determining the information required to specify the thermodynamic properties of materials was described. In this chapter we continue this line of reasoning with reference to a specific type of system: the single-phase simple system of a pure material. Single-phase simple systems of mixtures are treated separately in Chapter 8.

7.1 Gibbs Energy Representation of the Fundamental Equation

For a single-phase simple system of a pure material, the independent variables for the energy representation are $\underline{S}$, $\underline{V}$, N, and for the Gibbs energy representation, they are T, P, N. The latter set is usually more convenient because T and P are the variables normally under the control of the experimentalist.

The Fundamental Equation for a single component in the Gibbs energy form is

$$\underline{G} = f_G(T, P, N) \tag{7-1}$$

and, in differential form,

$$d\underline{G} = -\underline{S}\, dT + \underline{V}\, dP + \mu\, dN \tag{7-2}$$

The equations of state, obtained from the first-order partial derivatives of Eq. (7-1), are[1]

[1]Note that the equations of state obtained from the $\underline{U}$, $\underline{H}$, $\underline{A}$, and $\underline{G}$ representations are equivalent. Given the three equations of state for any one representation, we could solve them simultaneously to obtain the three equations of state for any other representation. For example, compare Eqs. (7-3) to (7-5) with Eqs. (5-5) to (5-7).

$$-\underline{S} = g_1(T, P, N) = G_T \tag{7-3}$$

$$\underline{V} = g_2(T, P, N) = G_P \tag{7-4}$$

and

$$\mu = g_3(T, P, N) = G_N \tag{7-5}$$

The three independent second-order partial derivatives are[2]

$$\frac{\partial^2 \underline{G}}{\partial T^2} = -\left(\frac{\partial \underline{S}}{\partial T}\right)_{P,N} = g_{11}(T, P, N) = G_{TT} \tag{7-6}$$

$$\frac{\partial^2 \underline{G}}{\partial P^2} = \left(\frac{\partial \underline{V}}{\partial P}\right)_{T,N} = g_{22}(T, P, N) = G_{PP} \tag{7-7}$$

$$\frac{\partial^2 \underline{G}}{\partial T\, \partial P} = -\left(\frac{\partial \underline{S}}{\partial P}\right)_{T,N} = \left(\frac{\partial \underline{V}}{\partial T}\right)_{P,N} = g_{12}(T, P, N) = G_{TP} \tag{7-8}$$

Equations (7-6) through (7-8) can be simplified because each of the g_{ij} functions are first order in mass. Thus,[3]

$$-\left(\frac{\partial S}{\partial T}\right)_P = -\frac{1}{N}\left(\frac{\partial \underline{S}}{\partial T}\right)_{P,N} = g_{11}(T, P) \tag{7-9}$$

$$\left(\frac{\partial V}{\partial P}\right)_T = \frac{1}{N}\left(\frac{\partial \underline{V}}{\partial P}\right)_{T,N} = g_{22}(T, P) \tag{7-10}$$

and

$$\left(\frac{\partial V}{\partial T}\right)_P = -\left(\frac{\partial S}{\partial P}\right)_T = \frac{1}{N}\left(\frac{\partial \underline{V}}{\partial T}\right)_{P,N} = g_{12}(T, P) \tag{7-11}$$

These second-order partial derivatives are related to three widely used properties: *the heat capacity at constant pressure*, C_p; *the isothermal compressibility*, κ_T; and *the coefficient of thermal expansion*, α_p. By definition,

$$C_p \equiv T\left(\frac{\partial S}{\partial T}\right)_P \tag{7-12}$$

$$\kappa_T \equiv -\frac{1}{V}\left(\frac{\partial V}{\partial P}\right)_T \tag{7-13}$$

and

$$\alpha_p \equiv \frac{1}{V}\left(\frac{\partial V}{\partial T}\right)_P \tag{7-14}$$

[2]When expressing $\underline{G}$ in extensive form, there are three more partial derivatives in addition to Eqs. (7-6) through (7-8). These derivatives, however, are either zero or are redundant with Eqs. (7-3) through (7-5). Specifically,

$$\left(\frac{\partial^2 \underline{G}}{\partial N^2}\right) = \left(\frac{\partial \mu}{\partial N}\right)_{T,P} = 0 \qquad \text{(see Section 5.2)}$$

$$\left(\frac{\partial^2 \underline{G}}{\partial T\, \partial N}\right) = -\left(\frac{\partial \underline{S}}{\partial N}\right)_{T,P} = -S = \frac{g_1}{N} \qquad \text{[see Eq. (7-3)]}$$

and

$$\left(\frac{\partial^2 \underline{G}}{\partial P\, \partial N}\right) = \left(\frac{\partial \underline{V}}{\partial N}\right)_{T,P} = V = \frac{g_2}{N} \qquad \text{[see Eq. (7-4)]}$$

[3]In the derivatives on the left-hand sides of Eqs. (7-9) through (7-11), note that the subscript N has been omitted. Of course, $n + 1$ variables must always be held constant in partial differentiation. It is, however, common practice to omit the mole numbers when expressing partial derivatives of intensive variables.

These three properties were among the first thermodynamic properties ever reported, principally because they can be measured by relatively simple experiments.

Let us now determine which data sets have the information equivalent to Eq. (7-1). We shall approach this problem once again by determining what data are required to reconstruct the Fundamental Equation.

Applying Euler's theorem to Eq. (7-2), we have the integrated form of the Fundamental Equation in the Gibbs energy representation:

$$\underline{G} = \mu N \tag{7-15}$$

Since Eq. (7-1) can be recovered by substituting $g_3(T, P)$ from Eq. (7-5) into Eq. (7-15), we need specify only one equation of state, namely, Eq. (7-5), to obtain the Fundamental Equation. This conclusion is not in contradiction to the discussion of Section 5.3, in which it was shown that two equations of state, Eqs. (5-5) and (5-6), were necessary to reconstruct the Fundamental Equation to within an arbitrary constant. As shown below, we must in fact know Eqs. (7-3) and (7-4) in order to obtain Eq. (7-5).

Let us first express Eq. (7-5) in differential form. In Section 5.4 it was shown that the Gibbs–Duhem equation for a pure material is

$$d\mu = -S\, dT + V\, dP \tag{7-16}$$

To evaluate g_3 of Eq. (7-5), we must express S and V as functions of T and P, substitute into Eq. (7-16), and integrate. The desired expressions for S and V are given by Eqs. (7-3) and (7-4) when written in intensive form. Thus,

$$d\mu = g_1(T, P)\, dT + g_2(T, P)\, dP \tag{7-17}$$

and, by integration from a reference state at $T = T^\circ$ and $P = P^\circ$, where μ is set equal to an arbitrary value of μ°,

$$\mu = \mu^\circ + \int_{T^\circ}^{T} [g_1(T, P)]_{P^\circ}\, dT + \int_{P^\circ}^{P} [g_2(T, P)]_T\, dP \tag{7-18}$$

Let us assume for the moment that Eqs. (7-3) and (7-4) are not known. In this case, we can evaluate g_1 and g_2 in the following manner.

Since S and V are properties, we can start with the exact differentials.

$$dS = \left(\frac{\partial S}{\partial T}\right)_P dT + \left(\frac{\partial S}{\partial P}\right)_T dP \tag{7-19}$$

and

$$dV = \left(\frac{\partial V}{\partial T}\right)_P dT + \left(\frac{\partial V}{\partial P}\right)_T dP \tag{7-20}$$

Of course, the partial derivatives in Eqs. (7-19) and (7-20) are related to the second-order partial derivatives of Eq. (7-1). Therefore,

$$dS = \frac{C_p}{T}\, dT - \alpha_p V\, dP \tag{7-21}$$

and

$$dV = \alpha_p V\, dT - \kappa_T V\, dP \tag{7-22}$$

Integrating from a reference state at T° and P° for which $S = S^\circ$ and $V = V^\circ$, we have

$$S = S^\circ + \int_{T^\circ}^{T} \left(\frac{C_p}{T}\right)_{P^\circ} dT - \int_{P^\circ}^{P} (\alpha_p V)_T \, dP \tag{7-23}$$

and

$$V = V^\circ + \int_{T^\circ}^{T} (\alpha_p V)_{P^\circ} \, dT - \int_{P^\circ}^{P} (\kappa_T V)_T \, dP \tag{7-24}$$

Equations (7-23) and (7-24) are equivalent to Eqs. (7-3) and (7-4), respectively. Substitution into Eq. (7-18) yields

$$\mu = \mu^\circ - S^\circ(T - T^\circ) + V^\circ(P - P^\circ) - g'(T, P, T^\circ, P^\circ) + g''(T, P, T^\circ, P^\circ) \tag{7-25}$$

where

$$g'(T, P, T^\circ, P^\circ) = \int_{T^\circ}^{T} \left[\int_{T^\circ}^{T} \left(\frac{C_p}{T}\right)_{P^\circ} dT - \int_{P^\circ}^{P} (\alpha_p V)_T \, dP \right]_{P^\circ} dT \tag{7-26}$$

and

$$g''(T, P, T^\circ, P^\circ) = \int_{P^\circ}^{P} \left[\int_{T^\circ}^{T} (\alpha_p V)_{P^\circ} \, dT - \int_{P^\circ}^{P} (\kappa_T V)_T \, dP \right]_T dP \tag{7-27}$$

[Note that the second term in Eq. (7-26) vanishes because the upper limit of integration, the dummy variable P, is held constant at P° in the second integration for dT.] In Eq. (7-25), μ° and S° can be assigned arbitrary values; the absolute value of volume, however, has physical significance and, therefore, V° must be the actual volume corresponding to T° and P°.

In summary, we have seen that the information content of the Fundamental Equation is contained in the data set consisting of the three second-order partial derivatives C_p, α_p, and κ_T in addition to one value of the specific volume in a reference state.

It should be clear that an equally valid data set consists of C_p and the P–V–T relationship, Eq. (7-4), since the latter contains the information content of α_p, κ_T, and the absolute value of volume. The P–V–T relationship is so commonly used that it is usually called *the* equation of state (as opposed to the equations of state, which is the general terminology for all the first-order derivatives of the Fundamental Equation). The data set of C_p and P–V–T is most commonly used to solve a wide variety of problems because this information is most readily available. In Section 7.2 we describe a method of obtaining any other derivatives involving thermodynamic properties directly from this data set.

Before concluding this discussion of the Gibbs energy representation, it is interesting to compare the results with those for the internal energy. From Eq. (7-25) for a change in state from T_1, P_1, to T_2, P_2,

$$\Delta G = \Delta\mu = -S^\circ(T_2 - T_1) + V^\circ(P_2 - P_1) - g'(T_2, P_2, T_1, P_1) + g''(T_2, P_2, T_1, P_1) \tag{7-28}$$

Whereas ΔU for the corresponding change in state would have a value independent of the reference values assigned to S° and μ°, the value of ΔG, as given by Eq. (7-28) depends on the value of S°. Since S° is an arbitrary constant, there can be no direct physical significance associated with ΔG for the general change in state under consideration.[4]

7.2 Transformation of Partial Derivatives

We have seen in Section 7.1 that the three independent second-order partial derivatives of the Gibbs energy representation are a convenient set because they can be measured readily. There are, nevertheless, many cases in which we must determine other partial derivatives. We shall now demonstrate that any partial derivative for a pure material can be expressed in terms of S, T, P, and V and also that any partial involving only these four variables can be expressed in terms of second-order partial derivatives of one of the four representations of the Fundamental Equation. Since we have already demonstrated in Section 5.5 that any second-order partial derivative of one representation of the Fundamental Equation can be expressed as a function of the three independent second-order partials of any other representation, it follows that any partial derivative can be expressed as a function of C_p, κ_T, and, α_p or their equivalent.

A partial derivative may involve intensive and extensive variables. For a *pure* material, only $n + 1 = 2$ intensive variables are independent; hence, a partial derivative involving only intensive variables can be expressed as $(\partial b/\partial c)_d$, where it is implied that N is constant. That is,

$$\left(\frac{\partial b}{\partial c}\right)_{d,N} = \left(\frac{\partial b}{\partial c}\right)_d \tag{7-29}$$

We now show that for a pure material, any partial derivative involving extensive variables can always be reduced to expressions involving partial derivatives of entirely intensive variables.

Consider the derivative $(\partial b/\partial c)_{d,e}$ where one of the four variables is extensive.

1(i). If c is extensive, then

$$\left(\frac{\partial b}{\partial \underline{c}}\right)_{d,e} = 0 \tag{7-30}$$

The proof follows from applying Euler's theorem to $b = f(\underline{c}, d, e)$:

$$0 = \left(\frac{\partial b}{\partial \underline{c}}\right)_{d,e} \underline{c}$$

1(ii). If b is extensive, then

$$\left(\frac{\partial \underline{b}}{\partial c}\right)_{d,e} = \frac{1}{(\partial c/\partial \underline{b})_{d,e}} = \frac{1}{0} = \infty \tag{7-31}$$

[4]Note that for isothermal processes, ΔG is no longer a function of S°, and Eq. (7-16) reduces to $\Delta G = \int_{P_1}^{P_2} V\,dP$. In this case, we can attach physical significance to the value of ΔG. Specifically, it represents the negative of the reversible flow work for the isothermal process (see Section 4.8).

1(iii). If d (or e) is extensive, then

$$\left(\frac{\partial b}{\partial c}\right)_{\underline{d},e} = \left(\frac{\partial b}{\partial c}\right)_e \tag{7-32}$$

Since

$$\left(\frac{\partial b}{\partial c}\right)_{\underline{d},e} = -\frac{(\partial \underline{d}/\partial c)_{b,e}}{(\partial \underline{d}/\partial b)_{c,e}} = -\frac{(\partial \underline{d}/\partial N)_{b,e}(\partial N/\partial c)_{b,e}}{(\partial \underline{d}/\partial N)_{c,e}(\partial N/\partial b)_{c,e}}$$

and

$$\left(\frac{\partial \underline{d}}{\partial N}\right)_{b,e} = \left(\frac{\partial \underline{d}}{\partial N}\right)_{c,e} = d$$

then

$$\left(\frac{\partial b}{\partial c}\right)_{\underline{d},e} = -\frac{(\partial N/\partial c)_{b,e}}{(\partial N/\partial b)_{c,e}} = \left(\frac{\partial b}{\partial c}\right)_{e,N} = \left(\frac{\partial b}{\partial c}\right)_e$$

Thus, if only one variable is extensive, the partial derivative is finite and nonzero only if d or e is extensive. The extensive variable may be deleted to yield a partial involving only intensive variables.

Now, consider the three cases in which two of the four variables are extensive.

2(i). If b and c are extensive, then

$$\left(\frac{\partial \underline{b}}{\partial \underline{c}}\right)_{d,e} = \frac{b}{c} \tag{7-33}$$

which follows directly by applying Euler's theorem to $\underline{b} = f(\underline{c}, d, e)$:

$$\underline{b} = \left(\frac{\partial \underline{b}}{\partial \underline{c}}\right)_{d,e} \underline{c} \quad \text{and} \quad \underline{b} = Nb, \qquad \underline{c} = Nc$$

2(ii). If b and d (or e) are extensive, then

$$\left(\frac{\partial \underline{b}}{\partial c}\right)_{\underline{d},e} = N\left[\left(\frac{\partial b}{\partial c}\right)_e - \frac{b}{d}\left(\frac{\partial d}{\partial c}\right)_e\right] \tag{7-34}$$

Since, expanding $\underline{b} = Nb$,

$$\left(\frac{\partial \underline{b}}{\partial c}\right)_{\underline{d},e} = N\left(\frac{\partial b}{\partial c}\right)_{\underline{d},e} + b\left(\frac{\partial N}{\partial c}\right)_{\underline{d},e} \tag{7-35}$$

Eq. (7-32) may be used to reduce $(\partial b/\partial c)_{\underline{d},e}$. The last term is reduced as follows:

$$\left(\frac{\partial N}{\partial c}\right)_{\underline{d},e} = \frac{(\partial \underline{d}/\partial c)_{N,e}}{(\partial \underline{d}/\partial N)_{c,e}} = -\frac{N(\partial d/\partial c)_e}{d}$$

2(iii). If d and e are extensive, then

$$\left(\frac{\partial b}{\partial c}\right)_{\underline{d},\underline{e}} = -\frac{(\partial \underline{d}/\partial c)_{b,\underline{e}}}{(\partial \underline{d}/\partial b)_{c,\underline{e}}} \tag{7-36}$$

The numerator and denominator can each be reduced by applying the results of case 2(ii).

Any partial derivative involving three extensive variables can now be reduced to partials involving two extensive variables by using Eq. (7-35) followed by one or more of the steps illustrated above. Similarly, partials involving four extensive variables can be reduced to three, etc. The net result is that *any partial derivative for a pure material can be expressed in terms of partials involving only three intensive variables.*

Let us now consider derivatives in intensive variables only and show that they can be reduced to expressions involving only $S, T. P, V$.

In general, a partial derivative of intensive variables may involve U, H, A, G, S, T, P, V. We have excluded N because it is extensive; μ is also excluded because μ is equal to G for a pure material.

The first step in reducing the general derivative $(\partial b/\partial c)_d$ is to eliminate any of the four potential functions U, H, A, or G, which may appear as b, c, or d. The three possibilities are treated as follows:

1. If b is U, H, A, or G, eliminate it by using the differential form of the transform. For example, with Eq. (5-41),

$$\left(\frac{\partial G}{\partial S}\right)_T = -S\left(\frac{\partial T}{\partial S}\right)_T^{\nearrow 0} + V\left(\frac{\partial P}{\partial S}\right)_T \tag{7-37}$$

2. If c is U, H, A, or G, invert the derivative by using the identity

$$\left(\frac{\partial c}{\partial b}\right)_d = \frac{1}{(\partial b/\partial c)_d} \tag{7-38}$$

   and then proceed by step 1.

3. If d is U, H, A, or G, bring the potential function into the brackets by using the relation

$$\left(\frac{\partial b}{\partial c}\right)_d = -\frac{(\partial d/\partial c)_b}{(\partial d/\partial b)_c} \tag{7-39}$$

   and then proceed by step 1.

Using these three steps, we can reduce any partial derivative involving only S, V, T, and P to a second-order derivative of one of the four forms of the Fundamental Equation. This is illustrated in Example 7.1.

Example 7.1

For a pure material there are 24 partial derivatives involving S, T, P, and V. Suggest a technique to relate any such derivative to second-order derivatives of the Gibbs energy and, therefore, to C_p, κ_T, and α_p.

Solution

Of the 24 derivatives, 12 are simply inverses. The others are as follows: (1) $(\partial S/\partial T)_p$, (2) $(\partial S/\partial T)_V$, (3) $(\partial S/\partial P)_T$, (4) $(\partial S/\partial P)_V$, (5) $(\partial S/\partial V)_T$, (6) $(\partial S/\partial V)_p$, (7) $(\partial V/\partial P)_T$, (8) $(\partial V/\partial P)_S$, (9) $(\partial V/\partial T)_p$, (10) $(\partial V/\partial T)_S$, (11) $(\partial P/\partial T)_S$, (12) $(\partial P/\partial T)_V$. Number (3) is equal to the negative of (9) from the Maxwell reciprocity theorem; similarly for (4) and (10). Number (5) is equal to (12) and (6) to (11). Thus, there remain (1) through (8). Three of these are second-order derivatives of the Gibbs energy:

(1) $(\partial S/\partial T)_p = -G_{TT} = C_p/T$
(3) $(\partial S/\partial P)_T = -G_{TP} = -G_{PT} = -(\partial V/\partial T)_p = -V\alpha_p$
(7) $(\partial V/\partial P)_T = G_{PP} = -V\kappa_T$

In addition, derivatives (2), (5), and (10) are related by the *X-Y-Z*-1 rule, as are (4), (6), and (8). [The sets (1), (3), (11) and (7), (9), (12) are also so related, but these derivatives have already been eliminated.] Thus, it is only necessary to relate (2), (4), and (6) to derivatives in $\underline{G}$ to allow all derivatives to be so expressed. The reduction of (2) has been introduced as Problem 5.12. For (4) and (6), we define our basis function $\underline{G} = y^{(0)} = f(T, P, N, \ldots)$, where the x–ξ values are shown below. For (6), we define $y^{(1)} = \underline{H} = f(\underline{S}, P, N_1, \ldots)$ and the x–ξ values are also shown below.

$y^{(0)}$			$y^{(1)}$	
x	ξ		x	ξ
1. T	$-\underline{S}$		1. $-\underline{S}$	$-T$
2. P	$\underline{V}$		2. P	$\underline{V}$
3. N_1	μ_1, etc.		3. N_1	μ_1, etc.

Thus, since $\underline{V} = y_2^{(1)}$, then

$$\left(\frac{\partial \underline{S}}{\partial \underline{V}}\right)_{P,N} = \left(\frac{\partial S}{\partial V}\right)_P = (-y_{12}^{(1)})^{-1}$$

But, by Table 5.1,

$$y_{12}^{(1)} = \frac{y_{12}^{(0)}}{y_{11}^{(0)}} = \frac{G_{TP}}{G_{TT}} = -\frac{\alpha_P VT}{C_P}$$

and $(\partial S/\partial V)_P = C_P/\alpha_P VT$.

For derivative (4), using the same basis function, we need to use a double transform; thus, $y^{(2)} = \underline{U} = f(\underline{S}, \underline{V}, N_1, \ldots)$ with

$y^{(2)}$	
x	ξ
1. $-\underline{S}$	$-T$
2. $\underline{V}$	$-P$
3. N_1	μ_1, etc.

and $(\partial \underline{S}/\partial P)_{\underline{V},N} = N(\partial S/\partial P)_V = N(y_{12}^{(2)})^{-1}$, since $P = -y_2^{(2)}$. From Table 5.2,

$$y_{12}^{(2)} = \frac{y_{12}^{(0)}}{y_{11}^{(0)}y_{22}^{(0)} - (y_{12}^{(0)})^2} = \frac{G_{TP}}{G_{TT}G_{PP} - G_{TP}^2}$$

and

$$\left(\frac{\partial S}{\partial P}\right)_V = \frac{G_{TT}G_{PP} - G_{TP}^2}{G_{TP}} = \frac{C_P\kappa_T}{T\alpha_P} - V\alpha_P$$

Although the procedure described above is always valid and unequivocal, there are shortcut methods which, when applicable, are less tedious. Many involve the use of the chain rule,

$$\left(\frac{\partial b}{\partial c}\right)_d = \left(\frac{\partial b}{\partial e}\right)_d \left(\frac{\partial e}{\partial c}\right)_d \tag{7-40}$$

Equation (7-40) is most useful when d is a property of the desired representation (e.g., T or P for G), but b and c are not; then e can be chosen as the second independent property in the desired representation.

Example 7.2

Evaluate the following partial derivatives as functions of P, V, T, their partial derivatives, and C_p: (a) $(\partial S/\partial P)_G$; (b) $(\partial A/\partial G)_T$.

Solution

(*a*) Using Eqs. (7-37) and (7-39), we find that

$$\left(\frac{\partial S}{\partial P}\right)_G = -\frac{(\partial G/\partial P)_S}{(\partial G/\partial S)_P} = \frac{-S(\partial T/\partial P)_S + V}{S(\partial T/\partial S)_P} \tag{7-41}$$

Using Eq. (7-39) to eliminate $(\partial T/\partial P)_S$,

$$\left(\frac{\partial T}{\partial P}\right)_S = -\frac{(\partial S/\partial P)_T}{(\partial S/\partial T)_P} = \frac{(\partial V/\partial T)_P}{C_p/T} \tag{7-42}$$

so that

$$\left(\frac{\partial S}{\partial P}\right)_G = \frac{VC_p}{TS} - \left(\frac{\partial V}{\partial T}\right)_P \tag{7-43}$$

The entropy, S, in Eq. (7-43) can be expressed as a function of the desired variables, as demonstrated in Section 7.1 [see Eq. (7-23)].

(*b*) This partial derivative can be reduced by the following shortcut procedure

$$\left(\frac{\partial A}{\partial G}\right)_T = \frac{(\partial A/\partial V)_T}{(\partial G/\partial V)_T} = -\frac{P}{V(\partial P/\partial V)_T} \tag{7-44}$$

in which Eq. (7-40) was employed. The insertion of V was a convenient but not an arbitrary choice. Although P could have been used in place of V with equal simplicity, S would have been less convenient. The choice was guided by the fact that $A = f(T, V)$ and $G = f(T, P)$; since T was the constant in the differentiation, either T, V or T, P would be a convenient set of independent variables.

Example 7.3

A well-insulated vessel is divided into two compartments by a partition. The volume of each compartment is 0.1 m³. One compartment initially contains 400 moles of argon at 294 K, and the other compartment is initially evacuated. The partition is then removed, and the gas is allowed to equilibrate. What is the final temperature?

Note: Under these conditions, argon is not an ideal gas. Assume that the van der Waals equation of state is valid:

$$\left(P + \frac{a}{V^2}\right)(V - b) = RT \tag{7-45}$$

where $a = 0.1362$ J m³/mol² and $b = 3.215 \times 10^{-5}$ m³/mol. Assume that $C_v = 12.56$ J/mol K, independent of temperature or pressure. Assume that the mass of the walls can be neglected.

Solution

If we choose the gas as the system, the process occurs at constant energy. Furthermore, we know the initial and final volume of the system. Since we are concerned with a one-component simple system, we can evaluate the change in any property if the changes of two other properties are known. Let us disregard the actual path between the initial and final conditions because it was clearly irreversible. If the process were carried out reversibly, we can relate the temperature variation to the volume and energy by

$$dT = \left(\frac{\partial T}{\partial V}\right)_U dV + \left(\frac{\partial T}{\partial U}\right)_V dU \tag{7-46}$$

Equation (7-46) is, of course, an exact differential. If we choose a reversible path in which U is constant, then it can be integrated:

$$T_f - T_i = \int_{V_i}^{V_f} \left(\frac{\partial T}{\partial V}\right)_U dV \tag{7-47}$$

The problem can be solved by expressing $(\partial T/\partial V)_U$ in terms of the available data. Thus,

$$\left(\frac{\partial T}{\partial V}\right)_U = -\frac{(\partial U/\partial V)_T}{(\partial U/\partial T)_V} = \frac{-T(\partial S/\partial V)_T + P}{T(\partial S/\partial T)_V} \tag{7-48}$$

We can evaluate $(\partial S/\partial V)_T$ in terms of P–V–T derivatives by applying the reciprocity relation to the first Legendre transform of $\underline{U}$ [i.e., $y^{(1)} = \underline{A}$] and

$$dy^{(1)} = d\underline{A} \tag{7-49}$$

and

$$\left(\frac{\partial S}{\partial V}\right)_T = \left(\frac{\partial P}{\partial T}\right)_V \tag{7-50}$$

so that Eq. (7-48) becomes

$$\left(\frac{\partial T}{\partial V}\right)_U = \frac{-T(\partial P/\partial T)_V + P}{C_v} \tag{7-51}$$

From the van der Waals equation, we find that

$$\left(\frac{\partial P}{\partial T}\right)_V = \frac{R}{V - b}$$

so that Eq. (7-51) reduces to

$$\left(\frac{\partial T}{\partial V}\right)_U = -\frac{a}{C_v V^2} \tag{7-52}$$

Substituting Eq. (7-52) into Eq. (7-47), we obtain

$$T_f = T_i + \frac{a}{C_v}\left(\frac{1}{V_f} - \frac{1}{V_i}\right) \tag{7-53}$$

or

$$T_f = T_i - \frac{a}{2C_v V_i} \tag{7-54}$$

$$= 294 - \frac{0.1362}{(2)(12.56)(0.1/400)} = 294 - 22$$

$$= 272 \text{ K}$$

7.3 P–$\underline{V}$–T–N Equation of State

As noted in Section 7.1, a knowledge of the P–$\underline{V}$–T–N behavior of a pure material allows one to obtain values of the isothermal compressibility κ_T, and the coefficient of thermal expansion α_p. With these parameters, and C_p, one may then calculate all derivatives containing three of the four variables S, T, V, and P.

The P–$\underline{V}$–T–N behavior of a material is often characterized by the compressibility factor Z, where

$$Z = \frac{P\underline{V}}{NRT} \tag{7-55}$$

For an ideal gas, $Z = 1.0$. For liquids, Z is a small number; for real gases Z can be either larger or smaller than unity. At the critical point, Z_c is normally in the range 0.27 to 0.29, but values can be less for highly polar materials.

The theory of nonideal gases is well developed on a mathematical basis, but to provide an analytical P–$\underline{V}$–T–N relation, some assertion must be made about the intermolecular energies between the molecules comprising the system. At the present time, only approximations have been possible to delineate such energies. Theory does, however, lead to a useful concept known as the *Theorem of Corresponding States*. This theorem suggests that values of thermodynamic properties of different materials can be compared when the property is divided by its value at the critical point. Applied to Eq. (7-55), this would suggest that

$$Z = f(Z_c, P_r, T_r) \tag{7-56}$$

where P_r is the reduced pressure, P/P_c, and T_r is the reduced temperature, T/T_c. The $f(\ )$ is then universal. If Z_c is now assumed to be approximately the same for all materials, then $Z = f(P_r, T_r)$. Generalized compressibility plots showing Z as a function of T_r and P_r are common in most textbooks in thermodynamics. The accuracy of such correlations is found to be greatly improved when Z_c is also employed as a third correlating variable.[5]

An even more accurate form of Eq. (7-56) was given by Pitzer et al.,[6] who substituted a different parameter, the acentric factor, for Z_c. In their relation

$$Z = Z^{(0)}(T_r, P_r) + \omega Z^{(1)}(T_r, P_r) \tag{7-57}$$

The acentric factor, ω, was developed as a measure of the difference in structure between the material of interest and a spherically symmetric gas such as argon (where ω is essentially zero). ω is defined as

$$\omega = -\log_{10} P_{vp_r} - 1 \tag{7-58}$$

[5]O. Hougen, K. M. Watson, and R. A. Ragatz, *Chemical Process Principles*, Pt. II (New York: Wiley, 1959).

[6]K. S. Pitzer, D. Z. Lippmann, R. F. Curl, C. M. Huggins, and D. E. Peterson, *J. Am. Chem. Soc.*, **77**, 3433 (1955).

where P_{vp_r} is the reduced vapor pressure at $T_r = 0.7$. Values of ω have been tabulated for many materials.[7]

The $Z^{(0)}$ and $Z^{(1)}$ functions in Eq. (7-57) depend only on T_r and P_r. The original tables of these functions given by Pitzer et al. were prepared from an analysis of experimental data. Most subsequent papers that presented analytical forms of a P–$\underline{V}$–T–N equation of state were, in fact, developed to match the Pitzer et al. tables and not experimental data!

Although the Hougen et al. or Pitzer et al. compressibility factor tabulations are very useful, they have essentially been replaced by analytical equations of state which are far more convenient for machine computation. Almost all such equations are explicit in pressure and are of the form

$$P = f(T, \underline{V}, N) \tag{7-59}$$

The famous van der Waals equation introduced in Example 7.3 is of this type.

As there are a great many different equations of state illustrated generally by Eq. (7-59), each with its own adherents, it is not possible to comment even briefly on all the more common. We have, therefore, selected three that are illustrative of the forms now being used.

The first given is the well-known *Redlich–Kwong*[8] *relation*, as it is the precursor of many later versions. Written in intensive form,[9] it is

$$P = \frac{RT}{V - b'} - \frac{a'}{T^{0.5}V(V + b')} \tag{7-60}$$

where a' and b' are constants specific for a given material. These parameters may be generalized by applying the stability criteria [Eq. (9-26)] at the critical point to yield

$$a' = \frac{\Omega_a R^2 T_c^{2.5}}{P_c} \tag{7-61}$$

$$b' = \frac{\Omega_b R T_c}{P_c} \tag{7-62}$$

where

$$\Omega_a = [(9)(2^{1/3} - 1)]^{-1} = 0.42748\ldots \tag{7-63}$$

$$\Omega_b = \frac{2^{1/3} - 1}{3} = 0.08664\ldots \tag{7-64}$$

A more accurate form of Eq. (7-60) that can be applied to both the liquid and vapor regions was given by Peng and Robinson,[10]

$$P = \frac{RT}{V - b} - \frac{a(\omega, T_r)}{V(V + b) + b(V - b)} \tag{7-65}$$

[7]See, for example, C. A. Passut and R. P. Danner, *Ind. Eng. Chem. Process Des. Dev.*, **12**, 365 (1973), or R. C. Reid, J. M. Prausnitz, and T. K. Sherwood, *Properties of Gases and Liquids*, 3rd ed. (New York: McGraw-Hill, 1977).

[8]O. Redlich and J. N. S. Kwong, *Chem. Rev.*, **44**, 233 (1949).

[9]To convert Eq. (7-60) to an extensive form, substitute $V = \underline{V}/N$ in the equation.

[10]D. -Y. Peng and D. B. Robinson, *Ind. Eng. Chem. Fundam.*, **15**, 59 (1976).

The parameters a and b in Eq. (7-65) are not the same as the a' and b' in Eq. (7-60). Also, in Eq. (7-65), the a parameter depends on both the acentric factor and reduced temperature.[11]

Again using Eq. (9-26) at the critical point, Peng and Robinson have shown that

$$a(\omega, T_r) = a(T_c)\alpha(\omega, T_r) \tag{7-66}$$

with

$$a(T_c) = \frac{0.45724R^2T_c^2}{P_c} \tag{7-67}$$

$$\alpha(\omega, T_r) = [1 + \kappa(1 - T_r^{1/2})]^2 \tag{7-68}$$

$$\kappa = 0.37464 + 1.54226\omega - 0.26992\omega^2 \tag{7-69}$$

and

$$b = \frac{0.07780RT_c}{P_c} \tag{7-70}$$

The Peng–Robinson relation is used as an example of an equation of state to calculate thermodynamic functions for pure materials (see Section 7.4), and also, in Chapters 8 and 10 to treat mixtures.

The last P–$\underline{V}$–T–N relation to be considered is the virial form,

$$P = \frac{RT}{\underline{V}} + \frac{BRT}{\underline{V}^2} + \frac{CRT}{\underline{V}^3} + \cdots \tag{7-71}$$

Equation (7-71) may be derived from statistical mechanics, and terms $B, C, \ldots$ are termed the second, third, . . . virial coefficients. If one possessed a valid theory to relate intermolecular energies to molecular separation distance, then $B, C, \ldots$ could be calculated. Lacking this, virial coefficients are normally correlated by some form of the Theorem of Corresponding States noted earlier.

The virial equation of state is normally truncated and simplified to

$$Z = \frac{P\underline{V}}{RT} = 1 + \frac{BP}{RT} \tag{7-72}$$

or

$$P = \frac{RT}{\underline{V} - B} \tag{7-73}$$

Equation (7-72) is obtained from Eq. (7-71) if $C, D, \ldots$ are set equal to zero, and Eq. (7-55) is employed to eliminate volume. After simplification, all terms in P^n where $n > 1$ are also dropped.

Equation (7-72) is quite easy to use, but it is limited to gases and its range of applicability is such that the reduced density should not exceed 0.5. Various techniques have been suggested to calculate B. The method developed by Abbott[12] is described here; it is similar to the form shown in Eq. (7-57).

[11] See footnote 9 to write Eq. (7-65) in extensive form.

[12] J. M. Smith and H. C. Van Ness, *Introduction to Chemical Engineering Thermodynamics*, 3rd ed. (New York: McGraw-Hill, 1975), p. 87.

$$B = \frac{RT_c}{P_c}(B^{(0)} + \omega B^{(1)}) \tag{7-74}$$

$$B^{(0)} = 0.083 - \frac{0.422}{T_r^{1.6}} \tag{7-75}$$

$$B^{(1)} = 0.139 - \frac{0.172}{T_r^{4.2}} \tag{7-76}$$

The three P–$\underline{V}$–T–N equations presented in this section should be considered only as illustrative examples. Many others could have been described, for example, the very well known Benedict–Webb–Rubin (B-W-R) equation that is widely used in the petroleum industry. It is again emphasized that such equations are explicit in pressure with V and T (or $\underline{V}, T, N$) as the independent variables. While such equations are of some value to determine say volumetric properties at a given pressure and temperature, their greatest use is to calculate isothermal changes in thermodynamic properties such as enthalpy or entropy—or for equilibrium phase studies. The application to thermodynamic properties of a pure material is discussed in the next section. The treatment of mixtures and phase equilibria is presented in Chapters 8 and 10.

7.4 Thermodynamic Properties

In this section we are concerned with developing relationships to determine changes in the thermodynamic properties of a pure substance. Initially we will limit ourselves to isothermal variations and then, later, expand the treatment to include temperature changes.

For isothermal variations, we are interested in so-called *departure functions.* As defined here a departure function is the difference between the property in the real state (as specified by T and V) and in an ideal gas state at the same temperature and at a volume V° which is determined from the real pressure (P) and temperature (T) by

$$V^\circ = \frac{RT}{P} \tag{7-77}$$

Thus, if the symbol B represents any thermodynamic property such as enthalpy, entropy, internal energy, Helmholtz energy, etc., the departure function is $B(V, T) - B^\circ(V^\circ, T)$.

The departure functions can be determined from a P–$\underline{V}$–T–N equation of state—or a P–V–T equation if the basis is 1 mole or a unit mass. Since, as shown in Section 7.3, P–V–T relations have V and T as independent variables, this suggests that the Helmholtz energy be used, since

$$A = U - TS = H - TS - PV$$
$$= f(V, T)$$

and

$$dA = -S\,dT - P\,dV \qquad \text{(constant } N\text{)} \tag{7-78}$$

At constant temperature,

$$dA = -P\,dV$$

Integrating between $A(T, V)$ and $A(T, \infty)$, we obtain

$$A(T, V) - A(T, \infty) = -\int_{\infty}^{V} P\,dV \tag{7-79}$$

When $V \longrightarrow \infty$, the material is assumed to behave as an ideal gas. Also, if we *assume* an ideal gas behavior from $V \longrightarrow \infty$ to $V = V^{\circ}$, then

$$A(T, \infty) - A^{\circ}(T, V^{\circ}) = -\int_{V^{\circ}}^{\infty} P\,dV \Big|_{\text{ideal gas}} \tag{7-80}$$

We add Eqs. (7-79) and (7-80) and, at the same time, add and subtract

$$\int_{\infty}^{V} \frac{RT}{V}\,dV$$

to the right-hand side. The final result, when $P_{\text{ideal gas}} = RT/V$, is

$$A(T, V) - A^{\circ}(T, V^{\circ}) = -\int_{\infty}^{V} \left(P - \frac{RT}{V}\right) dV + RT \ln \frac{V^{\circ}}{V} \tag{7-81}$$

In Eq. (7-81) the temperature is constant. Also, by substituting a pressure-explicit equation of state in the integral, the Helmholtz energy departure function is obtained (see Example 7.4). Also, from Eq. (7-78),

$$\left(\frac{\partial A}{\partial T}\right)_V = -S$$

So Eq. (7-81) may be differentiated with respect to temperature[13] at constant volume to obtain the entropy departure function,

$$S(T, V) - S^{\circ}(T, V^{\circ}) = \frac{\partial}{\partial T} \int_{\infty}^{V} \left(P - \frac{RT}{V}\right) dV - R \ln \frac{V^{\circ}}{V} \tag{7-82}$$

With the Helmholtz energy and entropy departure functions, others can be obtained by simple algebra; for example,

$$U(T, V) - U^{\circ}(T, V^{\circ}) = (A - A^{\circ}) + T(S - S^{\circ}) \tag{7-83}$$

$$H(T, V) - H^{\circ}(T, V^{\circ}) = (U - U^{\circ}) + PV - RT \tag{7-84}$$

$$G(T, V) - G^{\circ}(T, V^{\circ}) = (H - H^{\circ}) - T(S - S^{\circ}) \tag{7-85}$$

To reiterate, the departure functions shown in Eqs. (7-81) through (7-85) yield the difference in the pure-component property in the real state (V, T) from the value it would have in an ideal-gas state at the same temperature, T, and at a volume V° which is given by RT/P, with P the system pressure.

[13]When Eq. (7-81) was differentiated with respect to T with V constant and $N = 1$ mole, one should note that $(\partial A^{\circ}/\partial T)_V = -S^{\circ} - P(\partial V^{\circ}/\partial T)_V$ since $dA^{\circ} = -S^{\circ}\,dT - P\,dV^{\circ}$. However, in differentiating the right-hand side of Eq. (7-81) one also obtains the term $RT(\partial \ln V^{\circ}/\partial T)_V = (RT/V^{\circ})(\partial V^{\circ}/\partial T)_V = P(\partial V^{\circ}/\partial T)_V$, so the derivatives of V° with respect to T at constant V cancel and do not appear in Eq. (7-82).

With these departure equations, it is quite straightforward to obtain explicit equations once a P–V–T equation of state is chosen. This approach is illustrated in Example 7.4.

Example 7.4

Determine the departure functions shown in Eqs. (7-81) through (7-85) using the Peng–Robinson equation of state [Eq. (7-65)].

Solution

Inserting Eq. (7-65) for P in Eq. (7-81), we obtain

$$A(T, V) - A^\circ(T, V^\circ) = -\int_\infty^V \left(\frac{RT}{V-b} - \frac{RT}{V} - \frac{a}{V^2 + 2Vb - b^2}\right) dV + RT \ln \frac{V^\circ}{V} \tag{7-86}$$

$$= RT \ln \frac{V^\circ}{V-b} + \frac{a}{2\sqrt{2}b} \ln \frac{V + b(1-\sqrt{2})}{V + b(1+\sqrt{2})} \tag{7-87}$$

Then,

$$\begin{aligned} S(T, V) - S^\circ(T, V^\circ) &= -\frac{\partial}{\partial T}(A - A^\circ) \\ &= -R \ln\left(\frac{V^\circ}{V-b}\right) - \frac{1}{2\sqrt{2}b} \ln\left[\frac{V + b(1-\sqrt{2})}{V + b(1+\sqrt{2})}\right]\left(\frac{\partial a}{\partial T}\right)_V \end{aligned} \tag{7-88}$$

The parameter a is given in Eqs. (7-66) through (7-68). With these,

$$\left(\frac{\partial a}{\partial T}\right)_V = \frac{-a\kappa}{(\alpha T T_c)^{1/2}} \tag{7-89}$$

In Eq. (7-89), a is determined from Eq. (7-66), α from Eq. (7-68), and κ from Eq. (7-69). Then,

$$S(T, V) - S^\circ(T, V^\circ) = -R \ln \frac{V^\circ}{V-b} + \frac{a\kappa}{(\alpha T T_c)^{1/2} 2\sqrt{2}b} \ln \frac{V + b(1-\sqrt{2})}{V + b(1+\sqrt{2})} \tag{7-90}$$

The other departure functions are obtained by simple algebra; for example,

$$\begin{aligned} H(T, V) - H(T, V^\circ) &= (A - A^\circ) + T(S - S^\circ) + (PV - RT) \\ &= \left\{\frac{a}{2\sqrt{2}b} \ln\left[\frac{V + b(1-\sqrt{2})}{V + b(1+\sqrt{2})}\right]\right\}\left[1 + \frac{\kappa T}{(\alpha T T_c)^{1/2}}\right] + (PV - RT) \end{aligned} \tag{7-91}$$

In Eq. (7-91), the pressure in the last term would be found from the original equation of state [Eq. (7-65)] at the given V and T.

To determine the difference in a thermodynamic property between two states 1 and 2 that differ in temperature, we first form the isothermal departure functions and then vary temperature in the ideal-gas state. That is,

$$\begin{aligned} B(T_2, V_2) - B(T_1, V_1) &= [B(T_2, V_2) - B^\circ(T_2, V_2^\circ)] \\ &\quad - [B(T_1, V_1) - B^\circ(T_1, V_1^\circ)] + [B^\circ(T_2, V_2^\circ) - B^\circ(T_1, V_1^\circ)] \end{aligned} \tag{7-92}$$

The first two brackets on the right-hand side of the identity Eq. (7-92) represent the departure functions at T_2 and T_1, respectively. They are calculated as described earlier in this section. The third bracket represents the difference in property B between two *ideal-gas* states (T_2, V_2°) and (T_1, V_1°). The advantage of varying temperature in an ideal-gas state is that heat capacities of ideal gases can then be used and, as shown in Section 7.5, values of C_p° or C_v° are known for many pure materials or, if not, they can often be estimated from molecular structure.

For example, if B were the specific internal energy U, then

$$U^\circ(T_2, V_2^\circ) - U^\circ(T_1, V_1^\circ) = \int_{T_1}^{T_2} \left(\frac{\partial U^\circ}{\partial T}\right)_V dT = \int_{T_1}^{T_2} C_v^\circ \, dT \tag{7-93}$$

and no effect of the volume difference need be considered because, in an ideal-gas state, U° is a function of temperature only.

We also note that

$$B^\circ(T, V^\circ) = B^\circ(T, P) \tag{7-94}$$

where

$$P = \frac{RT}{V^\circ} \tag{7-95}$$

so that for the case where B is the enthalpy H,

$$\begin{aligned} H^\circ(T_2, V_2^\circ) - H^\circ(T_1, V_1^\circ) &= H^\circ(T_2, P_2) - H^\circ(T_1, P_1) \\ &= \int_{T_1}^{T_2} \left(\frac{\partial H^\circ}{\partial T}\right)_P dT = \int_{T_1}^{T_2} C_p^\circ \, dT \end{aligned} \tag{7-96}$$

As with U°, no volume (or pressure) term appears since H° is again only a function of temperature.[14]

For B as the entropy,

$$S^\circ(T_2, V_2^\circ) - S^\circ(T_1, V_1^\circ) = \int_{T_1}^{T_2} \left(\frac{\partial S^\circ}{\partial T}\right)_V dT + \int_{V_1^\circ}^{V_2^\circ} \left(\frac{\partial S^\circ}{\partial V}\right)_T dV \tag{7-97}$$

Since $(\partial S/\partial V)_T = (\partial P/\partial T)_V$, then, for an ideal gas, $(\partial S/\partial V)_T = R/V$. Thus Eq. (7-97) becomes

$$S^\circ(T_2, V_2^\circ) - S^\circ(T_1, V_1^\circ) = \int_{T_1}^{T_2} \frac{C_V^\circ}{T} \, dT + R \ln \frac{V_2^\circ}{V_1^\circ} \tag{7-98}$$

By using Eq. (7-95) and the fact that $C_p^\circ = C_v^\circ + R$ (see footnote 14), then an alternative form for Eq. (7-98) is

$$\begin{aligned} S^\circ(T_2, V_2^\circ) - S^\circ(T_1, V_1^\circ) &= S^\circ(T_2, P_2) - S^\circ(T_1, P_1) \\ &= \int_{T_1}^{T_2} \frac{C_p^\circ}{T} \, dT - R \ln \frac{P_2}{P_1} \end{aligned} \tag{7-99}$$

[14]Note that this would also imply that, in an ideal-gas state, $(\partial H^\circ/\partial T)_p = (\partial H^\circ/\partial T)_v = C_p^\circ$. Since $H^\circ = U^\circ + PV^\circ = U^\circ + RT$, then $C_p^\circ = C_v^\circ + R$.

If B were the Gibbs or the Helmholtz energy, departure functions can be readily calculated if a P–V–T relation is available. However, it is not possible to determine changes in these properties if the temperature is changed since derivatives with respect to temperature yield terms containing the absolute entropy, S. As the absolute value of this property is not known, nonisothermal variations in A or G (or, in fact, any Legendre transform where $\underset{\sim}{S}$ has been transformed to T) are without physical meaning. Also, if isothermal changes in G or A are desired, it is preferable to avoid departure functions, for it is more convenient to utilize the Fundamental Equation directly. For example, for the Gibbs energy, from Eq. (5-41), if $N =$ constant, then

$$\left(\frac{\partial G}{\partial P}\right)_{T,N} = V$$

and

$$G(T, P_2) - G(T, P_1) = \int_{P_1}^{P_2} V\,dP \tag{7-100}$$

In a similar manner,

$$A(T_1, V_2) - A(T_1, V_1) = -\int_{V_1}^{V_2} P\,dV \tag{7-101}$$

The difference between heat capacities in a real state at T, V and an ideal-gas state at T, V° can also be determined from an equation of state. For example, if the entropy departure function, Eq. (7-82) is differentiated with respect to temperature, at constant V, then with

$$\left(\frac{\partial S}{\partial T}\right)_V = \frac{C_v}{T}$$

one obtains

$$C_v(T, V) - C_v^\circ(T, V^\circ) = T\int_\infty^V \left(\frac{\partial^2 P}{\partial T^2}\right)_V dV \tag{7-102}$$

(See footnote 13 for the reason why there are no derivatives of V° with respect to T.)

It is less convenient to obtain $C_p - C_p^\circ$ from the departure functions developed above since

$$\left(\frac{\partial S}{\partial T}\right)_P = \frac{C_p}{T}$$

and a derivative at constant pressure is required, whereas the principal variables used in the present treatment use T and V. It is possible to show that

$$C_p - C_p^\circ = -T\int_0^P \left(\frac{\partial^2 V}{\partial T^2}\right)_P dP \tag{7-103}$$

which is similar in form to Eq. (7-102). However, since equations of state are explicit in pressure, the second derivative in the integral of Eq. (7-103) is not easily determined. It is probably more convenient to determine $C_v(T, V)$ by Eq.

(7-102) and then use the result of Problem 5.12, which states that

$$C_p = C_v + \frac{TV\alpha_p^2}{\kappa_T} \tag{7-104}$$

to relate C_p to C_v in the real state. α_p and κ_T are defined by Eqs. (7-13) and (7-14). After rearrangement,

$$C_p = C_v - \frac{T(\partial P/\partial T)_V^2}{(\partial P/\partial V)_T} \tag{7-105}$$

The derivatives in Eq. (7-105) are readily found from a pressure-explicit equation of state. Note that for an ideal gas, Eq. (7-105) reduces to

$$C_p^\circ - C_v^\circ = R \tag{7-106}$$

Thus, combining the results of Eqs. (7-102), (7-105), and (7-106) yields

$$C_p - C_p^\circ = T\int_\infty^V \left(\frac{\partial^2 P}{\partial T^2}\right)_V dV - \frac{T(\partial P/\partial T)_V^2}{(\partial P/\partial V)_T} - R \tag{7-107}$$

7.5 Ideal-Gas Heat Capacities

In Section 7.4, to calculate the change in thermodynamic properties U, H, and S between states that varied in temperature, departure functions were first employed to reach the ideal-gas state, and then temperature changes were effected in that state. Also, as pointed out in the same section, this procedure was developed so that *ideal-gas* heat capacities could be employed. The two most common are

$$C_v^\circ = \left(\frac{\partial U^\circ}{\partial T}\right)_V = T\left(\frac{\partial S^\circ}{\partial T}\right)_V \tag{7-108}$$

$$C_p^\circ = \left(\frac{\partial H^\circ}{\partial T}\right)_P = T\left(\frac{\partial S^\circ}{\partial T}\right)_P \tag{7-109}$$

where the $^\circ$ superscript on U, H, and S indicates these properties are to be evaluated in an ideal-gas state. Thus, in such a state,

$$H^\circ = U^\circ + PV^\circ = U^\circ + RT$$

and then

$$\frac{dH^\circ}{dT} = \frac{dU^\circ}{dT} + R = C_p^\circ = C_v^\circ + R \tag{7-110}$$

so that the restrictions of constant volume on $(\partial U^\circ/\partial T)$ or constant pressure on $(\partial H^\circ/\partial T)$ are not needed. This is true since U° and H° are functions only of temperature. The restriction of constant V or P on entropy in Eqs. (7-108) and (7-109) is, however, still necessary since the ideal-gas entropy is a function of temperature and volume (or pressure). Another interpretation to Eq. (7-110) is that neither C_p° nor C_v° is dependent on volume or pressure, but only on temperature. This assertion may also be proved in a formal sense, as for example, with C_p°,

$$\left(\frac{\partial C_p}{\partial P}\right)_T = \frac{\partial^2 H^\circ}{\partial P\,\partial T} = \frac{\partial}{\partial T}\left(\frac{\partial H^\circ}{\partial P}\right)_T$$

$$= \frac{\partial}{\partial T}\left[V^\circ - T\frac{\partial V^\circ}{\partial T}\right]_p$$

$$= \frac{\partial}{\partial T}\left[\frac{RT}{P} - T\frac{\partial}{\partial T}\left(\frac{RT}{P}\right)\right]_p$$

$$= \frac{\partial}{\partial T}(0) = 0$$

where $(\partial H/\partial P)_T$ was related to V and T by methods shown in Chapter 5 (see Problem 5.4b).

While C_v° and C_p° refer to an ideal-gas state at (T, V°), since they are only dependent on temperature, they could be expressed as C_v^* and C_p^* indicating a *real* state at T but as $P \longrightarrow 0$ or $V \longrightarrow \infty$. Although the physical interpretation of these states is quite different, numerical values of C_v° and C_v^* or C_p° and C_p^* are identical.

In most instances, experimental values of C_v or C_p have been obtained at low pressures ($P \leq 1$ bar) and, for engineering purposes, can be equated to C_v^* or C_p^*. However, most heat-capacity values available in the literature are not experimentally determined but have been obtained from theory and spectroscopic data. The essence of the method lies in expressing the internal energy of an ideal gas as a function of temperature; then, by differentiation, C_v^* is obtained. In this method, let us consider briefly the types of energies involved.

First, we rule out intermolecular energy since, by choosing an ideal gas, there are no interactions between molecules. Next, energies associated with electron motion, or intranuclear movement, or even energy associated with mass by relativistic considerations may be discarded since these are either constants or temperature dependent only at very high temperatures. For materials used normally in chemical engineering practice, none contribute to C_v^*.

Molecules do, however, move and have kinetic energy; in addition, these energies increase with temperature. By simple kinetic arguments it is easy to show that in each of the three possible directions of motion there is an energy of $mv^2/2$ or $kT/2$. k is Boltzmann's constant or R divided by Avogadro's number. This translational energy thus contributes $3(kT/2)$ per molecule or $3RT/2$ per mole. The translational contribution to C_v^* is then $3R/2$. Molecules having no other energy storage modes would have C_v^* values of about 12.5 J/mol K since R is 8.31 J/mol K. Such is the case for monatomic gases such as He, Ar, and Ne at low pressures. C_p^* is then about $12.5 + 8.3 = 20.8$ J/mol K.

More complex molecules can also store energy in other ways. Most are obvious from a mechanical visualization. Rotation of the molecule may occur both with the entire molecule and, to some degree, by rotation of certain segments in relation to one another. Individual atoms may also vibrate in relation to adjacent atoms. As the temperature increases, more energy can be stored in rotational and vibrational modes. To compute quantitatively the exact relation between energy and temperature is, however, not always a simple matter. Quantum mechanics enters the picture particularly with respect to vibrational energies, since vibrational energy quanta turn out to be large. Nevertheless, with sufficient experimental spectroscopic data on the characteristic vibrational levels, as well as the moments of inertia, the barriers to internal rotation, and any rotational–vibrational interactions, it is possible to calculate accurately C_v^*.

Engineers, however, have taken the C_v^* values calculated from theory and have developed simple, approximate atomic or group-additive techniques to allow rapid estimations of heat capacities as a function of temperature. Several methods to estimate C_p^* are given in Reid et al.[15]

PROBLEMS

7.1. A skeptical engineer questions whether the efficiency of a Carnot engine is always given by

$$\eta = \frac{-W_E}{Q_H} = \frac{T_H - T_C}{T_H}$$

where W_E = work of Carnot engine
Q_H = heat interaction with hot reservoir
T_H, T_C = temperature of hot (cold) reservoir

She agrees that the relation shown above is correct for a Carnot engine with an ideal gas as the working fluid but indicates that it may be in error if the working fluid is not an ideal gas. She suspects that the efficiency decreases as the working fluid becomes more nonideal.

How would you clarify the situation?

7.2. Our research laboratory has synthesized a new material and the properties of this substance are being studied under conditions where it is always a vapor. Two sets of experiments have been carried out and they are described below. Given the results of these experiments, the relationship $P\underline{V} = NCT$ is proposed to describe the P–$\underline{V}$–T properties of the material (C is a constant). If you agree with this proposal, show a rigorous proof. If you do not agree, either prove that the relationship cannot be applicable or describe clearly what additional experiments you would recommend, and demonstrate how you would use these data to show whether or not $P\underline{V} = NCT$.

Experiment A: Vapor is contained in subsystem I of a rigid, well-insulated container (Figure P7.2). Subsystem II is initially evacuated. The partition between I and II is broken and gas fills both I and II. Over a wide range of initial temperatures, pressures, and volumes of I, the final temperature, after expansion, equals the initial temperature.

[15] Reid et al., *op. cit.*, Chap. 7.

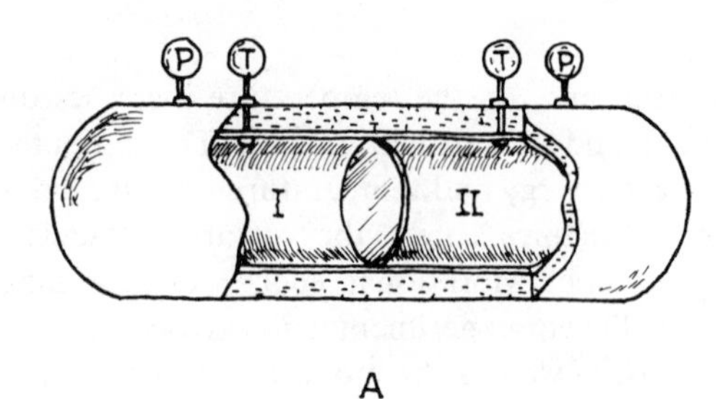

A

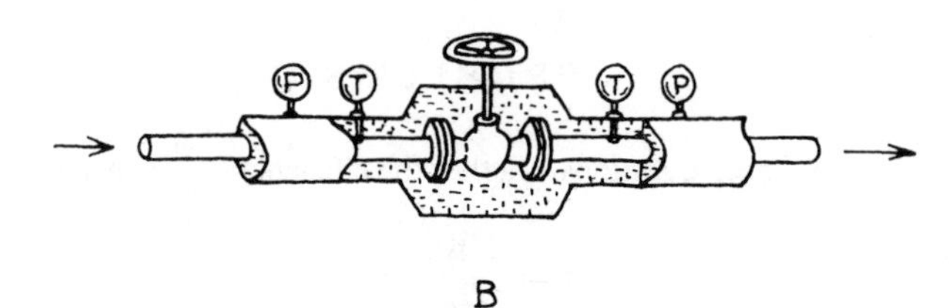

B

Figure P7.2

Experiment B: Vapor flows in an insulated pipe and through an insulated throttling valve wherein the pressure is reduced. Over a wide range of upstream temperatures and pressures, as well as downstream pressures, the temperature of the vapor does not change when the gas flows through the valve.

7.3. In Chapter 5 the concept of *equations of state* was introduced for the Fundamental Equation in energy representation. If we were, however, to use the Helmholtz energy representation of the Fundamental Equation, the three equations of state for a pure material would be

$$\underline{S} = f_1(T, \underline{V}, N) \tag{A}$$

$$P = f_2(T, \underline{V}, N) \tag{B}$$

$$\mu = f_3(T, \underline{V}, N) \tag{C}$$

Suppose that Eq. (B) were to be given by the extensive form of the Redlich–Kwong equation,

$$P = \frac{NRT}{\underline{V} - bN} - \frac{aN^2}{T^{1/2}\underline{V}(\underline{V} + bN)}$$

and the ideal-gas heat capacity were expressed as

$$C_v^* = A + BT + CT^2$$

(a) Derive an analytical expression for Eq. (A) assuming a reference state of an ideal gas at $T°$, $V°$.

(b) Since only two of the three equations of state are independent, derive Eq. (C) using the Redlich–Kwong equation and the result obtained for Eq. (A).

7.4. We have a constant-volume, closed vessel filled with vapor dichlorodifluoromethane. We plan to heat this vapor and would like to know how the specific entropy of the vapor varies with pressure. Derive a general relation to allow one to calculate the desired derivative, $(\partial S/\partial P)_V$ assuming that we know the total volume of the vessel, the moles of vapor, C_p as a function of T and P, and an equation of state.

Illustrate your result at the start of the heating process, where $T = 365.8$ K, $P = 16.5$ bar, and total volume $= 1.51 \times 10^{-3}$ m³. Assume that at this condition, $C_p = 94.9$ J/mol K and the Redlich–Kwong equation of state is applicable with

$$a = 20.839 \text{ J m}^3 \text{ K}^{1/2}/\text{mol}^2$$

$$b = 6.725 \times 10^{-5} \text{ m}^3/\text{mol}$$

7.5. Gas streams are often cooled either by expansion through a throttling valve or in a power turbine. To measure the relative efficiency of these two processes, the term μ_H/μ_S is often quoted. The throttling (or Joule–Thompson) coefficient is defined as $\mu_H = (\partial T/\partial P)_H$, whereas the isentropic coefficient is $\mu_S = (\partial T/\partial P)_S$.

Calculate (μ_H/μ_S) for CO_2 at 319.4 K over a range of reduced pressures from 0.1 to 20. What conclusions can you draw? Also determine the compressibility factor over the same pressure range at 319.4 K. Use the Peng–Robinson equation of state in your calculations. For CO_2, $T_c = 304.2$ K, $P_c = 73.76$ bar, and $\omega = 0.225$.

7.6. Sulfur dioxide vapor at 520 K and 100 bar fills one-half of a rigid, adiabatic cylinder. The other half is evacuated, and the two halves are separated by a metal diaphragm. If this should rupture, what would be the final temperature and pressure? Assume that the vapor is well mixed and that expansion is sufficiently rapid so that negligible heat transfer occurs between the walls and the SO_2 vapor.

For SO_2, $T_c = 430.8$ K, $P_c = 78.8$ bar, $V_c = 1.22 \times 10^{-4}$ m³/mol, $\omega =$ 0.251, and C_p^* (J/mol K) is given by

$$C_p^* = 23.852 + 6.699 \times 10^{-2}T - 4.961 \times 10^{-5}T^2 + 1.328 \times 10^{-8}T^3$$

with T in kelvins.

7.7. Saturated steam with a quality of one fills a rigid vessel at 472.1 K. It is desired to reduce the pressure and temperature of the steam simultaneously but to maintain always a saturated vapor. The cylinder is fitted with a piston that has negligible friction to allow for changes in volume. Heat transfer is allowed. Data for saturated steam are given below.

T (K)	P (bar)	V (m³/kg)	H (kJ/kg)	S (kJ/kg K)
469.3	14.321	0.1378	2789.3	6.4581
472.1	15.188	0.1300	2791.0	6.4372
474.9	16.097	0.1230	2792.6	6.4163

What is the magnitude of this heat interaction at the start of the pressure reduction?

7.8. Estimate the value of C_v for propylene vapor at 125°C at a density of 92.28 kg/m³ (48.28 bar). Perform the calculations using the experimental P–V–T data given below. Also, use the Peng–Robinson equation of state to carry out the estimation. Compare your results with the value of 1857 J/kg K reported by N. de Nevers and J.J. Martin [*AIChE J.*, **6**, 43 (1960)].

Data:

1. C_p^*, ideal gas: The isobaric heat capacity of propylene in the ideal-gas state is given by

$$C_p^* = 88.16 + 5.574T - 2.757 \times 10^{-3}T^2 + 5.239 \times 10^{-7}T^3 \quad \text{(J/kg K)}$$

2. P–V–T data [*Physica*, **19**, 287 (1953)]: The data are presented in terms of amagat units as a function of pressure and temperature. An amagat unit is the ratio of the true density to the density at 1 atm pressure and 273.15 K, which for propylene is 4.544×10^{-5} g-mol/cm³.

	Pressure (bar)		
Density (amagat)	100°C	125°C	150°C
1.000	1.3977	1.4932	1.5887
6.4976	8.6045	9.2626	9.9163
8.3409	10.8438	11.7045	12.5582
10.1832	12.9956	14.0657	15.1263
12.1074	15.1544	16.4500	17.7329
14.3997	17.6096	19.1839	20.7397
16.6494	19.8993	21.7574	23.5909
18.3557	21.5589	23.6388	25.6901
19.5100	22.6444	24.8775	27.0779
23.0078	25.7640	28.4762	31.1408
27.6638	29.5207	32.9048	36.2209
32.1907	32.7702	36.8431	40.8254
36.9366	35.7822	40.6151	45.3266
41.6597	38.4067	44.0276	49.4979
42.4108	38.7928	44.5451	50.1329
46.1992	40.6108	47.0201	53.2493
47.3154	41.1106	47.7175	54.1293
50.9531	42.6191	49.8806	56.9300

3. Properties for use in the Peng–Robinson equation of state:

$$T_c = 365.0 \text{ K}$$

$$P_c = 46.20 \text{ bar}$$

$$\omega = 0.148$$

$$M = 42.081$$

7.9. We wish to design a device to make dry ice (solid CO_2) which could be attached to the outlet tube of a cylinder of CO_2. Assume that the CO_2 in the cylinder consists of a liquid and gas in equilibrium at 300 K (67.01 bar), but that the outlet is connected only to the liquid phase by means of an eductor tube (Figure P7.9).

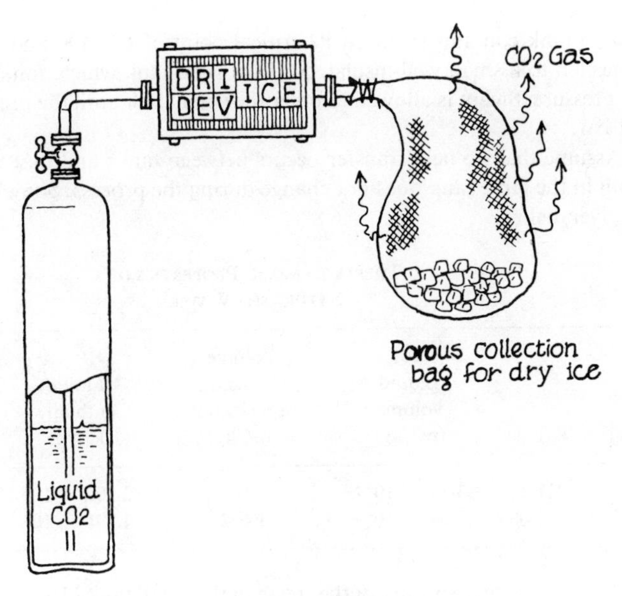

Figure P7.9

The enthalpy of vaporization of CO_2 at 300 K (67.01 bar) is 103.7 J/g and the enthalpy of sublimation at 1 bar is 570 J/g. $T_c = 304.2$ K, $P_c = 73.825$ bar, and the acentric factor is 0.225. The temperature of the solid–vapor mixture of CO_2 at 1 bar is 194.7 K. The heat capacity of CO_2 in the ideal-gas state may be approximated by

$$C_p \text{ (J/mol K)} = 26.62 + 0.0363T \text{ (K)}$$

between 194 and 300 K.

Suggest what you think would be a good engineering way to design such a device. How much dry ice at 1 bar do you think you could make per kilogram of liquid CO_2 drawn from the cylinder?

7.10. An isothermal compressor that operates essentially reversibly takes suction from a manifold containing CO_2 at 1.01 bar and 300 K. Gas is compressed and fed to a well-insulated 8.50-m³ storage tank originally containing CO_2 also at 1.01 bar and 300 K.

Assume that the CO_2 P–V–T properties may be correlated with the Redlich–Kwong equation of state with

$$a = 6.4648 \text{ J m}^3\text{ K}^{1/2}/\text{mol}^2$$
$$b = 2.9704 \times 10^{-5} \text{ m}^3/\text{mol}$$

The heat capacity of CO_2 in the ideal-gas state may be expressed as

$$C_p^* = 19.795 + 7.3436 \times 10^{-2}T - 5.6019 \times 10^{-5}T^2 + 1.7153 \times 10^{-8}T^3$$

with T in kelvins, and C_p^* is in J/mol K.

(a) What is the final temperature of the gas in the tank when the pressure reaches 48.3 bar?

(b) Plot the tank temperature as a function of tank pressure; on the same plot show the cumulative compressor work required.

(c) Repeat part (b) assuming that CO_2 is an ideal gas with a constant $C_p \sim$ 36.8 J/mol K.

7.11. A large tank contains steam at its critical point of 647.4 K and 221.2 bar. It is connected to a small well-insulated 0.1-m^3 rigid tank which, initially, has a very low pressure. Steam is allowed to enter the small tank until the pressure increases to 1 bar.

Assume that no heat transfer occurs between tanks and that the state of the steam in the large tank does not change during the process. Some data for steam are given below.

THERMODYNAMIC PROPERTIES OF SATURATED WATER

T (K)	P (bar)	Liquid volume (m^3/kg)	Volume change in vaporization (m^3/kg)	Liquid enthalpy (kJ/kg)	Enthalpy change in vaporization (kJ/kg)
647.4	221.2	3.17×10^{-3}	0	2.107×10^3	0
373.0	1.00	1.04×10^{-3}	1.672	4.191×10^2	2.257×10^3

(a) What is the temperature of the steam in the small tank? How many kilograms entered?

(b) Suppose that the "large" tank had contained CO_2 at 647.4 K and 221.2 bar instead of steam. What would have been the temperature in the 0.1-m^3 tank when the pressure had reached 1 bar? How many kilograms of CO_2 would have been transferred? In this case, use the Peng–Robinson equation of state to describe the properties of CO_2. For CO_2, the Peng–Robinson parameters are: $T_c = 304.2$ K, $P_c = 73.76$ bar, and $\omega = 0.225$; C_p^* as a function of temperature is given in Problem 7.10.

(c) If instead of either steam or CO_2, we had had helium gas in the large tank (again at 647.4 K and 221.2 bar) and had expanded this gas into the evacuated small tank, what would have been the temperature when the pressure attained 1 bar? How many kilograms would have been transferred? Assume helium to be an ideal gas at all temperatures and pressures; also let $C_p =$ 20.9 J/mol K = constant.

7.12. Two moles of a gas are in a closed vessel at 10 bar and 300 K. By some series of irreversible operations, the pressure increases to 20 bar and the temperature to 320 K. Three kilojoules of work is done *on* the gas. Assume that the P–V–T properties of the gas can be modeled by an equation of state of the form $P(V - b) = RT$, where $b = 5 \times 10^{-5}$ m^3/mol. $C_v = 0.1\, T$(K) at 10 bar over the temperature range of interest (J/mol K).

What is the net heat interaction of the system with the environment? What is the change of ΔH and ΔA for this change of state?

7.13. Memo to our thermodynamics Consultant:

As our thermo con-person, we need your help in developing a simple fire extinguisher. This device is nothing more than a small 30-liter vessel charged with liquid R-12 (dichlorodifluoromethane) at room temperature (assume 294. 3 K). We are now charging so that the total R-12 mass in the extinguisher initially is 38 kg. The initial pressure reads 5.853 bar, so I presume we have an equilibrium vapor–liquid mixture.

When used, liquid R-12 is discharged so that the exit port is on the bottom. I am trying to sell the unit to a client but she refuses to negotiate until we can show her some calculations as to how the pressure in the vessel will change as liquid is expelled. She has indicated to me that she believes the pressure drop will be so great, even for a small amount of liquid expelled, that the utility of the device will be quite limited (Figure P7.13). She wants me to add an inert gas to help

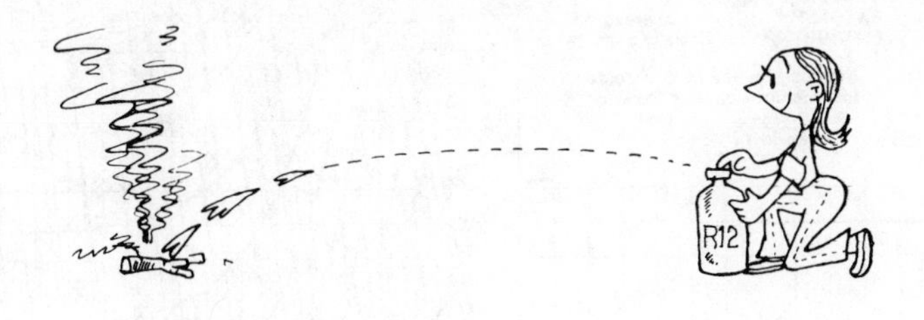

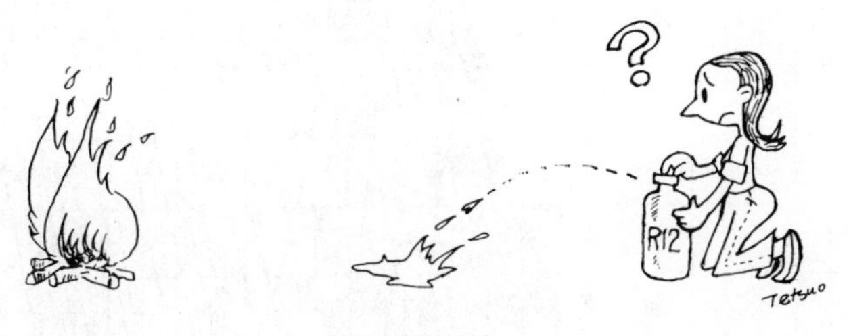

Figure P7.13

keep the pressure constant. I do not favor this idea, but, frankly, I am worried about the possible drop in pressure.

Some data for R-12 are shown below.

DATA FOR SATURATED LIQUID AND VAPOR DICHLORODIFLUOROMETHANE

T (K)	P (bar)	V^L (m^3/kg $\times 10^4$)	V^V (m^3/kg $\times 10^2$)	H^L (kJ/kg)	H^V (kJ/kg)	S^L (kJ/kg K)	S^V (kJ/kg K)
288.8	4.994	7.437	3.486	50.628	194.009	0.193	0.690
289.9	5.158	7.458	3.378	51.686	194.456	0.197	0.690
291.0	5.325	7.480	3.274	52.744	194.900	0.201	0.689
292.1	5.497	7.502	3.174	53.807	195.342	0.204	0.689
293.2	5.673	7.524	3.078	54.873	195.782	0.208	0.688
294.3	5.853	7.547	2.985	55.940	196.219	0.211	0.688
295.4	6.037	7.570	2.895	57.013	196.870	0.215	0.688

Please write me a concise memorandum detailing how you think the pressure and temperature vary as liquid is removed. Do we have to use an inert gas?

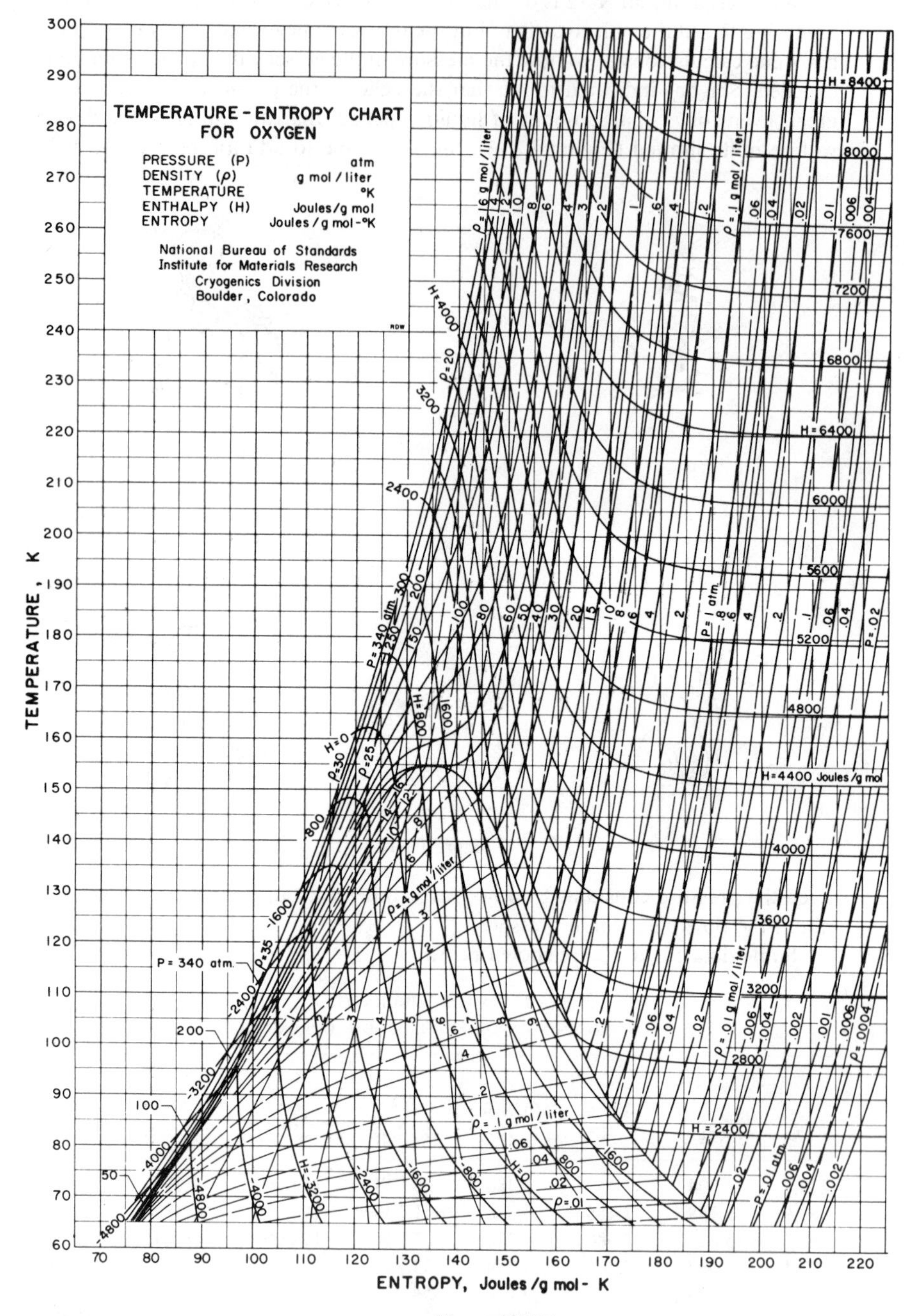
TEMPERATURE - ENTROPY CHART
FOR OXYGEN
PRESSURE (P) atm
DENSITY (ρ) g mol / liter
TEMPERATURE °K
ENTHALPY (H) Joules/g mol
ENTROPY Joules / g mol - °K
National Bureau of Standards
Institute for Materials Research
Cryogenics Division
Boulder, Colorado
TEMPERATURE, K
ENTROPY, Joules /g mol - K
H = 8400
H = 6400
H = 4400 Joules /g mol
H = 2400
P = 340 atm.
ρ = 1 g mol / liter
ρ = .1 g mol / liter
ρ = 4 g mol / liter
ρ = 16 g mol / liter
P = 1 atm
P = .01 atm

Figure P7.15

7.14. A 1-m³ tank is filled with supercritical CO_2 at 300 bar and 340 K. It is to be connected to a tank of identical size that contains, initially, CO_2 at 1 bar and 290 K. Both tanks are rigid and adiabatic. Assume no heat interactions between the tanks via the connecting line and neglect heat transfer to the tank walls. The valve in the interconnecting line is opened. Assume that the Peng–Robinson equation of state is applicable to CO_2. Critical data and C_p^* as a function of temperature are given in Problems 7.10 and 7.11.

What is the temperature and pressure in the "low-pressure" tank when the pressure in the "high-pressure" tank has dropped to 150 bar? What fraction of the CO_2 in the high-pressure tank has been transferred?

7.15. Oxygen gas at 150 K and 30.4 bar is to be compressed to 101.3 bar in a turbine with an efficiency of 80% (based on an isentropic process). If the flow rate is 5000 kg/h, what would be the power required? Carry our your calculations using:

(a) The temperature–entropy diagram in Figure P7.15.

(b) The Peng–Robinson equation of state. Data in this case are: $T_c = 154.6$ K, $P_c = 50.46$ bar, and $\omega = 0.021$. Also assume that C_p^* is 29.3 J/mol K, independent of temperature.

7.16. We plan to fill carbon dioxide cylinders from a supply manifold that is at 298 K and 1 bar (see Figure P7.16). Gas will be fed to the suction side of a compressor that operates in an adiabatic manner with an efficiency of 85%. Gas is discharged to a small heat exchanger and then to the cylinders.

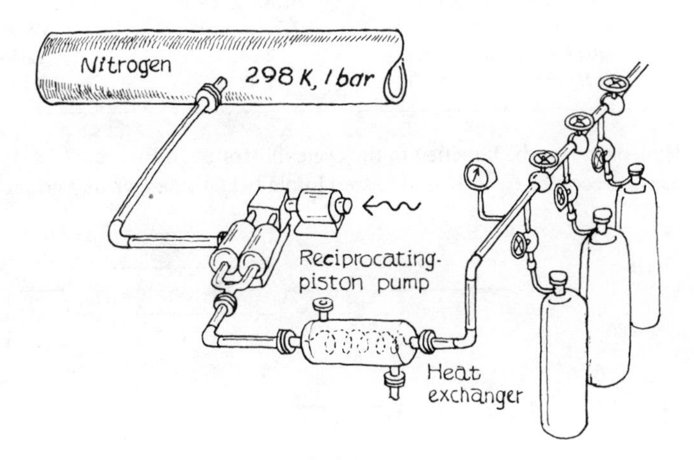

Figure P7.16

Initially, each cylinder contains CO_2 at 2 bar. The final pressure, after filling, is 140 bar. It may be assumed that the CO_2 leaves the heat exchanger at 298 K and, due to the high heat capacity of the bottle walls, the gas in the cylinder remains essentially at 298 K during filling. Each cylinder has a volume of 0.025 m³ (25 liters).

Determine the minimum power rating of the motor if we want to fill one cylinder in 3 min. Also, what is the total work required for a filling? Use the Peng–Robinson equation of state (see Problems 7.10 and 7.11 for data). The vapor pressure of liquid CO_2 at 298 K is 64.09 bar.

7.17. Differential scanning calorimeters (DSC) are now being employed to measure liquid heat capacities. Generally, liquid is placed inside a small ampule that has a locked cap to prevent losses. The energy added to effect a small temperature rise of the ampule and contents may be accurately ascertained with a DSC. In a separate experiment, the energy required to heat only the empty ampule (and cap) is also determined. The liquid and associated vapor are assumed to be in equilibrium at all times and no inert gas is present.

Prepare a rigorous analysis to indicate how one should carry out experiments to attain the highest accuracy in the measurement of the saturated liquid heat capacity, $(\partial H/\partial T)_{sat}$. In particular, comment on the effect of the initial volume fraction of vapor at the start of the experiment. Demonstrate your analysis for the case of isobutane at (a) 261.5 and (b) 316.5 K. Some property data are given below.

SAMPLE CASE: ISOBUTANE

	(a)	(b)
T (K)	261.5	316.5
P (bar)	1.103	5.738
V^V(sat vap) (m^3/kg)	3.549×10^{-1}	6.761×10^{-2}
V^L(sat liq) (m^3/kg)	1.685×10^{-3}	1.897×10^{-3}
ΔH_v (kJ/kg)	3.643×10^{2}	3.026×10^{2}
dV^L/dT (m^3/kg K)	3.146×10^{-6}	4.944×10^{-6}
dV^V/dT (m^3/kg K)	-1.236×10^{-2}	-1.708×10^{-3}
$(\partial H^L/\partial T)_{sat}$ (kJ/kg K)	2.345	2.544
$(\partial H^V/\partial T)_{sat}$ (kJ/kg K)	1.424	1.298

7.18. Hydrogen is to be liquefied in the cycle illustrated in Figure P7.18. High-pressure gas is precooled to 40 K and passed into a heat exchanger that employs saturated

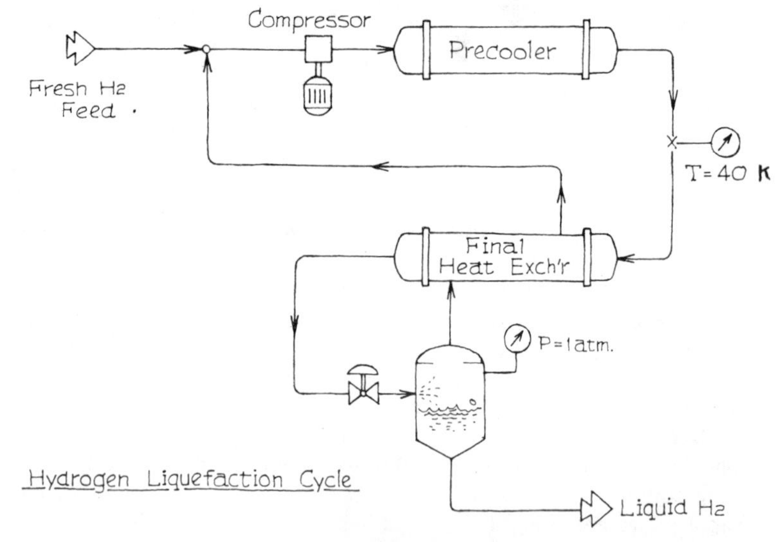

Figure P7.18

hydrogen recycle gas as the coolant. The high-pressure cold vapor is then expanded to 1 atm across an insulated expansion valve. The liquid fraction is separated and the vapor recycled as shown. The heat transfer area of this final exchanger is very large.

(a) At what pressure should the compressor be operated to maximize the amount of liquid formed per kg hydrogen flow?
(b) Sketch the state of hydrogen during the process on both T–S and H–P diagrams.
(c) Determine the minimum work that would be required if the process were carried out reversibly.

7.19.

Phone : 7-11711
Cable : Eureak

Mr. James Longthorne
c/o Faroutlake, Maine
Dear Jim:

Please forgive this intrusion upon your well-deserved vacation, but a problem has arisen in the office that demands immediate attention and everyone in the design section is very busy except you and me. I am, therefore, sending this problem to you in the hope that you can handle it as soon as possible.

A client has requested us to make rough calculations on his proposed process to liquefy monochlorodifluoromethane ($CHClF_2$) as per the attached flow sheet. The process gas enters at (1) at 1 atm (1.013 bar) and 20°C and is compressed adiabatically to (2). The compressor operates at about 95% of a reversible, adiabatic device and a similar efficiency is believed applicable for the electric drive motor. The hot, high-pressure gas at (2) is cooled to 400 K by heat transfer in exchanger A and in so doing makes saturated steam at 400 K. (The feed to A is water, liquid, saturated at 400 K.)

From (3) the gas is further cooled in exchanger B (which is essentially infinite in area), then passes through a Joule–Thompson valve and into a liquid–vapor separator at 1 atm. Liquid is removed and saturated vapor is used as the cooling medium in exchanger B. The exit vapor at (B) is cooled and recycled, but do not worry about this part of the process.

Now, the customer demands (although we have our doubts) that there be a maximum amount of liquefaction per pass.

Please submit a flow sheet showing me your recommendations for:

(a) All line sizes
(b) Motor power for the compressor
(c) Steam generated (kg/day)
(d) Liquid flow rate
(e) All pressures and temperatures in the process steams

The flow rate at (1) is 4.54 kg/min and the customary flow velocities of 60 m/s (gas), 15 m/s (gas–liquid), and 3 m/s (liquid) may be assumed.

Also attached are all the data on $CHClF_2$ that the client has available. Sorry it is not more; I have taken the liberty to ship you the text R. C. Reid, J. M. Prausnitz, and T. K. Sherwood, *Properties of Gases and Liquids*, 3rd ed., McGraw-Hill, New York, 1977, with this letter. I'm sure you can estimate the desired properties.

Hope to hear from you soon. Have a pleasant vacation.

With regards,

Edward Cel

Manager, Design Section

EC/jh

Attachment 1

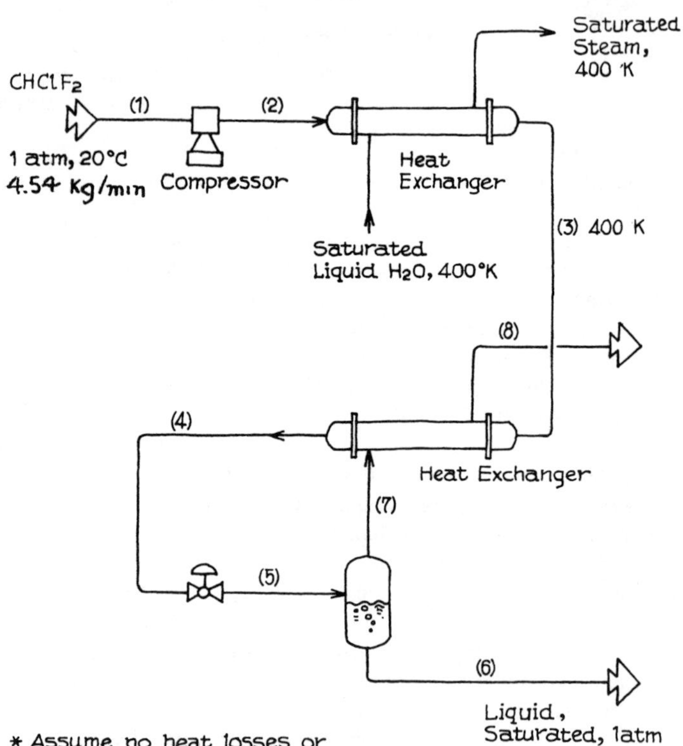

Attachment 2

Client Data on $CHClF_2$

Molecular weight = 86.48

Boiling point at 1 atm = 232.4 K

Melting point = 113 K

Flammability: nonflammable

Color: clear, water white

Odor: ethereal

Toxicity: Group 5-A Classification of Underwriters' Laboratory Report MH-3134

7.20. While making a conceptual process design for a plant to be built in the near future, you are requested to analyze a small section of the overall design. In the section of concern to you, pure ethylidine chloride (1,1-dichloroethane) vapor is obtained from a still at about 363 K and 2.55 bar (see Figure P7.20). This vapor

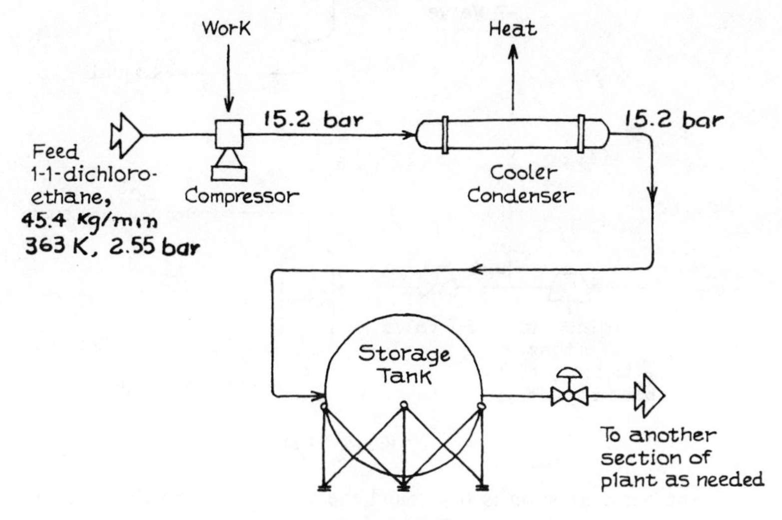

Figure P7.20

is to be compressed to 15.2 bar and then cooled and condensed to a saturated liquid. This hot liquid is to be stored as required for use later in the overall process. The estimated maximum flow rate is about 45.4 kg/min.

The compressor design has not yet been settled, but it is reasonable to assume that it operates adiabatically but with an efficiency of about 90% of theoretical. The pump drive unit is an electrical motor with an efficiency near 95%. The cooler–condenser is air-cooled. The storage tank is well insulated and must be capable of storing a 12-h flow of ethylidine chloride with a 10% ullage at the end of 12 h. A schematic flow sheet is given in Figure P7.20.

(a) What size motor do you recommend?
(b) What is the heat load in the cooler-condenser, in watts?
(c) What size storage tank is needed?
(d) What are the temperatures and specific volumes out of the compresser and cooler–condenser?
(e) If flow velocities are to be held near 7 m/s in the lines, what size piping do you recommend for all the connecting lines?

The *Handbook of Chemistry and Physics* lists a boiling point and liquid density at 20°C, and you may use these if desired. Use any correlation methods you desire.

7.21. We are faced with handling a stream of pure carbon monoxide at 300 atm (304 bar) and 197 K. We would like to liquefy as great a fraction as possible, but there seems to be some disagreement about the process to be used. One suggestion has been to expand this high-pressure fluid across a Joule–Thompson valve and take what liquid is formed [see Figure P7.21(a)].

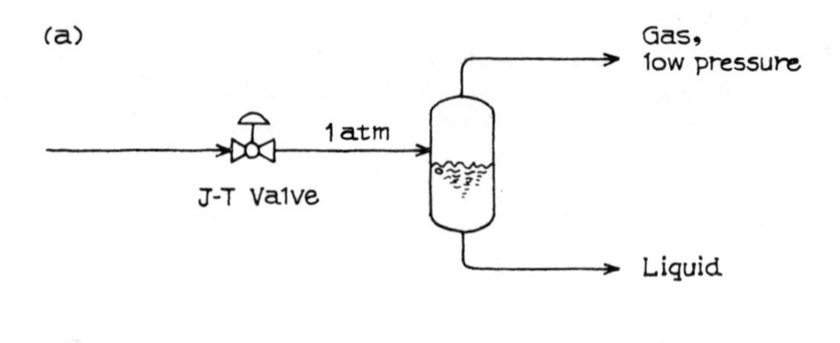

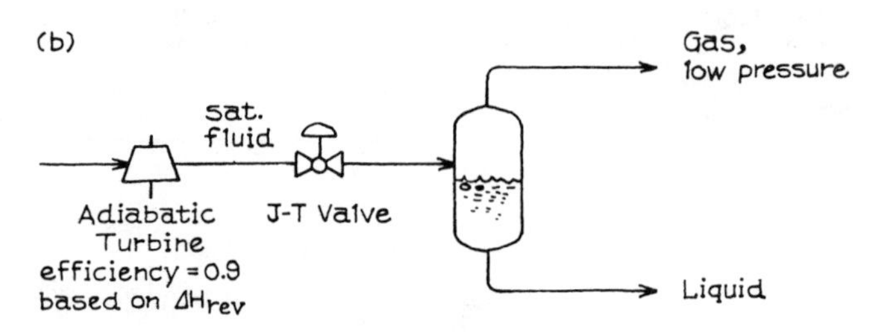

Figure P7.21

Another suggestion is to expand the vapor in an adiabatic turbine to the saturation curve and then follow with a Joule–Thompson expansion [see Figure P7.21(b)]. The two-step operation has been suggested to avoid erosion of the turbine blades with a two-phase mixture.

Evaluate these methods in order to indicate the fraction of initial gas one might be expected to liquefy in each. Also, suggest a better way to carry out the process to liquefy carbon monoxide that will give a larger fraction liquefied; draw a flow sheet, and calculate the fraction liquefied.

We wish we could supply you with thermodynamic property data but, unfortunately, we do not have access to any at the present time. Use any estimation methods you can find.

8 Single-Phase, Simple Systems of Mixtures

In Chapter 7 we developed thermodynamic relationships for pure materials. In this chapter we extend the treatment to cover single-phase, multicomponent simple mixtures. Multiphase systems are covered in Chapter 10, and reacting systems in Chapter 11. The mixtures may involve solid solutions (provided that they are in a stable equilibrium state), liquids, and gases. If dissociation of any component occurs (e.g., in ionization of salts in aqueous solutions), the degree of dissociation is assumed to be known.

To avoid repetitive derivations, we use a generalized property $\underline{B}$ (extensive) or B (intensive). Relationships for $\underline{B}$ and B are then applicable for properties such as internal energy, volume, entropy, or any Legendre transform of the energy (or entropy).

Two *model* mixtures are introduced: the *ideal-gas mixture* and the *ideal solution*. Deviations of real mixtures from these models lead to the use of fugacity, activity, and activity coefficient functions as measures of the *non-ideality* of the mixture.

In spite of the fact that there are few new concepts or principles, many relationships may appear unduly complex because the notation is often formidable. Yet, in the development of equations to describe mixtures, one must resort to detailed subscripting to denote the component involved or the conditions of restraint for partial differentiation.

8.1 Extensive and Intensive Thermodynamic Properties of Mixtures

Any extensive variable, $\underline{B}$, can be expanded in differential form as a function of $(n + 2)$ other variables and, for single-phase, simple systems there are no restrictions on the choice of variables. In Table 8.1, columns (b) and (c), we

TABLE 8.1
GENERAL EXTENSIVE PROPERTY, $\underline{B}$

(a) Pure material, Independent variables Y_1, Y_2, N	(b) Mixture, Independent variables $Y_1, Y_2, N_1, \ldots, N_n$	(c) Mixture, Independent variables $Y_1, Y_2, x_1, \ldots, x_{n-1}, N$
$\underline{B} = f(Y_1, Y_2, N)$ (8-1a)	$\underline{B} = f(Y_1, Y_2, N_1, \ldots, N_n)$ (8-1b)	$\underline{B} = f(Y_1, Y_2, x_1, \ldots, x_{n-1}, N)$ (8-1c)
$d\underline{B} = \left(\frac{\partial \underline{B}}{\partial Y_1}\right)_{Y_2,N} dY_1 + \left(\frac{\partial \underline{B}}{\partial Y_2}\right)_{Y_1,N} dY_2 + \left(\frac{\partial \underline{B}}{\partial N}\right)_{Y_1,Y_2} dN$ (8-2a)	$d\underline{B} = \left(\frac{\partial \underline{B}}{\partial Y_1}\right)_{Y_2,N_i} dY_1 + \left(\frac{\partial \underline{B}}{\partial Y_2}\right)_{Y_1,N_i} dY_2 + \sum_{i=1}^{n} \left(\frac{\partial \underline{B}}{\partial N_i}\right)_{Y_1,Y_2,N_j[i]} dN_i$ (8-2b)	$d\underline{B} = \left(\frac{\partial \underline{B}}{\partial Y_1}\right)_{Y_2,x,N} dY_1 + \left(\frac{\partial \underline{B}}{\partial Y_2}\right)_{Y_1,x,N} dY_2 + \sum_{i=1}^{n-1} \left(\frac{\partial \underline{B}}{\partial x_i}\right)_{Y_1,Y_2,x[i,n],N} dx_i + \left(\frac{\partial \underline{B}}{\partial N}\right)_{Y_1,Y_2,x} dN$ (8-2c)
	Integration by Euler's theorem. Assuming Y_1 and Y_2 are extensive properties	
$\underline{B} = \left(\frac{\partial \underline{B}}{\partial Y_1}\right)_{Y_2,N} Y_1 + \left(\frac{\partial \underline{B}}{\partial Y_2}\right)_{Y_1,N} Y_2 + \left(\frac{\partial \underline{B}}{\partial N}\right)_{Y_1,Y_2} N$ (8-3a)	$\underline{B} = \left(\frac{\partial \underline{B}}{\partial Y_1}\right)_{Y_2,N_i} Y_1 + \left(\frac{\partial \underline{B}}{\partial Y_2}\right)_{Y_1,N_i} Y_2 + \sum_{i=1}^{n} \left(\frac{\partial \underline{B}}{\partial N_i}\right)_{Y_1,Y_2,N_j[i]} N_i$ (8-3b)	$\underline{B} = \left(\frac{\partial \underline{B}}{\partial Y_1}\right)_{Y_2,x,N} Y_1 + \left(\frac{\partial \underline{B}}{\partial Y_2}\right)_{Y_1,x,N} Y_2 + \left(\frac{\partial \underline{B}}{\partial N}\right)_{Y_1,Y_2,x} N$ (8-3c)
	Simplifications if Y_1 and Y_2 are intensive properties	
$\underline{B} = \left(\frac{\partial \underline{B}}{\partial N}\right)_{Y_1,Y_2} N$ (8-4a)	$\underline{B} = \sum_{i=1}^{n} \left(\frac{\partial \underline{B}}{\partial N_i}\right)_{Y_1,Y_2,N_j[i]} N_i$ (8-4b)	$\underline{B} = \left(\frac{\partial \underline{B}}{\partial N}\right)_{Y_1,Y_2,x} N$ (8-4c)
$\left(\frac{\partial \underline{B}}{\partial N}\right)_{Y_1,Y_2} = \frac{\underline{B}}{N} = B$ (8-5a)	$\left(\frac{\partial \underline{B}}{\partial N_i}\right)_{Y_1,Y_2,N_j[i]} \neq \frac{\underline{B}}{N_i}$ (8-5b)	$\left(\frac{\partial \underline{B}}{\partial N}\right)_{Y_1,Y_2,x} = \frac{\underline{B}}{N} = B$ (8-5c)
	With $Y_1 = T$ and $Y_2 = P$	
$\underline{B} = \left(\frac{\partial \underline{B}}{\partial N}\right)_{T,P} N = BN$ (8-6a)	$\underline{B} = \sum_{i=1}^{n} \left(\frac{\partial \underline{B}}{\partial N_i}\right)_{T,P,N_j[i]} N_i = \sum_{i=1}^{n} \bar{B}_i N_i$ (8-6b)	$\underline{B} = \left(\frac{\partial \underline{B}}{\partial N}\right)_{T,P,x} N = BN$ (8-6c)
Eq. (8-2a) becomes $d\underline{B} = \left(\frac{\partial \underline{B}}{\partial T}\right)_{P,N} dT + \left(\frac{\partial \underline{B}}{\partial P}\right)_{T,N} dP + B\,dN$ (8-7a)	Eq. (8-2b) becomes $d\underline{B} = \left(\frac{\partial \underline{B}}{\partial T}\right)_{P,N_i} dT + \left(\frac{\partial \underline{B}}{\partial P}\right)_{T,N_i} dP + \sum_{i=1}^{n} \bar{B}_i\,dN_i$ (8-7b)	Eq. (8-2c) becomes $d\underline{B} = \left(\frac{\partial \underline{B}}{\partial T}\right)_{P,x,N} dT + \left(\frac{\partial \underline{B}}{\partial P}\right)_{T,x,N} dP + \sum_{i=1}^{n-1} \left(\frac{\partial \underline{B}}{\partial x_i}\right)_{T,P,x[i,n],N} dx_i + B\,dN$ (8-7c)

illustrate two variable sets that are commonly employed for mixtures. For reference, the corresponding relations for pure materials are shown in column (a).

In column (b), the independent variables are n mole numbers plus two other properties. In column (c), we have $(n - 1)$ mole fractions, the total number of moles, and two other properties. In this latter set, we have eliminated one mole fraction (i.e., x_n) so that any of the other $(n - 1)$ mole fractions can be varied independently. Thus, when $\underline{B}$ is expressed as a partial derivative with respect to x_i, x_n must vary so that $dx_n = -dx_i$.

Equation (8-2a-c) represents chain-rule expansions of $\underline{B}$ in terms of the independent variables. One should note that the subscript N_i in column (b) denotes that *all* mole numbers are held constant. In column (c) the subscript N indicates that the total moles are constant. Of course, the combined x and N subscripts in column (c) are equivalent to the N_i subscript in column (b). Also, subscripts such as $N_j[i]$ denote that all N_j are held constant except N_i. Similarly, the subscript $x[i, n]$ means that all x_j are constant except x_i and x_n.

Equation (8-2a-c) may be integrated by using Euler's theorem as shown in Appendix C. If Y_1 and Y_2 are extensive, Eq. (8-3a-c) is obtained; if they are intensive, the equations simplify to the form of Eq. (8-4a-c).

If Y_1 and Y_2 are intensive variables, the form of Eq. (8-4b) indicates that the extensive property $\underline{B}$ may be expressed simply as a weighted average of the partial derivatives $(\partial \underline{B}/\partial N_i)_{Y_1, Y_2, N_j[i]}$. In such cases it has been found that T and P form a convenient set of Y_1 and Y_2, and we define the derivative as a *partial molar property*, $\bar{B}_i$:

$$\bar{B}_i \equiv \left(\frac{\partial \underline{B}}{\partial N_i}\right)_{T, P, N_j[i]} \tag{8-8}$$

Note that partial molar properties are intensive and depend on the temperature, pressure, and composition of the system. The corresponding property for a pure material is, of course, the specific property, B.

With the definition of $\bar{B}_i$, Eqs. (8-2b) and (8-4b) are transformed to Eqs. (8-7b) and (8-6b), as shown in Table 8.1. Although not shown, many of the derivatives in column (c) can also be rewritten to employ partial molar properties. For example, let us express the derivative $(\partial \underline{B}/\partial x_i)_{T, P, x[i, n], N}$ of Eq. (8-7c) in terms of partial molar quantities. Starting with Eq. (8-7b), divide all differentials by dx_j and impose upon these derivatives the restraints that T, P, $x[j, n]$, N be constant. The first two terms on the right-hand side vanish; thus,

$$\left(\frac{\partial \underline{B}}{\partial x_j}\right)_{T, P, x[j, n], N} = \sum_{i=1}^{n} \bar{B}_i \left(\frac{\partial N_i}{\partial x_j}\right)_{T, P, x[j, n], N} \tag{8-9}$$

Since

$$N_i = x_i N \tag{8-10}$$

$$\left(\frac{\partial N_i}{\partial x_j}\right)_{T, P, x[j, n], N} = \begin{Bmatrix} 0 & i \neq j, n \\ 1 & i = j \\ -1 & i = n \end{Bmatrix} N \tag{8-11}$$

so that Eq. (8-9) becomes

$$\left(\frac{\partial \underline{B}}{\partial x_j}\right)_{T, P, x[j, n], N} = N(\bar{B}_j - \bar{B}_n) \tag{8-12}$$

Example 8.1

A ternary mixture of 50 mole% *n*-propanol, 25 mole% *n*-pentanol, and 25 mole% *n*-heptane is prepared by a two-step process in which pentanol and heptane are mixed in vat 1 and this mixture is then added to vat 2, containing the propanol. Each vat is equipped with stirrer and heating coils; the addition is slow enough so that the process can be considered isothermal at 294 K. What is the cooling load required for each vat?

Data:

Mole fractions		Partial molar enthalpies (−J/g-mol)		
Heptane	Propanol	Heptane	Propanol	Pentanol
0.00	0.00	—	—	0.0
0.00	0.25	—	67.1	6.5
0.00	0.50	—	46.6	16.0
0.00	0.75	—	18.2	54.9
0.00	1.00	—	0.0	—
0.25	0.00	1153.4	—	47.5
0.25	0.25	1155.5	167.4	53.5
0.25	0.50	1165.5	136.2	74.5
0.25	0.75	1200.8	106.4	—
0.50	0.00	864.8	—	237.1
0.50	0.25	884.1	335.7	229.2
0.50	0.50	919.2	280.7	—
0.75	0.00	361.3	—	1155.8
0.75	0.25	425.6	1203.1	—
1.00	0.00	0.0	—	—

Solution

First consider vat 1; choosing the contents as the system and employing a basis of 1 mole of product, at constant pressure

$$đQ = d\underline{H} - h_e\, dn_e$$

$$Q = \underline{H}_{\text{final}} - \underline{H}_{\text{initial}} - h_{a5}n_{a5} - h_h n_h$$

where the subscripts $a5$ and h denote *n*-pentanol and heptane, respectively. $\underline{H}_{\text{initial}}$ is zero since the vat was initially empty. From Eq. (8-6b),

$$\underline{H}_{\text{final}} = \bar{H}_{a5}n_{a5} + \bar{H}_h n_h$$

Since $n_{a5} = n_h = 0.5$ mole and the enthalpy base for the pure components is zero (i.e., $h_{a5} = h_h = 0$), then

$$-Q = (0.5)(864.8) + (0.5)(237.1) = 551 \text{ J/mol}$$

For vat 2, using the same approach,

$$\underline{H}_{\text{final}} = n_{a3}\bar{H}_{a3} + n_{a5}\bar{H}_{a5} + n_h\bar{H}_h$$

$$\underline{H}_{\text{initial}} = 0, \text{ pure } a3 \text{ (}n\text{-propanol)}$$

$$h_e = -551 \text{ J/mol (from above)}$$

$$n_e = 0.5$$

$$-Q = (0.5)(136.2) + (0.25)(74.5) + (0.25)(1165.5) - (0.5)(551)$$

$$= 103.1 \text{ J/mol}$$

The general relations for an intensive property are given in Table 8.2; columns

TABLE 8.2
GENERAL INTENSIVE PROPERTY, B

(a) Pure material Independent variables Y_1, Y_2, N	(b) Mixture Independent variables $Y_1, Y_2, N_1, \ldots, N_n$	(c) Mixture Independent variables $Y_1, Y_2, x_1, \ldots, x_{n-1}, N$
$B = f(Y_1, Y_2, N)$ (8-13a)	$B = f(Y_1, Y_2, N_1, \ldots, N_n)$ (8-13b)	$B = f(Y_1, Y_2, x_1, \ldots, x_{n-1}, N)$ (8-13c)
$dB = \left(\frac{\partial B}{\partial Y_1}\right)_{Y_2,N} dY_1 + \left(\frac{\partial B}{\partial Y_2}\right)_{Y_1,N} dY_2 + \left(\frac{\partial B}{\partial N}\right)_{Y_1,Y_2} dN$ (8-14a)	$dB = \left(\frac{\partial B}{\partial Y_1}\right)_{Y_2,N_i} dY_1 + \left(\frac{\partial B}{\partial Y_2}\right)_{Y_1,N_i} dY_2 + \sum_{i=1}^{n} \left(\frac{\partial B}{\partial N_i}\right)_{Y_1,Y_2,N_j[i]} dN_i$ (8-14b)	$dB = \left(\frac{\partial B}{\partial Y_1}\right)_{Y_2,x,N} dY_1 + \left(\frac{\partial B}{\partial Y_2}\right)_{Y_1,x,N} dY_2 + \sum_{i=1}^{n-1} \left(\frac{\partial B}{\partial x_i}\right)_{Y_1,Y_2,x[i,n],N} dx_i + \left(\frac{\partial B}{\partial N}\right)_{Y_1,Y_2,x} dN$ (8-14c)
Integration by Euler's theorem Assuming that Y_1 and Y_2 are extensive properties		
$0 = \left(\frac{\partial B}{\partial Y_1}\right)_{Y_2,N} Y_1 + \left(\frac{\partial B}{\partial Y_2}\right)_{Y_1,N} Y_2 + \left(\frac{\partial B}{\partial N}\right)_{Y_1,Y_2} N$ (8-15a)	$0 = \left(\frac{\partial B}{\partial Y_1}\right)_{Y_2,N} Y_1 + \left(\frac{\partial B}{\partial Y_2}\right)_{Y_1,N} Y_2 + \sum_{i=1}^{n} \left(\frac{\partial B}{\partial N_i}\right)_{Y_1,Y_2,N_j[i]} N_i$ (8-15b)	$0 = \left(\frac{\partial B}{\partial Y_1}\right)_{Y_2,x,N} Y_1 + \left(\frac{\partial B}{\partial Y_2}\right)_{Y_1,x,N} Y_2 + \left(\frac{\partial B}{\partial N}\right)_{Y_1,Y_2,x} N$ (8-15c)
Simplifications if Y_1 and Y_2 are intensive properties		
$0 = \left(\frac{\partial B}{\partial N}\right)_{Y_1,Y_2}$ (8-16a)	$0 = \sum_{i=1}^{n} \left(\frac{\partial B}{\partial N_i}\right)_{Y_1,Y_2,N_j[i]} N_i$ $\left[\text{but} \left(\frac{\partial B}{\partial N_i}\right)_{Y_1,Y_2,N_j[i]} \neq 0\right]$ (8-16b)	$0 = \left(\frac{\partial B}{\partial N}\right)_{Y_1,Y_2,x}$ (8-16c)

Table 8.2 (Continued)
General Intensive Property, B

(a) Pure material Independent variables Y_1, Y_2, N	(b) Mixture Independent variables $Y_1, Y_2, N_1, \ldots, N_n$	(c) Mixture Independent variables $Y_1, Y_2, x_1, \ldots, x_{n-1}, N$
With $Y_1 = T$ and $Y_2 = P$		
$0 = \left(\frac{\partial B}{\partial N}\right)_{T,P}$ (8-17a)	$0 = \sum_{i=1}^{n} \left(\frac{\partial B}{\partial N_i}\right)_{T,P,N_j[i]} N_i$ (8-17b)	$0 = \left(\frac{\partial B}{\partial N}\right)_{T,P,x}$ (8-17c)
Eq. (8-14a) becomes $dB = \left(\frac{\partial B}{\partial T}\right)_P dT + \left(\frac{\partial B}{\partial P}\right)_T dP$ (8-18a)	Eq. (8-14b) becomes $dB = \left(\frac{\partial B}{\partial T}\right)_{P,N_i} dT + \left(\frac{\partial B}{\partial P}\right)_{T,N_i} dP + \sum_{i=1}^{n} \left(\frac{\partial B}{\partial N_i}\right)_{T,P,N_j[i]} dN_i$ (8-18b)	Eq. (8-14c) becomes $dB = \left(\frac{\partial B}{\partial T}\right)_{P,x} dT + \left(\frac{\partial B}{\partial P}\right)_{T,x} dP + \sum_{i=1}^{n-1} \left(\frac{\partial B}{\partial x_i}\right)_{T,P,x[i,n]} dx_i$ (8-18c)

Note that we could have expressed B as a function of $(n + 1)$ intensive variables plus one extensive variable, e.g., $(T, P, x_1, \ldots, x_{n-1}, N)$ or $(T, P, x_1, \ldots, x_{n-1}, \underline{Y})$, where $\underline{Y}$ is extensive, In differential form, then

$$dB = \left(\frac{\partial B}{\partial T}\right)_{P,x,\underline{Y}} dT + \left(\frac{\partial B}{\partial P}\right)_{T,x,\underline{Y}} dP + \sum_{i=1}^{n-1} \left(\frac{\partial B}{\partial x_i}\right)_{T,P,x[i,n],\underline{Y}} dx_i + \left(\frac{\partial B}{\partial \underline{Y}}\right)_{T,P,x} d\underline{Y}$$

Applying Euler's theorem, $(\partial B/\partial \underline{Y})_{T,P,x} = 0$, regardless of what $\underline{Y}$ is chosen. It then follows that all derivatives such as $(\partial B/\partial T)_{P,x,\underline{Y}}$ can be written as $(\partial B/\partial T)_{P,x}$ where it is then implied that some extensive variable, $\underline{Y}$, is held constant.

(b) and (c) refer to a mixture and column (a) to a pure material. Note that when Y_1 and Y_2 are intensive variables, the sets in columns (a) and (c) involve $(n + 1)$ intensive and one extensive variable. Since any intensive variable can be expressed as a function of any other $(n + 1)$ *intensive properties*, if follows that $(\partial B/\partial N)$ in Eqs. (8-16a) and (8-16c) must vanish.

Equations (8-18b) and (8-18c) are the differential forms of the two most commonly used sets for expressing an intensive property of a mixture. They contain the terms $(\partial B/\partial N_i)_{T,P,N_j[i]}$ and $(\partial B/\partial x_i)_{T,P,x[i,n]}$, each of which can be expressed in terms of partial molar quantities. Since $(\partial B/\partial N_i)_{T,P,N_j[i]}$ already contains the partial molar set of properties $T, P, N_1, \ldots, N_n$, the transformation follows directly from the substitution of $B = \underline{B}/N$:

$$\left(\frac{\partial B}{\partial N_i}\right)_{T,P,N_j[i]} = \left(\frac{\partial \underline{B}/N}{\partial N_i}\right)_{T,P,N_j[i]} = \frac{\bar{B}_i}{N} - \frac{\underline{B}}{N^2}\left(\frac{\partial N}{\partial N_i}\right)_{T,P,N_j[i]} = \frac{1}{N}(\bar{B}_i - B) \quad (8\text{-}19)$$

In a similar manner, $(\partial B/\partial x_i)_{T,P,x[i,n]}$ can be related to $(\partial \underline{B}/\partial x_i)_{T,P,x[i,n]}$:

$$\left(\frac{\partial B}{\partial x_i}\right)_{T,P,x[i,n]} = \left(\frac{\partial \underline{B}/N}{\partial x_i}\right)_{T,P,x[i,n]} = \frac{1}{N}\left(\frac{\partial \underline{B}}{\partial x_i}\right)_{T,P,x[i,n]} = \bar{B}_i - \bar{B}_n \quad (8\text{-}20)[1]$$

With Eqs. (8-19) and (8-20), Eqs. (8-18b) and (8-18c) of Table 8.2 become

$$dB = \left(\frac{\partial B}{\partial T}\right)_{P,N_i} dT + \left(\frac{\partial B}{\partial P}\right)_{T,N_i} dP + \frac{1}{N}\sum_{i=1}^{n} \bar{B}_i\, dN_i - \frac{B}{N}\, dN \quad (8\text{-}21)$$

$$dB = \left(\frac{\partial B}{\partial T}\right)_{P,x} dT + \left(\frac{\partial B}{\partial P}\right)_{T,x} dP + \sum_{i=1}^{n-1} (\bar{B}_i - \bar{B}_n)\, dx_i \quad (8\text{-}22)$$

Expanding the sum in Eq. (8-22), we obtain

$$dB = \left(\frac{\partial B}{\partial T}\right)_{P,x} dT + \left(\frac{\partial B}{\partial P}\right)_{T,x} dP + \sum_{i=1}^{n} \bar{B}_i\, dx_i \quad (8\text{-}23)$$

Example 8.2

Relate the partial derivative $(\partial B/\partial x_i)_{T,P,x[i,k]}$ to $(\partial B/\partial x_i)_{T,P,x[i,n]}$.

Solution

From Eq. (8-20),

$$\left(\frac{\partial B}{\partial x_i}\right)_{T,P,x[i,n]} = \bar{B}_i - \bar{B}_n$$

$$\left(\frac{\partial B}{\partial x_i}\right)_{T,P,x[i,k]} = \bar{B}_i - \bar{B}_k$$

Then

$$\left(\frac{\partial B}{\partial x_i}\right)_{T,P,x[i,k]} - \left(\frac{\partial B}{\partial x_i}\right)_{T,P,x[i,n]} = \bar{B}_n - \bar{B}_k = \left(\frac{\partial B}{\partial x_n}\right)_{T,P,x[n,k]}$$

[1]Note that N is held constant during differentiation because $(\partial/\partial x_i)_{T,P,x[i,n]}$ implies that N is constant. See the note at the bottom of Table 8.2.

8.2 Partial Molar Properties

A partial molar property was defined in Eq. (8-8) and is related to a mixture property by Eq. (8-6b). It follows from the definition that if $\underline{W}$ and $\underline{Z}$ are any two extensive properties, and Y is either temperature or pressure, and if $\underline{W} = Y\underline{Z}$, then $\bar{W}_i = Y\bar{Z}_i$. Thus, the conjugate sets $T\underline{S}$ and $P\underline{V}$, when differentiated as indicated in Eq. (8-8), become $T\bar{S}_i$ and $P\bar{V}_i$. It also follows that any derivative of the form $(\partial \underline{Z}/\partial Y_1)_{Y_2,N_i}$, where Y_1 and Y_2 are T and P, when operated on by $(\partial/\partial N_i)_{T,P,N_j[i]}$, becomes $(\partial \bar{Z}_i/\partial Y_1)_{Y_2,N_i}$ because the order of differentiation is immaterial when the independent variables are consistent. For example,

$$\frac{\partial}{\partial N_i}\left[\left(\frac{\partial \underline{H}}{\partial P}\right)_{T,N_i}\right]_{T,P,N_j[i]} = \frac{\partial}{\partial P}\left[\left(\frac{\partial \underline{H}}{\partial N_i}\right)_{T,P,N_j[i]}\right]_{T,N_i} = \left(\frac{\partial \bar{H}_i}{\partial P}\right)_{T,N_i} \tag{8-24}$$

Hence, from $\underline{H} = \underline{U} + P\underline{V}$, we could immediately write

$$\bar{H}_i = \bar{U}_i + P\bar{V}_i \tag{8-25}$$

Also, from $C_p = T(\partial S/\partial T)_{P,x}$, we could define $\underline{C}_p = NC_p = T(\partial \underline{S}/\partial T)_{P,N_i}$, and it then follows that

$$\bar{C}_{p_i} = \frac{\partial}{\partial N_i}[NC_p]_{T,P,N_j[i]} = T\left(\frac{\partial \bar{S}_i}{\partial T}\right)_{P,N_i} \tag{8-26}$$

Transformations of potential functions require more effort. Let us apply the partial molar operator to

$$d\underline{G} = -\underline{S}\,dT + \underline{V}\,dP + \sum_{m=1}^{n} \mu_m\,dN_m$$

Then

$$d\bar{G}_i = -\bar{S}_i\,dT + \bar{V}_i\,dP + \sum_{m=1}^{n}\left(\frac{\partial \mu_m}{\partial N_i}\right)_{T,P,N_j[i]} dN_m + \sum_{m=1}^{n} \mu_m d\left[\left(\frac{\partial N_m}{\partial N_i}\right)_{T,P,N_j[i]}\right] \tag{8-27}$$

Since

$$\left(\frac{\partial N_m}{\partial N_i}\right)_{T,P,N_j[i]} = \begin{Bmatrix} 0 & m \neq i \\ 1 & m = i \end{Bmatrix} \tag{8-28}$$

the last term reduces to $d(1) = 0$. The third term on the right-hand side can be left unchanged or transformed to a mole fraction derivative.

Partial molar quantities can be evaluated by several methods. If the property $\underline{B}$ can be conveniently measured, directly or indirectly, $\bar{B}_i$ can be found by measuring the change in $\underline{B}$ upon addition of a small amount of component i to a mixture while holding the temperature and pressure constant. From the definition, Eq. (8-8),

$$\bar{B}_i \equiv \left(\frac{\partial \underline{B}}{\partial N_i}\right)_{T,P,N_j[i]} = \lim_{\Delta N_i \to 0}\left(\frac{\Delta \underline{B}}{\Delta N_i}\right)_{T,P,N_j[i]} \tag{8-29}$$

It is essential to keep in mind the fact that $\bar{B}_i$ is a property of the mixture and not simply a property of component i. Thus, $\bar{B}_i$ will generally vary with mixture composition. To emphasize this point, we note that $\bar{B}_i$ is an intensive mixture property and, as such, we could apply any of the equations in Table 8.2 to this property. For example, expressing $\bar{B}_i = f(T, P, x_1, \ldots, x_{n-1})$, Eq. (8-18c) becomes

$$d\bar{B}_i = \left(\frac{\partial \bar{B}_i}{\partial T}\right)_{P,x} dT + \left(\frac{\partial \bar{B}_i}{\partial P}\right)_{T,x} dP + \sum_{j=1}^{n-1} \left(\frac{\partial \bar{B}_i}{\partial x_j}\right)_{T,P,x[j,n]} dx_j \tag{8-30}$$

If experimental data or analytical expressions for $\underline{B}$ or B are available, $\bar{B}_i$ can be evaluated directly. Consider three possible cases:

(1) $\underline{B} = f(T, P, N_1, \ldots, N_n)$

(2) $B = f(T, P, N_1, \ldots, N_n)$

(3) $B = f(T, P, x_1, \ldots x_{i-1}, x_{i+1}, \ldots, x_n)$

For case (1) $\bar{B}_i$ can be evaluated by differentiation using Eq. (8-8). For case (2) we can obtain $(\partial B/\partial N_i)_{T,P,N_j[i]}$ directly from the data and use Eq. (8-19) to solve for $\bar{B}_i$:

$$\bar{B}_i = B + N\left(\frac{\partial B}{\partial N_i}\right)_{T,P,N_j[i]} \tag{8-31}$$

For case (3) we can obtain $(\partial B/\partial x_j)_{T,P,x[j,i]}$ directly from the data and then relate this partial to one of the forms above. For example, let us express $(\partial B/\partial N_i)_{T,P,N_j[i]}$ in Eq. (8-31) as a function of $(\partial B/\partial x_j)_{T,P,x[j,i]}$. We first express B as a differential in terms of $T, P, x_1, \ldots, x_{i-1}, x_{i+1}, \ldots, x_n$:

$$dB = \left(\frac{\partial B}{\partial T}\right)_{P,x} dT + \left(\frac{\partial B}{\partial P}\right)_{T,x} dP + \sum_{j \neq i} \left(\frac{\partial B}{\partial x_j}\right)_{T,P,x[j,i]} dx_j \tag{8-32}$$

Dividing by dN_i and imposing the restraint of constant T, P, and $N_j[i]$, we obtain

$$\left(\frac{\partial B}{\partial N_i}\right)_{T,P,N_j[i]} = -\frac{1}{N} \sum_{j \neq i} x_j \left(\frac{\partial B}{\partial x_j}\right)_{T,P,x[j,i]} \tag{8-33}$$

Substituting into Eq. (8-31) yields

$$\bar{B}_i = B - \sum_{j \neq i} x_j \left(\frac{\partial B}{\partial x_j}\right)_{T,P,x[j,i]} \tag{8-34}$$

In Eq. (8-34) we have obtained $\bar{B}_i$ from a data set in which x_i was eliminated. This equation cannot be used to obtain $\bar{B}_k$ unless we transform the given data to a set in which x_k is eliminated. Alternatively, we can use Eq. (8-20) (with k substituted for n) to solve for $\bar{B}_k$ from the data set of case (3). Thus,

$$\bar{B}_k = \bar{B}_i + \left(\frac{\partial B}{\partial x_k}\right)_{T,P,x[k,i]} \tag{8-35}$$

or, using Eq. (8-34),

$$\bar{B}_k = B + \left(\frac{\partial B}{\partial x_k}\right)_{T,P,x[k,i]} - \sum_{j \neq i} x_j \left(\frac{\partial B}{\partial x_j}\right)_{T,P,x[j,i]} \tag{8-36}$$

For a binary system of A and C, Eq. (8-36) reduces to

$$\bar{B}_A = B - y_C \left(\frac{\partial B}{\partial y_C}\right)_{T,P}$$

$$\bar{B}_C = B - y_A \left(\frac{\partial B}{\partial y_A}\right)_{T,P} \tag{8-37}$$

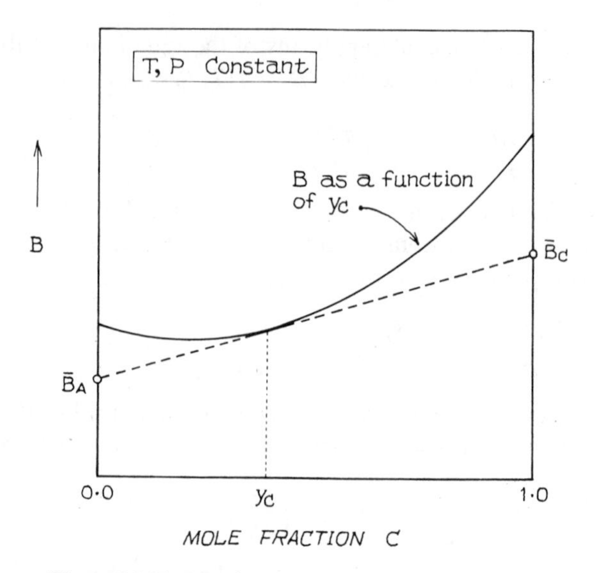

Figure 8.1 Tangent intercept rule for the binary A–C.

Equation set (8-37) can be easily visualized as shown in Figure 8.1 in which B is plotted as a function of y_C at constant T and P. At any y_C, a tangent to the curve, when extrapolated, intersects the $y_C = 0$ axis at $\bar{B}_A$ and the $y_C = 1$ axis at $\bar{B}_C$.[2] The application of Eq. (8-36) to a ternary system is shown in Example 8.3. Even in this relatively simple system, the data required to calculate partial molar properties are considerable—and only rarely available.

Example 8.3

Apply Eq. (8-36) to a ternary system of A, D, and C to obtain $\bar{B}_A$.

Solution

For a ternary system of A, D, and C, there are several ways to express $\bar{B}_A$. Suppose that C is the component "eliminated"; then from Eq. (8-36),

$$\bar{B}_A = B + \left(\frac{\partial B}{\partial y_A}\right)_{T,P,y_D} - y_A\left(\frac{\partial B}{\partial y_A}\right)_{T,P,y_D} - y_D\left(\frac{\partial B}{\partial y_D}\right)_{T,P,y_A}$$

$$= B + (1 - y_A)\left(\frac{\partial B}{\partial y_A}\right)_{T,P,y_D} - y_D\left(\frac{\partial B}{\partial y_D}\right)_{T,P,y_A}$$

On the other hand, if D were eliminated, we would have

$$\bar{B}_A = B + (1 - y_A)\left(\frac{\partial B}{\partial y_A}\right)_{T,P,y_C} - y_C\left(\frac{\partial B}{\partial y_C}\right)_{T,P,y_A}$$

[2]There are occasions when, in determining partial molar quantities, they become either very large or very small at low concentrations (e.g., partial molar entropy or Gibbs energy), and this intercept technique is not particularly accurate. An alternative technique for handling such cases is found in H. C. van Ness and R. V. Mrazek, "Treatment of Thermodynamic Data for Homogeneous Binary Systems," *AIChE J.*, **5**, 209 (1959).

The two expressions for $\bar{B}_A$ can be shown to be equivalent by equating them, using Eq. (8-20) while noting that $(1 - y_A) = y_D + y_C$. The choice between the two ways of expressing $\bar{B}_A$ depends on the data available; in each case, derivatives are required with the composition of different components held constant.

8.3 Duhem Relations for Partial Molar Quantities

For an n-component system, there are n partial molar quantities representing any extensive property. The set of $n + 2$ variables, $T, P, \bar{B}_1, \ldots, \bar{B}_n$ are all intensive and, therefore, only $n + 1$ variables from this set are independent: one can be expressed as a function of the other $n + 1$ variables.

The dependency can be expressed in a form analogous to the Gibbs–Duhem equation [see Eq. (5-45)]. Any property $\underline{B}$, for which $\bar{B}_i$ is employed as in Eq. (8-6b), can be expanded as a function of $T, P, N_1, \ldots, N_n$:

$$d\underline{B} = \left(\frac{\partial \underline{B}}{\partial T}\right)_{P,N_i} dT + \left(\frac{\partial \underline{B}}{\partial P}\right)_{T,N_i} dP + \sum_{i=1}^{n} \bar{B}_i \, dN_i \tag{8-38}$$

But $d\underline{B}$ can also be expressed by differentiating Eq. (8-6b):

$$d\underline{B} = \sum_{i=1}^{n} N_i \, d\bar{B}_i + \sum_{i=1}^{n} \bar{B}_i \, dN_i \tag{8-39}$$

Subtracting Eq. (8-39) from Eq. (8-38) yields

$$\sum_{i=1}^{n} N_i \, d\bar{B}_i = \left(\frac{\partial \underline{B}}{\partial T}\right)_{P,N_i} dT + \left(\frac{\partial \underline{B}}{\partial P}\right)_{T,N_i} dP \tag{8-40}$$

or, dividing by total moles, N, we get

$$\sum_{i=1}^{n} x_i \, d\bar{B}_i = \left(\frac{\partial B}{\partial T}\right)_{P,x} dT + \left(\frac{\partial B}{\partial P}\right)_{T,x} dP \tag{8-41}$$

Equations (8-40) and (8-41) are the most common forms of the *Duhem relation* for partial molar quantities. When integrated, they permit evaluation of any $\bar{B}_i$ in terms of the other $(n - 1)$ values of $\bar{B}_j$.

An alternative form of the Duhem relation can be obtained as follows. Noting that

$$\left(\frac{\partial B}{\partial T}\right)_{P,x} = \left[\frac{\partial\left(\sum_{i=1}^{n} \bar{B}_i x_i\right)}{\partial T}\right]_{P,x} = \sum_{i=1}^{n} x_i \left(\frac{\partial \bar{B}_i}{\partial T}\right)_{P,x} \tag{8-42}$$

and

$$\left(\frac{\partial B}{\partial P}\right)_{T,x} = \sum_{i=1}^{n} x_i \left(\frac{\partial \bar{B}_i}{\partial P}\right)_{T,x} \tag{8-43}$$

Eq. (8-41) can be rewritten as

$$\sum_{i=1}^{n} x_i \left[d\bar{B}_i - \left(\frac{\partial \bar{B}_i}{\partial T}\right)_{P,x} dT - \left(\frac{\partial \bar{B}_i}{\partial P}\right)_{T,x} dP\right] = 0 \tag{8-44}$$

Substituting Eq. (8-30) for $d\bar{B}_i$ in Eq. (8-44) yields

$$\sum_{i=1}^{n} x_i \left[\sum_{j \neq k} \left(\frac{\partial \bar{B}_i}{\partial x_j} \right)_{T,P,x[j,k]} dx_j \right] = 0 \tag{8-45}$$

or, changing the order of the summations,

$$\sum_{j \neq k} \left[\sum_{i=1}^{n} x_i \left(\frac{\partial \bar{B}_i}{\partial x_j} \right)_{T,P,x[j,k]} \right] dx_j = 0 \tag{8-46}$$

Since the brackets are coefficients of $(n - 1)$ terms in dx_j, all of which are independent (because x_k has been eliminated), it follows that each term in brackets must vanish. Thus, an equivalent Duhem relation is

$$\sum_{i=1}^{n} x_i \left(\frac{\partial \bar{B}_i}{\partial x_j} \right)_{T,P,x[j,k]} = 0 \tag{8-47}$$

Equation (8-47) could also have been obtained by dividing Eq. (8-41) by dx_j at constant T, P, $x[j, k]$.

We shall have occasion to use this form in treating fugacities of phases in equilibrium, as described in Chapter 10.

Example 8.4

For a binary of components 1 and 2, if the partial molar enthalpy $\bar{H}_1$ were available as a function of mole fraction x_1, show how $\bar{H}_2$ and the mixture enthalpy H can be determined. The data $\bar{H}_1 = f(x_1)$ are at constant T and P.

Solution

From Eq. (8-47),

$$x_1 \left(\frac{\partial \bar{H}_1}{\partial x_1} \right)_{T,P} + x_2 \left(\frac{\partial \bar{H}_2}{\partial x_1} \right)_{T,P} = 0$$

Separating and integrating between x_1^0 and x_1, we obtain

$$(\bar{H}_2)_{x_1} - (\bar{H}_2)_{x_1^0} = -\int_{x_1^0}^{x_1} \frac{x_1}{x_2} \left(\frac{\partial \bar{H}_1}{\partial x_1} \right) dx_1$$

Let x_1^0 be 0 (i.e., pure component 2). Then $(\bar{H}_2)_{x_1^0} = H_2$. If this is substituted in the above, the right-hand side may be found from the $\bar{H}_1 = f(x_1)$ data and $\bar{H}_2$ is then related to the pure component enthaply H_2. To obtain H, Eq. (8-6b) divided by N yields

$$H = x_1 \bar{H}_1 + x_2 \bar{H}_2$$

We have already seen that all $n(\bar{B}_i)$s can be found from a data set of the form $B = f(T, P, x_1, \ldots, x_{n-1})$. We have also shown that any $\bar{B}_i$ can be found—to within an arbitrary constant—from the other $(n - 1)(\bar{B}_j)$s. It is then possible to reconstruct B from these $(\bar{B}_i)s$ by dividing Eq. (8-6b) by N,

$$B = \sum_{i=1}^{n} \bar{B}_i x_i \tag{8-48}$$

Thus, $(n - 1)(\bar{B}_i)$s have an information content equivalent to B.

Frequently, data are reported in the literature for all $(n)(\bar{B}_i)$s (most commonly for binary systems in which $\bar{B}_1$ and $\bar{B}_2$ are measured independently). The redundant information can be used to check the consistency of the data. At any given T, P, and concentration, we can calculate the partial derivative $(\partial \bar{B}_i/\partial x_j)_{T,P,x[j,k]}$ for each component, and then use Eq. (8-47) to verify that the sum does indeed vanish.

In some cases, reported data will not give a satisfactory consistency check, but these may be the only available data. In that case, it is always possible to smooth the data in order to obtain a set of partial molar quantities that are consistent. The procedure is simply to reconstruct B from Eq. (8-48) and then apply an equation of the form of Eq. (8-34) or (8-36) to obtain the partial molar quantities. As shown in the following example, this set will always be consistent.

Example 8.5

Using Eqs. (8-34) and (8-36), prove that a set of partial molar properties obtained only from data of the form, $B = f(T, P, x_1, \ldots, x_{i-1}, x_{i+1}, \ldots, x_n)$ will always form a consistent set.

Solution

Multiply Eq. (8-36) by x_k and sum over all k except $k = i$.

$$\sum_{k\neq i} x_k\bar{B}_k = \sum_{k\neq i} x_k\left[B + \left(\frac{\partial B}{\partial x_k}\right)_{T,P,x[i,k]} - \sum_{j\neq i} x_j\left(\frac{\partial B}{\partial x_j}\right)_{T,P,x[i,j]}\right]$$

Multiply Eq. (8-34) by x_i and add to the above.

$$\sum_{k=1}^{n} x_k\bar{B}_k = B(x_i + \sum_{k\neq i} x_k) + \sum_{k\neq i} x_k\left(\frac{\partial B}{\partial x_k}\right)_{T,P,x[i,k]}$$
$$- (\sum_{k\neq i} x_k + x_i)\left[\sum_{j\neq i} x_j\left(\frac{\partial B}{\partial x_j}\right)_{T,P,x[i,j]}\right]$$

The second and third terms on the right-hand side cancel, and the result reduces to

$$\sum_{k=1}^{n} x_k\bar{B}_k = B$$

Thus, the set of $\bar{B}_k$ obtained from $B = f(T, P, x_1, \ldots, x_{i-1}, x_{i+1}, \ldots, x_n)$ will always form a consistent set.

8.4 *P–$\underline{V}$–T–N* Relations for Mixtures

In Section 7.3 we presented a brief introduction to *P–$\underline{V}$–T–N* relations for pure components. Three such relations, or equations of state, were used as illustrative examples. Essentially all pure-component equations of state can also be applied to mixtures; the difficulty of application lies in delineating rules to obtain mixture parameters which now depend on the mixture composition. No general theory has yet been developed (except for the virial equation of state) to determine appropriate *mixing rules;* thus they are ordinarily formulated in a manner to yield the least error when the predictions of the equation of state are compared with experiment.

Consider first the Redlich–Kwong equation of state given in Eq. (7-60). It is rewritten below, in intensive form in Eq. (8-49a) and in extensive form in Eq. (8-49b).

$$P = \frac{RT}{V - b_m} - \frac{a_m}{T^{1/2}V(V + b_m)} \tag{8-49a}$$

$$P = \frac{NRT}{\underline{V} - Nb_m} - \frac{N^2 a_m}{T^{1/2}\underline{V}(\underline{V} + Nb_m)} \tag{8-49b}$$

where a_m and b_m are the mixture constants. N is the total moles in the mixture and

$$N = \sum_{i=1}^{n} N_i \tag{8-50}$$

The pure component parameters a_i and b_i were given in Eqs. (7-61) and (7-62) in terms of the critical constants of component i. Redlich and Kwong suggested the mixing rules shown in Eqs. (8-51) and (8-52) to determine a_m and b_m.

$$a_m^{1/2} = \sum_{i=1}^{n} y_i a_i^{1/2} \tag{8-51}$$

$$b_m = \sum_{i=1}^{n} y_i b_i \tag{8-52}$$

where y_i is the mole fraction of component i.

With these rules, then, for Eq. (8-49b),

$$N^2 a_m = \left(\sum_{i=1}^{n} N_i a_i^{1/2}\right)^2 \tag{8-53}$$

$$Nb_m = \sum_{i=1}^{n} N_i b_i \tag{8-54}$$

Equations (8-53) and (8-54) are convenient forms should partial derivatives with respect to mole numbers be required, as illustrated in Example 8.6.

Example 8.6

Derive an expression for the partial molar volume, $\bar{V}_k$, using the Redlich–Kwong equation of state.

Solution

Using Eq. (8-8),

$$\bar{V}_k = \left(\frac{\partial \underline{V}}{\partial N_k}\right)_{T,P,N_j[k]}$$

Since the Redlich–Kwong equation of state is explicit in pressure with $\underline{V}$, T, N_1, . . . as the independent variables, the defining derivative for $\bar{V}_k$ is better rewritten as

$$\bar{V}_k = \left(\frac{\partial \underline{V}}{\partial N_k}\right)_{T,P,N_j[k]} = -\frac{(\partial P/\partial N_k)_{T,\underline{V},N_j[k]}}{(\partial P/\partial \underline{V})_{T,N}}$$

The denominator is readily found using Eq. (8-49b),

$$\left(\frac{\partial P}{\partial \underline{V}}\right)_{T,N} = \frac{-NRT}{(\underline{V} - Nb_m)^2} + \frac{N^2 a_m(2\underline{V} + Nb_m)}{T^{1/2}\underline{V}^2(\underline{V} + Nb_m)^2}$$

To obtain the numerator from Eq. (8-49b), one must also use Eqs. (8-53) and (8-54); that is, the derivatives of $N^2 a_m$ and Nb_m are necessary.

$$\frac{\partial}{\partial N_k}(N^2 a_m)_{T,\underline{V},N_j[k]} = \frac{\partial}{\partial N_k}\left[\left(\sum_{i=1}^{n} N_i a_i^{1/2}\right)^2\right]_{T,\underline{V},N_j[k]}$$

$$= 2\left(\sum_{i=1}^{n} N_i a_i^{1/2}\right) a_k^{1/2}$$

$$= 2Na_m^{1/2} a_k^{1/2}$$

$$\frac{\partial}{\partial N_k}(Nb_m)_{T,\underline{V},N_j[k]} = \frac{\partial}{\partial N_k}\left(\sum_{i=1}^{n} N_i b_i\right)_{T,\underline{V},N_j[k]}$$

$$= b_k$$

Therefore,

$$\left(\frac{\partial P}{\partial N_k}\right)_{T,\underline{V},N_j[k]} = RT\left[\frac{1}{\underline{V} - Nb_m} - \frac{Nb_k}{(\underline{V} - Nb_m)^2}\right] - \frac{1}{T^{1/2}\underline{V}}\left[\frac{2Na_m^{1/2}a_k^{1/2}}{\underline{V} + Nb_m} - \frac{N^2 a_m b_k}{(\underline{V} + Nb_m)^2}\right]$$

To evaluate $\bar{V}_k$, one must determine $\underline{V}$, a_m and b_m for the mixture at the given state of the material; then the derivatives with respect to P can be found as shown above.

For the more complex Peng–Robinson equation of state given in Eq. (7-65), the two parameters for a pure material were defined in Eqs. (7-66) through (7-70). For a mixture,

$$P = \frac{NRT}{\underline{V}_m - Nb_m} - \frac{N^2 a_m}{\underline{V}_m(\underline{V}_m + Nb_m) + Nb_m(\underline{V}_m - Nb_m)} \tag{8-55}$$

with

$$a_m = \sum_i \sum_j y_i y_j a_{ij} \tag{8-56}$$

$$b_m = \sum_i y_i b_i \tag{8-57}$$

and

$$a_{ij} = (1 - \delta_{ij})(a_i a_j)^{1/2} \qquad i \neq j \tag{8-58}$$

$$a_{ii} = a_i \tag{8-59}$$

The term δ_{ij} is a *binary interaction parameter* that is specific for the *i–j* binary. In most cases, δ_{ij} is assumed to be independent of the composition, system temperature, or system pressure. It is usually obtained from experimental data with the *i–j* binary; then it is used in multicomponent mixtures with the implied assumption that ternary or higher-order interactions are negligible. The introduction of binary interaction parameters is characterstic of most equations of state introduced in the past several years. While δ_{ij} is ordinarily a small number (ca. 0.1), the final result is quite sensitive to the value chosen. Note that if $\delta_{ij} = 0$, the mixing rule for a_m in the Peng–Robinson relation reduces to that for the Redlich–Kwong equation of state, Eq. (8-51).

Finally, for the virial equation of state, Eq. (7-71), the expression for the virial coefficients in a mixture can be developed from theory; that for the second virial is

$$B_m = \sum_i \sum_j y_i y_j B_{ij} \tag{8-60}$$

Assuming that little error is introduced in using the same value of B_m for the truncated form of the virial [Eq. (7-72)], then values of B_{ii} (B_i) and B_{jj} (B_j) can be calculated as shown earlier in Eqs. (7-74) through (7-76). To determine the *interaction* virial coefficient, B_{ij} ($i \neq j$), it would be convenient to use Eqs. (7-74) through (7-76), but *mixture* critical temperatures, critical pressures, and acentric factors are required. Prausnitz[3] suggests that

$$T_{c_{ij}} = (1 - k_{ij})(T_{c_i} T_{c_j})^{1/2} \tag{8-61}$$

$$\omega_{ij} = \frac{\omega_i + \omega_j}{2} \tag{8-62}$$

and

$$P_{c_{ij}} = \frac{RT_{c_{ij}}(Z_{c_i} + Z_{c_j})/2}{[(V_{c_i}^{1/3} + V_{c_j}^{1/3})/2]^3} \tag{8-63}$$

where Z_{c_i} and V_{c_i} are the critical compressibility factor and critical volume of component i. The term k_{ij} is a binary interaction parameter similar to the δ_{ij} introduced in the Peng–Robinson equation of state. (Note that k_{ij} does not equal δ_{ij}.) Values of k_{ij} for many binaries are available.[4] As with the pure-component virial equation of state, the mixture form is limited to the gas phase under conditions where the mixture density is less than one-half of the critical mixture density.[5]

In this brief discussion of mixture P–$\underline{V}$–T–N equations of state, it can be seen that the basic forms of the equations do not change when treating mixtures rather than pure components. This would, of course, be expected since the mixture equations of state must reduce to the pure-component form, as all mole fractions but one approach zero. However, in any application of mixture equations of state, mixing rules to relate characteristic equation parameters to composition must be available. Most mixing rules now in use are conceptually simple, yet algebraically complex results are inevitable when operations require that derivatives be taken with respect to specific mole numbers or to compositions.

The *departure* functions for mixtures are calculated in the same manner as for pure components except that mixture parameters must be used. Thus, Eqs. (7-81) through (7-85) are still applicable, but they apply to a mixture with a specific composition. When considering variations in temperature in the ideal-

[3] J. M. Prausnitz, *Molecular Thermodynamics of Fluid-Phase Equilibria* (Englewood Cliffs, N. J.: Prentice-Hall, 1969).

[4] P. L. Chueh and J. M. Prausnitz, *Ind. Eng. Chem. Fundam.*, **6**, 492 (1967).

[5] An estimate of the mixture critical density can be obtained from

$$\rho_{c_m} = V_{c_m}^{-1} \approx \left[\frac{\sum \sum y_i y_j (V_{c_i}^{1/3} + V_{c_j}^{1/3})^3}{8}\right]^{-1}$$

gas state [e.g., Eqs. (7-93), (7-96), and (7-97)], values of C_v° or C_p° for the mixture are calculated as

$$C_{v_m}^\circ = \sum_i y_i C_{v_i}^\circ \tag{8-64}$$

or

$$C_{p_m}^\circ = \sum_i y_i C_{p_i}^\circ \tag{8-65}$$

That is, in the ideal-gas state, heat capacities are mole-fraction averages of the pure-component values.

8.5 Ideal-Gas Mixtures and Ideal Solutions

In Chapter 3 we defined an ideal gas as one that exhibited P–$\underline{V}$–T–N behavior as given by

$$P = \frac{NRT}{\underline{V}} \tag{8-66}$$

and, in addition, the internal energy of the gas was a function only of temperature and mass,

$$\underline{U} = f(T, N) \tag{8-67}$$

Further, we indicated that this hypothetical model gas was well simulated by real gases at low pressures. As a limiting case, as pressure approaches zero, all gases are assumed to behave as ideal gases. This concept was then utilized in Section 7.4 to develop methods to calculate differences in thermodynamic properties by determining isothermal departure functions and varying temperature in an ideal-gas state. In all the previous work, we have implicitly assumed that the systems only contained a pure component. In this section we want to examine the *ideal-gas* mixture[6] as a role model for a mixture of gases. Also, for mixtures, there is, in addition, a less restrictive model than the ideal-gas mixture; this is the *ideal solution*. The latter model can be used both for gaseous and for liquid mixtures.

Denbigh[7] has suggested that the definitions of an ideal gas (pure), an ideal-gas mixture, and an ideal solution may all be written in a similar form:

Pure-component; ideal gas:

$$G_i = \mu_i = RT \ln P + \lambda_i(T) \tag{8-68}$$

Ideal-gas mixture; for component i:

$$\bar{G}_i = \mu_i = RT \ln p_i + \lambda_i(T) \tag{8-69}$$

Ideal solution; for component i:

$$\bar{G}_i = \mu_i = RT \ln y_i + \Lambda_i(T, P) \tag{8-70}$$

[6]Note that an ideal-gas mixture is not necessarily the same as a *mixture* of *ideal gases*.

[7]K. Denbigh, *The Principles of Chemical Equilibrium*, 3rd ed. (Cambridge: Cambridge University Press, 1971).

In these definitions, G_i is the Gibbs energy of pure i while $\bar{G}_i$ is the partial molar Gibbs energy of i in a mixture. P is the system pressure, y_i is the mole fraction of i, and p_i is the partial pressure of i,

$$p_i = y_i P \tag{8-71}$$

$\lambda_i(T)$ is a function of temperature and is specific for component i. $\Lambda_i(T, P)$ is a similar function for i, but it depends on both temperature and pressure.

Illustrating that Eq. (8-68) satisfies the criteria stated above for a pure-component, ideal gas, from Eq. (8-68)

$$\left(\frac{\partial G_i}{\partial P}\right)_T = \frac{RT}{P}$$

but, from Eq. (5-41), with N constant,

$$\left(\frac{\partial G_i}{\partial P}\right)_T = V_i$$

Thus, Eq. (8-66) is obtained. To prove the second criterion, we first note, from Eq. (5-40), that

$$\underline{G} = \underline{U} + P\underline{V} - T\underline{S} = \underline{H} - T\underline{S}$$

so

$$\frac{\partial}{\partial T}\left(\frac{\underline{G}}{T}\right)_{P,N} = \frac{\partial}{\partial T}\left(\frac{\underline{H}}{T} - \underline{S}\right)_{P,N} = -\frac{\underline{H}}{T^2} \tag{8-72}$$

where the identity

$$\left(\frac{1}{T}\right)\left(\frac{\partial \underline{H}}{\partial T}\right)_{P,N} = \left(\frac{\partial \underline{S}}{\partial T}\right)_{P,N} = \frac{NC_P}{T}$$

was used.

Writing Eq. (8-72) in intensive form and using Eq. (8-68),

$$\frac{\partial}{\partial T}\left(\frac{G_i}{T}\right)_{P,N} = \frac{-H_i}{T^2} = \frac{d}{dT}\left(\frac{\lambda_i(T)}{T}\right)$$

Thus, H_i is only a function of temperature. Also, since as shown above, $V_i = RT/P$, then

$$U_i = H_i(T) - PV_i = H_i(T) - RT$$

so U_i depends only on temperature.

In treating an ideal-gas mixture, using Eqs. (8-69), (8-71), and (8-27) with constant N and composition,

$$\left(\frac{\partial \bar{G}_i}{\partial P}\right)_{T,y} = \bar{V}_i = \frac{RT}{P}$$

Therefore,

$$V = \sum_i y_i \bar{V}_i = \frac{RT}{P}$$

As expected, an ideal-gas mixture also follows Eq. (8-66). Next, differentiating Eq. (8-72) with respect to N_i, keeping T, P, and all, moles except i constant,

$$\frac{\partial}{\partial T}\left(\frac{\bar{G}_i}{T}\right)_{P,N} = -\frac{\bar{H}_i}{T^2} \tag{8-73}$$

Using Eq. (8-73) with Eq. (8-69), noting that since P and N are held constant,

p_i is not a variable,

$$\frac{\partial}{\partial T}\left(\frac{\bar{G}_i}{T}\right)_{P,N} = -\frac{\bar{H}_i}{T^2} = \frac{d}{dT}\left(\frac{\lambda_i(T)}{T}\right)_{P,N}$$

and thus $\bar{H}_i$ depends only on temperature. Again, with

$$\bar{U}_i = \bar{H}_i(T) - P\bar{V}_i = \bar{H}_i(T) - RT$$

$\bar{U}_i$ also varies only with temperature. Thus, in an ideal-gas mixture, $\bar{H}_i$ and $\bar{U}_i$ are functions only of temperature. As will be seen later, the fact that neither of these properties is a function of composition will lead to the conclusion that there is no enthalpy (or internal energy) change when pure-component ideal gases are mixed to form an ideal-gas mixture.

While the consequences of the defining ideal-gas mixture equation [Eq. (8-69)] are those expected when an analogy is made with pure-component ideal gases, there is one other result that is readily obtained and is very useful in later work.

In Section 6.4 we considered the equilibrium criteria for situations involving semipermeable membranes. The criteria derived were that the temperatures were equal on both sides of the membrane, as were the chemical potentials of all components for which the membrane was permeable. Therefore, for an *ideal-gas mixture*, the partial pressures of permeable components would be equal across the membrane.

To complete the discussion of ideal-gas mixtures, we note that real-gas mixtures at low pressures simulate ideal-gas mixtures and, as the pressure is reduced toward zero, *all* gas mixtures behave in an ideal fashion.

Example 8.7

Calculate the minimum work required to separate air into pure oxygen and nitrogen. Assume a steady-flow process operating at 300 K. The inlet air is at 1 bar and contains 80 mole % nitrogen. Air may be considered as an ideal-gas mixture and pure oxygen and nitrogen as ideal gases.

Solution

Visualize that the inlet air contacts two semipermeable membranes. One is permeable only to nitrogen and the other only to oxygen. Consider the nitrogen-permeable membrane first. The chemical potentials of nitrogen are equal across the membrane. Since the mixture is an ideal-gas mixture, the chemical potential equality may be replaced by an equality of partial pressures. As $p_{N_2} = Py_{N_2} = (1)(0.8) = 0.8$ bar, the pure nitrogen downstream of the membrane is 0.8 bar. If this stream is compressed back to 1 bar, the reversible, isothermal flow work is, per mole of nitrogen flowing,

$$W = -\int V\,dP = -RT\int d\ln P = (8.314)(300)\ln\frac{1}{0.8}$$

$$= -557 \text{ J/mol } N_2$$

Similarly, for oxygen,

$$W = -(8.314)(300)\ln\frac{1}{0.2} = -4010 \text{ J/mol } O_2$$

On a basis of 1 mole of air as feed,

$$W = (0.8)(-557) + (0.2)(-4010) = -1248 \text{ J/mol air}$$

Thus, the minimum work to separate 1 mole of air at 300 K, 1 bar into pure nitrogen and pure oxygen streams at 300 K, 1 bar, is 1248 J, assuming the air to be an ideal-gas mixture.

In the case of an ideal solution, as defined by Eq. (8-70), we first differentiate with respect to pressure while maintaining the temperature and composition constant.

$$\left(\frac{\partial \bar{G}_i}{\partial P}\right)_{T,y} = \left\{\frac{\partial[\Lambda_i(T, P)]}{\partial P}\right\}_{T,y} = \bar{V}_i$$

Since Λ_i does not depend on composition, neither does $\bar{V}_i$; therefore, in an ideal solution,

$$\bar{V}_i \neq f(\text{composition}) \tag{8-74}$$

In a similar manner, using Eqs. (8-70) and (8-73), we obtain

$$\frac{\partial}{\partial T}\left(\frac{\bar{G}_i}{T}\right)_{P,y_j} = \left\{\frac{\partial[\Lambda_i(T, P)/T]}{\partial T}\right\}_{P,y_j} = -\frac{\bar{H}_i}{T^2}$$

so that, in an ideal solution,

$$\bar{H}_i \neq f(\text{composition}) \tag{8-75}$$

If the mixture follows an ideal-solution behavior over the entire compositional range, then Eqs. (8-74) and (8-75) imply that

$$\bar{V}_i = V_i \tag{8-76}$$

and

$$\bar{H}_i = H_i \tag{8-77}$$

where V_i and H_i are pure-component specific volumes and enthalpies of i at the same temperature and pressure as the mixture and for the same phase (liquid, vapor, solid) as the mixture. Further, in such a case, it is evident from Eq. (8-70) that

$$\bar{G}_i = G_i + RT \ln y_i \tag{8-78}$$

where G_i is the pure-component Gibbs energy. Since

$$G_i = H_i - TS_i$$

and

$$\bar{G}_i = \bar{H}_i - T\bar{S}_i$$

then

$$\bar{S}_i = S_i - R \ln y_i \tag{8-79}$$

Properties of the ideal solution may then be written as

$$G = \sum_i y_i G_i + RT \sum_i y_i \ln y_i \tag{8-80}$$

$$S = \sum_i y_i S_i - R \sum_i y_i \ln y_i \tag{8-81}$$

$$H = \sum_i y_i H_i \tag{8-82}$$

Equations (8-76) through (8-82) apply only to a solution that is ideal at all compositions.

Later, when we compare real solutions with ideal solutions, a superscript ID will be used to denote those properties in the model ideal solution.

Example 8.8

A new process calls for an aqueous feed of ammonium nitrate at high pressure at 25°C. The pressure currently being considered is 10 kilobars. To complete our design, the solubility of NH_4NO_3 in the feedstream is desired. A literature search of this system has provided the data shown below. What is your best estimate of the solubility at 10 kilobars?

Data: At 25°C and 1 bar, the solubility of NH_4NO_3 in water is 67.63 wt% salt. At a pressure of about 1 bar, the chemical potentials of both components have been determined at 25°C as a function of composition. These data are shown in Figure 8.2. Note that the chemical potential of NH_4NO_3 is referenced to a saturated solution, whereas for water the reference is pure water.

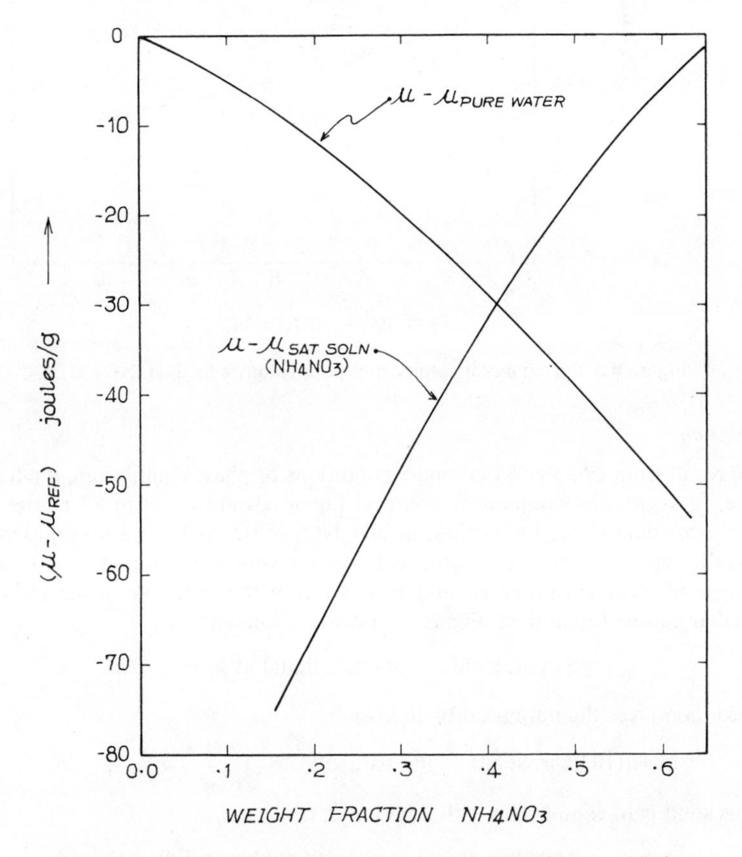

Figure 8.2 Chemical potentials for water and NH_4NO_3 in solution at 25°C, 1 bar.

Volumetric data for the NH_4NO_3–H_2O system at 25°C are given in Figure 8.3.

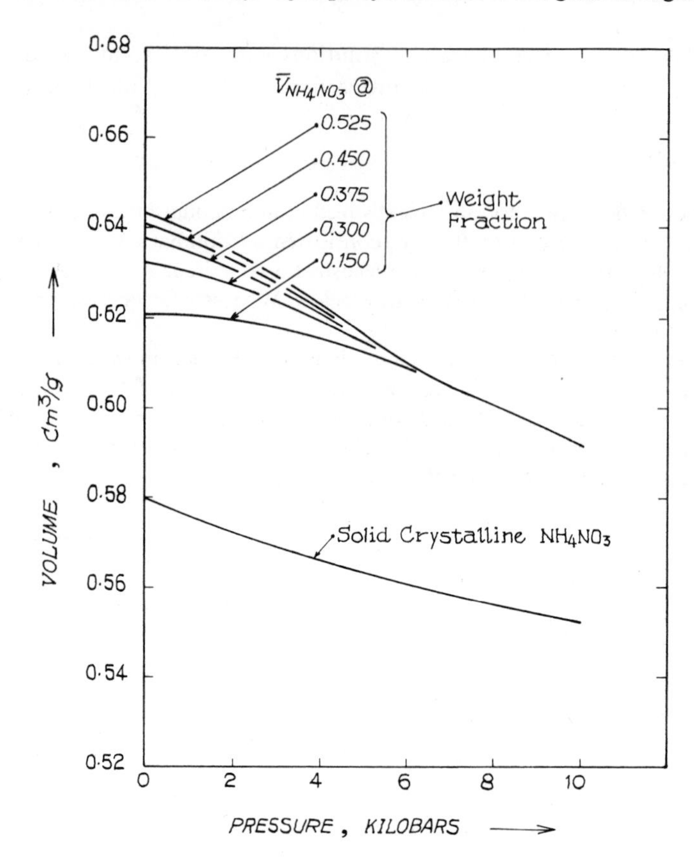

Figure 8.3 Partial molal volume and pure volumes for NH_4NO_3 at 25°C.

Solution

We recall from Chapter 6 that under conditions of phase equilibrium, the temperature, pressure, and component chemical potentials are equal in all phases. Let x be the solubility (weight fraction) of NH_4NO_3 in the high-pressure solution. Now visualize an isothermal cyclic process beginning with a solution at x, at 1 bar. The change in chemical potential of NH_4NO_3 from this solution to the solid phase at 1 bar can be found from Figure 8.2 once x is known:

$$\mu(\text{1 bar, solid}) - \mu(\text{1 bar, liquid at } x) = -f(x)$$

Next, compress the pure solid to 10 kbar.

$$\mu(\text{10 kbar, solid}) - \mu(\text{1 bar, solid}) = \int_{\text{1 bar}}^{\text{10 kbar}} V^S_{NH_4NO_3}\, dP$$

This solid is in equilibrium with liquid at x, i.e.,

$$\mu(\text{10 kbar, liquid at } x) - \mu(\text{10 kbar, solid}) = 0$$

Finally expand this liquid back to 1 bar,

$$\mu(1 \text{ bar, liquid at } x) - \mu(10 \text{ kbar, liquid at } x) = \int_{10 \text{ kbar}}^{1 \text{ bar}} \bar{V}^L_{NH_4NO_3}\, dP$$

Adding, we obtain

$$-f(x) = \int_{1 \text{ bar}}^{10 \text{ kbar}} [\bar{V}^L(x) - V^S]\, dP$$

By trial and error, with Figure 8.2 to obtain $f(x)$ and Figure 8.3 to obtain $\bar{V}^L(x)$ and V^S as a function of pressure, x is found to be 0.295. Thus, at 10 kbar the solubility of NH_4NO_3 is only about 29.5 wt%.

8.6 Mixing and Excess Functions

A mixture property is often related to the properties of a reference state, which can be real or hypothetical. The difference between the value of the actual and reference state properties is denoted by the symbol Δ and is called the *mixing or solution function*. The defining equation for Δ is

$$\Delta \underline{B} \equiv \underline{B}(T, P, N_1, \ldots, N_n) - \sum_{j=1}^{n} N_j \bar{B}_j^+(T^+, P^+, x_1^+, \ldots, x_{n-1}^+) \quad \text{(8-83)}[8]$$

or

$$\Delta B \equiv B(T, P, x_1, \ldots, x_{n-1}) - \sum_{j=1}^{n} x_j \bar{B}_j^+(T^+, P^+, x_1^+, \ldots, x_{n-1}^+) \quad \text{(8-84)}$$

where $\Delta \underline{B}$ and ΔB are the total and specific mixing (or solution) functions. For reasons that will become apparent later, the N_j or x_j (not x_j^+) within the summation is taken as the *actual mole number or mole fraction of the mixture*, and not that of the reference state.

It follows from the defining equations that the mixing functions are specified only when the reference states have been clearly delineated. (Note that a reference state must be defined for each component.) Reference states are chosen either for convenience (the properties are known) or for practicality (the deviations from the reference state are small).

If the mixing function is to be useful, it should depend on the properties of the mixture. That is, we desire

$$\Delta B = f(T, P, x_1, \ldots, x_{n-1}) \quad \text{(8-85)}$$

This requirement limits us to either one of two choices for each reference state variable; namely, $T^+, P^+, x_1^+, \ldots, x_{n-1}^+$ are either set equal to the actual mixture properties (hence, $\bar{B}_j^+$ varies as the mixture conditions change) or they can be set at some fixed conditions (hence, $\bar{B}_j^+$ is a contant).

In practice, the most common reference state is the pure component at the same temperature, pressure, and state of aggregation of the mixture. That

[8] Prior to this point, y was used to express composition. y is used primarily to express vapor mole fractions. Since mixing functions have their greatest use for liquid solutions, the change from y_j to x_j as a mole fraction has been made.

is, for component j, $T^+ = T$, $P^+ = P$, $x_j^+ = 1$, $x_i^+ = 0\ (i \neq j)$ and, therefore, $\bar{B}_j^+ = B_j(T, P)$. In this case, the temperature and pressure vary as the state of the mixture changes, but the reference-state mole fractions are fixed. Note that this definition of the reference state satisfies Eq. (8-85). Whenever the pure components exist in the same state of aggregation of the mixture at T and P, this reference state is *real*. There are, however, a number of cases in which the stable state of the pure material is different from that of the mixture and hence the reference state as defined above is *hypothetical* (e.g., an inorganic salt in an aqueous solution or a liquid mixture above the critical temperature of one of the components). In these cases, it is not uncommon to select other reference states (e.g., the saturated salt solution or an infinitely dilute solution).

The use of the mixture mole numbers in the summation of Eq. (8-83) results in a convenient definition of the partial molar mixing function. Substituting Eq. (8-6b),

$$\underline{B} = \sum_{j=1}^{n} N_j \bar{B}_j$$

into Eq. (8-83) yields

$$\Delta\underline{B} = \sum_{j=1}^{n} N_j(\bar{B}_j - \bar{B}_j^+) \tag{8-86}$$

Applying the partial molar operator, we get

$$\overline{\Delta B_j} = \left[\frac{\partial(\Delta\underline{B})}{\partial N_j}\right]_{T,P,N_i[j]} = (\bar{B}_j - \bar{B}_j^+) \tag{8-87}$$

or, rewriting Eq. (8-86),

$$\Delta\underline{B} = \sum_{j=1}^{n} N_j\, \overline{\Delta B_j} \tag{8-88}$$

Dividing both sides of Eq. (8-88) by N yields

$$\Delta B = \sum_{j=1}^{n} x_j\, \overline{\Delta B_j} \tag{8-89}$$

It also follows from Eq. (8-85) that

$$d(\Delta B) = \left(\frac{\partial(\Delta B)}{\partial T}\right)_{P,x} dT + \left(\frac{\partial(\Delta B)}{\partial P}\right)_{T,x} dP + \sum_{j\neq i}\left(\frac{\partial(\Delta B)}{\partial x_j}\right)_{T,P,x[i,j]} dx_j \tag{8-90}$$

Note that Eqs. (8-87), (8-89), and (8-90) are completely analogous to Eqs. (8-8), (8-48), and (8-32), with B replaced by ΔB. Since these three equations formed the basis for all the partial molar quantity relationships developed in Sections 8.2 and 8.3, it follows that all relationships developed for B also apply to ΔB. Thus, the analog to Eq. (8-34) is

$$\overline{\Delta B_i} = \Delta B - \sum_{j\neq i} x_j\left(\frac{\partial(\Delta B)}{\partial x_j}\right)_{T,P,x[j,i]} \tag{8-91}$$

and, hence, the slope-intercept method can be used to evaluate partial molar mixing functions from specific mixing functions. Similarly, the analogs to Eqs. (8-41) and (8-47) are

$$\sum_{i=1}^{n} x_i\, d(\overline{\Delta B_i}) = \left(\frac{\partial(\Delta B)}{\partial T}\right)_{P,x} dT + \left(\frac{\partial(\Delta B)}{\partial P}\right)_{T,x} dP \tag{8-92}$$

and

$$\sum_{i=1}^{n} x_i \left(\frac{\partial(\overline{\Delta B_i})}{\partial x_j}\right)_{T,P,x[j,k]} = 0 \tag{8-93}$$

and thus the Duhem relation can be used to check consistency of data or to find one partial molar mixing function from a set of $(n - 1)$ others.

Example 8.9

If 1 mole of pure sulfuric acid is diluted with N_w moles of water, the heat evolved is about equal to

$$Q(kJ) = \frac{74.78 N_w}{N_w + 1.7983} \qquad (18°\text{C},\ N_w < 20)$$

What is the differential enthalpy of solution for water and acid for a solution containing 40 mole% acid?

Solution

The heat evolved for 1 mole of acid is given in the problem statement. If we wished to obtain Q for N_A moles of acid and N_w moles of water in kilojoules,

$$Q = \frac{N_A(74.78\, N_w/N_A)}{(N_w/N_A) + 1.7983}$$

Then the integral heat of solution, $\Delta \underline{H}$, is $-Q$ and therefore

$$\overline{\Delta H_w} = \left(\frac{\partial \Delta \underline{H}}{\partial N_w}\right)_{T,P,N_A}$$

$$= \frac{-134 x_A^2}{(1 + 0.798 x_A)^2}$$

$$= -12.35 \text{ kJ/mol } H_2O$$

$$\overline{\Delta H_A} = \left(\frac{\partial \Delta \underline{H}}{\partial N_A}\right)_{T,P,N_w}$$

$$= \frac{74.78 x_w^2}{(1 + 0.798 x_A)^2}$$

$$= -15.50 \text{ kJ/mol acid}$$

Mixing functions can also be defined for the ideal solution, introduced in Section 8.5. In this case the reference state is invariably taken as *the pure components at the same temperature, pressure, and state of aggregation as that of the mixture*. Thus, the superscripts in Eq. (8-84) can be deleted:

$$\Delta B = B - \sum_{j=1}^{n} x_j B_j \tag{8-94}$$

For an ideal solution,

$$\Delta B^{\text{ID}} = B^{\text{ID}} - \sum_{j=1}^{n} x_j B_j \tag{8-95}$$

For the Gibbs energy, entropy, and enthalpy of an ideal solution, substituting Eqs. (8-80), (8-81), and (8-82), respectively, into Eq. (8-95) yields

$$\Delta G^{\text{ID}} = RT \sum_{j=1}^{n} x_j \ln x_j \tag{8-96}$$

$$\Delta S^{\text{ID}} = -R \sum_{j=1}^{n} x_j \ln x_j \tag{8-97}$$

$$\Delta H^{\text{ID}} = 0 \tag{8-98}$$

or

$$\overline{\Delta G_j^{\text{ID}}} = RT \ln x_j \tag{8-99}$$

$$\overline{\Delta S_j^{\text{ID}}} = -R \ln x_j \tag{8-100}$$

$$\overline{\Delta H_j^{\text{ID}}} = 0 \tag{8-101}$$

The deviation of a mixture from ideal solution behavior is commonly denoted by the *excess function*, $\underline{B}^{\text{EX}}$ or B^{EX}:

$$\underline{B}^{\text{EX}} \equiv \underline{B} - \underline{B}^{\text{ID}} \qquad \text{or} \qquad B^{\text{EX}} = B - B^{\text{ID}} \tag{8-102}$$

From Eq. (8-6b) for $\underline{B}$ and $\underline{B}^{\text{ID}}$,

$$\underline{B} = \sum_{j=1}^{n} N_j \bar{B}_j$$

$$\underline{B}^{\text{ID}} = \sum_{j=1}^{n} N_j \bar{B}_j^{\text{ID}} \tag{8-103}$$

it follows that

$$\underline{B}^{\text{EX}} = \sum_{j=1}^{n} N_j (\bar{B}_j - \bar{B}_j^{\text{ID}}) \tag{8-104}$$

Applying the partial molar operator to Eq. (8-102) yields

$$\bar{B}_j^{\text{EX}} = \bar{B}_j - \bar{B}_j^{\text{ID}} \tag{8-105}$$

or

$$\underline{B}^{\text{EX}} = \sum_{j=1}^{n} N_j \bar{B}_j^{\text{EX}} \tag{8-106}$$

Thus, B^{EX} and $\bar{B}_j^{\text{EX}}$ are completely analogous to B and $\bar{B}_j$, and we could calculate $\bar{B}_j^{\text{EX}}$ from B^{EX} using the equations developed in Section 8.2 or we could use the Duhem equation as developed in Section 8.3.

By analogy to Eq. (8-102), excess mixing functions are commonly defined:

$$\Delta \underline{B}^{\text{EX}} = \Delta \underline{B} - \Delta \underline{B}^{\text{ID}} \tag{8-107}$$

or

$$\overline{\Delta B_j^{\text{EX}}} = \overline{\Delta B_j} - \overline{\Delta B_j^{\text{ID}}} \tag{8-108}$$

But, unlike $\Delta \underline{B}$ or $\Delta \underline{B}^{\text{ID}}$, an excess function does not depend directly on the choice of the reference state. This can be readily appreciated by expanding $\overline{\Delta B_j}$ and $\overline{\Delta B_j^{\text{ID}}}$ in Eq. (8-108):

$$\overline{\Delta B_j} = \bar{B}_j - \bar{B}_j^{+} \tag{8-108a}$$

$$\overline{\Delta B_j^{\text{ID}}} = \bar{B}_j^{\text{ID}} - \bar{B}_j^{+} \tag{8-108b}$$

Thus,

$$\overline{\Delta B}_j^{EX} = \bar{B}_j - \bar{B}_j^{ID} = \bar{B}_j^{EX} \tag{8-109}$$

Therefore, ΔB^{EX} and B^{EX} are identical, provided that the same reference state is used for ΔB and ΔB^{ID}. As mentioned above, the conventional reference state for ΔB^{ID} is taken as pure components at the same temperature, pressure, and state of aggregation of the mixture. Unless otherwise specified, the same reference state is used for ΔB whenever the excess function is applied.

Because the temperature and pressure of the ideal mixture are taken equal to those of the real mixture, it follows that if

$$W = YZ \tag{8-110}$$

where Y is T or P, then

$$W^{EX} = YZ^{EX} \tag{8-111}$$

Thus, it can be shown that

$$H^{EX} = U^{EX} + PV^{EX} \tag{8-112}$$

and

$$G^{EX} = H^{EX} - TS^{EX} \tag{8-113}$$

For an ideal solution, the excess functions are zero, by definition. In general, if the components of a mixture have similar force fields and are not significantly different in size and symmetry, ideal solution behavior is a good first approximation. For mixtures that are free of strong associative forces such as hydrogen bonding, solvation, or complexes, *regular solution behavior* is sometimes a more accurate approximation than ideal behavior. A regular solution is defined as one for which $\bar{S}_j^{EX}$ is zero for all components. Thus,

$$G^{EX} = H^{EX} = \Delta H = \Delta H^{EX} \tag{8-114}$$

or

$$\bar{G}_j^{EX} = \bar{H}_j^{EX} = \overline{\Delta H}_j^{EX} = \overline{\Delta H}_j \tag{8-115}$$

That is, the Gibbs energy of mixing can be synthesized *from knowledge of only the enthalpy of mixing.*

We mention in passing a third model that is less commonly employed; an *athermal* solution is defined as one for which $\bar{H}_j^{EX} \equiv 0$ for all j. In this case, $G^{EX} = -TS^{EX}$, and estimations of G^{EX} are made from liquid models that allow an estimate of the excess entropy of mixing.

8.7 Fugacity and Fugacity Coefficient

We have seen in Chapter 6 that chemical potentials play an important role in defining equilibrium states: Chemical potentials are equal in all phases for multiphase systems, chemical potentials are the same for all permeable components across a membrane, and for chemical equilibrium, the sum of the products of the chemical potentials times their respective stoichiometric multipliers is zero. Thus, it is most desirable to be able to calculate such chemical potentials.

There are, however, three obstacles. First, the numerical value of any chemical potential can only be determined within an arbitrary constant, which is related to a reference entropy. Second, chemical potentials become negatively infinite as the system pressure approaches zero. Third, the chemical potential of a component in a mixture also becomes negatively infinite as the concentration of that component approaches zero.

For these reasons, a new function, the fugacity, is introduced. As will become apparent, the fugacity function may be employed instead of the chemical potential to define phase, membrane, or chemical equilibrium. Also, the fugacity may be numerically determined and is a well-behaved function at both low pressures and/or small concentrations.

The fugacity of a component i in a mixture, $\hat{f}_i$, is defined as

$$\bar{G}_i = \mu_i = RT \ln \hat{f}_i + \lambda_i(T) \tag{8-116}$$

$\hat{f}_i$ is a function of temperature, pressure, and composition.[9] The function $\lambda_i(T)$ is the same as introduced earlier in Eqs. (8-68) and (8-69) to define ideal gases and ideal-gas mixtures. The fugacity of a pure component, f_i, depends on pressure and temperature and is defined as

$$G_i = \mu_i = RT \ln f_i + \lambda_i(T) \tag{8-117}$$

Clearly, comparing Eq. (8-68) with (8-117) or Eq. (8-69) with (8-116), the fugacity of a pure component is equal to the pressure if the gas is ideal, and the fugacity of a component in an ideal-gas mixture is equal to its partial pressure. Since all substances (pure or mixtures) are assumed to approach an ideal-gas state as the pressure is reduced toward zero ($P \longrightarrow P^*$) then equivalent statements would be

$$\lim_{P \to P^*} \frac{f_i}{P} = 1 \tag{8-118}$$

or

$$\lim_{P \to P^*} \frac{\hat{f}_i}{y_i P} = 1 \tag{8-119}$$

The variation of fugacity with pressure is readily determined from the defining equations [Eqs. (8-116) and (8-117)] and Eq. (5-41),

$$\left(\frac{\partial \ln \hat{f}_i}{\partial P}\right)_{T,N} = \frac{1}{RT}\left(\frac{\partial \bar{G}_i}{\partial P}\right)_{T,N} = \frac{\bar{V}_i}{RT} \tag{8-120}$$

$$\left(\frac{\partial \ln f_i}{\partial P}\right)_{T,N} = \frac{1}{RT}\left(\frac{\partial G_i}{\partial P}\right)_{T,N} = \frac{V_i}{RT} \tag{8-121}$$

To find the temperature effect on fugacity, it is convenient first to relate the Gibbs departure function to fugacity. Recalling from Sections 7.4 and 8.4 that a departure function represents the difference between the property in the

[9] $\hat{f}_i$ is not a partial molar property. The $\wedge$ signifies only that the property applies to a component in a mixture.

real state (T, $\underline{V}$ or T, P) and in an ideal-gas state at (T, $\underline{V}^\circ$), where $\underline{V}^\circ = RT/P$, then

$$G_i - G_i^\circ = RT \ln \frac{f_i}{P} \tag{8-122}$$

Equation (8-68) was used to represent G_i° and Eq. (8-117) for G_i. Dividing Eq. (8-122) by RT and then differentiating with respect to temperature at constant pressure, with Eq. (8-72), we obtain

$$\frac{\partial}{\partial T}[(G_i - G_i^\circ)/RT]_P = \left[\frac{\partial \ln (f_i/P)}{\partial T}\right]_P = -\frac{H_i - H_i^\circ}{RT^2} \tag{8-123}$$

Thus, the variation in $\ln f_i$ with temperature is simply the negative of the enthalpy departure function divided by RT^2. For an ideal gas, f_i would not be a function of temperature. In a similar fashion, for a component in a mixture, Eq. (8-122) becomes

$$\bar{G}_i - \bar{G}_i^\circ = RT \ln \frac{\hat{f}_i}{y_i P} \tag{8-124}$$

where Eqs. (8-116), (8-69), and (8-71) have been used. $\bar{G}_i^\circ$ is the partial molar Gibbs energy in an ideal-gas mixture at T, $\underline{V}^\circ$ and for the same composition as the real mixture. Again dividing by RT and differentiating with respect to temperature at constant pressure and composition, with Eq. (8-73), we get

$$\frac{\partial}{\partial T}[(\bar{G}_i - \bar{G}_i^\circ)/RT]_{P,y} = \left[\frac{\partial \ln (\hat{f}_i/y_i P)}{\partial T}\right]_{P,y} = -\frac{\bar{H}_i - \bar{H}_i^\circ}{RT^2} \tag{8-125}$$

But for an ideal-gas mixture, $\bar{H}_i^\circ$ is only a function of temperature and is independent of composition and pressure, so $\bar{H}_i^\circ = H_i^\circ$. Then

$$\left(\frac{\partial \ln \hat{f}_i}{\partial T}\right)_{P,y} = -\frac{\bar{H}_i - H_i^\circ}{RT^2} \tag{8-126}$$

To obtain numerical values of the fugacity, we will find that an equation of state is necessary. Since, as noted several times earlier, such P–$\underline{V}$–T–N relations are normally explicit in pressure, it will be convenient to formulate the problem with T, $\underline{V}$, N as the independent variables. This conclusion suggests that a Legendre transform of the energy into T, $\underline{V}$, N space would be appropriate. Such a transform is the Helmholz energy; $\underline{A}$. Recalling that

$$\underline{A} = \underline{U} - T\underline{S} \tag{8-127}$$

$$d\underline{A} = -\underline{S}\,dT - P\,d\underline{V} + \sum_i \mu_i\,dN_i \tag{8-128}$$

then

$$\left(\frac{\partial \underline{A}}{\partial N_i}\right)_{T,\underline{V},N_j[i]} = \mu_i \tag{8-129}$$

The Helmholz energy-departure function was developed in Eq. (7-81). Rewriting it in extensive form yields

$$\underline{A}(T, \underline{V}, N) - \underline{A}^\circ(T, \underline{V}^\circ, N) = -\int_\infty^{\underline{V}} \left(P - \frac{NRT}{\underline{V}}\right) d\underline{V} + NRT \ln \frac{\underline{V}^\circ}{\underline{V}} \tag{8-130}$$

We can now differentiate Eq. (8-130) with respect to N_i while maintaining T,

$\underline{V}$, $N_j[i]$ constant. Also, using Eq. (8-116) to substitute for μ_i and Eq. (8-69) for μ_i°,

$$RT \ln \frac{\hat{f}_i}{y_i P} = -\int_\infty^{\underline{V}} \left[\left(\frac{\partial P}{\partial N_i}\right)_{T,\underline{V},N_j[i]} - \frac{RT}{\underline{V}}\right] d\underline{V} + RT \ln \frac{\underline{V}^\circ}{\underline{V}}$$

With $\underline{V}^\circ = NRT/P$ and $\underline{V} = ZNRT/P$, then

$$RT \ln \phi_i = RT \ln \frac{\hat{f}_i}{y_i P} = -\int_\infty^{\underline{V}} \left[\left(\frac{\partial P}{\partial N_i}\right)_{T,\underline{V},N_j[i]} - \frac{RT}{\underline{V}}\right] d\underline{V} - RT \ln Z \tag{8-131}$$

Z is the compressibility factor of the mixture.

Note that the integration in Eq. (8-131) is to be carried out at constant T, N, and composition. The term ϕ_i,

$$\phi_i = \frac{\hat{f}_i}{y_i P} \tag{8-132}$$

is termed the *fugacity coefficient* of component i.

Equation (8-131) can also be used to determine the fugacity coefficient for a pure material, ν_i,

$$\nu_i = \frac{f_i}{P} \tag{8-133}$$

By letting $y_i \longrightarrow 1$, then

$$RT \ln \nu_i = RT \ln \frac{f_i}{P} = -\int_\infty^{\underline{V}} \left[\left(\frac{\partial P}{\partial N}\right)_{T,\underline{V}} - \frac{RT}{\underline{V}}\right] d\underline{V} - RT \ln Z \tag{8-134}$$

In either Eq. (8-131) or (8-134), one employs a pressure-explicit equation of state to determine $(\partial P/\partial N_i)_{T,\underline{V},N_j[i]}$ or $(\partial P/\partial N)_{T,\underline{V}}$.

Example 8.10

Determine ν_i for a pure material whose P–$\underline{V}$–T–N behavior can be represented by the Peng–Robinson equation of state, Eq. (7-65).

Solution

In this particular case, rather than use Eq. (8-134), we have already developed the Helmholz energy departure function in Eq. (7-87). Then,

$$\frac{\partial}{\partial N}[(\underline{A} - \underline{A}^\circ)]_{T,\underline{V}} = \frac{\partial}{\partial N}[N(A - A^\circ)]_{T,\underline{V}} = \mu - \mu^\circ = RT \ln \nu$$

So

$$RT \ln \nu = (A - A^\circ) + N\left[\frac{\partial}{\partial N}(A - A^\circ)_{T,\underline{V}}\right]$$

To evaluate the derivative, from Eq. (7-87),

$$\frac{\partial}{\partial N}[(A - A^\circ)]_{T,\underline{V}} = \frac{\partial}{\partial N}\left[RT \ln \frac{\underline{V}^\circ}{\underline{V} - Nb} + \frac{a}{2\sqrt{2}b} \ln \frac{\underline{V} + Nb(1 - \sqrt{2})}{\underline{V} + Nb(1 + \sqrt{2})}\right]_{T,\underline{V}}$$

$$= \frac{bRT}{N(V - b)} - \frac{aV}{N[V(V + b) + b(V - b)]}$$

$$= \frac{RT}{N}(Z - 1)$$

Therefore,

$$RT \ln \nu = (A - A^\circ) + RT(Z - 1)$$

where $A - A^\circ$ is given in Eq. (7-87) and Z is the compressibility factor for the material.

It may be shown that a similar treatment to determine the fugacity coefficient of a component in a mixture, using the Peng–Robinson equation, yields

$$\ln \phi_i = \ln \frac{\hat{f}_i}{y_i P} = \frac{b_i}{b_m}(Z - 1) - \ln (Z - B) + \frac{A}{2\sqrt{2}B}\left(\frac{2\sum_k y_k a_{ik}}{a_m} - \frac{b_i}{b_m}\right) \ln \frac{Z + B(1 - \sqrt{2})}{Z + B(1 + \sqrt{2})} \quad (8\text{-}135)$$

where

$$A = \frac{a_m P}{(RT)^2} \qquad B = \frac{b_m P}{RT}$$

a_m and b_m are as given in Eqs. (8-56) and (8-57).

We have seen that, for a pure component that behaves as an ideal gas, $f_i = P$, and, from Eq. (8-134), ν_i (ideal gas) $= 1.0$. Also, for a component in an ideal-gas mixture, $\hat{f}_i = y_i P$; thus by Eq. (8-131), ϕ_i(ideal-gas mixture) $= 1.0$. For the case where one wishes to model a mixture as an ideal solution over the entire range of composition, Eq. (8-78) relates $\bar{G}_i^{\text{ID}}$ to G_i. With Eqs. (8-116) and (8-117),

$$\overline{\Delta G}_i^{\text{ID}} = \bar{G}_i^{\text{ID}} - G_i = RT \ln \frac{\hat{f}_i^{\text{ID}}}{f_i} = RT \ln y_i \quad (8\text{-}136)$$

Thus,

$$\hat{f}_i^{\text{ID}} = f_i y_i$$

This expression is often termed the *Lewis and Randall rule*. The pure component fugacity must be evaluated at the same temperature and pressure as the mixture and in the same state of aggregation. For example, if the mixture were a gas at T, P, then f_i would be determined for pure i at T, P and as a gas—even though at this T and P, the stable state might be other than a gas (i.e., a liquid or solid). This problem most often occurs when the Lewis and Randall rule is used to estimate mixture fugacities in phase equilibrium calculations; a further discussion of this problem is given in Chapter 10.

The Gibbs–Duhem equation for chemical potentials was derived in Section 5.4 [see Eq. (5-45)]. A comparable relation may be obtained for fugacities. In this case it is more convenient to begin with the Fundamental Equation in entropy representation [see Problem 5.1(b)]:

$$\underline{S} = S(\underline{U}, \underline{V}, N_1, \ldots, N_n) \quad (8\text{-}137)$$

The conjugate coordinates in this case are $(T^{-1}, \underline{U})$, $(PT^{-1}, \underline{V})$, $(-\mu_i T^{-1}, N_i)$. The total Legendre transform of Eq. (8-137) is

$$-\underline{U}\, d\left(\frac{1}{T}\right) - \underline{V}\, d\left(\frac{P}{T}\right) + \sum_i N_i\, d\left(\frac{\mu_i}{T}\right) = 0 \quad (8\text{-}138)$$

Expanding the first two derivatives, and using the definition for enthalpy, $\underline{H} = \underline{U} + P\underline{V}$,

$$\frac{\underline{H}}{T^2}\,dT - \frac{\underline{V}}{T}\,dP + \sum_i N_i\, d\left(\frac{\mu_i}{T}\right) = 0 \tag{8-139}$$

From Eq. (8-116),

$$d\left(\frac{\mu_i}{T}\right) = R\, d\ln \hat{f}_i + d\left[\frac{\lambda_i(T)}{T}\right] \tag{8-140}$$

To simplify, we use Eq. (8-116) once again but differentiate with respect to temperature, keeping pressure and composition constant. Then, with Eqs. (8-73) and (8-126),

$$\frac{\partial}{\partial T}\left(\frac{\mu_i}{T}\right)_{P,y} = -\frac{\bar{H}_i}{T^2} = -\frac{\bar{H}_i - H_i^\circ}{T^2} + \frac{d}{dT}\left(\frac{\lambda_i(T)}{T}\right)$$

Thus

$$d\left(\frac{\lambda_i(T)}{T}\right) = -\frac{H_i^\circ}{T^2}\,dT \tag{8-141}$$

Substituting Eq. (8-141) into (8-140) and this result into Eq. (8-139) yields

$$\sum_i N_i\, d\ln \hat{f}_i = -\left(\frac{\underline{H} - \sum_i N_i H_i^\circ}{RT^2}\right) dT + \frac{\underline{V}}{RT}\,dP \tag{8-142}$$

$$= -\sum_i \frac{N_i(\bar{H}_i - H_i^\circ)}{RT^2}\,dT + \sum_i \frac{N_i \bar{V}_i}{RT}\,dP \tag{8-143}$$

Either of these equations represents the Gibbs–Duhem equation for the fugacity; they may be written in intensive form by dividing by the total moles, N.

In closing this section on fugacity, we note that one other fugacity could have been defined (i.e., the *mixture* fugacity, f_m). In an analogous fashion to Eq. (8-116) or (8-117),

$$G_m = RT\ln f_m + g(\text{composition}, T) \tag{8-144}$$

Although of little utility in itself, it is interesting to note that all the previous relationships for the general property B may be applied to the fugacity function if $\ln f_m$ is used for B and $\ln(\hat{f}_i/y_i)$ for $\bar{B}_i$. Then, for example, Eq. (8-48) becomes

$$\ln f_m = \sum_i y_i \ln \frac{\hat{f}_i}{y_i} \tag{8-145}$$

One could also calculate f_m by assuming the mixture to be a pseudo-pure component (i.e., by maintaining the composition constant). Then the equations developed earlier in this section to determine f_i could also be used to find f_m. In Section 10.5 we do introduce $N\ln f_m$ as an extensive parameter to allow us to treat the problem of supercritical components in multicomponent mixtures.

8.8 Activity and Activity Coefficient

To calculate the fugacity of a component in a mixture, Eq. (8-131) may be used provided that an equation of state applicable for mixtures is available. Equation (8-135) illustrates such a relation when the Peng–Robinson equation of state is employed. Although Eq. (8-131) is normally limited to cases where the mixture is a gas, it can be used in some cases for liquid mixtures.[10] An alternative method, applicable primarily to the liquid phase, involves the use of a new function, the *activity*.

To introduce this function, we must first define a standard or reference state for the mixture. In fact, this was also the case for treating fugacity and, in Eq. (8-124), the reference state chosen was an ideal-gas mixture. For activity, let us denote the reference state by the superscript + and define the activity in terms of a difference in the partial molar Gibbs energy between the real state and the reference state:

$$\bar{G}_i(T, P, x_1, \ldots, x_{n-1}) - \bar{G}_i^+(T, P^+, x_1^+, \ldots, x_{n-1}^+) = RT \ln a_i \tag{8-146}$$

where a_i is the activity of component i. Note that the reference-state temperature is equal to the system temperature, but that the other reference-state conditions, $P^+, x_1^+, \ldots, x_{n-1}^+$, can be chosen arbitrarily. x_i is the mole fraction of i in the real state and x_i^+ is the mole fraction of i in the reference state. (x_i is used in this section rather than y_i to emphasize that the activity concept is primarily of use in condensed phases.)

We may also use Eq. (8-116) to define a fugacity of i in the reference state,

$$\bar{G}_i^+(T, P^+, x_1^+, \ldots, x_{n-1}^+) = RT \ln \hat{f}_i^+ + \lambda_i(T) \tag{8-147}$$

Thus it follows that

$$\bar{G}_i - \bar{G}_i^+ = RT \ln \frac{\hat{f}_i}{\hat{f}_i^+} \tag{8-148}$$

or

$$a_i = \frac{\hat{f}_i}{\hat{f}_i^+} \tag{8-149}$$

A number of reference states are in common use. Probably the most common is that of a *pure material reference* state,[11] that is, pure i at T and P of the mixture *and* in the same state of aggregation as the mixture. For this case

$$\hat{f}_i^+ = f_i(T, P) \tag{8-150}$$

and

$$RT \ln a_i = \bar{G}_i - G_i = \overline{\Delta G}_i \tag{8-151}$$

If the solution were ideal, then

$$\overline{\Delta G}_i = \overline{\Delta G}_i^{\text{ID}} = RT \ln x_i \tag{8-152}$$

or

$$a_i = x_i \qquad \text{(ideal solution)} \tag{8-153}$$

[10] The Benedict–Webb–Rubin [M. Benedict, G. B. Webb, and L. C. Rubin, *Chem. Eng. Prog.*, **47**, 419 (1951)], the Soave [G. Soave, *Chem. Eng. Sci.*, **27**, 1197 (1972)], and the Peng–Robinson [D. -Y. Peng and D. B. Robinson, *Ind. Eng. Chem. Fundam.*, **15**, 59 (1976)] equations of state have been shown to be applicable in determining vapor *and* liquid component fugacities in light hydrocarbon mixtures. This technique is discussed in more detail in Section 10.4.

[11] Alternative standard states are discussed later in this section.

For most solutions, a_i/x_i is not unity; the difference of the ratio from unity is a measure of the nonideality. This ratio is called the *activity coefficient*, γ_i;

$$\gamma_i = \frac{a_i}{x_i} \tag{8-154}$$

or

$$\gamma_i = \frac{\hat{f}_i}{f_i x_i} \tag{8-155}$$

As written, Eq. (8-155) indicates that γ_i represents the deviation of i from that predicted by the Lewis and Randall rule for ideal solutions.

When the pure-solvent reference state is employed, the activity coefficient can be related to the excess Gibbs energy of mixing. Substituting Eq. (8-154) into Eq. (8-151) and using Eq. (8-108) for $\overline{\Delta G_i}$ and Eq. (8-99) for $\overline{\Delta G_i^{ID}}$, we obtain

$$RT \ln (\gamma_i x_i) = \overline{\Delta G_i} = \overline{\Delta G_i^{EX}} + \overline{\Delta G_i^{ID}} = \overline{\Delta G_i^{EX}} + RT \ln x_i$$

or

$$\overline{\Delta G_i^{EX}} = RT \ln \gamma_i \tag{8-156}$$

This simple relationship between $\overline{\Delta G_i^{EX}}$ and $\ln \gamma_i$ allows us to obtain the effects of temperature and pressure on $\ln \gamma_i$ from the derivatives of $\overline{\Delta G_i^{EX}}$:

$$\left(\frac{\partial \ln \gamma_i}{\partial T}\right)_{P,x} = \frac{1}{R}\left[\frac{\partial(\overline{\Delta G_i^{EX}}/T)}{\partial T}\right]_{P,x} = -\frac{\overline{\Delta H_i^{EX}}}{RT^2} = -\frac{\overline{\Delta H_i}}{RT^2} \tag{8-157}$$

$$\left(\frac{\partial \ln \gamma_i}{\partial P}\right)_{T,x} = \frac{1}{RT}\left[\frac{\partial(\overline{\Delta G_i^{EX}})}{\partial P}\right]_{T,x} = \frac{\overline{\Delta V_i^{EX}}}{RT} = \frac{\overline{\Delta V_i}}{RT} \tag{8-158}$$

Differential enthalpies or volumes of mixing ($\overline{\Delta H_i}$ or $\overline{\Delta V_i}$) are normally quite small, so that activity coefficients are weak functions of temperature or pressure.

Many other relations for $\ln \gamma_i$ may be derived by utilizing the equality between $\ln \gamma_i$ and $\overline{\Delta G_i^{EX}}/RT$; that is, all the previous generalized relations for $B = \overline{\Delta B_i^{EX}}$ are applicable. One important result is the *Gibbs–Duhem equation*, which can be found from the generalized Duhem relation, Eq. (8-41), with $\bar{B}_i = \overline{\Delta G_i^{EX}}/RT$ and $B = \Delta G^{EX}/RT$:

$$\sum_i x_i \, d\left(\frac{\overline{\Delta G_i^{EX}}}{RT}\right) = \sum_i x_i \, d \ln \gamma_i$$

$$= -\frac{\Delta H}{RT^2} dT + \frac{\Delta V}{RT} dP \tag{8-159}$$

When activity coefficients for all components in a mixture are calculated or reported in the literature, the data can be examined for consistency. That is, the Duhem equation, Eq. (8-159), is a relationship involving all n activity coefficients, and hence any one can be found if the other $n - 1$ activity coefficients are given. Hence all n activity coefficients are not independent; they must conform to the Duhem equation.

The method described for obtaining $\bar{B}_i$ from B as a function of composition [Eq. (8-36)] can also be used to relate $\ln \gamma_i$ to $\overline{\Delta G_i^{EX}}$. However, a more common method is to employ Eq. (8-156). If this equation is multiplied by N_i and summed over all $i = 1, \ldots, n$, then

$$\Delta \underline{G}^{\text{EX}} = \sum_i N_i \, \overline{\Delta G}_i^{\text{EX}} = RT \sum_i N_i \ln \gamma_i \tag{8-160}$$

Now, we differentiate with respect to N_k, keeping T, P, and all moles constant except k; then, since

$$RT \sum_i N_i \left(\frac{\partial \ln \gamma_i}{\partial N_k} \right)_{T,P,N_j[k]} = 0$$

by virtue of Eq. (8-159), we have

$$RT \ln \gamma_k = \left(\frac{\partial \, \Delta \underline{G}^{\text{EX}}}{\partial N_k} \right)_{T,P,N_j[k]} \tag{8-161}$$

To use Eq. (8-161), one usually proposes a model for the solution behavior that relates $\Delta \underline{G}^{\text{EX}}$ to T, P, and N_i. By simple differentiation, the activity coefficients may be determined. For example, the simplest, nontrivial relation for $\Delta \underline{G}^{\text{EX}}$ in a binary mixture of 1 and 2, which meets the obvious criteria that $\Delta \underline{G}^{\text{EX}} = 0$, as either x_1 or $x_2 \longrightarrow 1.0$ is

$$\frac{\Delta \underline{G}^{\text{EX}}}{RT} = NCx_1x_2 \tag{8-162}$$

where C is a not a function of composition. With Eq. (8-161)

$$\ln \gamma_1 = Cx_2^2 \tag{8-163a}$$

$$\ln \gamma_2 = Cx_1^2 \tag{8-163b}$$

These equations relating γ_1, γ_2 to composition are suitable only for very simple liquid solutions. Prausnitz[12] illustrates their use for the liquid binaries argon–oxygen and benzene–cyclohexane. Identical results are obtained if one assumes the solution to be regular (i.e., $\Delta \underline{S}^{\text{EX}} = 0$) and that $\Delta \underline{H} = \Delta \underline{H}^{\text{EX}}$ is symmetrical in composition. C must then be proportional to $(1/T)$.

$$\frac{\Delta \underline{H}}{RT} = NCx_1x_2 = \frac{\Delta \underline{G}^{\text{EX}}}{RT} \tag{8-164}$$

For this simple case [Eq. (8-162) or (8-164)], the important properties are shown as a function of x_1 in Figure 8.4. In Figure 8.4(a), $C = 0$ (i.e., the solution is ideal).

In Figure 8.4(b), $C = 1$ and in Figure 8.4(c), $C = 2$. Note the increasing degree of nonideality with increasing C, particularly as reflected either in the activity or activity coefficients. These cases are examples of *positive deviation* from ideal solution behavior because $\gamma_1 > 1$ or $\ln \gamma_1 > 0$. Negative deviations would refer to cases where $\gamma_1 < 1$ or $\ln \gamma_1 < 0$. In general, solutions that exhibit positive deviations tend to be partially miscible over a certain range of temperatures. The case shown in Figure 8.4(c) is, in fact, in a state that is at the limit of

[12]Prausnitz, *op. cit.*, p. 194.

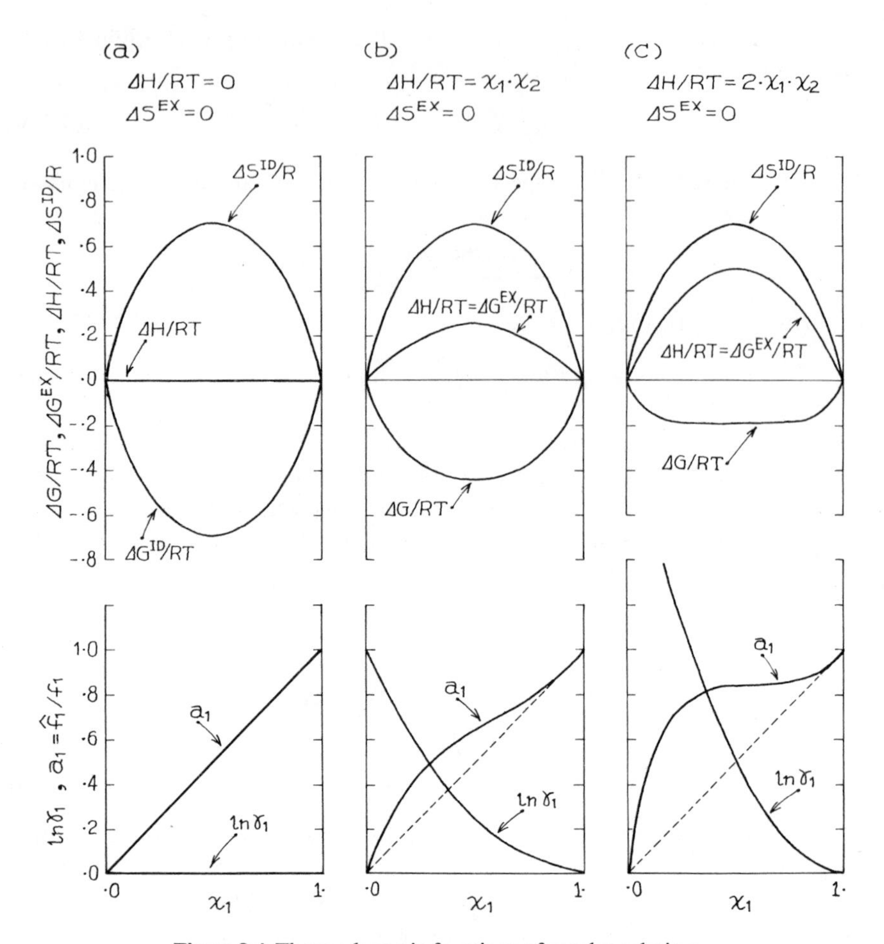

Figure 8.4 Thermodynamic functions of regular solutions.

stability, and any further increase in C (or T) would lead to immiscibility. This behavior is discussed later in Chapter 9.

Many liquid mixtures show very complex functionalities when enthalpies and excess entropies of mixing are plotted versus composition. This point is illustrated in Figure 8.5 for the two binary liquid solutions, ethanol–isooctane and ethanol–benzene. It is interesting to note, however, that in most cases there appear to be compensating factors in ΔH and $T\,\Delta S^{EX}$ such that ΔG^{EX} is only slightly asymmetrical. Thus, ΔG^{EX} can often be correlated with composition by expressions containing only two (or, at most, three) constants. Reid et al.[13] present tables showing the functional form for $\Delta G^{EX}(= \Delta \underline{G}^{EX}/N)$ for many solution models that have been proposed. Also, by using Eq. (8-161), the resultant activity coefficient expressions are given.

[13] R. C. Reid, J. M. Prausnitz, and T. K. Sherwood, *The Properties of Gases and Liquids*, 3rd ed. (New York: McGraw-Hill, 1977), Tables 8-3 and 8-8.

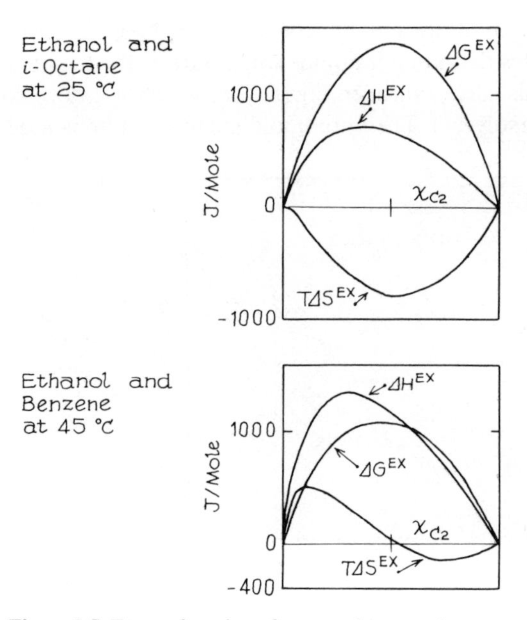

Figure 8.5 Excess functions for some binary mixtures.

Normally, one needs experimental data to be able to obtain the parameters in the ΔG^{EX} (or $\ln \gamma_i$) expressions, or, in some cases, group contribution estimation methods such as the UNIFAC technique developed by Prausnitz and his colleagues can be employed.[14] With values of $\ln \gamma_i$ as a function of composition, fugacities of components in the solution can be found from Eq. (8-155).

In the discussion above we have concentrated on the pure solvent activity coefficient, which is the form most commonly employed. As mentioned at the beginning of this section, there are other forms of activity coefficients that may be more convenient to use if one of the components does not exist as a pure liquid at the temperature and pressure in question.

In general, an activity coefficient can still be defined for any other standard state by the equation

$$\hat{f}_i = \hat{f}_i^+ \gamma_i^+ x_i \tag{8-165}$$

where $\hat{f}_i^+$ is the fugacity of i in the standard state. By analogy to Eq. (8-146),

$$RT \ln \gamma_i^+ + RT \ln x_i = \bar{G}_i(T, P, x_1, \ldots, x_{n-1}) - \bar{G}_i^+(T, P^+, x_1^+, \ldots, x_{n-1}^+) \tag{8-166}$$

Note again the requirement that the standard-state temperature be equal to that of the system, but the pressure, composition, and state of aggregation may be different. For comparison, Eq. (8-166), for the pure solvent standard state, is

$$RT \ln \gamma_i + RT \ln x_i = \bar{G}_i(T, P, x_1, \ldots, x_{n-1}) - G_i(T, P) \tag{8-167}$$

[14] *Ibid.*, p. 347. See also Section 10.4 for further discussion.

The infinite dilution activity coefficient, γ_i^{**}, is a particularly common choice for solutes that are not liquids when pure at the temperature in question. To visualize this state, consider Figure 8.6, which is drawn for a binary of A (solute) and B (solvent). The pure liquid standard state is used for the solvent;

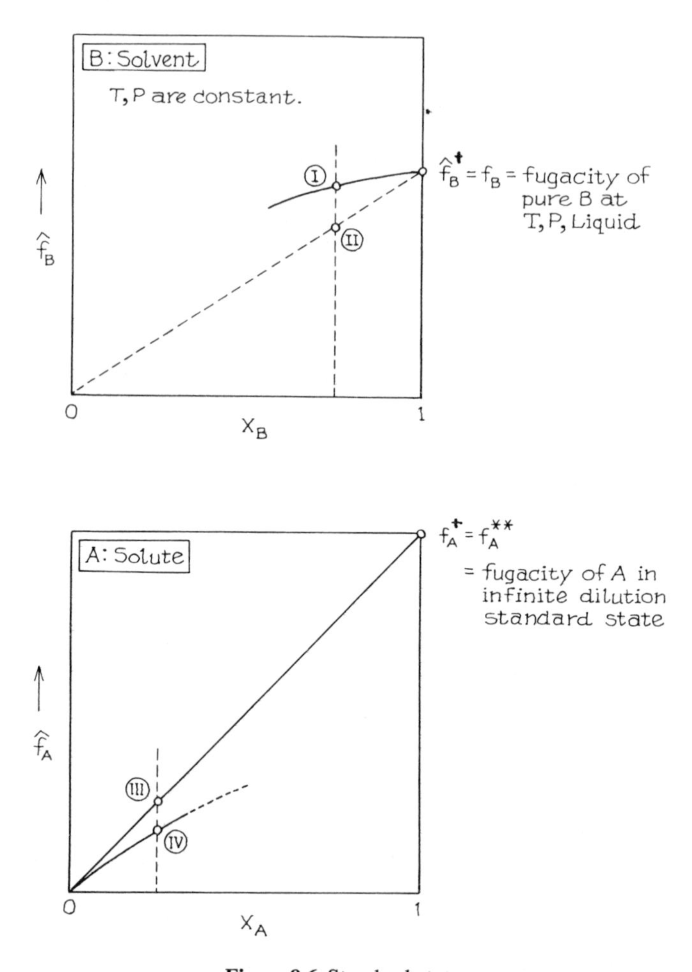

Figure 8.6 Standard states.

that is, $\hat{f}_B^+ = f_B = \hat{f}_B(x_B = 1)$. For the solute A, since data for $\hat{f}_A$ exist only for low concentrations of A, and pure liquid A does not exist at T, P, we define the infinite dilution standard state for A as the intersection of a tangent drawn to the $\hat{f}_A - x_A$ curve as $x_A \longrightarrow 0$ with the ordinate $x_A = 1$. Thus, to determine $\hat{f}_A^+ = \hat{f}_A^{**}$, data at a low concentration of A are all that are required.

In both cases one can express the activity coefficient in terms of the reference-state fugacity:

$$\gamma_B = \frac{\hat{f}_B}{x_B f_B} \tag{8-168}$$

$$\gamma_A^{**} = \frac{\hat{f}_A}{x_A f_A^{**}} \tag{8-169}$$

A pictorial representation of the activity coefficient may also be noted on Figure 8.6 as the ratios

$$\gamma_B = \frac{\hat{f}_B^{\text{I}}}{\hat{f}_B^{\text{II}}}, \qquad \gamma_A^{**} = \frac{\hat{f}_A^{\text{IV}}}{\hat{f}_A^{\text{III}}} \tag{8-170}$$

In these two cases, the use of a standard state as a pure solvent is often called *symmetrical normalization*, but if the infinite dilution standard state is used for the solute, we say that this component is *unsymmetrically normalized.* It is obvious that

$$\begin{aligned} &\text{symmetrical:} \quad &\gamma_i \longrightarrow 1, x_i \longrightarrow 1 \\ &\text{unsymmetrical:} \quad &\gamma_i^{**} \longrightarrow 1, x_i \longrightarrow 0 \end{aligned} \tag{8-171}$$

Other choices of standard states may be used. Many are discussed by Prausnitz.[15] The basic thermodynamic relations for all cases are similar, but care must be taken to denote the reference state. For example, if the infinite dilution reference state were chosen for component j, Eq. (8-157) would become

$$\left[\frac{\partial \ln \gamma_j^{**}}{\partial T}\right]_{P,x} = -\frac{\overline{\Delta H}_j}{RT^2} = -\frac{\bar{H}_j - \bar{H}_j^{**}}{RT^2} \tag{8-172}$$

where $\bar{H}_j^{**}$ is the partial molar enthalpy of j in an infinitely dilute solution. For binary mixtures, this simple definition suffices; for multicomponent mixtures, a more explicit definition of the infinite-dilution standard state is necessary since as $x_j \longrightarrow 0$, $\bar{H}_j^{**}$ can vary depending on the relative amounts of the other constituents.

PROBLEMS

8.1. A small-scale experiment requires the preparation of a 40 mole% sulfuric acid solution from pure water and 80 mole% sulfuric acid. The mixture is to be prepared by metering one fluid at a constant molar rate into a tank containing the second fluid. Cooling is to be provided to maintain the bath at a constant temperature.

At the temperature of operation, 298 K, partial molar enthalpies are listed below [*International Critical Tables*, Vol. 7, (New York: McGraw-Hill, 1930), p. 237]. The term $\bar{H}°(H_2SO_4)$ refers to a state of infinite dilution (i.e., where the mole fraction acid approaches zero). $H°(H_2O)$ is the enthalpy of pure water at 298 K. x is mole fraction and enthalpies are expressed in J/mol.

[15] Prausnitz, *op. cit.*, Chap. 6. The problem of choosing appropriate reference (standard) states for components that are not liquids, when pure, at the system temperature, is considered in more detail in Section 10.5.

Data:

$x_{H_2SO_4}$	$(\bar{H} - H^\circ)_{H_2O}$	$(\bar{H} - \bar{H}^\circ)_{H_2SO_4}$
0	0	0
0.05	−183	17,290
0.10	−1,228	32,360
0.15	−2,428	38,980
0.20	−4,187	46,850
0.25	−6,071	53,090
0.30	−7,997	58,490
0.35	−10,340	63,350
0.40	−12,810	67,660
0.45	−16,250	72,180
0.50	−20,310	76,660
0.55	−23,990	79,720
0.60	−26,380	81,770
0.65	−28,010	82,650
0.70	−29,350	83,360
0.75	−30,480	83,850
0.80	−31,360	84,150
0.85	−32,240	84,300
0.90	−32,950	84,380
0.95	−33,700	84,460
1.00	−34,440	84,570

(a) Which fluid should be charged to the reactor first in order to minimize the rate of heat release?
(b) What is the peak differential heat load per mole of fluid added?
(c) What is the total heat of mixing per mole of final solution for both methods?

8.2. We are faced with the problem of diluting a 90 wt % H_2SO_4 solution with water in the following manner. A tank contains 500 kg of pure water at 298 K; it is equipped with a cooling device to remove any heat of mixing. This cooling device operates with a boiling refrigerant reflux condenser system to maintain the temperature at 298 K. Because of the peculiarities of the system, the rate of heat transfer (W/m^2) must be constant. We wish to add 1500 kg of acid solution (at a variable rate) in 1 h. The acid is initially at 298 K. Enthalpy data are given in Problem 8.1.
(a) Plot the heat of solution (kJ/kg solution) versus weight fraction H_2SO_4 with the reference states as pure water and pure H_2SO_4, liquid, at 298 K.
(b) What is the total heat transferred in the dilution process described?
(c) Derive a differential equation to express the mass flow of 90 wt % acid, kg/min, as a function of the acid concentration in the solution.
(d) Using the result from part (c), determine the mass flow of 90 wt % acid when the overall tank liquid is 64.5 wt % acid.

8.3. In an experiment a mixture of helium and ammonia was prepared as follows. As shown in Figure P8.3, there are separate supply manifolds for the helium and

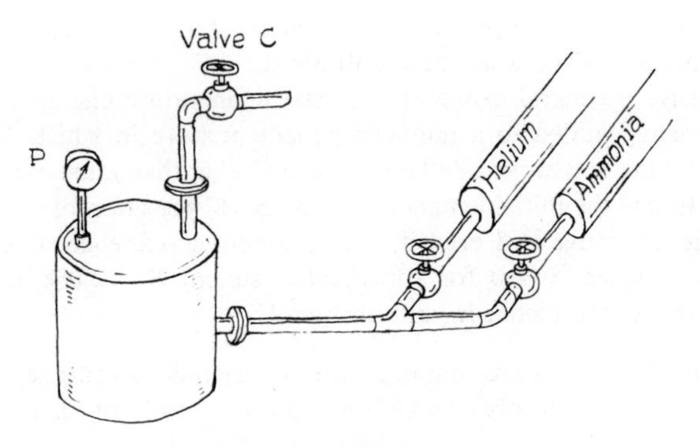

Figure P8.3

ammonia. The aluminum mixing tank is first evacuated to a very low pressure. Helium gas is then admitted very rapidly until the tank pressure is at 2 bar. The helium supply valve is then closed. Ten minutes later, the ammonia supply valve is opened to allow ammonia to flow rapidly into the tank. The valve is closed when the tank pressure reaches 3 bar. A day later, after diffusive mixing, the gas mixture is drawn off through valve C.

Data: Assume ideal gases. The heat capacities of helium and ammonia may be considered to be constants with the following values:

$$C_p(\text{He}) = 20.9 \text{ J/mol K}$$

$$C_p(\text{NH}_3) = 35.6 \text{ J/mol K}$$

The tank dimensions are: 0.3 m (inside diameter), 0.3 m tall. The wall thickness is 1.2 cm and the initial wall temperature is 310 K. C(aluminum) = 1 J/g K and the density of aluminum is about 2700 kg/m^3.

The helium in the manifold is at 310 K and 10 bar, but the ammonia manifold is at 310 K, 5 bar.

With only this description and the given data, what is your best engineering estimate of the composition of the mixture removed from valve C?

8.4. For a single-phase mixture of n components, it is often stated that

$$\sum_{k=1}^{n} N_k \left(\frac{\partial P}{\partial N_k}\right)_{T,\underline{V},N_i[k]} = \kappa_T^{-1}$$

where κ_T is the isothermal compressibility. Derive this relation. What is your physical interpretation of the term

$$N_k \left(\frac{\partial P}{\partial N_k}\right)_{T,\underline{V},N_i[k]}$$

for an ideal-gas mixture?

8.5. In a tank of steam at 2.068 bar and 477.6 K, the Keenan and Keyes steam tables show that the entropy is 7.510 J/g K. The reference entropy is 0.0 for saturated liquid water at 273.2 K. Suppose that you had a tank of steam at 2.068 bar and 477.6 K that contained 10 g of steam. You now injected a small quantity of additional steam (at 2.068 bar and 477.6 K) into the tank under conditions that the total entropy and pressure remained constant. The value of C_p for steam at 2.068 bar and 477.6 K is about 2.01 J/g K.

(a) Derive a general expression for the temperature change per mole of component j added to a multicomponent mixture in which the total entropy, pressure, and moles of all components (other than j) are maintained constant.

(b) Estimate the initial temperature change (K/per gram of steam). What would this derivative had been if, with a different reference state for entropy, we had chosen $S = 0$ for superheated steam, $P = 2.068$ bar, $T = 477.6$ K? How do you reconcile your answers?

8.6. Two simple systems are contained within a cylinder and are separated by a piston (Figure P8.6). Each subsystem is a mixture of $\frac{1}{2}$ mole of nitrogen and $\frac{1}{2}$ mole of

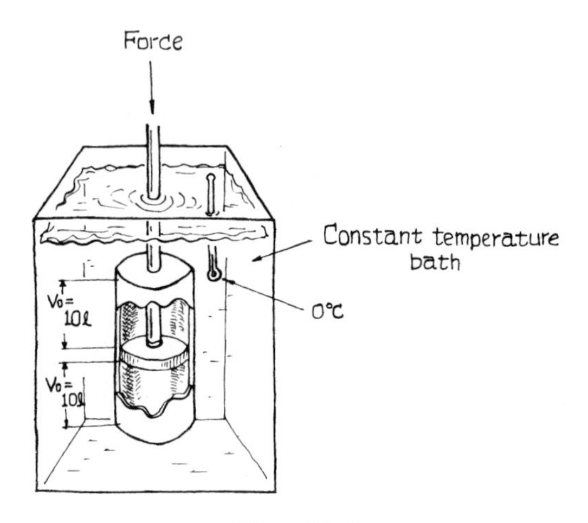

Figure P8.6

hydrogen (consider as ideal gases). The piston is in the center of the cylinder, each subsystem occupying a volume of 10 liters. The walls of the cylinder are diathermal and the system is in contact with a heat reservoir at a temperature of 0°C. The piston is permeable to H_2 but impermeable to N_2. How much work is required to push the piston to such a position that the volumes of the subsystems are 5 and 15 liters?

8.7. A constant-volume crystallizer vessel containing a saturated salt solution at 1 bar is being fed with an unsaturated 5 wt% sodium chloride solution at the same temperature as inside the pot. The water is vaporized and removed at 1 bar, and simultaneously solid salt is removed from the bottom. The heat of vaporization and crystallization is provided by an electric heater. The system is in steady state. How would you calculate:

(a) The entropy production rate of the universe?

(b) The rate of entropy production for the system, consisting of the liquid, solid, and vapor in the crystallizer?

8.8. We have a stream of n-butane and carbon dioxide at 15 MPa and 393.2 K flowing from a well. The composition is 50 mole% carbon dioxide. We would like to make a separation of this stream to end with one stream of 2 mole % CO_2 and one with 90 % CO_2—both streams still at 15 MPa and 393.2 K.

	CO_2		n-Butane
T_c (K)	304.2		425.2
P_c (MPa)	7.376		3.80
ω	0.225		0.193
δ_{12}		0.13	

(a) What is the minimum work required per mole of inlet feed? Use the Peng–Robinson equation of state.

(b) How does this value compare if all streams were assumed to be ideal-gas mixtures?

8.9. I am thinking about purchasing a cottage by the ocean. The location is, however, remote from most amenities of civilization and I would like a reliable power source for lights, electric blankets, etc.

There is a stream near the cottage which flows into the ocean and one of my friends, Rocky Jones, has told me I could obtain the power I need if I would only mix the fresh water in my stream with the ocean water in an appropriate manner. Simply allowing this clear, bubbly stream to cascade merrily into the ocean does seem wasteful in this day and age (Figure P8.9). But how to harness this source?

Figure P8.9

Rocky is willing to manufacture just what I need (so he says) but, friends through we are, his price is outrageous and I would be helpless to repair it since he will only sell sealed units. Before I seriously consider his offer, in any case, could you show me with appropriate detail just how much power I could expect if the unit worked perfectly? (Data are given below.)

Also, I would like some good ideas from an engineer as to just how I might build my own unit to obtain power from this mixing process.

Data:

The stream on my property is essentially at mean sea level; the average flow is 1 m^3/s. The average stream and ocean temperatures are 10°C.

There is a rapid ocean current near the shore to aid the mixing step, if desired.

The ocean averages ~3.5% NaCl and the partial pressure of water over this saline solution (at 10°C) is 1200 N/m^2, whereas for fresh water at 10°C, the vapor pressure is 1228 N/m^2. The densities of the ocean water and the fresh water are, respectively, 1025.5 and 1000 kg/m^3.

$\overline{\Delta H_w}$ and $\overline{\Delta V_w}$ for these dilute saline solutions are essentially zero, and both fresh water and ocean water may be assumed incompressible.

(a) Suggest at least one device (with dimensions, if relevant) and show that the power from this unit will give the same power as you calculated above.
(b) If I need about 10 kW of power, would I have to be very efficient, or is my hope an impossible dream?

8.10. We have a mixture of n components in a vessel that is maintained at constant temperature and pressure. Attached to this vessel is a small compartment that connects to the larger vessel by an ideal semipermeable membrane that is permeable only to one component, j. This small compartment is at the same temperature as the large one, but it is at a different pressure, P^+; at equilibrium it contains only component j.

Obtain relations for

$$\bar{S}_j(T, P, y_1, \ldots) - S_j(T, P^+)$$

when the mixture is:
(a) An ideal solution
(b) A regular solution
(c) An athermal solution
(d) An ideal-gas mixture

8.11. Calculate the osmotic pressure across an ideal semipermeable membrane (permeable to water) for the system NaCl-H_2O at 25°C at pressures of 1 and 500 bar. Consider the concentration range of 0 to 25 wt% NaCl.

Compare your results with the case wherein the solution of NaCl and water is assumed to form an ideal solution.

Osmotic pressures are often calculated by the relation

$$\text{Osmotic pressure} = CRT$$

where C is equal to the molar concentration of the solute. How is this expression derived?

Data are provided in Figures P8.11(a) through (d).

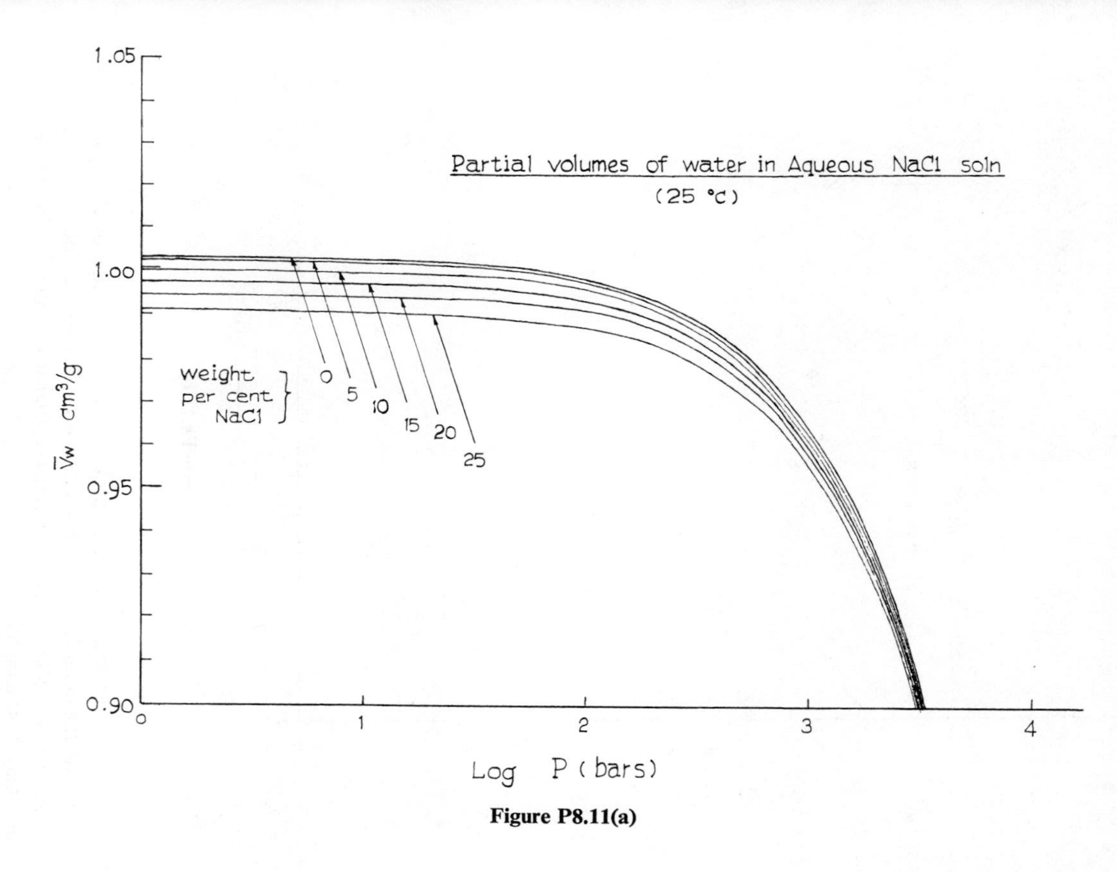

Figure P8.11(a)

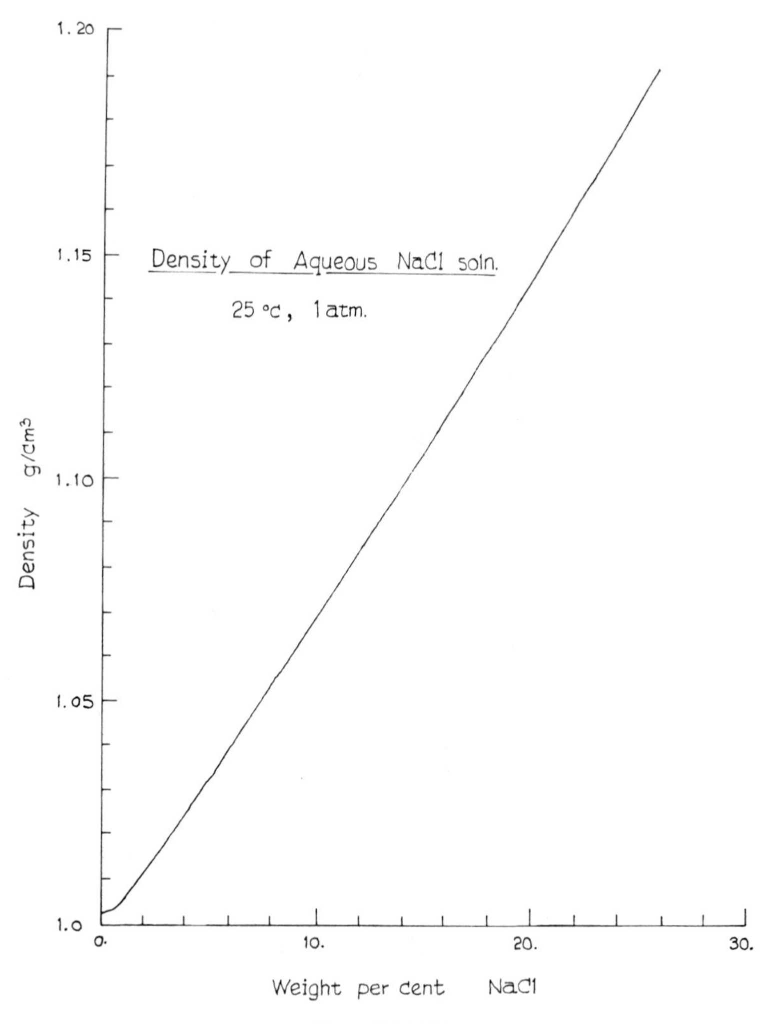

Figure P8.11(b)

8.12. Many industries face a common problem of concentrating dilute aqueous solutions with the minimum expenditure of energy. In the sugar industry, for example, dilute sucrose solutions must be concentrated before purification by crystallization may be attempted.

Numerous water-removal schemes have been suggested as alternatives to the evaporative methods now used, and we wish to evaluate these new schemes.

Data:

Temperature of separation 20°C = 293.2 K

Sucrose molecular weight = 342.30 ($C_{10}H_{22}O_{11}$)

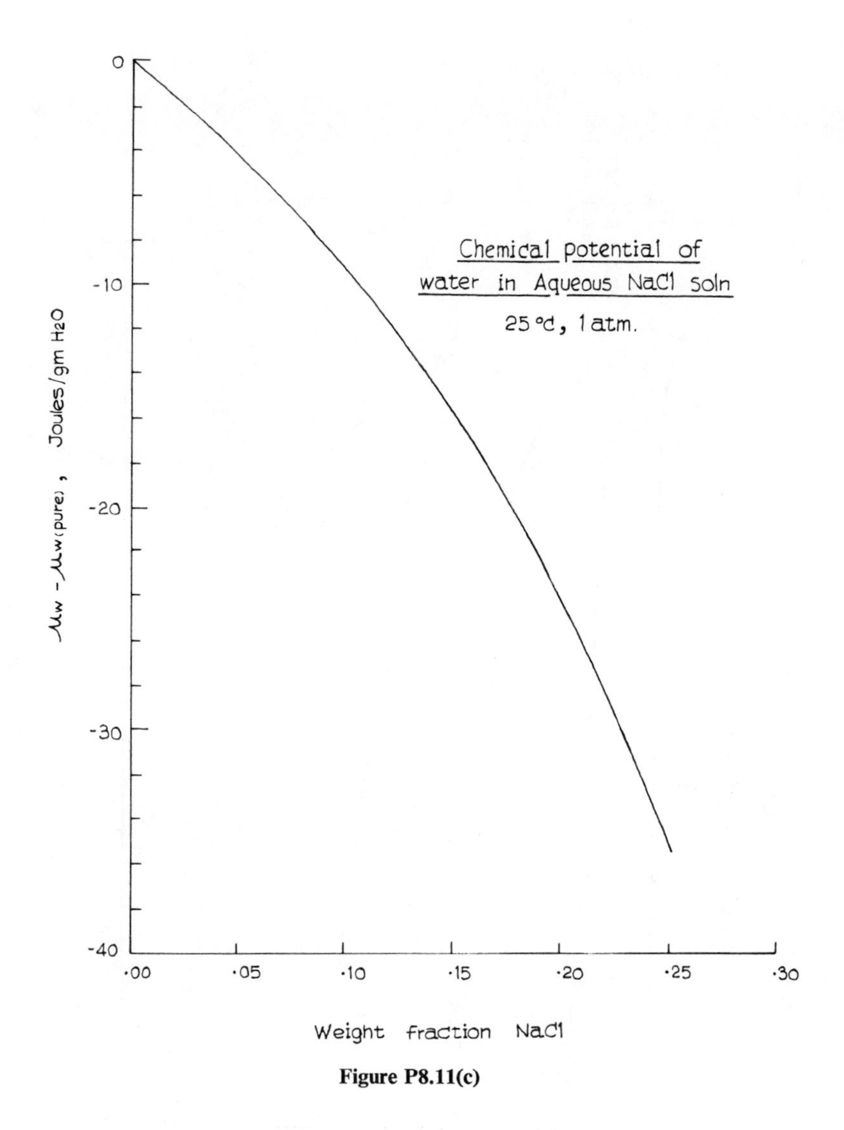

Figure P8.11(c)

The only reliable thermodynamic data for sucrose–water solutions show the activity of water as a function of molality (moles sucrose per kilogram of water). We show these data below and have added, for your convenience, the equivalent mole fraction water and weight fraction sucrose that correspond to the given molality.

The activity of water is defined as p_w/P_{VP_w} where p_w is the partial pressure of water over the sucrose solution and P_{VP_w} is the vapor pressure of pure water at 293.2 K.

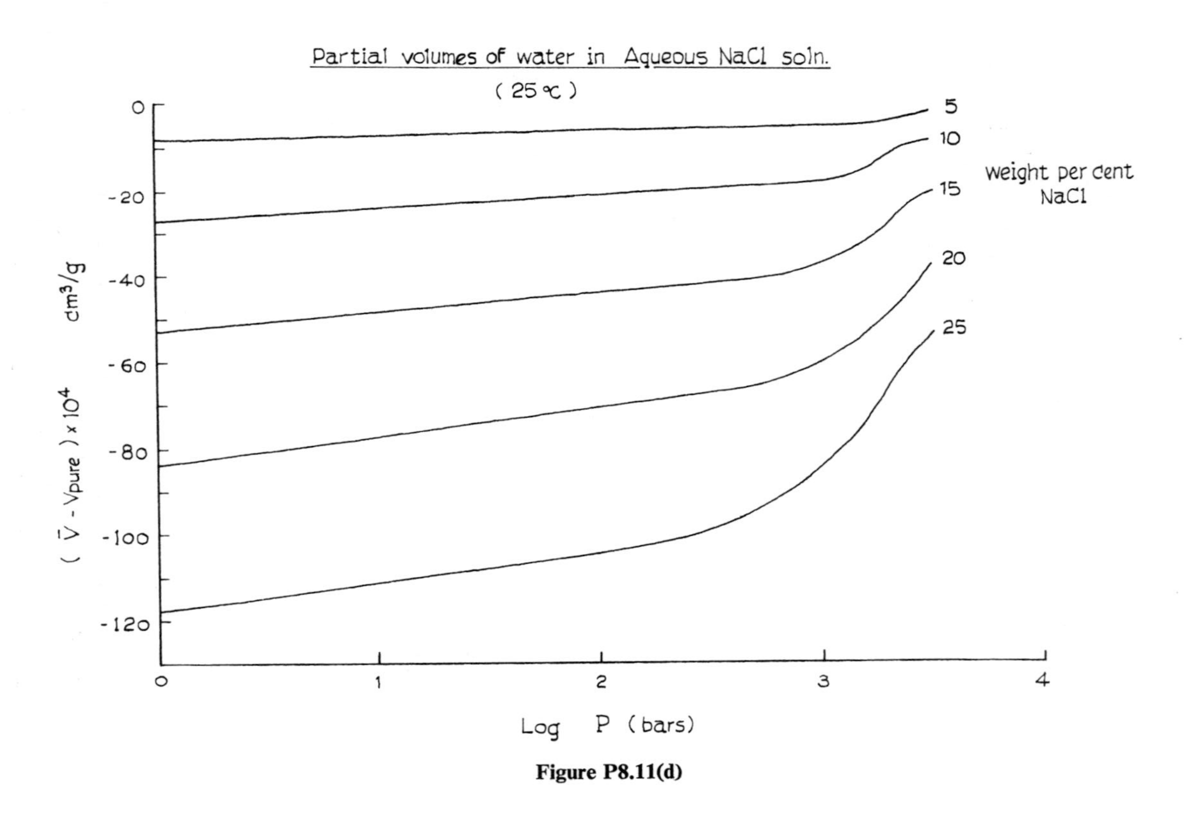

Figure P8.11(d)

Molality (moles sucrose/kg H_2O)	Mole fraction Water	Weight fraction sucrose	Activity of water
0.2	0.99641	0.064	0.99635
0.3	0.99463	0.093	0.99448
0.3249	0.99419	0.10	0.99393
0.4	0.99285	0.120	0.99259
0.5	0.99108	0.146	0.99067
0.6	0.98932	0.170	0.98871
0.8	0.98580	0.215	0.98470
1.0	0.98232	0.255	0.98057
1.2	0.97886	0.291	0.97635
1.253	0.97794	0.30	0.97596
1.5	0.97371	0.339	0.96972

(a) Calculate the *minimum* work required to concentrate a dilute sucrose solution. Assume that we begin with a 10 wt% solution of sucrose in water and we want to end with a 30 wt% sucrose solution. Pick a basis of 1 kg of original 10% solution and express your answer in joules.

(b) Compare your answer to that which would be obtained if you assumed the sucrose–water solution to be ideal (i.e., where the activities are set equal to the mole fraction).

References:

Robinson, R. A., and D. A. Sinclair, *J. Am. Chem. Soc.*, **56**, 1830 (1934).

Sinclair, D. A., *J. Phys. Chem.*, **37**, 495 (1933).

8.13. For the liquid mixture of benzene and cyclohexane, experimental data have shown that the activity coefficient of benzene may be expressed as

$$RT \ln \gamma_B = (3800 - 8T)(1 - x_B)^2$$

where $R = 8.314$ J/mol K
T in K
the subscript B represents benzene

Calculate the entropy and enthalpy of dilution when 1 mole of pure benzene is added to 2 moles of a solution containing 80 mole% cyclohexane at 300 K, 1 bar. The mixing process is isothermal and isobaric.

8.14. Experimental data for the binary liquid mixture of ethanol (E) and methylcyclohexane (M) are shown below.

(1) Enthalpy of mixing and *excess* Gibbs energy of mixing at 35°C.

Mole fraction E	ΔH (J/mol)	ΔG^{EX} (J/mol)
0	0	0
0.0742	452.9	544.2
0.1979	625.4	1036.9
0.3456	670.6	1336.6
0.5324	641.3	1387.2
0.8004	399.3	905.4
1.0	0	0

(2) Excess liquid heat capacities as a function of composition and temperature. Excess heat capacities are defined as

$$C_p^{EX} = C_{p_{mix}} - x_E C_{p_E} - x_M C_{p_M}$$

C_p^{EX} is expressed in J/mol K as

$$C_p^{EX} = \sum_{j=1}^{4} b_j T^{j-1} \qquad \text{with } b_j = f(\text{composition}), \quad T \text{ in K}$$

(See the accompanying table.)

(3) Heat capacities for pure liquid ethanol and methylcyclohexane as a function of temperature,

$$C_p = \sum_{j=1}^{5} b_j t^{j-1}$$

where C_p is in J/mol K and t is in *degrees Celsius.*

MOLE FRACTION ETHANOL

	0.0742	0.1979	0.3456
b_1	3.505 019 85 E+01	6.448 759 04 E+00	2.570 176 4 E+01
b_2	−3.449 617 30 E−01	3.470 872 13 E−02	−1.710 333 5 E−01
b_3	8.597 602 38 E−04	−8.545 534 82 E−04	−1.438 276 5 E−04
b_4	0	2.635 205 92 E−06	1.844 259 9 E−06

MOLE FRACTION ETHANAL

	0.5324	0.8004
b_1	1.457 230 3 E−01	−1.607 368 3 E+01
b_2	1.256 900 5 E−01	2.477 038 3 E−01
b_3	−1.262 525 8 E−03	−1.359 608 9 E−03
b_4	3.155 998 4 E−06	2.543 389 4 E−06

	Ethanol	Methylcyclohexane
b_1	1.036 929 3 E+02	1.749 883 1 E+02
b_2	2.986 102 6 E−01	3.870 387 7 E−01
b_3	1.878 244 3 E−03	8.778 553 5 E−04
b_4	5.558 858 1 E−06	0
b_5	1.200 221 1 E−08	0

We would like to use these data to aid in the design of a low-temperature separation process. Neglect any effect of pressure on the liquid-phase properties.

(a) Estimate the activity coefficients of both ethanol and methylcyclohexane at −85°C for a liquid mixture containing 80 mole% ethanol.

(b) Comment on the phase stability of the liquid mixture at this temperature of −85°C; do you see any evidence that the liquid mixture may split into two liquid phases? If so, indicate your best estimate of the phase compositions; if not, explain clearly why you feel there is no phase split at this temperature.

8.15. Given only the differential heat of solution of cyclohexane in methyl ethyl ketone at 18°C as determined by M. B. Donald and K. Ridgway ["The Binary System Cyclohexane–Methylethyl Ketone," *Chem. Eng. Sci.*, **5**, 188 (1956)], what is:

(a) The integral heat of solution as a function of the mole fraction cyclohexane?

(b) The differential heat of solution of methyl ethyl ketone in cyclohexane as a function of the mole fraction cyclohexane?

Mole fraction cyclohexane	Differential heat of solution of cyclohexane in methyl ethyl ketone (J/mol)
0.025	4831
0.095	4243
0.193	3541
0.400	2410
0.485	1935
0.614	1276
0.783	564
0.875	381

(c) Compare the result in part (b) with the following data. What conclusions do you reach?

Mole fraction cyclohexane	Differential heat of solution of methyl ethyl ketone in cyclohexane (J/mol)
0.124	124
0.238	263
0.381	438
0.491	894
0.598	1497
0.797	3034
0.895	4305
0.958	6651

8.16. Liquid anhydrous ammonia is now shipped in well-insulated tankers. A potential hazard exists if an accident should occur with loss of the liquid into the sea. The liquid ammonia exists at a temperature of about 239 K at 1 bar. In contacting ambient water, presumably some would dissolve and some would vaporize. A few small and medium-sized test spills have indicated, surprisingly, that the fraction of the spilled ammonia which ends up dissolved (as NH_4OH) is constant at about 0.70, independent of the quantity spilled or the exact conditions of the spill. It would be very useful to have an analytical model to describe an ammonia spill.

Develop such a model and show that it does lead to the prediction that a constant fraction of the liquid ammonia dissolves while the remaining fraction vaporizes to be dispersed downwind. Apply this model to a situation where 1000 kg of liquid ammonia contacts 293 K water at 1 bar.

Data:

At 1 bar, the enthalpy of a saturated solution of ammonia in water may be expressed empirically as:

$$H^L = 789.04 - 905.65 \sin (2.7129 - 2.045\, x_N)$$

where x_N is the weight fraction NH_3 and H^L is in J/g.

where $x_N = 1$, H^L (pure liquid NH_3 at 239 K, 1 bar) = 228.1 J/g.

where $x_N = 0$, H^L (pure water at 373 K, 1 bar) = 412.6 J/g.

H^L (pure water at 293 K) = 83.7 J/g.

It may also be assumed that the vapor in equilibrium with an ammonia–water solution is essentially pure NH_3. (This is true except for very dilute solutions.) The enthalpy of ammonia vapor at 1 bar is not a strong function of temperature and may be taken as a constant equal to 1570 J/g.

Stability 9

In this chapter we derive the criteria of stability for thermodynamic systems and determine the conditions when such systems become unstable. This concept of stable and unstable thermodynamic states is of particular importance when one deals with *metastable* states. As an example, let us select liquid water at 80°C and 1 bar as our system. If we were to heat this water isobarically to about 100°C, we would expect to see boiling begin. However, if we could supress all active nucleation sites, it could be possible to heat the *liquid* water well in excess of 100°C with no vaporization. In such a condition we refer to the liquid water as *superheated.* It is in a metastable state. The higher the temperature we attain, the easier it becomes to initiate nucleation and subsequent vaporization. Yet even in this superheated state, it is still *stable* in a thermodynamic sense. There is, however, a limit to the temperature that can be reached before *spontaneous* nucleation occurs. This state is one at the *limit of stability;* one of the important results of this chapter is the derivation of thermodynamic relations to allow one to predict these stability limits.

Superheated liquids are found in *bubble* chambers. Here a liquid (often hydrogen) exists in a state where the temperature exceeds that expected from the prevailing pressure. Nucleation in this case results from the penetration of the chamber by nuclear particles. The track of small bubbles is used to characterize the type and energy of the particle.

All phase changes can be associated with metastable states. Another device to detect and study high-energy nuclear particles is the *cloud chamber*. In this case a vapor is *subcooled;* that is, the pressure and temperature are such that, if equilibrium existed, one would have *expected* to find a condensed phase. But condensation requires a nucleation surface to initiate the formation of the liquid phase. (This process is described in Chapter 12.) Without an efficient nucleating surface, the vapor may be subcooled until the appearance of an energetic particle upon which condensation can take place—or, until the limit of stability of the subcooled vapor is reached. In the latter case, condensation is immediate.

Similar examples could be cited for supercooling in liquids, supersaturation in solutions, etc.

Another example of the use of the equations derived in this chapter relates to some very special cases when a system is both *at the limit of* stability and yet is still a *stable state*. These states are called *critical states*. Critical states do not occur for all phase transformations.

Considerable use is made of Legendre transform notation as developed in Chapter 5. In this manner it is possible to develop stability and critical point critieria in a very general manner. Selection of the specific forms to employ in any given problem is then dictated largely by the types of data or property correlations available.

9.1 Criteria of Stability

Let us examine a simple, isolated system with but a single phase. Let us further suppose that the system satisfies the criteria of equilibrium; that is, T, P, and μ_j are uniform throughout the system and hence $\delta \underline{S} = 0$. To determine if $\underline{S}$ is a maximum, we must show that the lowest-order nonvanishing variation in $\underline{S}$ is negative (see Section 6.2).

It might appear, at first glance, that no internal variations are possible for an isolated, single-phase system. Nevertheless, we can restructure our homogeneous system into one large portion α and a second small portion β (see Figure 9.1). That is, we conceptually insert a membrane enclosing some finite element β inside the system so that we may distinguish this element from the remainder of the system, α. We must, however, allow this membrane to be diathermal, nonrigid, and permeable to all components, so that the composite system is still a simple system. In the original state both subsystems have properties identical to those of the α-phase.

Employing the stability criteria of Eq. (6-7), for small variations, we should examine second-order inequalities first. Of course, if $\delta^2 \underline{S}$ should be zero, we must examine higher-order terms. Let us agree to write second-order derivatives in a shorthand notation such as $S^{\alpha}_{UV} \equiv (\partial^2 \underline{S}^{\alpha} / \partial \underline{U}^{\alpha} \partial \underline{V}^{\alpha})$, etc. Then, if the original system (i.e., the α-phase) is stable,

$$
\begin{aligned}
\delta^2 \underline{S} &= \delta^2(\underline{S}^{\alpha} + \underline{S}^{\beta}) = \delta^2 \underline{S}^{\alpha} + \delta^2 \underline{S}^{\beta} \\
&= S^{\alpha}_{UU}(\delta \underline{U}^{\alpha})^2 + 2S^{\alpha}_{UV}\, \delta \underline{U}^{\alpha}\, \delta \underline{V}^{\alpha} + S^{\alpha}_{VV}(\delta \underline{V}^{\alpha})^2 + \sum_{j=1}^{n} \sum_{k=1}^{n} S^{\alpha}_{N_j N_k}\, \delta N^{\alpha}_j\, \delta N^{\alpha}_k \\
&\quad + 2 \sum_{j=1}^{n} S^{\alpha}_{UN_j}\, \delta \underline{U}^{\alpha}\, \delta N^{\alpha}_j + 2 \sum_{j=1}^{n} S^{\alpha}_{VN_j}\, \delta \underline{V}^{\alpha}\, \delta N^{\alpha}_j \\
&\quad + \text{similar terms for subsystem } \beta < 0
\end{aligned}
\tag{9-1}
$$

Figure 9.1 Conceptual visualization of a subsystem β within a homogeneous system α.

Noting that in the Taylor expansion the second-order partial derivatives are evaluated at the initial conditions, and since the α-phase is identical to the β-phase at the outset of the perturbation, it follows that $N^\alpha S^\alpha_{XY} = N^\beta S^\beta_{XY}$, where X and Y may be $\underline{U}$, $\underline{V}$, or N_j. Furthermore, the α- and β-phase variations are related by the equations of isolation:

$$(\delta \underline{U}^\alpha)^2 = (\delta \underline{U}^\beta)^2 \tag{9-2}$$

$$(\delta \underline{V}^\alpha)^2 = (\delta \underline{V}^\beta)^2 \tag{9-3}$$

$$(\delta N_j^\alpha)^2 = (\delta N_j^\beta)^2 \tag{9-4}$$

With these substitutions, Eq. (9-1) simplifies to

$$\begin{aligned}\delta^2 \underline{S} = \frac{N}{N^\beta}[&S^\alpha_{\underline{U}\underline{U}}(\delta \underline{U}^\alpha)^2 + 2S^\alpha_{\underline{U}\underline{V}}\,\delta \underline{U}^\alpha\,\delta \underline{V}^\alpha + S^\alpha_{\underline{V}\underline{V}}(\delta \underline{V}^\alpha)^2 \\ &+ 2\sum_{j=1}^{n}(S^\alpha_{\underline{U}N_j}\,\delta \underline{U}^\alpha + S^\alpha_{\underline{V}N_j}\,\delta \underline{V}^\alpha)\,\delta N_j^\alpha + \sum_{j=1}^{n}\sum_{k=1}^{n} S^\alpha_{N_jN_k}\,\delta N_j^\alpha\,\delta N_k^\alpha] < 0\end{aligned} \tag{9-5}$$

Note that Eq. (9-5) contains only derivatives and variations for the α-phase. Although the β-phase was introduced to allow us to vary the parameters of the α-phase, we see that the stability of the composite reduces to determining the stability of the original α-phase. Thus, our stability analysis, when completed, will tell us whether or not the original system is stable. If it is unstable, the analysis will not tell us what other phase may form in its place, but it will show that some transformation, leading to a more stable condition, would occur.

Note that we could have eliminated the α terms instead of the β terms, in which case we would have obtained Eq. (9-5) with β superscripts instead of

α superscripts. Since the properties of the α- and β-phases are identical in the initial state, and since the Taylor series expansion is for deviations from that initial state, it is immaterial which superscript we retain.

It is more convenient to explore the consequences of stability in the energy representation instead of the entropy representation, because the $\underline{U}$ criteria can be readily modified to the $\underline{H}, \underline{A}, \underline{G}, \underline{U}'$, etc. forms using the Legendre transform technique discussed in Section 5.5. In the internal energy representation, the analog of Eq. (9-5) is

$$\begin{aligned}\delta^2\underline{U} = \frac{N}{N^\beta}\Big[& U_{SS}(\delta\underline{S})^2 + 2U_{SV}\,\delta\underline{S}\,\delta\underline{V} + U_{VV}(\delta\underline{V})^2 \\ & + 2\sum_{j=1}^{n}(U_{SN_j}\,\delta\underline{S} + U_{VN_j}\,\delta\underline{V})\,\delta N_j \\ & + \sum_{j=1}^{n}\sum_{k=1}^{n} U_{N_jN_k}\,\delta N_j\,\delta N_k\Big] > 0\end{aligned} \tag{9-6}$$

where the superscript α has been dropped.[1]

The conditions where $\delta^2\underline{U} > 0$ or $\delta^2\underline{S} < 0$ are called the criteria of *intrinsic stability*, since they relate to a single phase. The conditions where these criteria are first violated, beginning with a stable phase, are $\delta^2\underline{U} = 0$ or $\delta^2\underline{S} = 0$ and are called the *limits of intrinsic stability*. Systems at the limit of intrinsic stability may or may not be stable; to answer this question, $\delta^3\underline{U}$ or $\delta^3\underline{S}$ (or higher-order) expansions must be examined. We return to this point in Section 9.3.

Equation (9-6) is a general quadratic and may be written in a more condensed form by defining $\underline{U}$ as a basis function $y^{(0)}$. $\underline{U}$ is a function of $n + 2$ independent variables ($\underline{S}, \underline{V}, N_1, \ldots, N_n$).

$$\underline{U} = y^{(0)} = f(x_1, \ldots, x_m) \tag{9-7}$$

where $m = n + 2$ and the x's represent $\underline{S}, \underline{V}, N_1, \ldots, N_n$ with arbitrary ordering. Then Eq. (9-6) becomes

$$\delta^2 y^{(0)} = K\sum_{i=1}^{m}\sum_{j=1}^{m} y_{ij}^{(0)}\,\delta x_i\,\delta x_j > 0 \tag{9-8}$$

where K is a positive numerical constant. It is dropped from the subsequent treatment, as we are only interested in variations that affect the sign of $\delta^2 y^{(0)}$. For a stable system, $\delta^2 y^{(0)}$ is positive; for an unstable system, it is negative. Thus, the limit of intrinsic stability occurs when a system with an initial positive value of $\delta^2 y^{(0)}$ becomes zero.

Since the variations x_i and x_j may be either positive or negative, it is more convenient to rearrange the summations in a sum-of-squares form:

$$\sum_{i=1}^{m}\sum_{j=1}^{m} y_{ij}^{(0)}\,\delta x_i\,\delta x_j = \sum_{k=1}^{m} y_{kk}^{(k-1)}\,\delta Z_k^2 > 0 \tag{9-9}$$

with

$$\delta Z_k = \delta x_k + \sum_{j=k+1}^{m} y_{kj}^{(k)}\,\delta x_j \qquad k = 1, 2, \ldots, m-1 \tag{9-10}$$

$$= \delta x_m \qquad k = m \tag{9-11}$$

[1]To derive Eq. (9-6), Eqs. (9-3) and (9-4) were used. Instead of Eq. (9-2), however, one employs

$$(\delta\underline{S}^\alpha)^2 = (\delta\underline{S}^\beta)^2$$

The advantage of arranging the stability criteria in a sum-of-squares form [Eq. (9-9)] is that δZ_k^2 is always positive irrespective of the sign of δZ_k. Then, since Eq. (9-9) must be positive for *all possible variations* in $\delta \underline{S}^\alpha$, $\delta \underline{V}^\alpha$, and δN_j^α, or for any combination of these variables as given by δZ_k, the stability of the system is dictated by the signs of $y_{kk}^{(k-1)}$. For example, we could hypothesize some variation occurring in the α-β system such that all δZ_k values were zero except one, δZ_r. In this case, the sum in Eq. (9-9) simplifies to

$$y_{rr}^{(r-1)} \, \delta Z_r^2 > 0$$

Therefore,

$$y_{rr}^{(r-1)} > 0$$

Since $y_{rr}^{(r-1)}$ is determined by the state of the original α–β system and is independent of the proposed variation, $y_{rr}^{(r-1)}$ must be positive for a stable system for *any possible variation.* It follows logically that, since the index r was chosen arbitrarily, for a stable system the necessary conditions are that

$$y_{kk}^{(k-1)} > 0, \qquad k = 1, 2, \ldots, m-1 \tag{9-12}$$

Note that in Eq. (9-12), the index k did not include the case when $k = m$. For this special situation, we can show that $y_{mm}^{(m-1)}$ is identically zero. That is,

$$y_{mm}^{(m-1)} = \left(\frac{\partial^2 y^{(m-1)}}{\partial x_m^2}\right)_{\xi_1,\ldots,\xi_{m-1}} = \left(\frac{\partial \xi_m}{\partial x_m}\right)_{\xi_1,\ldots,\xi_{m-1}} = 0 \tag{9-13}$$

$y_{mm}^{(m-1)}$ must be zero since, by specifying $(m-1)$ intensive variables $(\xi_1, \ldots, \xi_{m-1})$, all other intensive variables in a system are fixed. That is, since the second derivative itself is a derivative of an intensive property with respect to an extensive property at constant $(m-1)$ intensive variables, it must be identically zero.

The limit of stability is defined as the state when any of the $y_{kk}^{(k-1)}$ derivatives in Eq. (9-12) becomes zero.

Example 9.1

What are the criteria for a stable ternary system composed of components B, C, and D?

Solution

For a ternary system, $m = n + 2 = 5$. The criteria satisfying Eq. (9-12) are that $y_{11}^{(0)}, y_{22}^{(1)}, y_{33}^{(2)}$, and $y_{44}^{(3)}$ are all positive. Note that $y_{55}^{(4)}$ is not included as, by Eq. (9-13), it is zero. If the basis function $\underline{U} = y^{(0)}$ is ordered in the manner,

$$\underline{U} = y^{(0)} = U(\underline{S}, \underline{V}, N_B, N_C, N_D)$$

then

$$y_{11}^{(0)} = U_{SS} > 0$$

$$y_{22}^{(1)} = A_{VV} > 0$$

$$y_{33}^{(2)} = G_{BB} > 0$$

$$y_{44}^{(3)} = G'_{CC} > 0$$

and

$$y_{55}^{(4)} = G''_{DD} = 0$$

It should be clear that a reordering of variables for this basis function will allow other stability criteria to be developed. For example, if the ordering of the variables $\underline{S}$ and $\underline{V}$ were inverted, then

$$y_{11}^{(0)} = U_{VV} > 0$$

$$y_{22}^{(1)} = H_{SS} > 0$$

but the second derivatives of $y^{(2)}$ and $y^{(3)}$ are not affected.

Additional stability criteria may be formulated by using the step-down relations in Tables 5.2 and 5.3. As an illustration, for $y_{44}^{(3)}$ with the ordering of the basis function as shown originally,

$$y_{44}^{(3)} = \frac{\begin{vmatrix} y_{33}^{(2)} & y_{34}^{(2)} \\ y_{34}^{(2)} & y_{44}^{(2)} \end{vmatrix}}{y_{33}^{(2)}} = \frac{\begin{vmatrix} G_{BB} & G_{BC} \\ G_{BC} & G_{CC} \end{vmatrix}}{G_{BB}}$$

$$= \frac{\begin{vmatrix} y_{22}^{(1)} & y_{23}^{(1)} & y_{24}^{(1)} \\ y_{23}^{(1)} & y_{33}^{(1)} & y_{34}^{(1)} \\ y_{24}^{(1)} & y_{34}^{(1)} & y_{44}^{(1)} \end{vmatrix}}{\begin{vmatrix} y_{22}^{(1)} & y_{23}^{(1)} \\ y_{23}^{(1)} & y_{33}^{(1)} \end{vmatrix}} = \frac{\begin{vmatrix} A_{VV} & A_{VB} & A_{VC} \\ A_{VB} & A_{BB} & A_{BC} \\ A_{VC} & A_{BC} & A_{CC} \end{vmatrix}}{\begin{vmatrix} A_{VV} & A_{VB} \\ A_{VB} & A_{BB} \end{vmatrix}}$$

$$= \frac{\begin{vmatrix} y_{11}^{(0)} & y_{12}^{(0)} & y_{13}^{(0)} & y_{14}^{(0)} \\ y_{12}^{(0)} & \cdot & \cdot & \cdot \\ y_{13}^{(0)} & & & \cdot \\ y_{14}^{(0)} & \cdot\ \cdot\ \cdot & & y_{44}^{(0)} \end{vmatrix}}{\begin{vmatrix} y_{11}^{(0)} & y_{12}^{(0)} & y_{13}^{(0)} \\ y_{12}^{(0)} & y_{22}^{(0)} & y_{23}^{(0)} \\ y_{13}^{(0)} & y_{23}^{(0)} & y_{33}^{(0)} \end{vmatrix}} = \frac{\begin{vmatrix} U_{SS} & U_{SV} & U_{SB} & U_{SC} \\ U_{SV} & U_{VV} & U_{VB} & U_{VC} \\ U_{SB} & U_{VB} & U_{BB} & U_{BC} \\ U_{SC} & U_{VC} & U_{BC} & U_{CC} \end{vmatrix}}{\begin{vmatrix} U_{SS} & U_{SV} & U_{SB} \\ U_{SV} & U_{VV} & U_{VB} \\ U_{SB} & U_{VB} & U_{BB} \end{vmatrix}}$$

Thus, there are four ways to express the fact that $y_{44}^{(3)} > 0$. (In fact, there are many more if the ordering of the basis function were changed.) Similar step-down relations for $y_{33}^{(2)}$ and $y_{22}^{(1)}$ are shown below.

$$y_{33}^{(2)} = \frac{\begin{vmatrix} y_{22}^{(1)} & y_{23}^{(1)} \\ y_{23}^{(1)} & y_{33}^{(1)} \end{vmatrix}}{y_{22}^{(1)}} = \frac{\begin{vmatrix} A_{VV} & A_{VB} \\ A_{VB} & A_{BB} \end{vmatrix}}{A_{VV}}$$

$$= \frac{\begin{vmatrix} y_{11}^{(0)} & y_{12}^{(0)} & y_{13}^{(0)} \\ y_{12}^{(0)} & y_{22}^{(0)} & y_{23}^{(0)} \\ y_{13}^{(0)} & y_{23}^{(0)} & y_{33}^{(0)} \end{vmatrix}}{\begin{vmatrix} y_{11}^{(0)} & y_{12}^{(0)} \\ y_{12}^{(0)} & y_{22}^{(0)} \end{vmatrix}} = \frac{\begin{vmatrix} U_{SS} & U_{SV} & U_{SB} \\ U_{SV} & U_{VV} & U_{VB} \\ U_{SB} & U_{VB} & U_{BB} \end{vmatrix}}{\begin{vmatrix} U_{SS} & U_{SV} \\ U_{SV} & U_{VV} \end{vmatrix}}$$

$$y_{22}^{(1)} = \frac{\begin{vmatrix} y_{11}^{(0)} & y_{12}^{(0)} \\ y_{12}^{(0)} & y_{22}^{(0)} \end{vmatrix}}{y_{11}^{(0)}} = \frac{\begin{vmatrix} U_{SS} & U_{SV} \\ U_{SV} & U_{VV} \end{vmatrix}}{U_{SS}}$$

It is clear from Example 9.1 that there are many ways to express criteria for a stable thermodynamic system. The necessary conditions are, however, still given by Eqs. (9-12). If these $(m - 1)$ derivatives (or any equivalent step-down form) are positive for any given ordering of the basis function, the system is stable. One does not have to test all possible forms of the criteria to answer the question of stability.

Although Eqs. (9-12) are necessary to define a stable system, it is possible to show that, if a system is initially in a stable state, as conditions change and the limit of intrinsic stability is approached, the particular derivative $y^{(m-2)}_{(m-1)(m-1)}$ goes to zero before any other. The proof of this statement depends on the use of the step-down relation shown in Table 5.3. In particular, it can be shown that

$$y^{(k-1)}_{kk} = y^{(k-2)}_{kk} - \frac{(y^{(k-2)}_{k(k-1)})^2}{y^{(k-2)}_{(k-1)(k-1)}} \tag{9-14}$$

From the stability criteria, Eqs. (9-12), both $y^{(k-1)}_{kk}$ and $y^{(k-2)}_{(k-1)(k-1)}$ must be positive. It is then clear from Eq. (9-14) that $y^{(k-2)}_{kk}$ must also be positive. It may also be noted that $y^{(k-2)}_{kk}$ would have been the coefficient of δZ_k if the ordering of variables x_{k-1} and x_k were to have been reversed. Thus, it is reasonable to expect no unusual behavior from this term as either $y^{(k-1)}_{kk}$ or $y^{(k-2)}_{(k-1)(k-1)}$ is reduced to simulate a system that is approaching the limit of intrinsic stability. An important, logical conclusion may then be drawn. Suppose that $y^{(k-2)}_{(k-1)(k-1)}$ decreases toward zero. Equation (9-14) then indicates that as $y^{(k-2)}_{(k-1)(k-1)}$ decreases, before it can reach the value zero, $y^{(k-1)}_{kk}$ becomes negative. Thus the positive nature of $y^{(k-1)}_{kk}$ is always violated before that of $y^{(k-2)}_{(k-1)(k-1)}$. Generalizing, one can state that the *necessary* and *sufficient criterion* of stability is

$$y^{(m-2)}_{(m-1)(m-1)} > 0 \tag{9-15}$$

The state of a system at the limit of stability is then expressed as

$$y^{(m-2)}_{(m-1)(m-1)} = 0 \tag{9-16}$$

Equations (9-15) and (9-16) may also be written in terms of lower-order Legendre transforms. A particularly convenient form is obtained using the relations in Table 5.3. For a stable system,

$$\mathfrak{L}_i > 0 \tag{9-17}$$

and at the limit of stability,

$$\mathfrak{L}_i = 0 \tag{9-18}$$

where $0 \leq i \leq m - 2$ and $\mathfrak{L}_i$ is a determinant defined as

$$\mathfrak{L}_i \equiv \begin{vmatrix} y^{(i)}_{(i+1)(i+1)} & y^{(i)}_{(i+1)(i+2)} & \cdots & y^{(i)}_{(i+1)(m-1)} \\ y^{(i)}_{(i+2)(i+1)} & y^{(i)}_{(i+2)(i+2)} & \cdots & y^{(i)}_{(i+2)(m-1)} \\ \vdots & \vdots & & \vdots \\ y^{(i)}_{(m-1)(i+1)} & y^{(i)}_{(m-1)(i+2)} & \cdots & y^{(i)}_{(m-1)(m-1)} \end{vmatrix} \tag{9-19}$$

Actually, with Table 5.3 one obtains the general relation

$$y^{(m-2)}_{(m-1)(m-1)} = \frac{\mathfrak{L}_i}{\prod_{r=i}^{m-3} y^{(r)}_{(r+1)(r+1)}} \tag{9-20}$$

But, since $y^{(r)}_{(r+1)(r+1)}$, $r < m - 2$, are always positive for a stable system and remain positive as $y^{(m-2)}_{(m-1)(m-1)}$ approaches zero, one can employ only the $\mathfrak{L}_i$ determinant to test stability. Equation (9-20) was, in fact, illustrated in Example 9.1. In this case of a ternary system, $m = 5$, and it was shown that

$$\begin{aligned} \mathfrak{L}_3 = y^{(m-2)}_{(m-1)(m-1)} = y^{(3)}_{44} &= \frac{\mathfrak{L}_2}{y^{(2)}_{33}} \\ &= \frac{\mathfrak{L}_1}{y^{(2)}_{33} y^{(1)}_{22}} \\ &= \frac{\mathfrak{L}_0}{y^{(2)}_{33} y^{(1)}_{22} y^{(0)}_{11}} \end{aligned}$$

9.2 Applications to Thermodynamic Systems

Pure substances

In this case

$$\underline{U} = U(\underline{S}, \underline{V}, N) = U(x_1, x_2, x_3) \tag{9-21}$$

To apply the general criterion of stability, Eq. (9-17), one must first choose the ordering of $\underline{S}$, $\underline{V}$, and N. For example, if $\underline{S} = x_1$, $\underline{V} = x_2$, $N = x_3$, then $y^{(0)} = \underline{U}$, $y^{(1)} = \underline{A}$, and $m = 3$. Only two forms of Eq. (9-17) are possible, that is, when $i = 0$ and $i = 1$.

$i = 0$:

$$\mathfrak{L}_0 = \begin{vmatrix} U_{SS} & U_{SV} \\ U_{SV} & U_{VV} \end{vmatrix} > 0 \tag{9-22}$$

$i = 1$:

$$\mathfrak{L}_1 = A_{VV} > 0 \tag{9-23}$$

Equations (9-22) and (9-23) are completely equivalent criteria. With the latter,

$$A_{VV} = \left(\frac{\partial^2 \underline{A}}{\partial \underline{V}^2}\right)_{T,N} = -\left(\frac{\partial P}{\partial \underline{V}}\right)_{T,N} = -\frac{1}{N}\left(\frac{\partial P}{\partial V}\right)_T > 0 \tag{9-24}$$

Equation (9-24) expresses the well-known fact that, for *stable* systems of pure materials, the pressure increases as the volume decreases if the system is constrained to be isothermal and of constant mass.

Yet even Eqs. (9-22) and (9-23) are not unique ways to express the stability criterion for pure materials. Different ordering of the x_1, x_2, x_3 (that is, $\underline{S}$, $\underline{V}$, N) set produces equivalent criteria. In fact, there are six equivalent forms of Eq. (9-15), and these are listed in Table 9.1. If a pure component system were to reach the limit of intrinsic stability, all of these criteria would be violated (i.e., equal zero) simultaneously.

TABLE 9.1 EQUIVALENT STABILITY CRITERIA FOR A PURE MATERIAL[a]

Ordering of (x_1, x_2, x_3)	Stability criterion
$(\underline{S}, \underline{V}, N)$	$A_{VV} = -(\partial P/\partial \underline{V})_{T,N} > 0$
$(\underline{S}, N, \underline{V})$	$A_{NN} = (\partial \mu/\partial N)_{T,\underline{V}} > 0$
$(\underline{V}, \underline{S}, N)$	$H_{SS} = (\partial T/\partial \underline{S})_{P,N} > 0$
$(\underline{V}, N, \underline{S})$	$H_{NN} = (\partial \mu/\partial N)_{P,\underline{S}} > 0$
$(N, \underline{S}, \underline{V})$	$U'_{SS} = (\partial T/\partial \underline{S})_{\mu,\underline{V}} > 0$
$(N, \underline{V}, \underline{S})$	$U'_{VV} = -(\partial P/\partial \underline{V})_{\mu,\underline{S}} > 0$

[a]Following the convention introduced in Chapter 5, potential functions in which one mole number is transformed to a chemical potential are denoted by a prime, those in which two mole numbers are transformed are denoted by a double prime, etc. Thus, $\underline{U}' = f(\underline{S}, \underline{V}, \mu_1, N_2, \ldots, N_n)$.

The criteria shown in Table 9.1 indicate that when a system is stable, all $y_{kk}^{(k-1)}$ must be positive. In the pure-component case, the only other positive term is $y_{11}^{(0)}(U_{SS}, U_{VV}$, or $U_{NN})$. For example, $U_{SS} = (\partial^2 \underline{U}/\partial \underline{S}^2)_{\underline{V},N} = (\partial T/\partial \underline{S})_{\underline{V},N} = T/NC_v > 0$. C_v is then positive. As noted previously, but repeated again for emphasis, U_{SS} never attains a zero or negative value as the system becomes unstable since higher-order criteria (Table 9.1) approach zero values before lower-order terms.

One criterion of stability for a pure material shown in Table 9.1, $A_{VV} > 0$, can be visualized with the aid of Figure 9.2. This is a typical P–V plot for the liquid and gaseous regions of a pure substance; also shown is an isotherm at temperature T. Reducing the pressure at constant temperature, one follows the isotherm to point A, where the liquid is called a saturated liquid and would be in equilibrium with saturated vapor at C. The isotherm for the vapor phase thus normally emerges at C. At both A and C, $(\partial P/\partial V)_T < 0$ and both phases are stable.

Suppose, however, that we propose an experiment in which a liquid under conditions at A is further expanded isothermally to some lower pressure noted by B. The criterion $(\partial P/\partial V)_T < 0$ is still satisfied and B is a stable equilibrium state, at least with respect to small variations in volume. We stress the fact that at B the variation must be small since it is not difficult to show that there are possible variations that could lead to a state of higher entropy; that is,

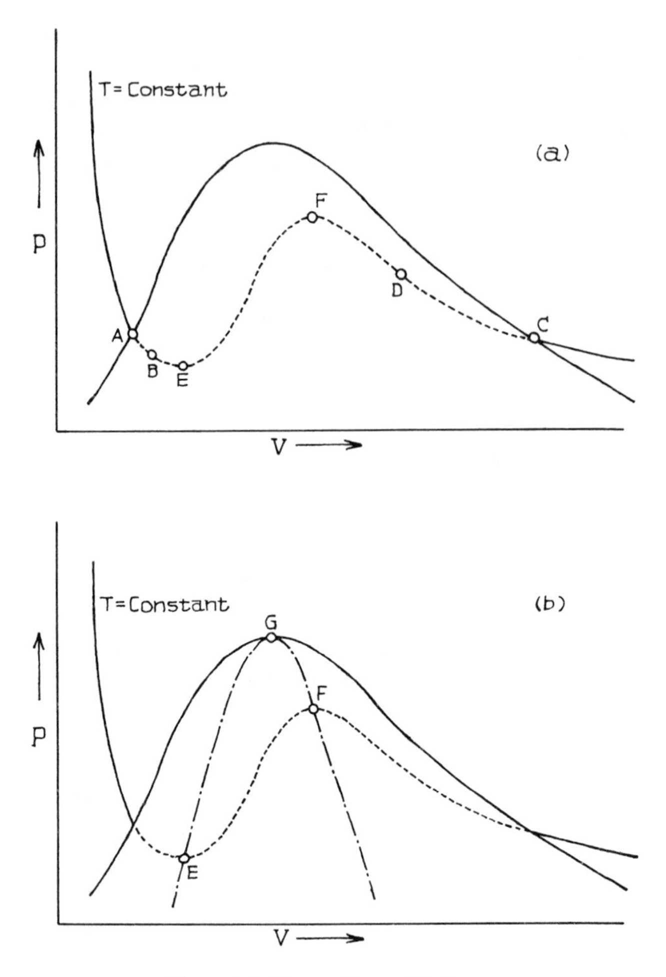

Figure 9.2 Metastable equilibrium.

with regard to some variations, B is not a stable equilibrium state. One obvious variation fitting this description would be to allow some vapor to form. State B is, therefore, metastable. To ever attain this state we would need to start with a very clean, degassed liquid, free of nucleation motes to prevent vapor from forming.

In a similar manner, we could have begun with the saturated vapor at C and compressed isothermally to D. Again, the latter state is metastable.

It is generally assumed that A, B, C, and D lie on a continuous curve. If this is true, the curve would have the general shape of A–B–E–F–D–C in Figure 9.2(a). At E and F it is obvious that $(\partial P/\partial V)_T = 0$, which is the limit of intrinsic stability for the temperature in question. These points fall on the spinodal curve shown by the dot–dashed curve in Figure 9.2(b). States on the spinodal curve are defined by the condition $(\partial P/\partial V)_T = 0$.

To examine more closely the stability of states on the spinodal curve, since we are interesed in variations along an isotherm, we can conveniently apply the Helmholtz energy extrema principle. Equations (6.30) and (6.29), written in the form

$$\delta \underline{A} = 0$$

$$\delta^2 \underline{A} \geq 0, \qquad \delta^m \underline{A} > 0$$

where $\delta^m \underline{A}$ is the lowest-order nonvanishing variation. The significance of $(\delta \underline{A})_{T,\underline{V},N} = 0$ leads to the equality of pressures and chemical potentials throughout the system. By similar reasoning, when there is a variation in volume at constant temperature and mass, the criterion $\delta^2 \underline{A}$ reduces to

$$\delta^2 \underline{A} = -\left(\frac{\partial P}{\partial \underline{V}}\right)_{T,N} \delta \underline{V}^2$$

As we have indicated, for points E and F, $(\partial P/\partial V)_T = 0$, and thus the third-order variation of $\underline{A}$ must be considered. We must then show that

$$\delta^3 \underline{A} = -\left(\frac{\partial^2 P}{\partial \underline{V}^2}\right)_{T,N} \delta \underline{V}^3 > 0$$

or, if this equals zero, then $\delta^4 \underline{A} > 0$, etc.

Now, at point E on Figure 9.2, $(\partial^2 P/\partial V^2)_T$ is clearly positive; thus, one would conclude that for variations in volume that reduce the volume (i.e., $\delta \underline{V}^3 < 0$), the phase is intrinsically stable. If, however, the volume were increased, $\delta \underline{V}^3 > 0$, and the phase would be unstable. Similar reasoning applied to F would indicate that the phase is stable if $\delta \underline{V}^3 > 0$ but not if $\delta \underline{V}^3 < 0$. The arguments may be illustrated on an $\underline{A}$–$\underline{V}$ diagram as shown in Figure 9.3. Points A and C represent saturated liquid and vapor phases at the given temperature. These points have a common tangent, the slope of which is -P. E and F represent states in which $\partial^2 \underline{A}/\partial \underline{V}^2 = 0$ and they correspond to points E and F, respectively, in Figure 9.2. For any system between E and F, $\partial^2 \underline{A}/\partial \underline{V}^2 < 0$ and this system is unstable. Points E and F represent stable states only for particular variations (i.e., for E, $\delta \underline{V} < 0$ and for F, $\delta \underline{V} > 0$).

At higher temperatures, states comparable to E and F on the spinodal curve begin to approach each other and, at the critical point, they coincide. That is, the two phases become indistinguishable (see Figure 9.2, point G) as the intensive properties of each phase (e.g., density, specific enthalpy, etc.) become identical. For a pure substance, if a liquid in equilibrium with its vapor is heated until the meniscus disappears, we denote the vapor pressure at this critical temperature as the critical pressure. At this unique point, from the

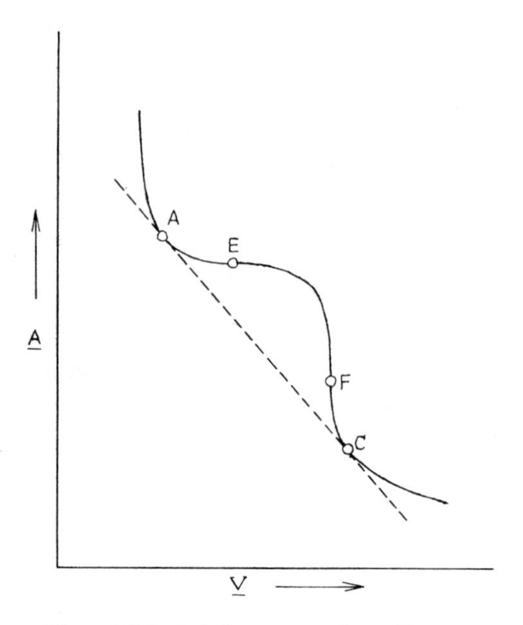

Figure 9.3 Helmholtz energy-volume diagram.

arguments shown above, it is clear that

$$\begin{aligned} \delta \underline{A} &= 0 \\ \delta^2 \underline{A} &= 0 \\ \delta^3 \underline{A} &= 0 \\ \delta^4 \underline{A} &\geq 0 \end{aligned} \tag{9-25}$$

or

$$\begin{aligned} \left(\frac{\partial P}{\partial \underline{V}}\right)_{T,N} &= 0 \\ \left(\frac{\partial^2 P}{\partial \underline{V}^2}\right)_{T,N} &= 0 \\ \left(\frac{\partial^3 P}{\partial \underline{V}^3}\right)_{T,N} &\leq 0 \end{aligned} \tag{9-26}$$

Above the critical point, $(\partial P/\partial V)_T$ is everywhere negative, and all such systems are intrinsically stable.

This treatment has touched only briefly on the extraordinary behavior of matter at the critical point between liquids and gases. Yet the study of matter in this region yields a number of interesting and unusual results. Critical phenomena are considered in more detail in Section 9.3.

Example 9.2

Determine the limits of stability for both liquid *and* vapor *n*-hexane over a wide range of pressures.

Solution

We can use any of the criteria given in Table 9.1. If, however, we select an equation of state such as the Peng–Robinson [Eq. (7-65)] to represent both the vapor and liquid properties of *n*-hexane, the first criterion employing A_{VV} is the most convenient.

The Peng–Robinson equation of state is

$$P = \frac{RT}{V - b} - \frac{a(\omega, T_r)}{V(V + b) + b(V - b)} \tag{A}$$

At the limit of stability, $A_{VV} = 0$ or $(\partial P/\partial \underline{V})_{T,N} = 0$. With N constant,

$$\left(\frac{\partial P}{\partial V}\right)_T = 0 = \frac{-RT}{(V - b)^2} + \frac{2a(\omega, T_r)(V + b)}{[V(V + b) + b(V - b)]^2} \tag{B}$$

The parameters for the Peng–Robinson equation are defined in Eqs. (7-66) through (7-70). For *n*-hexane: $T_c = 507.4$ K, $P_c = 29.7$ bar, and $\omega = 0.296$. Thus, with Eq. (7-67), $a(T_c) = 2.739$ J m^3/mol^2; with Eq. (7-70), $b = 1.105 \times 10^{-4}$ m^3/mol; with Eq. (7-69), $\kappa = 0.807$; and

$$\begin{aligned}\alpha(\omega, T_r) &= \frac{a(\omega, T_r)}{a(T_c)} \\ &= [1 + \kappa(1 - T_r^{1/2})]^2\end{aligned}$$

Equations (A) and (B) may be solved simultaneously to eliminate one of the variable set (P, V, T). Below the critical point there are three roots for V. The smallest represents the limit of stability for liquid phase, and the largest yields the volume at the limit of stability for the subcooled vapor. The intermediate value is not meaningful, as it represents a state in the unstable region.

The results are shown in Figures 9.4 and 9.5. In Figure 9.4, the liquid spinodal curve is drawn from a negative pressure of -10 bar to a pressure near the critical. (Negative pressures simply indicate that the liquid is in tension.) Also shown is the saturated liquid volume curve. In this case, since the liquid is in equilibrium with vapor, negative pressure is not possible. A single isotherm is drawn in Figure 9.4. For this temperature (467 K), at pressures above about 20 bar, the liquid hexane is subcooled. Below about 20 bar, the liquid is superheated. The limit of stability for liquid hexane at 467 K is about 0.8 to 0.9 bar. Another way of interpreting this last point is to visualize the heating of liquid hexane under a pressure of 1 bar. The normal boiling point is reached at about 342 K, but if nucleation from surfaces can be suppressed, the results shown above indicate that it is possible to heat this liquid hexane to 467 K before spontaneous nucleation occurs.

Note that the 467 K isotherm has a slope of zero when it intersects the spinodal curve. This is necessary to match the $A_{VV} = 0$ criterion at the limit of stability.

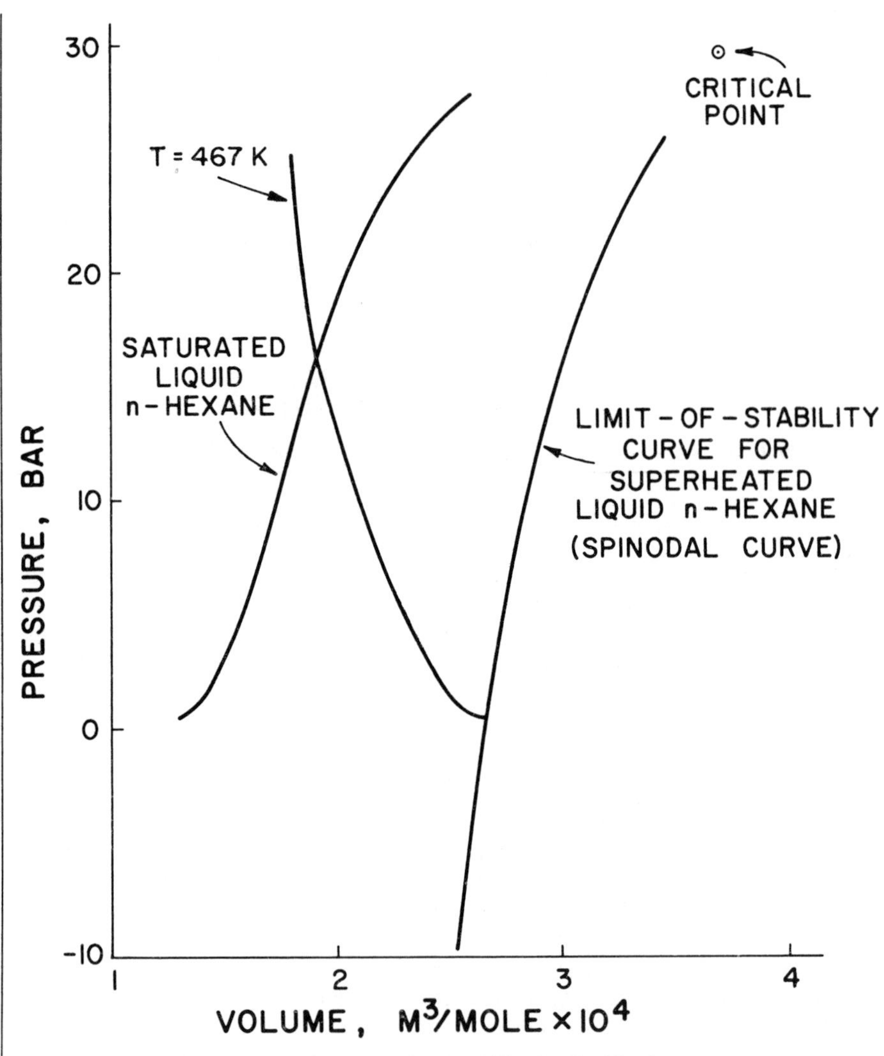

Figure 9.4 Estimated limit of stability for liquid *n*-hexane.

Another way to show the limit of stability for liquid hexane is as in Figure 9.5. Here the spinodal curve is compared against the saturated vapor pressure curve. Both intersect at the critical point. The spinodal curve is nearly linear and extends into the negative-pressure region. Also shown in Figure 9.5 are experimental data for the limit of stability of *n*-hexane as measured by Skripov and Ermakov [*Zh. Fiz. Khim.*, **38**, 396 (1964)]. The agreement is quite good considering that the Peng–Robinson equation of state does not predict liquid-phase properties with high accuracy.

For the vapor spinodal, refer to Figure 9.6. Here the limit of stability for sub-

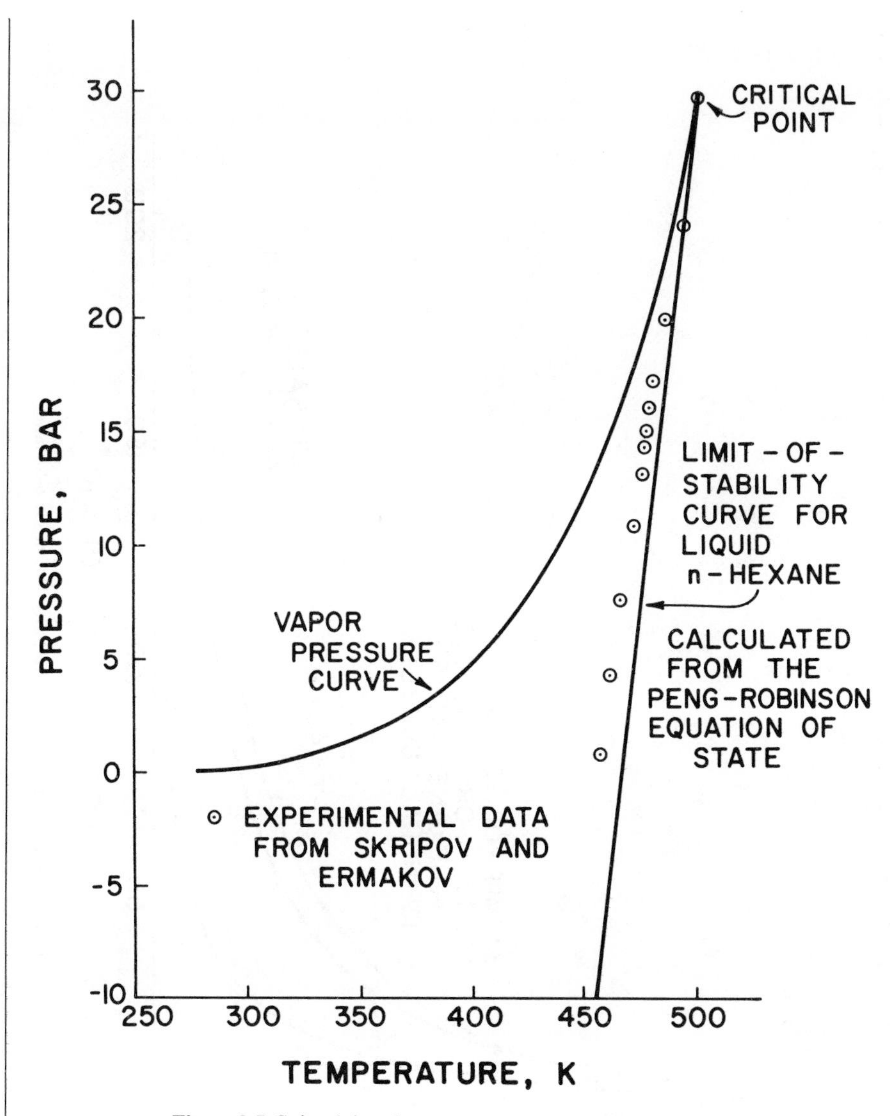

Figure 9.5 Spinodal and vapor pressure curve for *n*-hexane.

cooled *n*-hexane is shown to the left of the curve designating saturated vapor. Between these two curves we have a metastable region with subcooled vapor. Also plotted in this graph is the 467 K isotherm. At pressures below about 16 bar, states on this isotherm are in the superheated *vapor* region. Above 16 bar, we have subcooled vapor. The limit of stability for this isotherm is about 20 bar.

Matching Figures 9.4 and 9.6, one would find the spinodal curves to be continuous and to pass through the critical point.

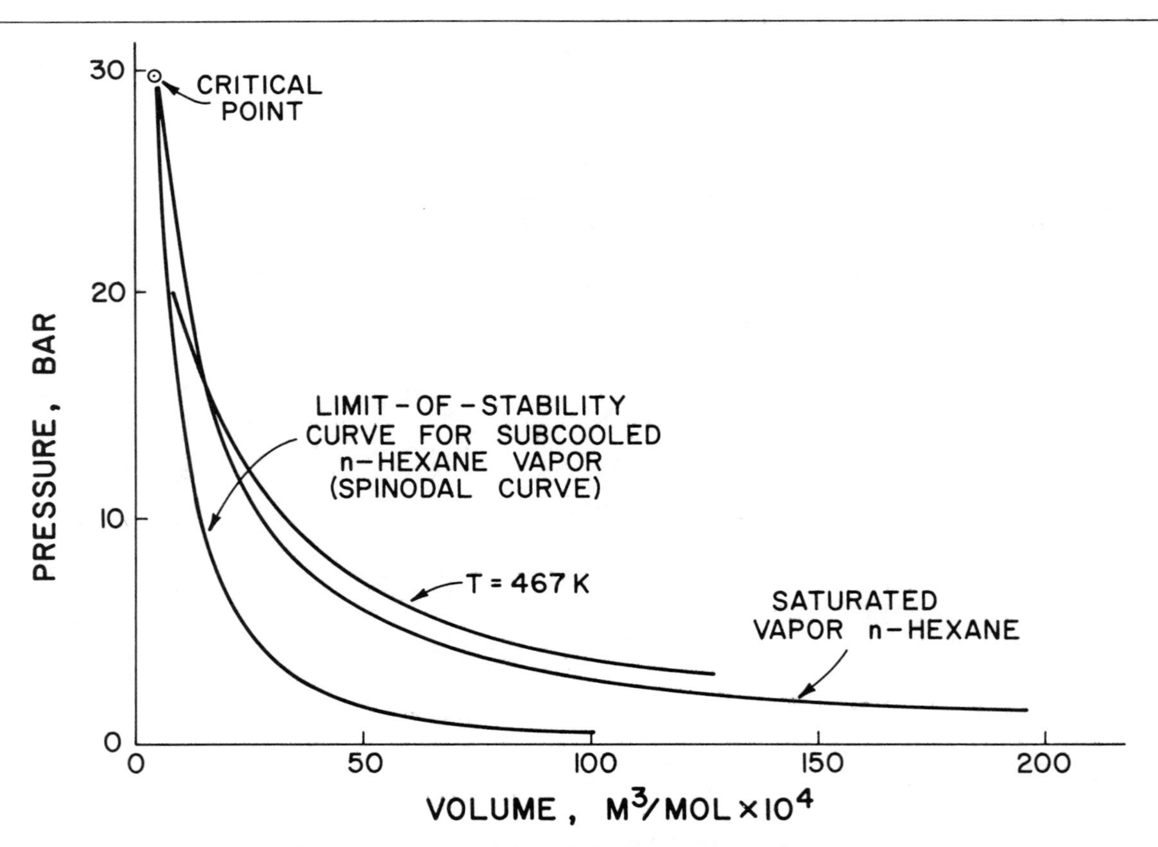

Figure 9.6 Estimated limit-of-stability for vapor *n*-hexane.

Binary systems

For a system composed of B and C,

$$\underline{U} = U(\underline{S}, \underline{V}, N_B, N_C) \tag{9-27}$$

Choosing the $(x_1, \ldots, x_4)$ ordering as given in Eq. (9-27), the stability limit criterion is simply

$$y_{33}^{(2)} = G_{BB} = \left(\frac{\partial^2 G}{\partial N_B^2}\right)_{T,P,N_C} > 0 \tag{9-28}$$

Or, in a more common form,

$$\left(\frac{\partial \mu_B}{\partial N_B}\right)_{T,P,N_C} > 0 \tag{9-29}$$

To visualize this inequality, examine Figure 9.7, in which, for a hypothe-

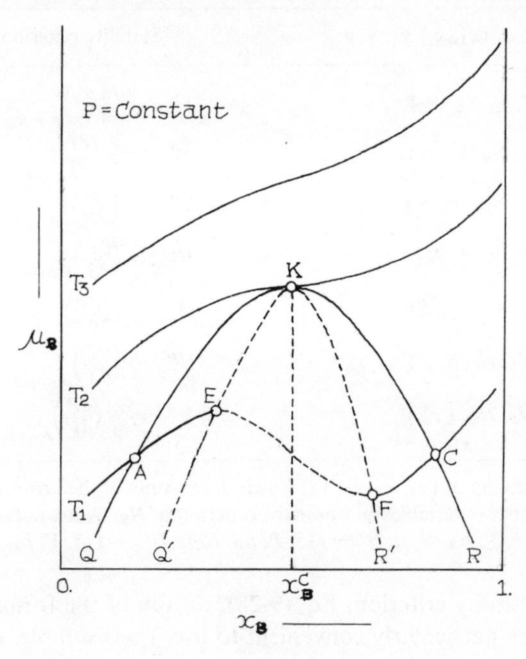

Figure 9.7 Chemical potential as a function of composition for a binary system.

tical binary system at constant pressure, μ_B is plotted against x_B at several different temperatures. QKR defines the locus of all phases that are in equilibrium (i.e., a phase at A is in phase equilibrium with a phase at C at T_1). At these points $(\partial \mu_B / \partial x_B)$ is positive and the phase is stable. A continuation of the T_1 isotherm from A to E or from C to F would define metastable equilibrium states that could be attained experimentally only if the phase transitions were inhibited. Such curves are quite similar to those in Figures 9.2 and 9.4, which were developed earlier to treat mechanical stability. States between E and F are unstable and at E and F,

$$\left(\frac{\partial \mu_B}{\partial x_B}\right)_{P,T} = 0 \tag{9-30}$$

Equation (9-30) defines states on the spinodal curve $Q'KR'$. Both the equilibrium phase envelope (QKR) and the spinodal curve become tangent at K, the critical point for this binary. The critical temperature is at T_2 and x_B^c for this particular pressure. It is obvious that Eq. (9-30) would still apply at point K.

Different stability criteria may be determined if the ordering of the independent variables in Eq. (9-27) is changed. These are shown in Table 9.2. As for pure component systems, *all* the criteria are identical, and if a system should reach the limit of intrinsic stability, *all* are violated simultaneously.

TABLE 9.2
STABILITY CRITERIA FOR BINARY SYSTEMS OF COMPONENTS B AND C

Ordering of (x_1, x_2, x_3, x_4)[a]	Stability criterion[b]
$(\underline{S}, \underline{V}, N_B, N_C)$	$G_{BB} = \left(\frac{\partial \mu_B}{\partial N_B}\right)_{T,P,N_C} > 0$
$(\underline{S}, N_B, \underline{V}, N_C)$	$A'_{VV} = -\left(\frac{\partial P}{\partial \underline{V}}\right)_{T,\mu_B,N_C} > 0$
$(\underline{S}, N_B, N_C, \underline{V})$	$A'_{CC} = \left(\frac{\partial \mu_C}{\partial N_C}\right)_{T,\mu_B,\underline{V}} > 0$
$(\underline{V}, N_B, \underline{S}, N_C)$	$H'_{SS} = \left(\frac{\partial T}{\partial \underline{S}}\right)_{P,\mu_B,N_C} > 0$
$(\underline{V}, N_B, N_C, \underline{S})$	$H'_{CC} = \left(\frac{\partial \mu_C}{\partial N_C}\right)_{P,\mu_B,\underline{S}} > 0$
$(N_B, N_C, \underline{S}, \underline{V})$	$U''_{SS} = \left(\frac{\partial T}{\partial \underline{S}}\right)_{\mu_B,\mu_C,\underline{V}} > 0$
$(N_B, N_C, \underline{V}, \underline{S})$	$U''_{VV} = -\left(\frac{\partial P}{\partial \underline{V}}\right)_{\mu_B,\mu_C,\underline{S}} > 0$

[a] Any ordering of (x_1, x_2, x_3, x_4) which differs only in the arrangement of the first two variables and/or in the ordering of N_B, N_C are not shown.
[b] $\underline{A}' = f(T, \underline{V}, \mu_B, N_C)$; $\underline{H}' = f(\underline{S}, P, \mu_B, N_C)$; $\underline{U}'' = f(\underline{S}, \underline{V}, \mu_B, \mu_C)$.

The binary stability criterion, Eq. (9-28), or any of the forms in Table 9.2, may not always be particularly convenient to use. For example, P–V–T data or correlations are normally expressed in a pressure explicit equation of state; that is,

$$P = f(\underline{V}, T, N_B, N_C) \tag{9-31}$$

In this case, criteria employing the Helmholtz energy are more desirable. With Eq. (9-17) and for $i = 1, y^{(1)} = \underline{A}, x_1 = \underline{S}, x_2 = \underline{V}, x_3 = N_B, x_4 = N_C$,

$$\mathfrak{L}_1 = \begin{vmatrix} y_{22}^{(1)} & y_{23}^{(1)} \\ y_{23}^{(1)} & y_{33}^{(1)} \end{vmatrix} = \begin{vmatrix} A_{VV} & A_{VB} \\ A_{VB} & A_{BB} \end{vmatrix} > 0$$

where

$$A_V = \left(\frac{\partial \underline{A}}{\partial \underline{V}}\right)_{T,N} = -P$$

and

$$A_{VV} = \left(\frac{\partial^2 \underline{A}}{\partial \underline{V}^2}\right)_{T,N} = -\left(\frac{\partial P}{\partial \underline{V}}\right)_{TN},$$

$$A_{VB} = \frac{\partial^2 \underline{A}}{\partial \underline{V}\, \partial N_B} = -\left(\frac{\partial P}{\partial N_B}\right)_{T,\underline{V},N_C}$$

and[2]

$$A_{BB} = \left(\frac{\partial^2 \underline{A}}{\partial N_B^2}\right)_{T,\underline{V},N_C} = \left(\frac{\partial \mu_B}{\partial N_B}\right)_{T,\underline{V},N_C}$$

$$= \int_{\underline{V}}^{\infty} \left(\frac{\partial^2 P}{\partial N_B^2}\right)_{T,\underline{V},N_C} d\underline{V} + \frac{RT}{N_B} \tag{9-32}$$

Thus all the A_{ij} derivatives may be found given an equation of state of the form shown in Eq. (9-31).

As an example of the use of the $\mathfrak{L}_i$ determinant form, consider Figure 9.8.

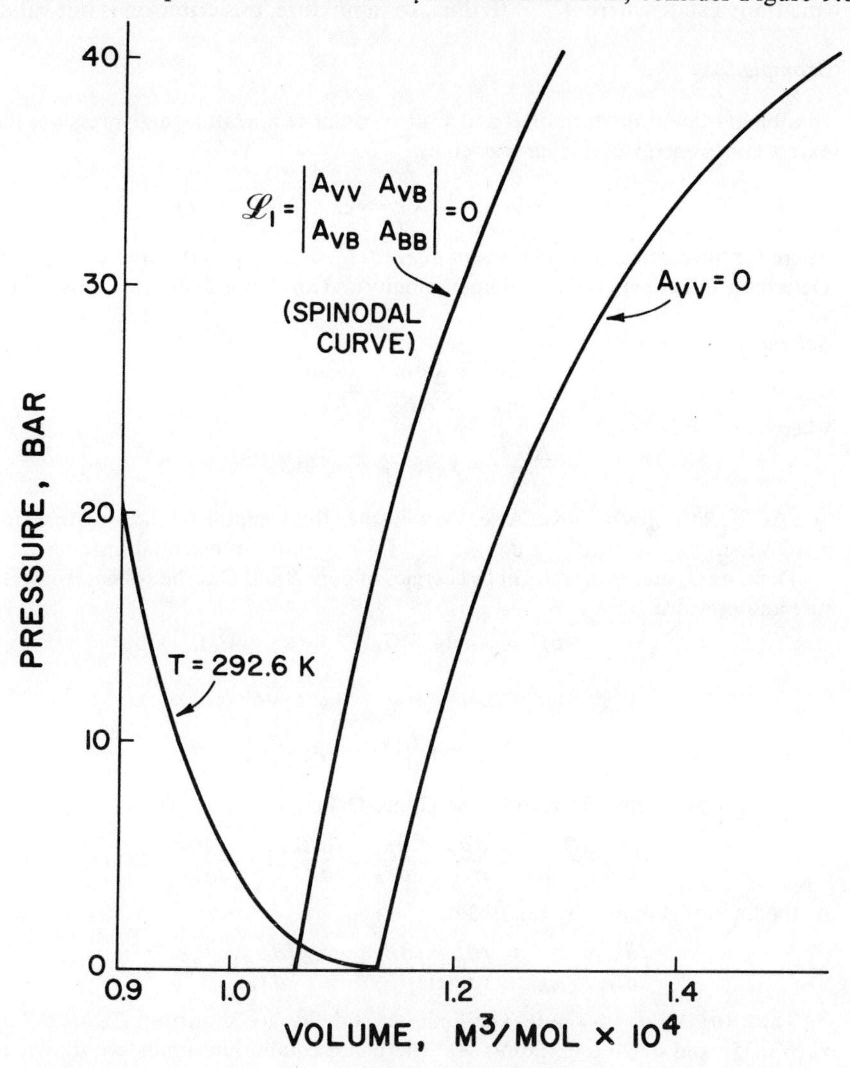

Figure 9.8 Spinodal curve for a liquid mixture of 80% ethane and 20% *n*-butane.

[2]The derivation of Eq. (9-32) is given as Problem 9.2.

On this graph we have plotted the locus where $\mathcal{L}_1 = 0$ for a liquid mixture consisting of 80 mole % ethane and 20 mole % *n*-butane. The derivatives A_{VV}, A_{VB}, and A_{BB} were determined from the Peng–Robinson equation of state. Also shown is one isotherm (292.6 K). To the left of the spinodal curve ($\mathcal{L}_1 = 0$), we have a stable or metastable liquid mixture (the binodal or saturated liquid locus is not shown). The liquid mixture would reach the limit of stability when the thermodynamic state is on the spinodal curve. At 292.6 K, the system pressure would be about 1 bar. We note that if the binary liquid had been treated as a *pure* component, the criterion would have been $A_{VV} = 0$. This locus is also shown in Figure 9.8. Clearly, the *mixture* becomes unstable before it can attain a state where $A_{VV} = 0$; thus, for a mixture, this criterion is not valid.

Example 9.3

In a binary liquid mixture of B and C at constant temperature and pressure, the excess Gibbs energy of mixing is given by

$$\frac{\Delta G^{EX}}{RT} = x_B x_C[k_1 + k_2(x_B - x_C) + k_3(x_B - x_C)^2]$$

where for the particular conditions of interest, $k_1 = 2.0$, $k_2 = 0.2$, and $k_3 = -0.8$. Determine if there are regions of immiscibility and any limits of essential instability.

Solution

$$\frac{\Delta G}{RT} = \frac{\Delta G^{ID}}{RT} + \frac{\Delta G^{EX}}{RT}$$

where

$$\frac{\Delta G^{ID}}{RT} = x_B \ln x_B + x_C \ln x_C$$

and $\Delta G^{EX}/RT$ is given above. Also, we will solve this example by showing that, for a ΔG versus x_B curve, $(\partial^2 \Delta G/\partial x_B^2)_{T,P} = 0$ for any states of essential instability.

Defining G_B and G_C as the Gibbs energies of pure B and C at the system temperature and pressure,

$$\Delta G = x_B(\mu_B - G_B) + x_C(\mu_C - G_C)$$

$$\left(\frac{\partial\, \Delta G}{\partial x_B}\right)_{T,P} = (\mu_B - G_B) - (\mu_C - G_C) + x_B\left(\frac{\partial \mu_B}{\partial x_B}\right) + x_C\left(\frac{\partial \mu_C}{\partial x_B}\right)$$

The last two terms are zero by the Gibbs–Duhem equation. Then

$$\left(\frac{\partial^2\, \Delta G}{\partial x_B^2}\right)_{T,P} = \frac{\partial \mu_B}{\partial x_B} - \frac{\partial \mu_C}{\partial x_B} = \frac{\partial \mu_B}{\partial x_B}\left(1 + \frac{x_B}{x_C}\right)$$

At the limit of stability, by Eq. (9-29),

$$\left(\frac{\partial \mu_B}{\partial N_B}\right)_{T,P,N_C} = \left(\frac{\partial \mu_B}{\partial x_B}\right)\left(\frac{\partial x_B}{\partial N_B}\right) = \frac{x_C}{N}\left(\frac{\partial \mu_B}{\partial x_B}\right) = 0$$

Thus, $(\partial^2\, \Delta G/\partial x_B^2)_{T,P} = 0$. The limits of stability are shown on Figure 9.9 at $x_B = 0.351$ and 0.690 (i.e., points A). The phase equilibrium limits are shown at points B, that is, where a tangent line is common to the curve.

Point C is $\overline{\Delta G}_C/RT$ and D is $\overline{\Delta G}_B/RT$.

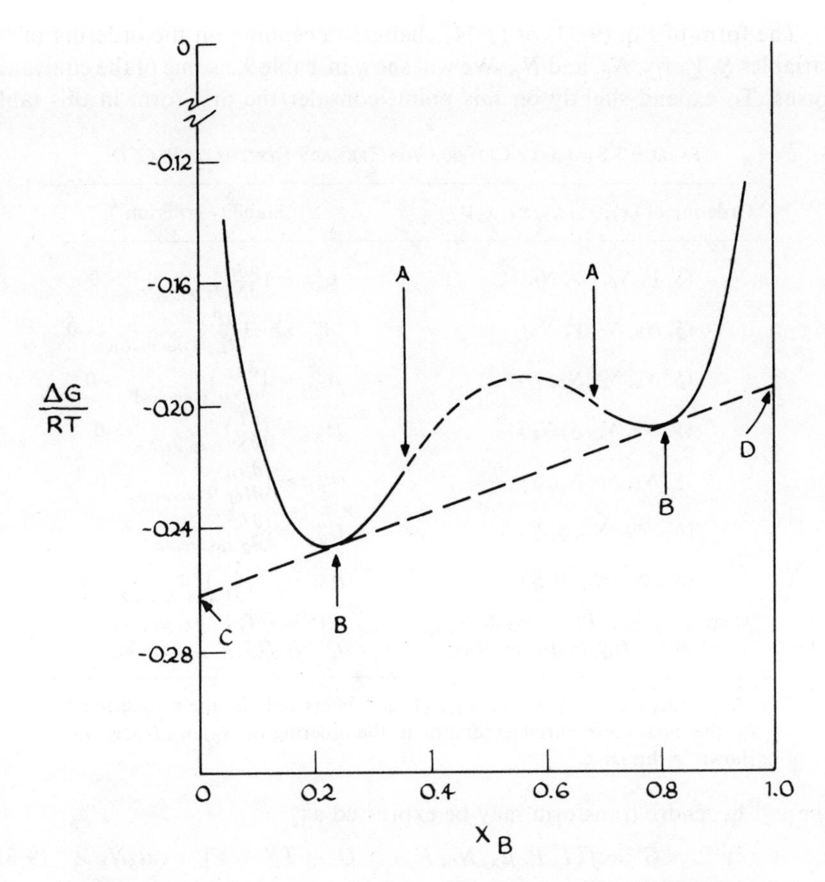

Figure 9.9 Gibbs energies of mixing for a system that exhibits phase splitting.

Ternary systems

The final example is a ternary system of B, C, and D. Extrapolation to multicomponent systems is readily accomplished.

For the system to be stable, with $m = 5$, Eq. (9-15) indicates that

$$y_{44}^{(3)} > 0 \tag{9-33}$$

or if one should desire to express this criterion using derivatives of the Helmholtz energy, then with Eq. (9-19), $i = 1$,

$$\mathfrak{L}_1 = \begin{vmatrix} A_{VV} & A_{VB} & A_{VC} \\ A_{VB} & A_{BB} & A_{BC} \\ A_{VC} & A_{BC} & A_{CC} \end{vmatrix} > 0 \tag{9-34}$$

The second-order derivatives were defined earlier and may be readily determined from a pressure explicit equation of state.

The form of Eq. (9-33) or (9-34) changes depending on the ordering of the variables $\underline{S}$, $\underline{V}$, N_B, N_C, and N_D. We will show in Table 9.3 some of the equivalent forms. To expand slightly on this point, consider the first form in this table.

TABLE 9.3 STABILITY CRITERIA FOR TERNARY SYSTEMS OF B, C, D

Ordering of $(x_1, x_2, x_3, x_4, x_5)$[a]	Stability criterion
$(\underline{S}, \underline{V}, N_B, N_C, N_D)$	$G'_{CC} = \left(\frac{\partial \mu_C}{\partial N_C}\right)_{T,P,\mu_B,N_D} > 0$
$(\underline{S}, N_B, N_C, \underline{V}, N_D)$	$A''_{VV} = -\left(\frac{\partial P}{\partial \underline{V}}\right)_{T,\mu_B,\mu_C,N_D} > 0$
$(\underline{S}, N_B, N_C, N_D, \underline{V})$	$A''_{DD} = \left(\frac{\partial \mu_D}{\partial N_D}\right)_{T,\mu_B,\mu_C,\underline{V}} > 0$
$(\underline{V}, N_B, N_C, \underline{S}, N_D)$	$H''_{SS} = \left(\frac{\partial T}{\partial \underline{S}}\right)_{P,\mu_B,\mu_C,N_D} > 0$
$(\underline{V}, N_B, N_C, N_D, \underline{S})$	$H''_{DD} = \left(\frac{\partial \mu_D}{\partial N_D}\right)_{P,\mu_B,\mu_C,\underline{S}} > 0$
$(N_B, N_C, N_D, \underline{S}, \underline{V})$	$U'''_{SS} = \left(\frac{\partial T}{\partial \underline{S}}\right)_{\mu_B,\mu_C,\mu_D,\underline{V}} > 0$
$(N_B, N_C, N_D, \underline{V}, \underline{S})$	$U'''_{VV} = -\left(\frac{\partial P}{\partial \underline{V}}\right)_{\mu_B,\mu_C,\mu_D,\underline{S}} > 0$
where $\underline{G}' = f(T, P, \mu_B, N_C, N_D)$	$\underline{A}'' = f(T, \underline{V}, \mu_B, \mu_C, N_D)$
$\underline{H}'' = f(\underline{S}, P, \mu_B, \mu_C, N_D)$	$\underline{U}''' = f(\underline{S}, \underline{V}, \mu_B, \mu_C, \mu_D)$

[a]Any ordering of $(x_1, x_2, x_3, x_4, x_5)$ that differs only in the arrangement of the first three variables and/or in the ordering of N_B, N_C, N_D is not shown in this table.

The $y^{(3)}$ Legendre transform may be expressed as

$$y^{(3)} = \underline{G}' = f(T, P, \mu_B, N_C, N_D) = \underline{U} - T\underline{S} + P\underline{V} - \mu_B N_B \tag{9-35}$$

$$dy^{(3)} = d\underline{G}' = -\underline{S}\,dT + \underline{V}\,dP - N_B\,d\mu_B + \mu_C\,dN_C + \mu_D\,dN_D$$

$$\left(\frac{\partial y^{(3)}}{\partial x_4}\right)_{\xi_1,\xi_2,\xi_3,x_5} = \left(\frac{\partial \underline{G}'}{\partial N_C}\right)_{T,P,\mu_B,N_D} = G'_C = \mu_C$$

and

$$y^{(3)}_{44} = \left(\frac{\partial^2 y^{(3)}}{\partial x_4^2}\right)_{\xi_1,\xi_2,\xi_3,x_5} = \left(\frac{\partial^2 \underline{G}'}{\partial N_C^2}\right)_{T,P,\mu_B,N_D} \tag{9-36}$$

$$= G'_{CC} = \left(\frac{\partial \mu_C}{\partial N_C}\right)_{T,P,\mu_B,N_D}$$

9.3 Critical States

The criterion for stable equilibrium is given by Eq. (9-15). An alternative form is

$$y^{(m-2)}_{(m-1)(m-1)} = \left(\frac{\partial \xi_{m-1}}{\partial x_{m-1}}\right)_{\xi_1,\xi_2,\ldots,\xi_{m-2},x_m} > 0 \tag{9-37}$$

Let us assume that we have a stable equilibrium state but by changes in some state variables we are approaching a state where $(\partial \xi_{m-1}/\partial x_{m-1})_{\xi_1,\ldots,\xi_{m-2},x_m}$ is near zero. That is, we are near the limit of essential instability on the spinodal curve. Returning to the hypothetical α-β systems introduced in Section 9.1, at the very limit of essential instability, to test the system for stability, the criterion indicates that we should perturb the x_{m-1} variable and determine how ξ_{m-1} behaves. Or, in different terms, as x_{m-1} is varied, how does $y^{(m-2)}_{(m-1)(m-1)}$ respond? The conditions of restraint indicate that

$$dx^{\alpha}_{m-1} + dx^{\beta}_{m-1} = 0 \tag{9-38}$$

On the spinodal curve (approached from a stable phase), $y^{(m-2)}_{(m-1)(m-1)} = 0$ in both α and β. This must be true, since α and β are identical in the original formulation. To determine the effect of interchanging additional x_{m-1} for a system on the spinodal curve, we must consider the derivative $y^{(m-2)}_{(m-1)(m-1)(m-1)}$. If this derivative were positive, and if $dx_{m-1}(\alpha) > 0$, then $y^{(m-2)}_{(m-1)(m-1)} > 0$ in subsystem α. That is, α is stable with respect to the transfer. However, with Eq. (9-38), $y^{(m-2)}_{(m-1)(m-1)}(\beta) < 0$, and subsystem β has become unstable and must form a new phase. Similar but opposite conclusions are reached if $dx_{m-1} < 0$.

Should there exist a state on the spinodal curve that satisfies Eq. (9-15) and, in addition, is a stable state, then

$$y^{(m-2)}_{(m-1)(m-1)(m-1)} = 0 \tag{9-39}$$

Stable states on the spinodal curve, that is, those which satisfy Eqs. (9-15) and (9-39), are termed *critical states*. If the critical state is to be a stable state, then

$$y^{(m-2)}_{(m-1)(m-1)(m-1)(m-1)} \geq 0 \tag{9-40}$$

and if $y^{(m-2)}_{(m-1)(m-1)(m-1)(m-1)}$ is zero, then the lowest even-order, nonvanishing derivative of $y^{(m-2)}$ must be positive, and all lower-order derivatives must be zero.

The second critical criterion [Eq. (9-39)] also may be written:

$$\left(\frac{\partial^2 \xi_{m-1}}{\partial x^2_{m-1}}\right)_{\xi_1, \xi_2, \ldots, \xi_{m-2}, x_m} = 0 \tag{9-41}$$

For example, with a ternary mixture of B, C, and D. Eq. (9-33) expresses the criterion for a stable system while at the critical point with $m = 5$,

$$\begin{aligned} y^{(3)}_{44} &= G'_{CC} = 0 \\ y^{(3)}_{444} &= G'_{CCC} = 0 \\ y^{(3)}_{4444} &= G'_{CCCC} \geq 0 \end{aligned} \tag{9-42}$$

With a pure material, $m = 3$ and at the critical point,

$$\begin{aligned} y^{(1)}_{22} &= A_{VV} = -\left(\frac{\partial P}{\partial \underline{V}}\right)_{T,N} = 0 \\ y^{(1)}_{222} &= A_{VVV} = -\left(\frac{\partial^2 P}{\partial \underline{V}^2}\right)_{T,N} = 0 \\ y^{(1)}_{2222} &= A_{VVVV} = -\left(\frac{\partial^3 P}{\partial \underline{V}^3}\right)_{T,N} \geq 0 \end{aligned} \tag{9-43}$$

To use lower-order Legendre transforms, the function $\mathfrak{L}_i$ [Eq. (9-19)] is set equal to zero and also

$$\mathfrak{M}_i = 0 \tag{9-44}$$[3]

where

$$\mathfrak{M}_i = \begin{vmatrix} y^{(i)}_{(i+1)(i+1)} & y^{(i)}_{(i+1)(i+2)} & \cdots & y^{(i)}_{(i+1)(m-1)} \\ y^{(i)}_{(i+2)(i+1)} & \vdots & \cdots & y^{(i)}_{(i+2)(m-1)} \\ \vdots & & & \vdots \\ y^{(i)}_{(m-2)(i+1)} & & \cdots & y^{(i)}_{(m-2)(m-1)} \\ \dfrac{\partial \mathfrak{L}_i}{\partial x_{i+1}} & & \cdots & \dfrac{\partial \mathfrak{L}_i}{\partial x_{m-1}} \end{vmatrix} \tag{9-45}$$

As an example, consider an n-component system and order the variables as $\underline{U} = f(\underline{S}, \underline{V}, N_B, N_C, \ldots, N_n)$. Choose $i = 2$. Then $y^{(2)} = \underline{G}$, the Gibbs energy, and with $m = n + 2$ the critical-point criteria are

$$\mathfrak{L}_2 = \begin{vmatrix} G_{BB} & G_{BC} & \cdots & G_{B,n-1} \\ G_{CB} & G_{CC} & \cdots & G_{C,n-1} \\ \vdots & \vdots & & \vdots \\ G_{n-1,B} & & \cdots & G_{n-1,n-1} \end{vmatrix} = 0 \tag{9-46}$$

$$\mathfrak{M}_2 = \begin{vmatrix} G_{BB} & G_{BC} & \cdots & G_{B,n-1} \\ G_{CB} & G_{CC} & \cdots & G_{C,n-1} \\ \vdots & & & \vdots \\ G_{n-2,B} & & \cdots & G_{n-2,n-1} \\ \dfrac{\partial \mathfrak{L}_2}{\partial N_B} & & \cdots & \dfrac{\partial \mathfrak{L}_2}{\partial N_{n-1}} \end{vmatrix} = 0 \tag{9-47}$$

where $G_{BB} = \partial^2 \underline{G}/\partial N_B^2$, etc.

Equations (9-46) and (9-47) are identical to the criteria given by Gibbs.[4] There are, however, many alternative but equivalent criteria, depending on the ordering of the independent variables and on the value of i.

Example 9.4

Ordering variables as $\underline{U} = U(\underline{S}, \underline{V}, N_B, N_C)$ for a binary of B and C, write the critical-point criteria in terms of derivatives of the Helmholtz and Gibbs energies.

Solution

Here $m = 4$, $n = 2$. With $i = 1$, Eqs. (9-18) and (9-44) become

$$\mathfrak{L}_1 = \begin{vmatrix} A_{VV} & A_{VB} \\ A_{BV} & A_{BB} \end{vmatrix} = 0$$

[3]The derivation of Eq. (9-44) is presented in Appendix D.

[4]J. W. Gibbs, "On the Equilibrium of Heterogeneous Substances," *Trans. Conn. Acad.*, **3**, 108 (1876); 343 (1878).

$$\mathfrak{M}_1 = \begin{vmatrix} A_{VV} & A_{VB} \\ \dfrac{\partial \mathfrak{L}_1}{\partial \underline{V}} & \dfrac{\partial \mathfrak{L}_1}{\partial N_B} \end{vmatrix} = 0$$

or

$$\mathfrak{L}_1 = A_{VV}A_{BB} - A_{VB}^2 = 0$$

$$\mathfrak{M}_1 = A_{BBB}A_{VV}^2 - A_{VVV}A_{BB}A_{VB} - 3A_{VBB}A_{VV}A_{VB} + 3A_{VVB}A_{VB}^2 = 0$$

With $i = 2$,

$$G_{BB} = G_{BBB} = 0$$

where

$$G_{BB} = \left(\frac{\partial^2 \underline{G}}{\partial N_B^2}\right)_{T,P,N_C} = \left(\frac{\partial \mu_B}{\partial N_B}\right)_{T,P,N_C}$$

$$G_{BBB} = \left(\frac{\partial^3 \underline{G}}{\partial N_B^3}\right)_{T,P,N_C} = \left(\frac{\partial^2 \mu_B}{\partial N_B^2}\right)_{T,P,N_C}$$

Example 9.5

Ordering variables as $\underline{U} = U(\underline{S}, N_B, N_C, \ldots, N_n, \underline{V})$, write the stability and critical-point criteria in terms of Helmholtz energies.

Solution

Here $m = n + 2$, and $i = 1$, Eqs. (9-18) and (9-44) become

$$\mathfrak{L}_1 = \begin{vmatrix} A_{BB} & A_{BC} & \cdots & A_{Bn} \\ A_{CB} & A_{CC} & & \vdots \\ A_{nB} & \cdots & \cdots & A_{nn} \end{vmatrix} = 0$$

$$\mathfrak{M}_1 = \begin{vmatrix} A_{BB} & A_{BC} & \cdots & A_{Bn} \\ A_{CB} & A_{CC} & & \vdots \\ \vdots & \vdots & & \vdots \\ \dfrac{\partial \mathfrak{L}_1}{\partial N_B} & \cdots & \cdots & \dfrac{\partial \mathfrak{L}_1}{\partial N_n} \end{vmatrix} = 0$$

By ordering $\underline{U}$ in such a manner $\underline{V}$ is a terminal variable, the $\mathfrak{L}_1$ and $\mathfrak{M}_1$ determinants contain only derivatives in mole numbers. No pressure derivatives (such as A_{VV}) appear. Heidemann and Khalil [R. A. Heidemann and A. M. Khalil, *AIChE J.*, **26**, 769 (1980)] utilized these determinants with the Soave modification of the Redlich–Kwong equation [G. Soave, *Chem. Eng. Sci.*, **27**, 1197 (1972)] to calculate critical points of multicomponent systems.

9.4 Indeterminacy

A stable equilibrium state is defined by Eq. (9-12). An argument was then presented, using Eq. (9-14), to show that the derivative $y_{(m-1)(m-1)}^{(m-2)}$ in the equation set (9-12) was the first to attain a value zero as the spinodal curve was approached from a stable region. Equation (9-16) resulted and has been employed to delineate when a stable system reached the limit of stability.

There may, however, be cases where Eq. (9-14) is indeterminate. Rewriting Eq. (9-14) for $y^{(m-2)}_{(m-1)(m-1)}$, we obtain

$$y^{(m-2)}_{(m-1)(m-1)} = \frac{\begin{vmatrix} y^{(m-3)}_{(m-2)(m-2)} & y^{(m-3)}_{(m-2)(m-1)} \\ y^{(m-3)}_{(m-1)(m-2)} & y^{(m-3)}_{(m-1)(m-1)} \end{vmatrix}}{y^{(m-3)}_{(m-2)(m-2)}} = \frac{\mathfrak{L}_{m-3}}{y^{(m-3)}_{(m-2)(m-2)}} \tag{9-48}$$

If both the numerator and denominator attain a value zero simultaneously [as $y^{(m-3)}_{(m-2)(m-2)}$ is decreased to zero], the case is indeterminate. To visualize the problem in a different way, we note that in the *unstable* domain, there may be regions where

$$y^{(k-1)}_{kk} = 0 \qquad 1 \leq k \leq m-2 \tag{9-49}$$

For example, Heidemann[5] shows curves where $y^{(2)}_{33} = 0$ in the unstable region [as defined by $y^{(3)}_{44} = 0$] for a ternary system at constant temperature and pressure. Teja and Kropholler[6] show $y^{(1)}_{22} = 0$ curves in the unstable region [as defined by $y^{(2)}_{33} = 0$] for a binary system. As long as all curves defined by Eq. (9-49) are located in the unstable region [defined by Eq. (9-16)], no difficulties are encountered in numerical computations. Should, however, $y^{(m-3)}_{(m-2)(m-2)}$ become tangent to the spinodal curve at the critical (or any other) point, an indeterminacy results where $y^{(m-2)}_{(m-1)(m-1)}$ and $y^{(m-3)}_{(m-2)(m-2)}$ both become zero.

To remove the indeterminacy, the stability criterion on the spinodal curve may be modified to

$$y^{(m-2)}_{(m-1)(m-1)} y^{(m-3)}_{(m-2)(m-2)} = 0 \tag{9-50}$$

The only case where Eq. (9-50) would fail is when $y^{(m-4)}_{(m-3)(m-3)}$ and lower-order terms become zero simultaneously.

Equation (9-20) can be used to demonstrate this technique by writing the various products in terms of the $\mathfrak{L}$ determinant [see Eq. (9-19)].

$$\begin{aligned}
y^{(m-2)}_{(m-1)(m-1)} &= \mathfrak{L}_{m-2} = \frac{\mathfrak{L}_{m-3}}{y^{(m-3)}_{(m-2)(m-2)}} = 0 \\
y^{(m-2)}_{(m-1)(m-1)} y^{(m-3)}_{(m-2)(m-2)} &= \mathfrak{L}_{m-3} = \frac{\mathfrak{L}_{m-4}}{y^{(m-4)}_{(m-3)(m-3)}} = 0 \\
&\vdots \\
\prod_{r=1}^{m-2} y^{(r)}_{(r+1)(r+1)} &= \mathfrak{L}_i = \frac{\mathfrak{L}_{i-1}}{y^{(i-1)}_{ii}} = 0 \\
&\vdots \\
\prod_{r=0}^{m-2} y^{(r)}_{(r+1)(r+1)} &= \mathfrak{L}_0 = 0
\end{aligned} \tag{9-51}$$

The first equation of this set is Eq. (9-48). If it is indeterminate, the second form can be used—unless $\mathfrak{L}_{m-4}/y^{(m-4)}_{(m-3)(m-3)}$ is indeterminate. Continuing, if necessary, to the case where the product index r has dropped to zero, the criterion is simply $\mathfrak{L}_0 = 0$ and no indeterminacy can result.

[5] R. A. Heidemann, *AIChE J.*, **21**, 824 (1975).

[6] A. S. Teja and H. W. Kropholler, *Chem. Eng. Sci.*, **30**, 435 (1975).

The use of equation set (9-51) is rarely required. In essentially all cases the simple criterion for stability, Eq. (9-15), is sufficient.

Should $\mathfrak{L}_{m-3}/y^{(m-3)}_{(m-2)(m-2)}$ be indeterminate at the critical point, the criterion for criticality [Eq. (9-39)] should be multiplied by $[y^{(m-3)}_{(m-2)(m-2)}]^3$ go avoid numerical difficulties.

9.5 Use of Mole Fractions in the $\mathfrak{L}_i$ and $\mathfrak{M}_i$ Determinants

Throughout this chapter the independent set of variables for the internal energy $\underline{U}$ was total entropy $\underline{S}$, total volume $\underline{V}$, and all mole numbers $N_1, \ldots, N_n$. This set of independent properties delineated the appropriate variables that had to be maintained constant in any derivative of $\underline{U}[y^{(0)}]$ or other Legendre transform $y^{(i)}$. For example, in a ternary, with the variables ordered as $\underline{U} = U(\underline{S}, \underline{V}, N_1, N_2, N_3)$, then

$$\mathfrak{L}_3 = y^{(3)}_{44} = \left(\frac{\partial^2 G'}{\partial N_2^2}\right)_{T,P,\mu_1,N_3} \tag{9-52}$$

$$\mathfrak{L}_2 = \begin{vmatrix} y^{(2)}_{33} & y^{(2)}_{34} \\ y^{(2)}_{43} & y^{(2)}_{44} \end{vmatrix} = \begin{vmatrix} \left(\frac{\partial^2 \underline{G}}{\partial N_1^2}\right)_{T,P,N_2,N_3} & \left(\frac{\partial^2 \underline{G}}{\partial N_1\, \partial N_2}\right) \\ \left(\frac{\partial^2 \underline{G}}{\partial N_1\, \partial N_2}\right) & \left(\frac{\partial^2 \underline{G}}{\partial N_2^2}\right)_{T,P,N_1,N_3} \end{vmatrix} \tag{9-53}$$

etc.

Suppose, however, that the variables chosen were

$$\underline{U} = f(\underline{S}, \underline{V}, N_1, N_2, N) \tag{9-54}$$

where $N = \sum_j N_j$. With the same approach as used earlier, we would have reached the conclusion that $\mathfrak{L}_i = \mathfrak{M}_i = 0$ at the critical point. But the definitions of $\mathfrak{L}_i$ and $\mathfrak{M}_i$ are slightly modified. For example, Eq. (9-52) now becomes

$$\mathfrak{L}_3 = y^{(3)}_{44} = \left(\frac{\partial^2 \underline{G}^+}{\partial N_2^2}\right)_{T,P,\eta_1,N} \tag{9-55}$$

where

$$\underline{G}^+ = \underline{U} - T\underline{S} + P\underline{V} - \eta_1 N_1 \tag{9-56}$$

$$\eta_1 = \left(\frac{\partial \underline{U}}{\delta N_1}\right)_{\underline{S},\underline{V},N_2,N} = \left(\frac{\partial \underline{G}}{\partial N_1}\right)_{T,P,N_2,N} \tag{9-57}$$

Note the difference between η_1 and the chemical potential μ_1:

$$\mu_1 = \left(\frac{\partial \underline{U}}{\partial N_1}\right)_{\underline{S},\underline{V},N_2,N_3} = \left(\frac{\partial \underline{G}}{\partial N_1}\right)_{T,P,N_2,N_3} \tag{9-58}$$

$$\mu_1 - \mu_n = \eta_1 \tag{9-59}$$

Equation (9-55) could also be written as

$$\mathfrak{L}_3 = \frac{1}{N}\left(\frac{\partial^2 G^+}{\partial z_2^2}\right)_{T,P,\eta_1,N} \tag{9-60}$$

with

$$G^+ = \frac{\underline{G}^+}{N}$$

and z is a mole fraction. With N held *constant*, it is then a simple matter to convert this derivative to one employing mole fractions.

A more useful example of this method results when $\mathfrak{L}_2$ is determined using the ordering in Eq. (9-54):

$$\mathfrak{L}_2 = \begin{vmatrix} y_{33}^{(2)} & y_{34}^{(2)} \\ y_{43}^{(2)} & y_{44}^{(2)} \end{vmatrix} = \begin{vmatrix} \left(\frac{\partial^2 \underline{G}}{\partial N_1^2}\right)_{T,P,N_2,N} & \left(\frac{\partial^2 \underline{G}}{\partial N_1\, \partial N_2}\right) \\ \left(\frac{\partial^2 \underline{G}}{\partial N_1\, \partial N_2}\right) & \left(\frac{\partial^2 \underline{G}}{\partial N_2^2}\right)_{T,P,N_1,N} \end{vmatrix} \tag{9-61}$$

where the Legendre transform $\underline{G}$ (Gibbs energy) is the same in Eqs. (9-53) and (9-61). Equation (9-61) can be written as

$$\mathfrak{L}_2 = \frac{1}{N^2} \begin{vmatrix} \left(\frac{\partial^2 G}{\partial z_1^2}\right)_{T,P,z_2} & \left(\frac{\partial^2 G}{\partial z_1\, \partial z_2}\right) \\ \left(\frac{\partial^2 G}{\partial z_1\, \partial z_2}\right) & \left(\frac{\partial^2 G}{\partial z_2^2}\right)_{T,P,z_1} \end{vmatrix} \tag{9-62}$$

where, in each derivative, the mole fraction z_3 varies to maintain the requirement that $\sum_j dz_j = 0$.

The $\mathfrak{L}_i$ and $\mathfrak{M}_i$ determinants may always be converted to a mole fraction form [e.g., Eq. (9-60) or (9-62)], but only if the Legendre transform is $\underline{A}$, $\underline{H}$, or $\underline{G}$ does one avoid introducing modified chemical potentials η_j. If $\underline{A}$, $\underline{H}$, or $\underline{G}$ is used, the expansion shown in Eq. (9-56) may be employed. As shown in Section 9.1, the last variable in the ordering (N in this case) is always a constraint in every derivative used in stability theory. With such a constraint, for any extensive variable, it is easy to convert to an intensive form; that is, if $\underline{B}$ were an extensive variable, then, *with N constant*, $d\underline{B} = N dB$, $d^2\underline{B} = N d^2 B$, etc.

Finally, it is obvious that if one should wish to employ mole fractions as working variables to specify stability and critical-point cirteria, one should order variables as in Eq. (9-54), although $\underline{S}$ and $\underline{V}$ may be interchanged and the particular mole number eliminated is not important.

To conclude, we summarize the convenient stability and critical-point criteria for an n-component system when mole fractions are used. In all derivatives involving mole fractions, z_n is not constant:

$$\underline{U} = f(\underline{S}, \underline{V}, N_1, \ldots, N_{n-1}, N)$$

(ordering of N_i is arbitrary) and

$$\mathfrak{L}_2 = 0 = \frac{1}{N^{n-1}} \begin{vmatrix} \left(\frac{\partial^2 G}{\partial z_1^2}\right)_{z_2,\ldots,z_{n-1}} & \left(\frac{\partial^2 G}{\partial z_1\, \partial z_2}\right) & \left(\frac{\partial^2 G}{\partial z_1\, \partial z_{n-1}}\right) \\ \left(\frac{\partial^2 G}{\partial z_2\, \partial z_1}\right) & \left(\frac{\partial^2 G}{\partial z_2^2}\right)_{z_1,z_3,\ldots,z_{n-1}} & \left(\frac{\partial^2 G}{\partial z_2\, \partial z_{n-1}}\right) \\ \vdots & \vdots & \vdots \\ \left(\frac{\partial^2 G}{\partial z_{n-1}\, \partial z_1}\right) & \cdots & \left(\frac{\partial^2 G}{\partial z_{n-1}^2}\right)_{z_1,\ldots,z_{n-2}} \end{vmatrix} \tag{9-63}$$

(T, P, N) are constant in each derivative).

$\mathfrak{M}_2$ is the determinant constructed from $\mathfrak{L}_2$ by replacing the bottom row by $(\partial\mathfrak{L}_2/\partial z_j)_{z_1,\ldots,z_{j-1},z_{j+1},\ldots,z_{n-1}}$, where j is the jth column, and $\mathfrak{L}_2$ and $\mathfrak{M}_2$ are $(n-1)$ by $(n-1)$ determinants.[7] If $\mathfrak{L}_1$ were desired, with the ordering as shown, then

$$\mathfrak{L}_1 = 0 = \frac{1}{N^{n-1}}\begin{vmatrix} \left(\frac{\partial^2 A}{\partial V^2}\right)_z & \left(\frac{\partial^2 A}{\partial V\,\partial z_1}\right) & \cdots & \left(\frac{\partial^2 A}{\partial V\,\partial z_{n-1}}\right) \\ \left(\frac{\partial^2 A}{\partial z_1\,\partial V}\right) & \left(\frac{\partial^2 A}{\partial z_1^2}\right)_{V,z_2,\ldots,z_{n-1}} & \cdots & \left(\frac{\partial^2 A}{\partial z_1\,\partial z_{n-1}}\right) \\ \cdot & & & \cdot \\ \cdot & & & \cdot \\ \cdot & & & \cdot \\ \left(\frac{\partial^2 A}{\partial z_{n-1}\,\partial V}\right) & \left(\frac{\partial^2 A}{\partial z_{n-1}\,\partial z_1}\right) & \cdots & \left(\frac{\partial^2 A}{\partial z_{n-1}^2}\right)_{V,z_1,\ldots,z_{n-2}} \end{vmatrix} \tag{9-64}$$

with T also constant. $\mathfrak{M}_1 = \mathfrak{L}_1$ with the entry in the bottom row of the first column replaced by $(\partial\mathfrak{L}_1/\partial V)_{z_1,\ldots,z_{n-1}}$ and each bottom entry in the other $n-1$ columns replaced by $(\partial\mathfrak{L}_1/\partial z_j)_{V,z_1,\ldots,z_{j-1},z_{j+1},\ldots,z_{n-1}}$, where $1 \le j \le n-1$. In a similar manner, $\mathfrak{L}_0$ and $\mathfrak{M}_0$ may be written.

PROBLEMS

9.1. In Table 9.1 there is a listing of equivalent stability criteria for a pure material. Consider the first two entires:

$$A_{\underline{V}\underline{V}} = -\left(\frac{\partial P}{\partial \underline{V}}\right)_{T,N} > 0$$

$$A_{NN} = \left(\frac{\partial \mu}{\partial N}\right)_{T,\underline{V}} > 0$$

At the limit of stability it is stated that both $A_{\underline{V}\underline{V}}$ and A_{NN} go to zero simultaneously. If one had an applicable equation of state to express the P–$\underline{V}$–T–N relationship for a pure material, and wished to use the criterion $A_{NN} = 0$ at the limit of stability, how could this be accomplished? Compare your result with that obtained using the alternative form, $A_{\underline{V}\underline{V}} = 0$.

9.2. Derive Eq. (9-32).

9.3. We have liquid ethane at 1 bar pressure. What is the maximum temperature the *liquid* ethane can be heated before it reaches the limit of stability? Use the Peng–Robinson equation of state [Eq. (7-65)] with the following data:

$$T_c = 305.4 \text{ K}$$

$$P_c = 48.8 \text{ bar}$$

[7]The $\mathfrak{L}_2$ and $\mathfrak{M}_2$ determinants written in mole fraction form were used by Peng and Robinson [D.-Y. Peng and D. B. Robinson, *AIChE J.*, **23**, 137 (1977)] to predict critical properties of multicomponent systems. The Peng–Robinson equation of state [Eqs. (8-55) through (8-59)] was employed.

$$\omega = 0.098$$

$$T_b = 184.5 \text{ K (at 1.013 bar)}$$

(W. Porteous and M. Blander [*AIChE J.*, **21**, 560 (1976)] report an experimental value of 269 K at the limit of stability.)

9.4. In the storage of liquefied gases under pressure, the liquid is in equilibrium with the vapor in the ullage. In case of a severe accident in which the vessel fails and the pressure is reduced very rapidly to atmospheric, the bulk liquid is believed to undergo an isentropic expansion to the lower pressure. As the pressure drops, boiling will be initiated at the walls, but, due to the lack of nucleation sites within the bulk liquid, this liquid could become superheated. There is even the possibility that the limit-of-stability curve (the spinodal) may be reached during the pressure release. If this happens, a spontaneous generation of vapor would occur with shock waves to exacerbate the incident.

Demonstrate your ability to analyze an accident of this type by considering a cylinder containing both gas and liquid carbon dioxide. The temperature initially is at 294 K and the corresponding vapor pressure 58.4 bar. A weld seam in the vapor space fails and the pressure drops very rapidly to 1 bar. Neglecting any boiling on the cylinder walls, will the bulk liquid reach the limit of stability?

The Peng–Robinson equation of state may be used if desired. [See Eq. (7-65).] Some data for saturated liquid CO_2 are shown as follows if you find them of value.

Saturated Liquid Carbon Dioxide[a]

T (K)	P (bar)	S (J/mol K)	V (cm^3/mol)
270	32.034	135.93	46.502
272	33.801	136.68	47.077
274	35.638	137.44	47.687
276	37.549	138.21	48.336
278	39.533	138.98	49.029
280	41.595	139.77	49.773
282	43.737	140.56	50.576
284	45.960	141.37	51.449
286	48.269	142.21	52.404
288	50.665	143.06	53.460
290	53.152	143.94	54.641
292	55.734	144.86	55.981
294	58.415	145.83	57.531
296	61.198	146.87	59.376
298	64.090	148.00	61.661
300	67.095	149.31	64.690
302	70.220	150.95	69.297
304	73.475	154.16	81.703
304.21[b]	73.825	156.58	94.440

[a]*Source:* S. Angus, B. Armstrong, and K. M. deReuck, Eds. *Int. Thermodynamic Tables of the Fluid State Carbon Dioxide*, (New York: Pergamon Press, 1976).
[b]Critical state.

9.5. Prove for a stable thermodynamic system that $C_P > C_V$.

9.6. A ternary liquid mixture of B, C, and D is at a constant pressure and temperature. The excess Gibbs energy is given as

$$\frac{\Delta G^{EX}}{RT} = \alpha(x_B x_C + x_B x_D + x_C x_D)$$

Determine the limits of essential stability (i.e., the spinodal curves) and any critical points if $\alpha = 2.5$.

9.7. The stability criteria, Eq. (9-43), are often used to relate constants in an equation of state to critical constants. To illustrate, the well-known two-constant equation of state of Redich–Kwong [Eq. (7-60)],

$$P = \frac{RT}{V - b} - \frac{a}{T^{1/2}V(V + b)}$$

employs a and b as constants for a pure material.

To nondimensionalize these constants, we define new parameters, Ω_a and Ω_b:

$$a = \frac{\Omega_a R^2 T_c^{5/2}}{P_c}$$

$$b = \frac{\Omega_b R T_c}{P_c}$$

The terms Ω_a and Ω_b are dimensionless and can be evaluated from the stability criteria. Show that for this particular equation of state they are pure numbers and equal to

$$\Omega_a = [9(2^{1/3} - 1)]^{-1}$$

$$\Omega_b = \frac{2^{1/3} - 1}{3}$$

9.8. For the NRTL correlation for the excess Gibbs energy of mixing [H. Renon and J. M. Prausnitz, *AIChE J.*, **14**, 135 (1968)],

$$\frac{\Delta G^{EX}}{RT} = x_1 x_2 \left(\frac{\tau_{21} G_{21}}{x_1 + x_2 G_{21}} + \frac{\tau_{12} G_{12}}{x_2 + x_1 G_{12}} \right)$$

x_1 and x_2 are the liquid mole fractions of 1 and 2, τ_{12} and τ_{21} are binary parameters, and $G_{12} = \exp(-\alpha\tau_{12})$, $G_{21} = \exp(-\alpha\tau_{21})$. α is a third parameter for the mixture. Let $x_1 = x_2$ and choose a symmetric mixture where $\tau_{21} = \tau_{12}$. What is the largest value of α that can be used for this correlation if it is to show phase splitting?

9.9. Wilson [G. M. Wilson, *J. Am. Chem. Soc.*, **86**, 127, 133 (1964)] has developed an excess Gibbs energy relation for nonideal liquid mixtures that has been shown to be widely applicable. For a binary system of components B and C it may be written as

$$\frac{\Delta G^{EX}}{RT} = -x_B \ln(x_B + \Lambda_{BC} x_C) - x_C \ln(x_C + \Lambda_{CB} x_B)$$

Show that irrespective of the parameters Λ_{BC} and Λ_{CB}, this relation cannot be used for systems that can phase split into two liquid fractions.

9.10. Calculate the temperature at the limit of stability for a liquid mixture of 95% ethane and 5% n-butane at 1 bar. Use the Peng–Robinson equation of state and assume that the binary interaction parameter for ethane–n-butane is zero.

	Ethane	n-Butane
T_c (K)	305.4	425.2
P_c (bar)	48.8	38.0
ω	0.098	0.193

9.11. For a binary mixture of B and C, show that immiscibility occurs if

$$x_B\left(\frac{\partial \ln \gamma_B}{\partial x_B}\right)_{T,P} + 1 = 0$$

where γ_B is the activity coefficient of component B.

Phase Equilibrium 10

The criteria for coexistence of phases in equilibrium were developed in Section 6.5, and it was shown there that the temperatures, pressures, and component chemical potentials were equal in all phases. These criteria are valid even if chemical reactions occur in one or more phases. They may not, however, necessarily be valid if there are any constraints to the flow of mass or energy between phases. Thus, it is often stated that such criteria only apply in the general sense for *simple*, multiphase systems. Example 10.1, discussed later, illustrates a *nonsimple* system.

It is imperative to note that the equilibrium criteria stated above always form the starting point for further developments in phase equilibrium. This will become apparent in the treatment in this chapter.

10.1 The Phase Rule

In Section 2.7 it was stated in Postulate I that $(n + 2)$ *independently variable* properties characterize completely the stable equilibrium state of a simple system. We admitted that there might arise cases in which properties were *not* independently variable, although we have not yet met such cases (with the trivial exception of excluding one mole fraction from a set of n in a mixture).

We are now interested in extending our treatment to multiphase systems to determine the minimum set of variables to describe completely such systems both in extent and intensity. We will find even here that a general result may only be found for simple, composite systems. For such a system with π phases let us describe each phase separately and then relate all properties by invoking the criteria of phase equilibrium. If we were to apply Postulate I to each phase separately, we could choose any $(n+2)$ properties of each phase, provided that each set includes no more than $(n+1)$ intensive variables (see Section 5.2).

For each phase, a particularly convenient set of $(n+2)$ properties is

$$T^\gamma, P^\gamma, x_1^\gamma, \ldots, x_{n-1}^\gamma, N^\gamma \tag{10-1}$$

where superscript γ is used as a dummy index to denote a phase.

For a composite system containing π phases we have a set such as Eq. (10-1) for each phase, or $\pi(n+2)$ properties. To determine which of these $\pi(n+2)$ properties are not independent, we must apply the criteria of phase equilibria, which for this system are

$$T^\alpha = T^\beta = \cdots = T^\gamma = \cdots = T^\pi \tag{10-2}$$

$$P^\alpha = P^\beta = \cdots = P^\gamma = \cdots = P^\pi \tag{10-3}$$

$$\mu_j^\alpha = \mu_j^\beta = \cdots = \mu_j^\gamma = \cdots = \mu_j^\pi \qquad (j = 1, \ldots, n) \tag{10-4}$$

There are $(\pi - 1)$ equalities in each of Eqs. (10-2) and (10-3) and $n(\pi - 1)$ in Eq. (10-4) for a total of $(n+2)(\pi-1)$. If we transform this set of $(n+2)$ $(\pi - 1)$ equalities into relations containing only the properties in the $\pi(n+2)$ set of the kind given in Eq. (10-1), the $(n+2)(\pi-1)$ equalities can be used as restraining equations to determine the relationships between the $\pi(n+2)$ properties of the coexisting phases.

Let us carry out this transformation. We see that the set of variables in Eq. (10-1) is particularly convenient because we need transform only the chemical potentials. Specifically, we can expand these by relations of the kind that we have used in Chapter 8:

$$\mu_j^\gamma = g_j^\gamma(T^\gamma, P^\gamma, x_1^\gamma, \ldots, x_{n-1}^\gamma) \tag{10-5}$$

Thus, each of the equalities of Eq. (10-4) takes the form

$$g_j^\alpha(T^\alpha, P^\alpha, x_1^\alpha, \ldots, x_{n-1}^\alpha) = g_j^\beta(T^\beta, P^\beta, x_1^\beta, \ldots, x_{n-1}^\beta) \tag{10-6}$$

and we have $n(\pi - 1)$ such relations. The $(n+2)(\pi-1)$ restraining equations of Eqs. (10-2), (10-3), and (10-6) involve only $\pi(n+1)$ *intensive* variables of the $\pi(n+2)$ properties in Eq. (10-1). Clearly, then, the extensive variables included in the set of Eq. (10-1)(i.e., $N^\alpha, N^\beta, \ldots, N^\gamma, \ldots, N^\pi$) are not related by the criteria of phase equilibria and hence are independently variable.

Of the remaining $\pi(n+1)$ intensive variables of the set given by Eq. (10-1), we have $(n+2)(\pi-1)$ restraining equations so that the number of independently variable intensive properties is $\pi(n+1) - (n+2)(\pi-1) = (n+2-\pi)$. Thus, we conclude that to describe the composite simple system completely, there are $(n+2-\pi)$ independently variable intensive properties and π independently variable extensive properties for the total of $(n+2)$ variables required by Postulate I.

Although we have derived this result using specific sets of properties, the result is of general validity and not restricted to any particular set of intensive and extensive variables. For example, one could readily show that instead of the set of Eq. (10-1), we could have started with the set $T^\gamma, P^\gamma, \bar{S}_1^\gamma, \ldots, \bar{S}_{n-1}^\gamma, \underline{S}^\gamma$. In place of the expansion of Eq. (10-5), we would then have to use $\mu_j^\gamma = g_j(T^\gamma, P^\gamma, \bar{S}_1^\gamma, \ldots, \bar{S}_{n-1}^\gamma)$. Additional transformations are required if the chosen set contains intensive variables other than T^γ and P^γ, but the final result is unchanged.

The general conclusion, which is one of the most important results of thermodynamics of multiphase systems, can be stated as follows: *For a composite simple system containing π phases and n components in which chemical reactions do not occur, there are $n + 2 - \pi$ independently variable intensive properties, and therefore at least π extensive properties must be included in the set of $n + 2$ properties necessary to describe the composite system completely.*

This result was first expressed by J. Willard Gibbs in 1875 and is commonly referred to as the *Gibbs phase rule*. In most texts the number of *independently variable intensive* properties is referred to as the *variance* or degrees of freedom, denoted by $\mathcal{f}$, and the phase rule is written as

$$\mathcal{f} = n + 2 - \pi \tag{10-7}$$

In a binary system,

$$\mathcal{f} = 4 - \pi \tag{10-8}$$

For a single phase, $\mathcal{f} = 3$ (i.e., we must specify three intensive properties such as T, P, x to describe completely all other intensive properties). We have dealt with such systems in Chapter 8.

Although the phase rule appears to be very simple, the application in some cases leads to results that are difficult to interpret. For example, in many binary liquid–vapor systems, dew and bubble points are monotonic functions of composition. There is no reason to expect this always to be the case. In fact, we frequently encounter systems in which the pressure, for an isothermal section, or the temperature, for an isobaric section, attains a maximum or minimum value. These systems are called *azeotropic mixtures* and a minimum-boiling type is illustrated in Figure 10.1. The azeotrope is that state in which the concentrations in the liquid and vapor phases become identical. Thus, for this state the relative volatility $[\alpha_{AB} \equiv (y_A/x_A)/(y_B/x_B)]$ becomes unity and separation processes such as simple distillation become impossible.

As shown in Figure 10.1 in the isothermal section at $T = T_z$, the pressure at the azeotrope is noted as P_z. The isobaric section is drawn for $P = P_z$ and again the azeotropic temperature is T_z. The azeotropic concentrations are drawn to be the same in both figures.

The azeotropic mixture illustrates one of the limitations of the Gibbs phase rule. The variance of a two-phase binary is 2. There are, however, cases in which we cannot necessarily describe completely the intensive properties of the phases in a mixture simply by specifying the temperature and pressure. For example, in Figure 10.1(a), for the section at $T = T_z$, choose a pressure P' less than P_z. At T_z, P', there are *two* equally valid sets of coexisting liquid and vapor compositions: The Gibbs phase rule apparently breaks down. The problem is that we

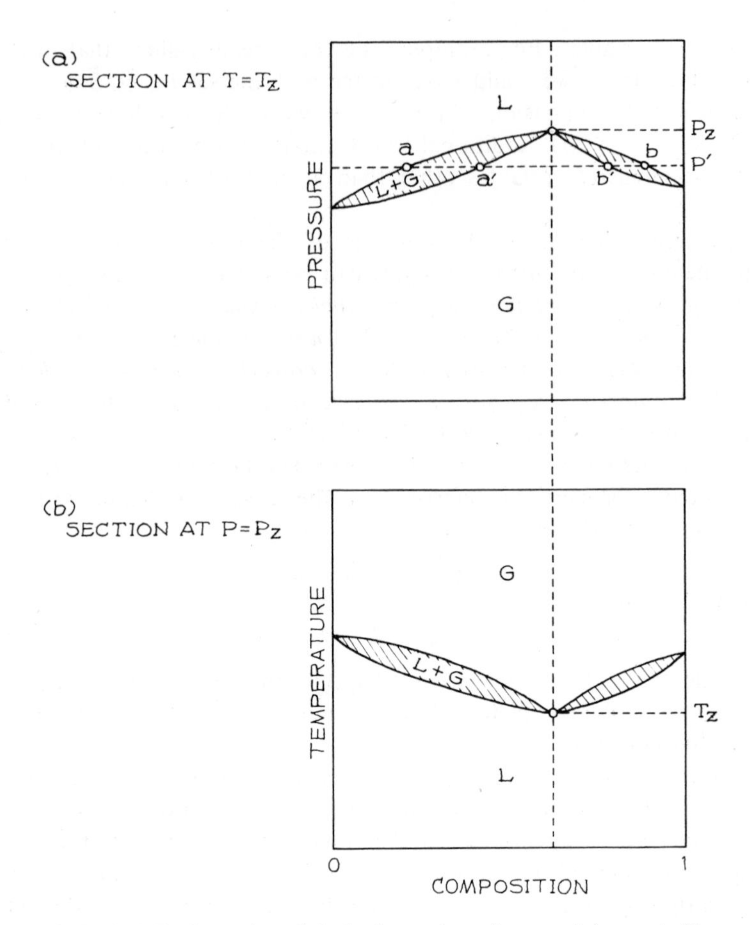

Figure 10.1 Isothermal and isobaric sections for a minimum boiling azeotropic system.

are expecting too much. As seen in the next section, the relation we obtain for $P = f(x)_T$ can be multivalued.

The apparent difficulty in interpreting the Gibbs phase rule is avoided if we reconsider the discussion earlier in this section. We stated that there are *certain* sets of intensive properties which, when selected, will completely define the system. But this statement did not imply that we could choose *any* set. Thus, for the case of an azeotropic system, we may not be able to use the pressure–temperature set. On the other hand, if we specify temperature and liquid composition, we always define a unique state for which pressure and vapor composition are determined.

It is also interesting to note that to specify the system completely (i.e., extensive as well as intensive properties), we could use pressure and temperature provided that we also include N_A and N_B in the set of $(n + 2)$ properties required by Postulate I. In this case, the average mole fractions of both components are known, and hence we can make an unambiguous choice regarding the branch of the azeotrope that is the correct one [e.g., in Figure 10.1(a), *a–a'* or *b–b'*].

Before concluding the discussion of the phase rule, let us note that it does not apply to composite systems that are not simple systems. For nonsimple systems, no general rule can be developed; each case must be analyzed separately with respect to the internal constraints that are present. Figure 10.2 and Example 10.1 illustrate this point.

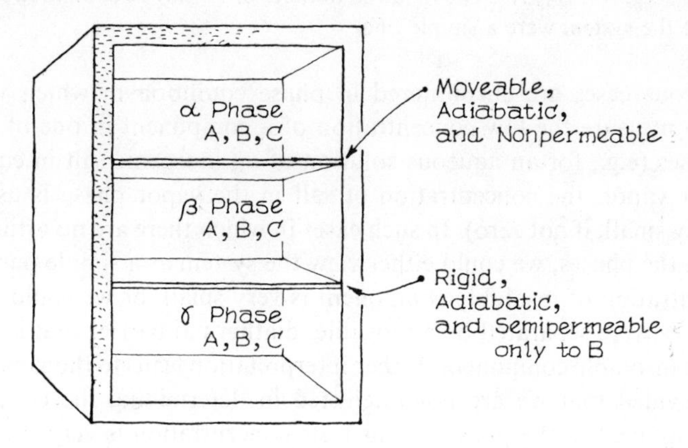

Figure 10.2

Example 10.1

The isolated composite system shown in Figure 10.2 contains three phases with components *A*, *B*, and *C* in each phase. Determine the minimum number of properties necessary to describe completely the composite system and suggest how these properties should be chosen.

Solution

Because the composite system is not a simple system, Postulate I does not apply. To determine the number of independently variable properties, we proceed by choosing $n + 2$ or five properties for each phase and use the particular set of equilibrium criteria that applies to the given internal restraints. Let us choose the following $\pi(n + 2)$ or 15 properties:

$$T^\alpha, T^\beta, T^\gamma, P^\alpha, P^\beta, P^\gamma, x_A^\alpha, x_A^\beta, x_A^\gamma, x_B^\alpha, x_B^\beta, x_B^\gamma, N^\alpha, N^\beta, N^\gamma$$

The criteria of equilibrium can be shown to be (see Sections 6.4 and 6.5)

$$T^\beta = T^\gamma$$

$$P^\alpha = P^\beta$$

$$\mu_B^\beta = \mu_B^\gamma$$

The last relation can be expanded:

$$\mu_B^\beta = g_B^\beta(T^\beta, P^\beta, x_A^\beta, x_B^\beta) = \mu_B^\gamma = g_B^\gamma(T^\gamma, P^\gamma, x_A^\gamma, x_B^\gamma)$$

The three restraining equations allow us to eliminate three variables from the set of 15. Nevertheless, we clearly cannot eliminate any arbitrary three. We can, however, eliminate one from each of the following three sets:

$$T^\beta, T^\gamma$$

$$P^\alpha, P^\beta$$

$$T^\beta, T^\gamma, P^\beta, P^\gamma, x_A^\beta, x_A^\gamma, x_B^\beta, x_B^\gamma$$

Thus, one satisfactory set of independently variable properties is T^α, P^α, P^γ, x_A^α, x_A^β, x_A^γ, x_B^α, x_B^β, x_B^γ, N^α, N^β, N^γ. The required number of 12 may be compared to the value of five if the system were a simple one.

Numerous cases are encountered in phase equilibria in which we find it difficult to measure the low concentration of a component in one of the coexisting phases (e.g., for an aqueous solution of an inorganic salt in equilibrium with water vapor, the concentration of salt in the vapor phase is usually immeasurably small, if not zero). In such cases in which there are no artificial walls separating the phases, we could either view the system as a simple one in which the concentration of insoluble component is very small, or we could treat it as if the phases were separated by a movable, diathermal barrier that is impermeable to the insoluble component. Either interpretation leads to the same practical result, provided that we are not interested in determining the concentration of the component in the phase in which its concentration is very small. In the first case (simple system), we would not use the restraining equation of equality of chemical potentials of the insoluble component because we are not interested in its concentration in one of the phases; in the second case (impermeable barrier), we do not have an equality of chemical potentials, and we assume that the concentration of the insoluble component is zero. Thus, from the pragmatic viewpoint, the two cases are equivalent.

10.2 The Differential Approach for Phase Equilibrium Relationships

We shall develop two approaches for evaluating the analytical relationships between properties of coexisting phases: the differential approach, described in this section, and the integral approach, treated in Sections 10.4 and 10.5. Although the integral approach is more commonly used, the differential approach is conceptually more elucidating. It is for this reason that the latter is developed here in some detail.

We first treat a simple binary system of components A and B with two coexisting phases α and β in which each component is present in each phase. The reasoning is then extended to three coexisting phases of a binary and to the general case of multicomponent, multiphase simple systems.

To develop analytical relationships, we use the theoretical basis described in Section 10-1 with one modification: For the criteria of equilibrium between coexisting phases, instead of employing chemical potentials [as in Eq. (10-4)], we use the equivalent criteria of equality of the fugacity of each component in each phase.[1] Thus, the criteria of equilibrium for a binary mixture coexisting as α and β phases are

$$T^\alpha = T^\beta \tag{10-9}$$

$$P^\alpha = P^\beta \tag{10-10}$$

$$\hat{f}_A^\alpha = \hat{f}_A^\beta \text{ or } \ln \hat{f}_A^\alpha = \ln \hat{f}_A^\beta \tag{10-11}$$

and

$$\hat{f}_B^\alpha = \hat{f}_B^\beta \text{ or } \ln \hat{f}_B^\alpha = \ln \hat{f}_B^\beta \tag{10-12}$$

To obtain differential equations involving P, T, and phase compositions, we take the total derivatives of Eq. set (10-9) through (10-12) and expand all fugacities in terms of T, P, and the pertinent phase compositions. In view of Eqs. (10-9) and (10-10), no phase designation need be made for T and P.

For example, suppose that we choose as the $(n + 1)$ variables in each phase, T, P, and x_A. Thus, for component A, we have

$$\begin{aligned} d \ln \hat{f}_A^\alpha &= \left(\frac{\partial \ln \hat{f}_A^\alpha}{\partial T}\right)_{P,x^\alpha} dT + \left(\frac{\partial \ln \hat{f}_A^\alpha}{\partial P}\right)_{T,x^\alpha} dP + \left(\frac{\partial \ln \hat{f}_A^\alpha}{\partial x_A^\alpha}\right)_{T,P} dx_A^\alpha \\ &= -\frac{\bar{H}_A^\alpha - H_A^*}{RT^2} dT + \frac{\bar{V}_A^\alpha}{RT} dP + \left(\frac{\partial \ln \hat{f}_A^\alpha}{\partial x_A^\alpha}\right)_{T,P} dx_A^\alpha \end{aligned} \tag{10-13}$$

$$\begin{aligned} d \ln \hat{f}_A^\beta &= \left(\frac{\partial \ln \hat{f}_A^\beta}{\partial T}\right)_{P,x^\beta} dT + \left(\frac{\partial \ln \hat{f}_A^\beta}{\partial P}\right)_{T,x^\beta} dP + \left(\frac{\partial \ln \hat{f}_A^\beta}{\partial x_A^\beta}\right)_{T,P} dx_A^\beta \\ &= -\frac{\bar{H}_A^\beta - H_A^*}{RT^2} dT + \frac{\bar{V}_A^\beta}{RT} dP + \left(\frac{\partial \ln \hat{f}_A^\beta}{\partial x_A^\beta}\right)_{T,P} dx_A^\beta \end{aligned} \tag{10-14}$$

where H_A^* refers to the enthalpy of A in an ideal-gas state at the temperature of the system. Note that the expansions of Eqs. (10-13) and (10-14) are written separately for each phase. As discussed in Chapters 5 and 8, these expansions are always valid for each phase, regardless of whether or not the phase is in equilibrium with other phases.

The fugacity expansions for component B are more convenient if we use the same set of intensive variables. Thus, the equations corresponding to Eqs. (10-13) and (10-14) for component B are

$$d \ln \hat{f}_B^\alpha = -\frac{\bar{H}_B^\alpha - H_B^*}{RT^2} dT + \frac{\bar{V}_B^\alpha}{RT} dP + \left(\frac{\partial \ln \hat{f}_B^\alpha}{\partial x_A^\alpha}\right)_{T,P} dx_A^\alpha \tag{10-15}$$

$$d \ln \hat{f}_B^\beta = -\frac{\bar{H}_B^\beta - H_B^*}{RT^2} dT + \frac{\bar{V}_B^\beta}{RT} dP + \left(\frac{\partial \ln \hat{f}_B^\beta}{\partial x_A^\beta}\right)_{T,P} dx_A^\beta \tag{10-16}$$

[1] Note that whereas equality of component fugacities is completely equivalent to equality of chemical potentials, equality of component activities is not necessarily equivalent. The discrepancy exists because different reference states of *unit activity* are sometimes chosen for a component in different phases. For fugacities, the reference state is always chosen as an ideal gas at the system temperature.

Equating differentials of $\ln \hat{f}_i$, we get

$$-\frac{\bar{H}_A^\alpha - \bar{H}_A^\beta}{RT^2}\,dT + \frac{\bar{V}_A^\alpha - \bar{V}_A^\beta}{RT}\,dP + \left(\frac{\partial \ln \hat{f}_A^\alpha}{\partial x_A^\alpha}\right)_{T,P} dx_A^\alpha - \left(\frac{\partial \ln \hat{f}_A^\beta}{\partial x_A^\beta}\right)_{T,P} dx_A^\beta = 0 \tag{10-17}$$

and

$$-\frac{\bar{H}_B^\alpha - \bar{H}_B^\beta}{RT^2}\,dT + \frac{\bar{V}_B^\alpha - \bar{V}_B^\beta}{RT}\,dP + \left(\frac{\partial \ln \hat{f}_B^\alpha}{\partial x_A^\alpha}\right)_{T,P} dx_A^\alpha - \left(\frac{\partial \ln \hat{f}_B^\beta}{\partial x_A^\beta}\right)_{T,P} dx_A^\beta = 0 \tag{10-18}$$

where we must remember that $\bar{H}_i^\alpha$ and $\bar{V}_i^\alpha$ are evaluated at x_A^α, T, P, whereas $\bar{H}_i^\beta$ and $\bar{V}_i^\beta$ are at x_A^β, T, P.

Equations (10-17) and (10-18) are two differential equations that must be satisfied simultaneously. We can solve by eliminating any one of the differentials in T, P, x_A^α, and x_A^β, thereby obtaining one differential equation in three variables. Integration of that equation yields the desired result.

Thus, the relationship $P = f(T, x_A^\beta)$ can be obtained by eliminating dx_A^α simultaneously from Eqs. (10-17) and (10-18). This final result may be simplified by noting that the coefficients of the two dx_A^α terms and the two dx_A^β are related by the Duhem equation for component fugacities, that is, for a binary system of phase π,

$$x_A^\pi \left(\frac{\partial \ln \hat{f}_A^\pi}{\partial x_A^\pi}\right)_{T,P} + x_B^\pi \left(\frac{\partial \ln \hat{f}_B^\pi}{\partial x_A^\pi}\right)_{T,P} = 0 \tag{10-19}$$

Thus, multiplying Eq. (10-17) by x_A^α and multiplying Eq. (10-18) by x_B^α, adding the two equations, and applying Eq. (10-19) to each phase, the result simplifies to

$$-\frac{x_A^\alpha(\bar{H}_A^\alpha - \bar{H}_A^\beta) + x_B^\alpha(\bar{H}_B^\alpha - \bar{H}_B^\beta)}{RT^2}\,dT + \frac{x_A^\alpha(\bar{V}_A^\alpha - \bar{V}_A^\beta) + x_B^\alpha(\bar{V}_B^\alpha - \bar{V}_B^\beta)}{RT}\,dP - \left[\left(x_A^\alpha - \frac{x_A^\beta x_B^\alpha}{x_B^\beta}\right)\left(\frac{\partial \ln \hat{f}_A^\beta}{\partial x_A^\beta}\right)_{T,P}\right] dx_A^\beta = 0 \tag{10-20}$$

Solving explicitly for dP,

$$dP = \frac{1}{T}\,\frac{x_A^\alpha(\bar{H}_A^\alpha - \bar{H}_A^\beta) + x_B^\alpha(\bar{H}_B^\alpha - \bar{H}_B^\beta)}{x_A^\alpha(\bar{V}_A^\alpha - \bar{V}_A^\beta) + x_B^\alpha(\bar{V}_B^\alpha - \bar{V}_B^\beta)}\,dT + \frac{RT(x_A^\alpha - x_A^\beta x_B^\alpha/x_B^\beta)(\partial \ln \hat{f}_A^\beta/\partial x_A^\beta)_{T,P}}{x_A^\alpha(\bar{V}_A^\alpha - \bar{V}_A^\beta) + x_B^\alpha(\bar{V}_B^\alpha - \bar{V}_B^\beta)}\,dx_A^\beta \tag{10-21}$$

Since P is a property, and since T and x_A^β form an independent set for P under conditions of two-phase equilibrium, Eq. (10-21) must be an exact differential equation. Therefore, it follows immediately that

$$\left(\frac{\partial P}{\partial T}\right)_{x_A^\beta, [\alpha-\beta]} = \frac{1}{T}\,\frac{x_A^\alpha(\bar{H}_A^\alpha - \bar{H}_A^\beta) + x_B^\alpha(\bar{H}_B^\alpha - \bar{H}_B^\beta)}{x_A^\alpha(\bar{V}_A^\alpha - \bar{V}_A^\beta) + x_B^\alpha(\bar{V}_B^\alpha - \bar{V}_B^\beta)} \tag{10-22}$$

and

$$\left(\frac{\partial P}{\partial x_A^\beta}\right)_{T, [\alpha-\beta]} = \frac{RT(x_A^\alpha - x_A^\beta x_B^\alpha/x_B^\beta)(\partial \ln \hat{f}_A^\beta/\partial x_A^\beta)_{T,P}}{x_A^\alpha(\bar{V}_A^\alpha - \bar{V}_A^\beta) + x_B^\alpha(\bar{V}_B^\alpha - \bar{V}_B^\beta)} \tag{10-23}$$

where the subscript $[\alpha-\beta]$ denotes conditions under which phases α and β coexist at equilibrium. Note that the partial derivatives in the left-hand side of Eqs. (10-22) and (10-23) do not imply that x_A^α is constant; x_A^α will in fact change as T is varied under conditions of $\alpha-\beta$ phase equilibrium and constant x_A^β. Similarly, x_A^α will change as x_A^β is varied with phase equilibrium and constant T. (Recall that for this case the variance is 2, so that x_A^α can be expressed as a function of T and x_A^β.)

Since

$$\left(\frac{\partial T}{\partial x_A^\beta}\right)_{P,[\alpha-\beta]}\left(\frac{\partial P}{\partial T}\right)_{x_A^\beta,[\alpha-\beta]}\left(\frac{\partial x_A^\beta}{\partial P}\right)_{T,[\alpha-\beta]} = -1 \tag{10-24}$$

it follows that

$$\left(\frac{\partial T}{\partial x_A^\beta}\right)_{P,[\alpha-\beta]} = -RT^2\frac{(x_A^\alpha - x_A^\beta x_B^\alpha/x_B^\beta)(\partial \ln \hat{f}_A^\beta/\partial x_A^\beta)_{T,P}}{x_A^\alpha(\bar{H}_A^\alpha - \bar{H}_A^\beta) + x_B^\alpha(\bar{H}_B^\alpha - \bar{H}_B^\beta)} \tag{10-25}$$

Example 10.2

Prove that for a minimum-boiling azeotrope at constant temperature the pressure maximizes at the azeotropic concentration.

Solution

An azeotrope occurs when $x_A^V = x_A^L$, at which point the term $(x_A^V - x_A^L x_B^V/x_B^L)$ in Eq. (10-25) vanishes. By inspection of Eq. (10-23) it is clear that $(\partial P/\partial x_A^L)_{T,[L-V]}$ also vanishes at the azeotropic concentration. To prove that a temperature minimum corresponds to a pressure maximum, all we need show is that $(\partial T/\partial x_A^L)_{P,[L-V]}$ and $(\partial P/\partial x_A^L)_{T,[L-V]}$ have opposite signs. The ratio of these two derivatives is $-[(\partial P/\partial T)_{x_A^L,[L-V]}$ by Eq. (10-24). Equation (10-22) shows that with $\alpha = V$ and $\beta = L$, the enthalpy and volume changes are always positive. Therefore, $(\partial P/\partial T)_{x_A^L,[L-V]}$ is always positive and, hence, $(\partial T/\partial x_A^L)_{P,[L-V]}$ and $(\partial P/\partial x_A^L)_{T,[L-V]}$ always have opposite signs.

In the procedure described above we evaluated the partial derivatives involving T, P, and x_A^β by eliminating the dx_A^α terms in solving Eqs. (10-17) and (10-18). Similarly, we could have worked with T, P, and x_A^α by eliminating dx_A^β [the results would be identical to Eqs. (10-22), (10-23), and (10-25) with x_A^α replacing x_A^β], or T, x_A^α, and x_A^β by eliminating dP, or P, x_A^α, and x_A^β by eliminating dT.

For example, to obtain the partial derivatives in T, x_A^α, and x_A^β, eliminating dP from Eqs. (10-17) and (10-18) yields

$$\begin{aligned}\frac{1}{T}\left[\frac{\bar{H}_A^\alpha - \bar{H}_A^\beta}{\bar{V}_A^\alpha - \bar{V}_A^\beta} - \frac{\bar{H}_B^\alpha - \bar{H}_B^\beta}{\bar{V}_B^\alpha - \bar{V}_B^\beta}\right] dT \\ - RT\left[\frac{(\partial \ln \hat{f}_A^\alpha/\partial x_A^\alpha)_{T,P}}{\bar{V}_A^\alpha - \bar{V}_A^\beta} - \frac{(\partial \ln \hat{f}_B^\alpha/\partial x_A^\alpha)_{T,P}}{\bar{V}_B^\alpha - \bar{V}_B^\beta}\right] dx_A^\alpha \\ + RT\left[\frac{(\partial \ln \hat{f}_A^\beta/\partial x_A^\beta)_{T,P}}{\bar{V}_A^\alpha - \bar{V}_A^\beta} - \frac{(\partial \ln \hat{f}_B^\beta/\partial x_A^\beta)_{T,P}}{\bar{V}_B^\alpha - \bar{V}_B^\beta}\right] dx_A^\beta = 0\end{aligned} \tag{10-26}$$

Simplifying Eq. (10-26) by appying the Duhem equation for fugacity for each phase, we obtain

$$-\frac{(\bar{H}_A^\alpha - \bar{H}_A^\beta)(\bar{V}_B^\alpha - \bar{V}_B^\beta) - (\bar{H}_B^\alpha - \bar{H}_B^\beta)(\bar{V}_A^\alpha - \bar{V}_A^\beta)}{RT^2}\,dT$$
$$+\frac{x_B^\alpha(\bar{V}_B^\alpha - \bar{V}_B^\beta) + x_A^\alpha(\bar{V}_A^\alpha - \bar{V}_A^\beta)}{x_B^\alpha}\left(\frac{\partial \ln \hat{f}_A^\alpha}{\partial x_A^\alpha}\right)_{T,P} dx_A^\alpha \tag{10-27}$$
$$-\frac{x_B^\beta(\bar{V}_B^\alpha - \bar{V}_B^\beta) + x_A^\beta(\bar{V}_A^\alpha - \bar{V}_A^\beta)}{x_B^\beta}\left(\frac{\partial \ln \hat{f}_A^\beta}{\partial x_A^\beta}\right)_{T,P} dx_A^\beta = 0$$

From Eq. (10-27), it follows that

$$\left(\frac{\partial x_A^\beta}{\partial x_A^\alpha}\right)_{T,[\alpha-\beta]} = \frac{x_B^\alpha(\bar{V}_B^\alpha - \bar{V}_B^\beta) + x_A^\alpha(\bar{V}_A^\alpha - \bar{V}_A^\beta)}{x_B^\beta(\bar{V}_B^\alpha - \bar{V}_B^\beta) + x_A^\beta(\bar{V}_A^\alpha - \bar{V}_A^\beta)}\left(\frac{x_B^\beta}{x_B^\alpha}\right)\frac{(\partial \ln \hat{f}_A^\alpha/\partial x_A^\alpha)_{T,P}}{(\partial \ln \hat{f}_A^\beta/\partial x_A^\beta)_{T,P}} \tag{10-28}$$

$$\left(\frac{\partial T}{\partial x_A^\alpha}\right)_{x_A^\beta,[\alpha-\beta]} = \frac{x_B^\alpha(\bar{V}_B^\alpha - \bar{V}_B^\beta) + x_A^\alpha(\bar{V}_A^\alpha - \bar{V}_A^\beta)}{(\bar{H}_A^\alpha - \bar{H}_A^\beta)(\bar{V}_B^\alpha - \bar{V}_B^\beta) - (\bar{H}_B^\alpha - \bar{H}_B^\beta)(\bar{V}_A^\alpha - \bar{V}_A^\beta)} \times \frac{RT^2}{x_B^\alpha}\left(\frac{\partial \ln \hat{f}_A^\alpha}{\partial x_A^\alpha}\right)_{T,P} \tag{10-29}$$

and $(\partial T/\partial x_A^\beta)_{x_A^\alpha,[\alpha-\beta]}$ is given by Eq. (10-29) with x_A^α and x_A^β interchanged.

From the partial derivatives obtained from Eqs. (10-22), (10-23), (10-25), (10-28), and (10-29), we could construct a variety of isoplethal, isothermal, or isobaric cross sections. To evaluate an isopleth [e.g., $P = f(T)$ at constant x_A^β], we must integrate $(\partial P/\partial T)_{x_A^\beta,[\alpha-\beta]}$. To carry out the integration of Eq. (10-22), we must express the right-hand side as a function of P, T, and x_A^β, where x_A^β is held constant during integration. From physical property data of individual phases, we can express $\bar{H}_i^\alpha$, $\bar{V}_i^\alpha$ and $\bar{H}_i^\beta$, $\bar{V}_i^\beta$ as a function of P, T, x_A^α and P, T, x_A^β, respectively. However, x_A^α is not a constant during the integration since it varies with T. Therefore, in addition to physical property data, we must know x_A^α as a function of T and x_A^β under a $\alpha - \beta$ phase equilibrium. This relationship can be obtained by integration of Eq. (10-28), provided that we can express the right-hand side of the equation as a function of x_A^α, x_A^β, and T. But here we run into the same problem: the properties are functions of P also, and P is a floating variable in Eq. (10-28), and, therefore, P must be expressed as a function of x_A^α, x_A^β, and T. This functionality we originally sought by integration of Eq. (10-22). Thus, to obtain a rigorous solution to an isopleth requires simultaneous solution of Eqs. (10-22) and (10-28). Although the procedure is complicated for the rigorous case, in practice it can usually be simplified considerably by using judicious approximations. One example is given below.

Example 10.3

A binary system of components A and B coexists in liquid–vapor equilibrium at 448 K and at low pressure. The vapor phase can be considered an ideal mixture of ideal gases; the liquid-phase activity coefficients can be approximated by the van Laar equations:

$$\ln \gamma_A = \frac{A_{12}}{[1 + (A_{12}/A_{21})(x_A/x_B)]^2} \tag{10-30}$$

and

$$\ln \gamma_B = \frac{A_{21}}{[1 + (A_{21}/A_{12})(x_B/x_A)]^2} \tag{10-31}$$

where A_{12} and A_{21} are functions of temperature, and γ_A and γ_B can be considered independent of pressure.

(a) Determine the y–x relationship at a constant temperature of 448 K, and indicate how a $(P–x)_T$ diagram would be constructed.

(b) If isothermal P–x data were available instead and activity coefficients were not known, would it be possible to construct the y–x relationship? If so, indicate the procedure one would follow. *Note*: For a liquid phase obeying the van Laar equations, the following limiting law can be shown to pertain:

$$\lim_{x_A \to 0} \frac{y_A}{x_A} = \left(\frac{P_{vp_A}}{P_{vp_B}}\right) e^{A_{12}} \tag{10-32}$$

where P_{vp_A} and P_{vp_B} are the vapor pressures of pure A and B at the temperature of the system. For the system in question at 448 K, $P_{vp_A} = 5.65$ bar and $P_{vp_B} = 8.98$ bar.

Solution

(a) The y–x relationships at constant T can be determined from Eq. (10-28) if we let α = liquid and β = vapor. Thus,

$$\left(\frac{\partial y_A}{\partial x_A}\right)_{T,[\alpha-\beta]} = \frac{x_A(\bar{V}_A^V - \bar{V}_A^L) + x_B(\bar{V}_B^V - \bar{V}_B^L)}{y_A(\bar{V}_A^V - \bar{V}_A^L) + y_B(\bar{V}_B^V - \bar{V}_B^L)}\left(\frac{y_B}{x_B}\right)\frac{(\partial \ln \hat{f}_A^L/\partial x_A)_{T,P}}{(\partial \ln \hat{f}_A^V/\partial y_A)_{T,P}} \tag{10-33}$$

For an ideal vapor mixture of ideal gases, the following simplifications are applicable:

$$\bar{V}_i^V = V_i^V = \frac{RT}{P} \tag{10-34}$$

$$\left(\frac{\partial \ln \hat{f}_A^V}{\partial y_A}\right)_{T,P} = \frac{1}{y_A} \tag{10-35}$$

For conditions far removed from the critical point, there is an additional simplification:

$$\bar{V}_i^V = \frac{RT}{P} \gg \bar{V}_i^L \tag{10-36}$$

Expressing the liquid fugacity as a function of activity coefficient,

$$\left(\frac{\partial \ln \hat{f}_A^L}{\partial x_A}\right)_{T,P} = \frac{1}{x_A} + \left(\frac{\partial \ln \gamma_A}{\partial x_A}\right)_{T,P} \tag{10-37}$$

Substituting Eqs. (10-34) through (10-37) into Eq. (10-33) and simplifying, we obtain

$$\left(\frac{\partial y_A}{\partial x_A}\right)_{T,[\alpha-\beta]} = \frac{y_A y_B}{x_A x_B}\left[1 + x_A\left(\frac{\partial \ln \gamma_A}{\partial x_A}\right)_{T,P}\right] \tag{10-38}$$

or

$$\int \frac{dy_A}{y_A(1-y_A)} = \int \frac{dx_A}{x_A(1-x_A)} + \int \frac{1}{1-x_A}\left(\frac{\partial \ln \gamma_A}{\partial x_A}\right)_{T,P} dx_A \quad (T\ \text{constant}) \tag{10-39}$$

Equation (10-39) can be integrated directly by evaluating the activity coefficient partial derivative from Eq. (10-30). It is, however, more convenient to rearrange terms prior to integration. Since

$$\frac{1}{1-x_A} = \frac{1}{x_B} = \frac{1 - x_B + x_B}{x_B} = \frac{x_A}{x_B} + 1$$

$$\frac{1}{1-x_A}\left(\frac{\partial \ln \gamma_A}{\partial x_A}\right)_{T,P} = \frac{x_A}{x_B}\left(\frac{\partial \ln \gamma_A}{\partial x_A}\right)_{T,P} + \left(\frac{\partial \ln \gamma_A}{\partial x_A}\right)_{T,P} \tag{10-40}$$

From the Duhem expression for γ,

$$\frac{x_A}{x_B}\left(\frac{\partial \ln \gamma_A}{\partial x_A}\right)_{T,P} = -\left(\frac{\partial \ln \gamma_B}{\partial x_A}\right)_{T,P}$$

Thus,

$$\int \frac{1}{1-x_A}\left(\frac{\partial \ln \gamma_A}{\partial x_A}\right)_{T,P} dx_A = \int \left[\frac{\partial \ln (\gamma_A/\gamma_B)}{\partial x_A}\right]_{T,P} dx_A \tag{10-41}$$

Substituting Eq. (10-41) into Eq. (10-39), and using the indefinite integral method, we obtain

$$\ln \frac{y_A}{1-y_A} = \ln \frac{x_A}{1-x_A} + \ln \frac{\gamma_A}{\gamma_B} + C \tag{10-42}$$

The constant of integration can be evaluated using the limiting condition given in the problem statement. That is, as $x_A \longrightarrow 0$ and $y_A \longrightarrow 0$, $\gamma_A \longrightarrow e^{A_{12}}$, $\gamma_B \longrightarrow 1$; thus, from Eqs. (10-32) and (10-42),

$$\ln\left(\lim_{x_A \to 0} \frac{y_A}{x_A}\right) = \lim_{x_A \to 0}\left(\ln \frac{y_A}{x_A}\right) = \ln \frac{e^{A_{12}} P_{vp_A}}{P_{vp_B}} = A_{12} + C \tag{10-43}$$

and

$$C = \ln \frac{P_{vp_A}}{P_{vp_B}} \tag{10-44}$$

so

$$\frac{y_A}{1-y_A} = \frac{x_A}{1-x_A}\left(\frac{P_{vp_A}}{P_{vp_B}}\right)\frac{\gamma_A}{\gamma_B} \qquad (T \text{ constant}) \tag{10-45}$$

Substituting Eqs. (10-30) and (10-31) into Eq. (10-45) and simplifying, we obtain the desired result:

$$\frac{y_A}{1-y_A} = g(x_A) = \frac{P_{vp_A}}{P_{vp_B}}\left(\frac{x_A}{x_B}\right) \exp\left[\frac{A_{12}A_{21}(A_{21}x_B^2 - A_{12}x_A^2)}{(A_{12}x_A + A_{21}x_B)^2}\right] \qquad (T \text{ constant}) \tag{10-46}$$

or

$$y_A = \frac{g(x_A)}{1 + g(x_A)} \qquad (T \text{ constant}) \tag{10-47}$$

To construct a P–x diagram at constant T, we must integrate Eq. (10-23). Substituting the simplifying assumptions given above into Eq. (10-23) (with β = liquid), we have

$$\left(\frac{\partial P}{\partial x_A}\right)_{T,[\alpha-\beta]} = \frac{RT(y_A - x_A y_B/x_B)[(1/x_A) + (\partial \ln \gamma_A/\partial x_A)_{T,P}]}{RT/P} \tag{10-48}$$

or

$$\left(\frac{\partial \ln P}{\partial x_A}\right)_{T,[\alpha-\beta]} = \left(y_A - \frac{x_A y_B}{x_B}\right)\left\{\frac{1}{x_A} + x_B\left[\frac{\partial \ln (\gamma_A/\gamma_B)}{\partial x_A}\right]_{T,P}\right\} \tag{10-49}$$

Thus,

$$\int d\ln P = \int \frac{1}{1-x_A}\left[\frac{g(x_A)}{1+g(x_A)} - x_A\right]\left\{\frac{1}{x_A} + x_B\left[\frac{\partial \ln(\gamma_A/\gamma_B)}{\partial x_A}\right]_{T,P}\right\} dx_A \qquad (10\text{-}50)$$

(T constant)

The integration is complex; the final result is much easier to obtain by the integral approach described in Section 10.4. The point to note is that simplifications resulting from the facts that one phase is a vapor (i.e., $\bar{V}^V \gg \bar{V}^L$) and that the vapor is ideal results in a procedure which, although complex, is manageable. Furthermore, we have seen that in the frequently occurring cases in which these assumptions are valid, knowledge of the condensed phase activity coefficient is all that is required to generate $(y\text{–}x)_T$ and $(P\text{–}x)_T$ diagrams.

(b) If we did not have an expression for the activity coefficient, and if we did have isothermal P–x data, we should be able to reverse the procedure above to construct the y–x relationship. Clearly, if we assumed that the liquid phase behaved as a van Laar liquid, all terms in the right-hand side of Eq. (10-50) could be expressed as functions of x_A and the van Laar constants, A_{12} and A_{21}. Thus, these constants could be evaluated by curve-fitting Eq. (10-50) to the experimental P–x data, and these constants could then be used in Eq. (10-46) to obtain the y–x relationship. Alternatively, we could use other applicable activity coefficient–liquid composition correlations and, in an analogous procedure, evaluate the coefficients in the expansion from the P–x data. Similarly, if the vapor phase were not ideal, an appropriate real-gas equation of state would have to be used for the vapor properties. Thus, it is possible to obtain y–x equilibrium information from the relatively simple measurement of equilibrium pressure without recourse to measurement of vapor phase composition.

We have seen in the discussion of Eqs. (10-13) through (10-18) that in order to generate isothermal, isobaric, and isoplethal cross sections, we must have knowledge of the temperature, pressure, and concentration dependency of the fugacities (or chemical potentials). If, however, we did not have knowledge of all property relations, we could develop some diagrams from knowledge of other cross sectional diagrams. The minimum number of such diagrams needed to specify all others is equivalent to determining the minimum number of independent partial derivatives involving temperature, pressure, and concentration.

For a binary system involving two phases, we require the minimum number of independent partials involving the variables, T, P, y_A, and x_A. In Section 7.2 we faced a similar problem for the four variables S, T, V, and P. There we found that all but four partials could be eliminated as independent by mathematical manipulation, and one of these four could be eliminated by thermodynamic reasoning [e.g., the Maxwell reciprocity relation was obtained from the fact that these variables satisfied the Fundamental Equation, $U = f(S, V)$]. Since the reciprocity condition does not apply here, it follows that there are four independent partials involving T, P, y_A, and x_A, but that no more than two can be chosen from any one set of the combinations of (T, P, x_A^α), (T, P, x_A^β), $(T, x_A^\alpha, x_A^\beta)$, and $(P, x_A^\alpha, x_A^\beta)$. Thus, any cross section could be obtained, for example, for a vapor–liquid binary system, from T–x, T–y, P–x, and P–y diagrams. Alternatively, in the general case in which simplifying assumptions are not applicable, we cannot obtain a y–x diagram from P–x data alone.

Let us now generalize the results obtained for the two-phase binary system to additional phases and components.

For a binary system involving the three phases α, β, and γ, the criteria of equilibrium become

$$T^\alpha = T^\beta = T^\gamma \tag{10-51}$$

$$P^\alpha = P^\beta = P^\gamma \tag{10-52}$$

$$\ln \hat{f}_A^\alpha = \ln \hat{f}_A^\beta = \ln \hat{f}_A^\gamma \tag{10-53}$$

$$\ln \hat{f}_B^\alpha = \ln \hat{f}_B^\beta = \ln \hat{f}_B^\gamma \tag{10-54}$$

These criteria are identical to solving simultaneously the two cases of α–β and β–γ phase equilibria. For the α–β equilibrium case, the relationships developed previously [namely, Eqs. (10-17), (10-18), (10-21), and (10-27)] are still valid. For the β–γ equilibrium, we would obtain identical equations with α replaced by γ. We would then solve the two sets simultaneously.

For example, by eliminating x_A^α from Eqs. (10-17) and (10-18), we obtained Eq. (10-21), which we can write as

$$dP = \left(\frac{\partial P}{\partial T}\right)_{x_A^\beta, [\alpha-\beta]} dT + \left(\frac{\partial P}{\partial x_A^\beta}\right)_{T, [\alpha-\beta]} dx_A^\beta \tag{10-55}$$

where the partials are given by Eqs. (10-22) and (10-23). The analogous relation for the β–γ equilibrium in which we eliminate x_A^γ would then be

$$dP = \left(\frac{\partial P}{\partial T}\right)_{x_A^\beta, [\beta-\gamma]} dT + \left(\frac{\partial P}{\partial x_A^\beta}\right)_{T, [\beta-\gamma]} dx_A^\beta \tag{10-56}$$

From Eq. (10-55) and (10-56), we could solve simultaneously to obtain P as a function of T by eliminating dx_A^β, P as a function of x_A^β by eliminating T, or T as a function of x_A^β by eliminating P. For example, let us eliminate x_A^β. Thus,

$$\left(\frac{\partial x_A^\beta}{\partial P}\right)_{T,[\beta-\gamma]}\left[dP - \left(\frac{\partial P}{\partial T}\right)_{x_A^\beta,[\beta-\gamma]} dT\right] = \left(\frac{\partial x_A^\beta}{\partial P}\right)_{T,[\alpha-\beta]}\left[dP - \left(\frac{\partial P}{\partial T}\right)_{x_A^\beta,[\alpha-\beta]} dT\right] \tag{10-57}$$

or

$$dP = \frac{(\partial P/\partial T)_{x_A^\beta,[\alpha-\beta]}(\partial x_A^\beta/\partial P)_{T,[\alpha-\beta]} - (\partial P/\partial T)_{x_A^\beta,[\beta-\gamma]}(\partial x_A^\beta/\partial P)_{T,[\beta-\gamma]}}{(\partial x_A^\beta/\partial P)_{T,[\alpha-\beta]} - (\partial x_A^\beta/\partial P)_{T,[\beta-\gamma]}}\, dT$$

or

$$dP = -\left[\frac{(\partial x_A^\beta/\partial T)_{P,[\alpha-\beta]} - (\partial x_A^\beta/\partial T)_{P,[\beta-\gamma]}}{(\partial x_A^\beta/\partial P)_{T,[\alpha-\beta]} - (\partial x_A^\beta/\partial P)_{T,[\beta-\gamma]}}\right] dT \tag{10-58}$$

The bracket in Eq. (10-58) is clearly $(\partial P/\partial T)_{[\alpha-\beta-\gamma]}$, which can be written as $(\partial P/\partial T)$ for the monovariant system of three phases in equilibrium. Equation (10-58) represents, in effect, a locus of triple points. Note that for each set of

T and P that satisfies Eq. (10-58), x_A^β will vary. We could have obtained this variation of x_A^β with T, for example, by eliminating dP from Eqs. (10-55) and (10-56). Other variables, such as x_A^α and x_A^γ, could be expressed as functions of T, P, or x_A^β in an analogous manner.

In the general case of n components distributed between π phases, we obtain, *for each component*, $\pi - 1$ equations of the form

$$-\frac{\bar{H}_i^\alpha - \bar{H}_i^\beta}{RT^2}\,dT + \frac{\bar{V}_i^\alpha - \bar{V}_i^\beta}{RT}\,dP + \sum_{j \neq k}^{n}\left[\left(\frac{\partial \ln \hat{f}_i^\alpha}{\partial x_j^\alpha}\right)_{T,P,x_i^\alpha[x_j^\alpha, x_k^\alpha]} dx_j^\alpha - \left(\frac{\partial \ln \hat{f}_i^\beta}{\partial x_j^\beta}\right)_{T,P,x_i^\beta[x_j^\beta, x_k^\beta]} dx_j^\beta\right] = 0 \tag{10-59}$$

Thus, we have $n(\pi - 1)$ equations of this form relating the $[(n - 1)(\pi) + 2]$ variables involving T, P, and x. Solving these equations simultaneously, we can eliminate $n(\pi - 1) - 1$ variables, resulting in a differential equation involving $n + 3 - \pi$ variables. From this equation, the differential of any one variable is expressed as a function of the remaining $n + 2 - \pi$ variables.[2]

10.3 Pressure–Temperature Relations

It is well known that in phase equilibria the system pressure is often a strong function of temperature. Also it turns out that the derivative dP/dT is related to the enthalpy (or entropy) and volume changes during a phase transformation. Since the latter properties are often of considerable interest to engineers, we shall illustrate a general approach by examining a few simple, but common systems, all of which involve a vapor phase in addition to one or more condensed phases.

The simplest case encountered is a pure liquid in equilibrium with its vapor. At equilibrium, we have $f^V = f^L$ in addition to temperature and pressure equalities. Thus,

$$d \ln f^V = d \ln f^L \tag{10-60}$$

Since the fugacity of a pure material is a function only of T and P, expanding we obtain

$$\left(\frac{\partial \ln f^V}{\partial T}\right)_P dT + \left(\frac{\partial \ln f^V}{\partial P}\right)_T dP = \left(\frac{\partial \ln f^L}{\partial T}\right)_P dT + \left(\frac{\partial \ln f^L}{\partial P}\right)_T dP \tag{10-61}$$

Substituting Eqs. (8-121) and (8-123) into Eq. (10-61) and collecting terms,

$$\left(\frac{dP}{dT}\right)_{[L\text{-}V]} = \frac{H^V - H^L}{T(V^V - V^L)} = \frac{\Delta H^{\text{vap}}}{T\,\Delta V^{\text{vap}}} \tag{10-62}$$

[2]Note that the Duhem equations (one for each phase for a total of π) can be used to simplify the $n(\pi - 1)$ equations of the form of Eq. (10-59) but not to reduce the number of variables. That is, the Duhem equations reduce the number of coefficients of the dx_j terms and therefore reduce the amount of physical property data required in integrations to obtain the final result.

Equation (10-62) is commonly called the Clausius–Clapeyron equation. Expressing $\Delta V^{vap} = (RT/P)(Z^V - Z^L) = (RT/P)\,\Delta Z^{vap}$, then

$$\left[\frac{d \ln P}{d(1/T)}\right]_{[L\text{-}V]} = \frac{-\Delta H^{vap}}{R\,\Delta Z^{vap}} \tag{10-63}$$

The ratio $(\Delta H^{vap}/\Delta Z^{vap})$ is a weak but essentially linear function of temperature except near the critical point. Often, at low pressures, the assumption is made that over a nominal temperature range, the ratio is constant and thus Eq. (10-63) is readily integrated. When such an assumption is not valid or when high accuracy is desired, other integration techniques are available.[3]

Next, let us consider the case in which we have a nonvolatile solute, such as an inorganic salt, dissolved in a volatile solvent such as water. The vapor above the solution is then essentially pure water. Since the system is divariant, the pressure–temperature relation for such a system is not unique unless we place a further restriction on the system. Two kinds of restricted systems will be illustrated.

Let the salt concentration in the liquid be constant as the system temperature (or pressure) is varied. For the volatile component (denoted by subscript w),

$$d \ln f_w^V = d \ln \hat{f}_w^L \tag{10-64}$$

Expanding in terms of T and P for pure vapor and in terms of T, P, and x_w for the liquid mixture, we obtain

$$\begin{aligned}\left(\frac{\partial \ln f_w^V}{\partial T}\right)_P dT &+ \left(\frac{\partial \ln f_w^V}{\partial P}\right)_T dP \\ &= \left(\frac{\partial \ln \hat{f}_w^L}{\partial T}\right)_{P,x} dT + \left(\frac{\partial \ln \hat{f}_w^L}{\partial P}\right)_{T,x} dP + \left(\frac{\partial \ln \hat{f}_w^L}{\partial x_w}\right)_{T,P} dx_w\end{aligned} \tag{10-65}$$

Substituting for the partial derivatives of fugacity and noting that $dx_w = 0$ for this case of constant liquid composition, after simplification, one finds

$$\left(\frac{\partial P}{\partial T}\right)_{x,[L\text{-}V]} = \frac{H_w^V - \bar{H}_w^L}{T(V_w^V - \bar{V}_w^L)} \tag{10-66}$$

The numerator represents the enthalpy change in vaporizing 1 mole of water from the solution at constant composition. The quantity H_w^V is not the enthalpy of saturated water vapor since, at a given T, the system pressure does not correspond to the equilibrium vapor pressure for pure water. The pressure correction, however, is ordinarily small and usually neglected. In fact, the numerator if often expanded by adding and subtracting H_w^L, the enthalpy of pure *liquid* water at the system temperature and pressure. Then[4]

$$H_w^V - \bar{H}_w^L = (H_w^V - H_w^L) + (H_w^L - \bar{H}_w^L) \tag{10-67}$$

[3] See, for example, R. C. Reid, J. M. Prausnitz, and T. K. Sherwood, *The Properties of Gases and Liquids*, 3rd ed. (New York: McGraw-Hill, 1977), Chap. 6.

[4] Note that this state may not be a thermodynamically stable state if, at T, $P < P_{vp_w}$.

When pressure corrections to the enthalpy of pure water are neglected, the first term in Eq. (10-67) is the enthalpy of vaporization of pure water at the system temperature; the second term is then the partial molar enthalpy of mixing, $\overline{\Delta H}_w^L$.

Then, if the additional assumptions that $V_w^V \gg \bar{V}_w^L$ and $V_w^V = RT/P$ are invoked,

$$\left[\frac{\partial \ln P}{\partial(1/T)}\right]_{x,[L\text{-}V]} = \frac{-(\Delta H_w^{\text{vap}} - \overline{\Delta H}_w^L)}{R} \tag{10-68}$$

In theory, then, from P–T data at constant liquid composition, one could determine partial molar enthalpies of mixing. Both $\overline{\Delta H}_w^L$ and $\overline{\Delta H}_s^L$ can be found, the former from Eq. (10-68) and the latter from $\overline{\Delta H}_w^L$ and the Duhem equation. Nevertheless, neat as this approach appears, it is unfortunately not a very useful one. Even with very accurate $(P\text{–}T)_x$ data, differentiation leads to some error and since, usually $\overline{\Delta H}_w^L < \Delta H_w^{\text{vap}}$, it is difficult to extract accurate values of $\overline{\Delta H}_w^L$ from the data. Direct calorimetric determination of enthalpies of mixing is usually the preferred way to measure this property.

Instead of keeping the salt solution concentration constant, one could impose the restriction that the solution is at all times saturated with the nonvolatile solute (i.e., there is always some undissovled salt present). In this case since $\pi = 3$ and $n = 2$, we have a univariant system.

Following the same general treatment, we have a liquid mixture, a pure water vapor phase, and a pure salt solid phase. Thus,

$$d \ln f_w^V = d \ln \hat{f}_w^L \tag{10-69}$$

and

$$d \ln f_s^S = d \ln \hat{f}_s^L \tag{10-70}$$

Expanding the fugacity in the pure phases in terms of T and P and the liquid phase in terms of T, P, and x_w, and substituting for the fugacity partials with respect to T and P, we obtain

$$-\frac{H_w^V - \bar{H}_w^L}{RT^2}\,dT + \frac{V_w^V - \bar{V}_w^L}{RT}\,dP = \left(\frac{\partial \ln \hat{f}_w^L}{\partial x_w}\right)_{T,P} dx_w \tag{10-71}$$

and

$$-\frac{H_s^S - \bar{H}_s^L}{RT^2}\,dT + \frac{V_s^S - \bar{V}_s^L}{RT}\,dP = \left(\frac{\partial \ln \hat{f}_s^L}{\partial x_w}\right)_{T,P} dx_w \tag{10-72}$$

To determine the P–T relation, we must solve Eqs. (10-71) and (10-72) simultaneously for dx_w and then use the Duhem relation to eliminate one of the fugacity partials with respect to concentration. A simple procedure is to multiply Eqs. (10-71) and (10-72), respectively, by x_w and x_s. Upon adding the resultant equations, we see that terms in dx_w drop out by virtue of the Duhem equation. Thus,

$$\left(\frac{\partial P}{\partial T}\right)_{[S\text{-}L\text{-}V]} = \frac{x_w(H_w^V - \bar{H}_w^L) + x_s(H_s^S - \bar{H}_s^L)}{T[x_w(V_w^V - \bar{V}_w^L) + x_s(V_s^S - \bar{V}_s^L)]} \tag{10-73}$$

To compare this case to Eq. (10-68), assume that the predominant term in the denominator is $x_w V_w^V$ and that $V_w^V = RT/P$; then

$$\left[\frac{\partial \ln P}{\partial(1/T)}\right]_{[S\text{-}L\text{-}V]} = -\frac{x_w(H_w^V - \bar{H}_w^L) + x_s(H_s^S - \bar{H}_s^L)}{Rx_w} \tag{10-74}$$

The term $d \ln P/d(1/T)$ is again related to an enthalpy change (i.e., the change in evaporating x_w moles of water while simultaneously crystallizing x_s moles of salt). The division by x_w simply yields the result as the enthalpy change per mole of water evaporated.

We might carry Eq. (10-74) one step further. The enthalpy of mixing of water and salt is often of interest. This ΔH^{mix} is defined as

$$\Delta H^{\text{mix}} = x_w(\bar{H}_w^L - H_w^L) + x_s(\bar{H}_s^L - H_s^S) \tag{10-75}$$

where H_w^L is the enthalpy of he pure water at the system T and P. If one adds and subtracts $x_w H_w^L$ to the right-hand side of Eq. (10-74), then

$$\left[\frac{d \ln P}{d(1/T)}\right]_{[S\text{-}L\text{-}V]} = \frac{\Delta H^{\text{mix}}}{Rx_w} - \frac{\Delta H_w^{\text{vap}}}{R} \tag{10-76}$$

Thus, from P–T data for saturated solutions and the heat of vaporization of the volatile component, one can determine, at least approximately, heats of mixing.

Now let us consider a binary liquid–vapor system in which both components exist in each phase. For this divariant system we shall use the additional restriction of constant liquid composition.

The pressure–temperature relation in this case follows directly from Eq. (10-22), is we denote liquid phase by β and vapor phase by α. Thus,

$$\left(\frac{\partial P}{\partial T}\right)_{x,[L\text{-}V]} = \frac{1}{T}\frac{y_A(\bar{H}_A^V - \bar{H}_A^L) + y_B(\bar{H}_B^V - \bar{H}_B^L)}{y_A(\bar{V}_A^V - \bar{V}_A^L) + y_B(\bar{V}_B^V - \bar{V}_B^L)} \tag{10-77}$$

Assuming that $\bar{V}_A^V \gg \bar{V}_A^L$, $\bar{V}_B^V \gg \bar{V}_B^L$, and noting that

$$y_A\bar{V}_A^V + y_B\bar{V}_B^V = V^V = \frac{RT}{P} \tag{10-78}$$

$$y_A\bar{H}_A^V + y_B\bar{H}_B^V = H^V \tag{10-79}$$

then

$$\left[\frac{\partial \ln P}{\partial(1/T)}\right]_{x,[L\text{-}V]} = -\frac{H^V - y_A\bar{H}_A^L - y_B\bar{H}_B^L}{R} \tag{10-80}$$

As expected, the slope of $\ln P$ versus $1/T$ corresponds to an enthalpy change. This enthalpy change is, however, somewhat unusual. The term H^V represents the enthalpy of the saturated vapor mixture at y_A, y_B. The terms $\bar{H}_A^L$ and $\bar{H}_B^L$ represent the partial molar enthalpies of the saturated liquid at x_A and x_B. To visualize the situation, consider Figure 10.3. The enthalpy of the saturated vapor and liquid of a binary mixture of A and B is shown as a function of liquid

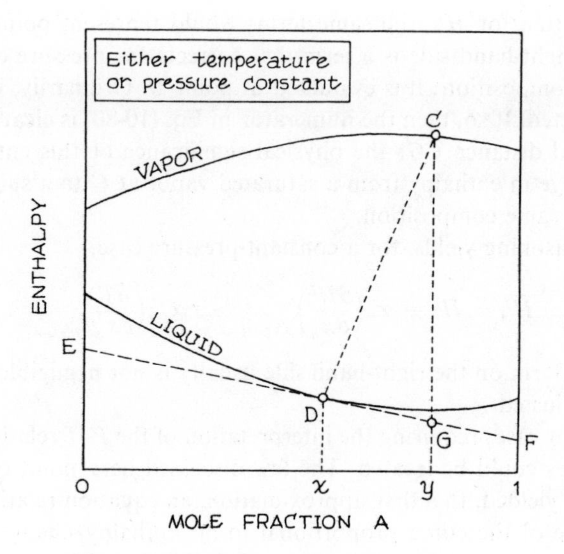

Figure 10.3 Enthalpy–concentration diagram.

and vapor mole fraction. Point C represents the enthalpy of a saturated vapor at y_A. This vapor is in equilibrium with a liquid of composition x_A with enthalpy H^L(point D). The diagram is drawn for either constant T or P. Consider a tangent to the lower curve at D. The tangent intersects the left-hand ordinate (pure B) at E and the right-hand ordinate (pure A) at F. Let us consider the significance of these points E and F.

If Figure 10.3 is drawn at constant temperature, then

$$dH^L = \left(\frac{\partial H^L}{\partial P}\right)_{T, x_A} dP + \left(\frac{\partial H^L}{\partial x_A}\right)_{P,T} dx_A \tag{10-81}$$

Dividing Eq. (10-81) by dx_A and placing the restriction of saturated solution yields

$$\left(\frac{\partial H^L}{\partial x_A}\right)_{T,[L\text{-}V]} = \left(\frac{\partial H^L}{\partial P}\right)_{T, x_A} \left(\frac{\partial P}{\partial x_A}\right)_{T,[L\text{-}V]} + \left(\frac{\partial H^L}{\partial x_A}\right)_{P,T} \tag{10-82}$$

but

$$\bar{H}^L_A = H^L + x_B\left(\frac{\partial H^L}{\partial x_A}\right)_{T,P} \tag{10-83}$$

so by multiplying Eq. (10-82) by x_B and subtracting Eq. (10-83) from the result, we obtain

$$\bar{H}^L_A = H^L + x_B\left(\frac{\partial H^L}{\partial x_A}\right)_{T,[L\text{-}V]} - x_B\left(\frac{\partial H^L}{\partial P}\right)_{T, x_A}\left(\frac{\partial P}{\partial x_A}\right)_{T,[L\text{-}V]} \tag{10-84}$$

A similar relation may be written for $\bar{H}^L_B$.

The $\bar{H}^L_A$ and $\bar{H}^L_B$ terms are those required for Eq. (10-80). The first two terms on the right-hand side of Eq. (10-84) yield point F on Figure 10.3. (If the equation were written for $\bar{H}_B^L$, the same terms would represent point E). The last term on the right-hand side is a term that corrects for pressure changes in the system with composition; it is evaluated at point D. Ordinarily, it is small and usually neglected. If so, then the numerator in Eq. (10-80) is clearly represented by the vertical distance $\overline{CG}$; the physical significance of this enthalpy term is then the change in enthalpy from a saturated vapor at C to a subcooled liquid at G with the same composition.

Similar reasoning yields, for a constant-pressure case,

$$\bar{H}_A^L = H^L + x_B\left(\frac{\partial H^L}{\partial x_A}\right)_{P,[L\text{-}V]} - x_B C_p^L\left(\frac{\partial T}{\partial x_A}\right)_{P,[L\text{-}V]} \tag{10-85}$$

Since the last term on the right-hand side usually is not negligible, point G will have to be adjusted.

Many other cases requiring the interpretation of the P–T relations of liquid–vapor mixtures could be treated. The few discussed here point out the typical approach; all yielded, to a first approximation, an equation relating $\ln P$ to $1/T$ with the slope of the curve proportional to an enthalpy change between the phases.

10.4 The Integral Approach to Phase Equilibrium Relationships

The integral approach proceeds from the same starting point as the differential approach, namely, equating component fugacities in each coexisting phase. Instead of differentiating these relationships, however, we treat them directly in the integral form. In general, the procedure is to equate the fugacities of each component

$$\hat{f}_i^\alpha = \hat{f}_i^\beta \tag{10-86}$$

and then to expand these fugacities as functions of temperature, pressure, and composition in each phase. Two approaches are used. One utilizes an equation of state to determine component fugacities in *both* phases, whereas in the other, an equation of state is employed only for the gas phase (if present) and fugacities in condensed phases are correlated in terms of activity coefficients. Both methods are described below.

Equation-of-state approach

Component fugacities are expressed in terms of a fugacity coefficient as given in Eq. (8-132).

$$\begin{aligned} \hat{f}_i^\alpha &= \phi_i^\alpha P x_i^\alpha \qquad (\alpha\text{-phase}) \\ \hat{f}_i^\beta &= \phi_i^\beta P x_i^\beta \qquad (\beta\text{-phase}) \end{aligned} \tag{10-87}$$

and the fugacity coefficients ϕ_i are related to P–$\underline{V}$–T–N properties of the mixture by Eq. (8-131). To illustrate these relations for a case where α = vapor and β = liquid, one obtains

$$\frac{y_i}{x_i} = \frac{\phi_i^L}{\phi_i^v} \tag{10-88}$$

ϕ_i^v is the fugacity coefficient of component i in the vapor phase and is a function of T, P, and vapor composition $(y_1, \ldots, y_i, \ldots, y_{n-1})$.

In Eq. (8-135) we show a relation for ϕ_i as determined from the Peng–Robinson equation of state, Eq. (8-55). For a n-component system, the phase rule requires that n-intensive variables be specified. Suppose that these were $(n - 1)$ liquid-phase mole fractions and the system temperature. There are n equations of the form of Eq. (10-88) to calculate the unknown $(n - 1)$ vapor-phase mole fractions and the system pressure. [Note in Eq. (8-135) that compressibility factors are also present. These would be determined separately for both phases, at the system temperature and pressure, at the appropriate phase composition.]

Clearly, the solution of the n equations in Eq. (10-88) requires machine computation as the expressions for ϕ_i are highly nonlinear. The success of this method depends, however, on the choice of an appropriate mixture equation of state. Few have been shown to be of much practical value to calculate phase equilibria. The Benedict–Webb-Rubin equation of state (or one of its many modifications) has been used for a number of years to determine phase equilibrium behavior of mixtures of light hydrocarbons. More recently, phase behavior has been correlated with two-constant modifications of the Redlich–Kwong equation of state such as suggested by Soave[5] or by Peng and Robinson [Eq. (8-55)]. For systems that are essentially nonpolar, these correlations have often been quite successful. In Figure 10.4 we show experimental and calculated pressures for the isobutane–carbon dioxide system at 311 K where the Peng–Robinson equation of state was used to determine phase volumes and fugacity coefficients. The sensitivity of the computed pressures to the correct choice of the isobutane–carbon dioxide interaction parameter [see Eq. (8-58)] is well illustrated.

Another application of the equation-of-state approach lies in computing solubilities of relatively nonvolatile solids in supercritical fluids.[6] Returning to Eq. (10-86), let α be the supercritical phase and β be the solid phase, with i the solute. Then $\hat{f}_i^\alpha$ is determined from an equation of state as shown in Eq. (10-87). $\hat{f}_i^\beta$ is now the fugacity of pure solid solute at the temperature and pressure of the system (assuming no solvent gas dissolves in the solid). Therefore,

[5]G., Soave, *Chem. Eng. Sci.*, **27**, 1197 (1972); see also M. L. Michelsen [*Fluid Phase Eq.*, **4**, 1 (1980)] for a review of computer algorithms using the Soave equation of state and I. Wichterle [*Fluid Phase Eq.*, **2**, 59 (1978)] for a general survey.

[6]The term *supercritical fluid* indicates the solvent phase is at a temperature and pressure above the critical-state values of the pure solvent.

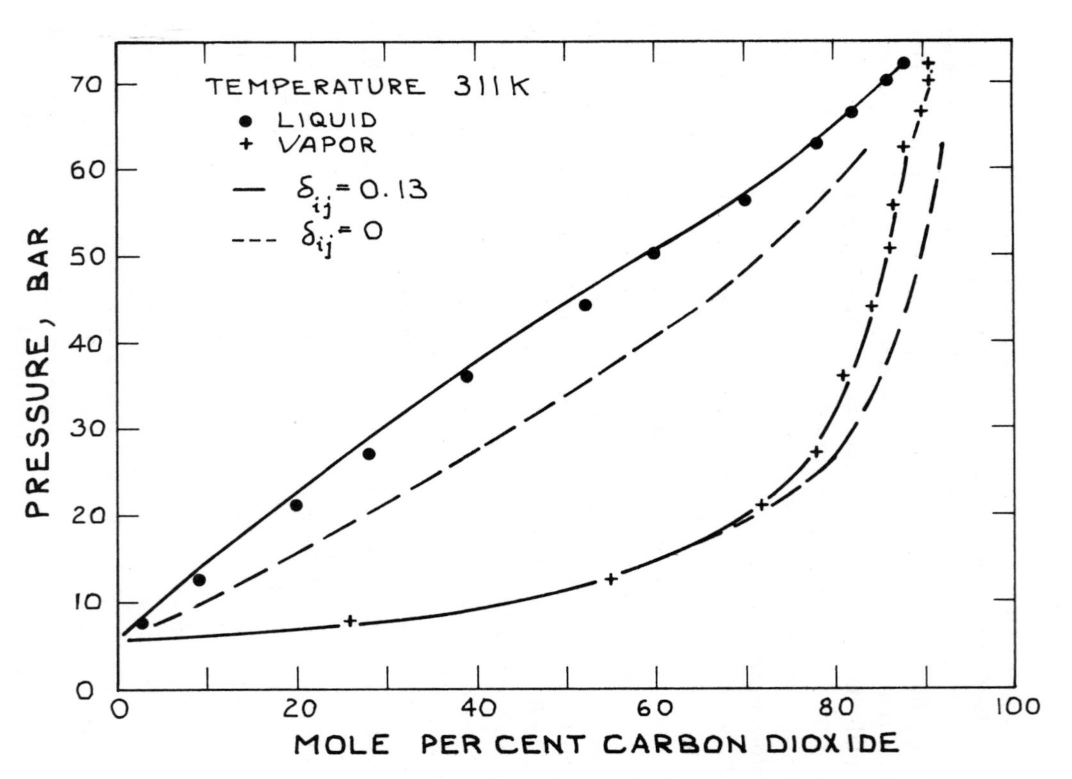

Figure 10.4 Estimation of the phase behavior of isobutane–carbon dioxide system using the Peng–Robinson equation of state. [Adapted from D.-Y. Peng and D. B. Robinson, *Ind. Eng. Chem. Fundam.*, **15**, 59 (1976)]

$$\hat{f}_i^\alpha = \phi_i^\alpha(P, T, y_i)Py_i = f_i^\beta(P, T)$$
$$= P_{vp_i}\nu_i(P_{vp_i}, T)\exp\left(\int_{P_{vp_i}}^{P} \frac{V_i^\beta}{RT}\,dP\right) \tag{10-89}$$

The fugacity coefficient for vapor above pure solid i at its vapor pressure, ν_i, is normally very close to unity. Assuming that the solid is incompressible $[V_i^\beta \neq f(P)]$, Eq. (10-89) can then be written as

$$y_i = \left(\frac{P_{vp_i}}{P}\right)\frac{1}{\phi_i}\left\{\exp\left[\frac{V_i^\beta}{RT}(P - P_{vp_i})\right]\right\} \tag{10-90}$$

Thus, the mole fraction of the solute in the supercritical fluid is given as a product of three terms. The first is called the *ideal solubility* and would simply indicate that, at constant temperature, y_i is proportional to P^{-1}. The third term is the Poynting correction and measures the effect of pressure on the solid fugacity. This term increases with pressure. The second term reflects the nonideality of the fluid-phase mixture of solvent and solute. ϕ_i is close to unity at low pressures, but it can become very small (to give large values of y_i) at high pressures. To illustrate Eq. (10-90), consider Figure 10.5. In this figure the solubility of solid 2,6-dimethylnaphthalene (2,6-DMN) in supercritical carbon dioxide is shown as a function of pressure. The computed curves were made with the Peng–Robinson equation of state to determine ϕ for the 2,6-DMN.[7] The interaction parameter, δ_{ij}, used was 0.10. At low pressures y (2,6-DMN) decreases with pressure, reflecting the first term in Eq. (10-90). When the pressure exceeds about 20 bar, ϕ(2,6-DMN) begins to decrease and, in so doing, causes the solubility to pass through a minimum and then increase rapidly with pressure. The Poynting correction term, althought not insignificant, does not play an important role in the rapid rise in solubility with pressure. For example, at 318 K and 244 bar, the solubility of 2,6-DMN is about 6.3×10^{-3} mole fraction. At 318 K, the vapor pressure of pure 2,6-DMN is about 3.5×10^{-5} bar. Thus, for Eq. (10-90), $P_{vp}/P \sim 1.4 \times 10^{-7}$, the Poynting correction is about 4.2 and ϕ^{-1} is 10,700. Clearly, an ideal-gas assumption of $\phi = 1$ is grossly incorrect.

In Figure 10.5, there is seen to be an inversion in solubility with temperature at pressures around 80 to 100 bar, that is, an increase in temperature results in a decrease in solubility. By the use of the differential approach described in Sections 10.2 and 10.3, one can show that in this region, the dissolution process is exothermic. That is, to dissolve 2,6-DMN in supercritical CO_2 under isobaric and isothermal conditions, one must remove heat from the system.

We have indicated only a few examples of the equation-of-state approach for determining phase equilibrium behavior. As noted, the success or failure of this method hinges on having an appropriate mixture equation of state. Progress in this area will most certainly lead to a wider use of this technique, even to the computation of phase behavior in liquid–liquid systems.

[7] R. T. Kurnik, S. J. Holla, and R. C. Reid, *J. Chem. Eng. Data.*, **26**, 47 (1981).

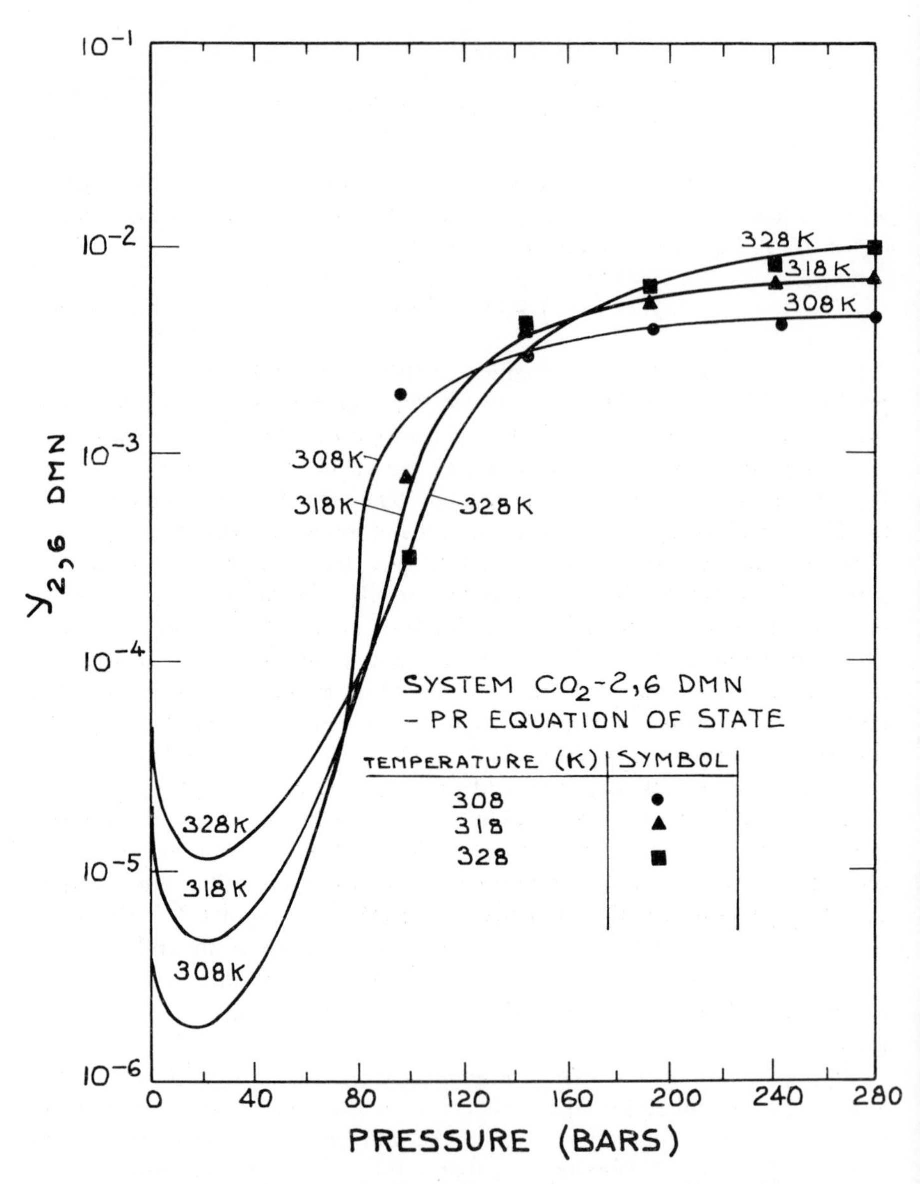

Figure 10.5 Solubility of 2,6-dimethylnaphthalene in supercritical carbon dioxide. [Adapted from R. T. Kurnik, S. J. Holla, and R. C. Reid, *J. Chem. Eng. Data*, **26**, 47 (1981)]

Activity coefficient approach for condensed phases

The component fugacity is given as shown in Eq. (8-165),

$$\hat{f}_i^\sigma = \hat{f}_i^+ \gamma_i^+ x_i^\sigma \tag{10-91}$$

$\hat{f}_i^+$ is the fugacity of i in a reference state defined by T^+, P^+, $x_1^+, \ldots, x_{n-1}^+$ and a state of aggregation which may be different from that of the mixture. Since $\hat{f}_i^\sigma$ and x_i are properties of the actual mixture in phase (σ), it follows that γ_i^+, the activity coefficient, is a function of the actual mixture properties as well as the reference-state properties, that is,

$$\gamma_i^+ = g_i(T, P, x_1^\sigma, \ldots, x_{n-1}^\sigma, T^+, P^+, x_1^+, \ldots, x_{n-1}^+) \tag{10-92}$$

Although we have the option of defining reference states in a very general manner, in actual fact, there are few that are ever used. In every case T^+ is defined to be equal to the system temperature T and in *almost* all cases, the composition in the reference state is chosen to be the pure component at T, P^+, and in the same condensed state (σ) as the mixture.

Equation (10-91) then becomes

$$\hat{f}_i^\sigma = f_i^\sigma(T, P^+)\gamma_i(T, P^+, x_1^\sigma, \ldots, x_{n-1}^\sigma)x_i^\sigma \tag{10-93}$$

For systems where the temperaure does not exceed the critical temperature of any component,[8] P^+ is usually set equal to the system pressure P. It is then convenient to relate $f_i^\sigma(T, P)$ to the fugacity of i at its vapor pressure,

$$\ln\left(\frac{f_{i,P}^\sigma}{f_{i,P_{vp}}^\sigma}\right) = \int_{P_{vp_i}}^{P} \left(\frac{\partial \ln f_i}{\partial P}\right)_T dP = \frac{1}{RT}\int_{P_{vp_i}}^{P} V_i^\sigma \, dP$$

where the volume in the integral is the molar volume of pure i at T in the same condensed state as the mixture.

$$f_{i,P}^\sigma = f_{i,P_{vp}}^\sigma \exp\left(\int_{P_{vp_i}}^{P} \frac{V_i^\sigma}{RT} dP\right) \tag{10-94}$$

The exponential term in Eq. (10-94) is called the *Poynting correction factor*. It can usually be neglected for pressures within an order of magnitude of P_{vp} and at temperatures not near the critical. [For example, if $(P - P_{vp}) = 10$ bar, $T = 400$ K, and $V_i^\sigma = 100$ cm³/mol, the Poynting term is 1.031.]

Also, for pure condensed i at T, P_{vp},

$$f_i^\sigma(T, P_{vp}) = f_i^v(T, P_{vp}) \tag{10-95}$$

$$= P_{vp_i}\nu_i(T, P_{vp}) \tag{10-96}$$

where $\nu_i(T, P_{vp})$ is the fugacity coefficient of pure i vapor [see Eqs. (8-133) and (8-134)].

With Eqs. (10-91), (10-94), and (10-96), if we have a system consisting of a vapor and liquid phase,

$$\phi_i^v P y_i = P_{vp_i}\nu_i\gamma_i x_i \exp\left(\int_{P_{vp_i}}^{P} \frac{V_i^\sigma}{RT} dP\right) \tag{10-97}$$

Note that Eq. (10-87) was used to express the fugacity of i in the vapor phase.

[8]Section 10.5 covers systems with supercritical components.

At low pressures and at temperatures away from the critical point, both the Poynting correction term and ν_i may be set equal to unity. Then

$$\phi_i P y_i = P_{vp_i} \gamma_i x_i \tag{10-98}$$

Equation (10-98) has been widely used to correlate vapor–liquid equilibrium data. Note that if the gas phase were ideal ($\phi_i = 1.0$) and the solution exhibited ideal behavior ($\gamma_i = 1.0$), then Eq. (10-98) simplifies to

$$P y_i = P_{vp_i} x_i \tag{10-99}$$

which is termed *Raoult's law*.

For liquid–liquid equilibria, Eqs. (10-86) and (10-87) become

$$\gamma_i^\alpha x_i^\alpha = \gamma_i^\beta x_i^\beta \tag{10-100}$$

as the reference state has been chosen to be the same for each condensed phase.

In Eq. (10-98) or (10-100), the activity coefficients, besides being a function of composition, vary with the system temperature and pressure. The pressure effect is often neglected at low pressures, and only the temperature and composition are considered. A wide variety of forms of the function

$$\gamma^\sigma = g(T, x_1^\sigma, \ldots, x_{n-1}^\sigma) \tag{10-101}$$

have been suggested. As noted in Chapter 8, a solution model for the condensed phase is developed from theory and the examination of experimental data. The excess Gibbs energy of the mixture is obtained, and the component activity coefficients found from Eq. (8-161). An example was given in Chapter 8 to illustrate this method using the very simple relation $\overline{\Delta G^{EX}} = C x_1 x_2$ for a binary of 1 and 2. Other binary models are shown in Table 10.1.[9]

Note in Table 10.1 that the Margules and van Laar models assume that the constants in the ΔG^{EX} function are temperature independent so, for the activity-coefficient expressions, the linear temperature dependence enters simply from Eq. (8-161). This proportionality is only very approximate. For the Wilson, NRTL, and UNIQUAC models, the parameters are assumed to be functions of temperature.

For most accurate correlations of vapor–liquid equilibrium, the Wilson, NRTL, or UNIQUAC models are now employed. The last is particularly useful since a significant amount of work has been expended to develop an activity coefficient estimation scheme (UNIFAC) to allow UNIQUAC to be used on systems where no or very few experimental data exist.[10] In a comprehensive review of correlation and prediction methods for liquid–liquid equilibria, the UNIQUAC model is also recommended.[11]

[9] R. C. Reid, J. M. Prausnitz, and T. K. Sherwood, *The Properties of Gases and Liquids* (New York: McGraw-Hill, 1977), Chap. 8.

[10] Aa. Fredenslund, J. Gmehling, and P. Rasmussen, *Vapor–Liquid Equilibrium Using UNIFAC* (Amsterdam: Elsevier, 1977); see also *Ind. Eng. Chem. Proc. Des. Dev.*, **16**, 450 (1977).

[11] J. M. Sørensen, T. Magnussen, P. Rasmussen, and Aa. Fredenslund, *Fluid Phase Eq.*, **2**, 297 (1979); **3**, 47 (1979); **4**, 151 (1980).

TABLE 10.1

SOME MODELS FOR THE EXCESS GIBBS ENERGY AND SUBSEQUENT ACTIVITY COEFFICIENTS FOR BINARY SYSTEMS

Name	ΔG^{EX}	Binary parameters	$\ln \gamma_1$ and $\ln \gamma_2$
Two-suffix[a] Margules	$\Delta G^{EX} = Ax_1x_2$	A	$RT \ln \gamma_1 = Ax_2^2$ $RT \ln \gamma_2 = Ax_1^2$
Three-suffix[a] Margules	$\Delta G^{EX} = x_1x_2[A + B(x_1 - x_2)]$	A, B	$RT \ln \gamma_1 = (A + 3B)x_2^2 - 4Bx_2^3$ $RT \ln \gamma_2 = (A - 3B)x_1^2 + 4Bx_1^3$
van Laar	$\Delta G^{EX} = \dfrac{Ax_1x_2}{x_1(A/B) + x_2}$	A, B	$RT \ln \gamma_1 = A\left(1 + \dfrac{Ax_1}{Bx_2}\right)^{-2}$ $RT \ln \gamma_2 = B\left(1 + \dfrac{Bx_2}{Ax_1}\right)^{-2}$
Wilson	$\dfrac{\Delta G^{EX}}{RT} = -x_1 \ln (x_1 + \Lambda_{12}x_2) - x_2 \ln (x_2 + \Lambda_{21}x_1)$	$\Lambda_{12}, \Lambda_{21}$	$\ln \gamma_1 = -\ln (x_1 + \Lambda_{12}x_2) + x_2\left(\dfrac{\Lambda_{12}}{x_1 + \Lambda_{12}x_2} - \dfrac{\Lambda_{21}}{\Lambda_{21}x_1 + x_2}\right)$ $\ln \gamma_2 = -\ln (x_2 + \Lambda_{21}x_1) - x_1\left(\dfrac{\Lambda_{12}}{x_1 + \Lambda_{12}x_2} - \dfrac{\Lambda_{21}}{\Lambda_{21}x_1 + x_2}\right)$
Four-suffix[a] Margules	$\Delta G^{EX} = x_1x_2[A + B(x_1 - x_2) + C(x_1 - x_2)^2]$	A, B, C	$RT \ln \gamma_1 = (A + 3B + 5C)x_2^2 - 4(B + 4C)x_2^3 + 12Cx_2^4$ $RT \ln \gamma_2 = (A - 3B + 5C)x_1^2 + 4(B - 4C)x_1^3 + 12Cx_1^4$
NRTL[b]	$\dfrac{\Delta G^{EX}}{RT} = x_1x_2\left(\dfrac{\tau_{21}G_{21}}{x_1 + x_2G_{21}} + \dfrac{\tau_{12}G_{12}}{x_2 + x_1G_{12}}\right)$ where $\tau_{12} = \dfrac{\Delta G_{12}}{RT}$, $\tau_{21} = \dfrac{\Delta G_{21}}{RT}$ $\ln G_{12} = -\alpha_{12}\tau_{12}$, $\ln G_{21} = -\alpha_{12}\tau_{21}$	$\Delta G_{12}, \Delta G_{21}, \alpha_{12}$[c]	$\ln \gamma_1 = x_2^2\left[\tau_{21}\left(\dfrac{G_{21}}{x_1 + x_2G_{21}}\right)^2 + \dfrac{\tau_{12}G_{12}}{(x_2 + x_1G_{12})^2}\right]$ $\ln \gamma_2 = x_1^2\left[\tau_{12}\left(\dfrac{G_{12}}{x_2 + x_1G_{12}}\right)^2 + \dfrac{\tau_{21}G_{21}}{(x_1 + x_2G_{21})^2}\right]$

TABLE 10.1 (Continued)

SOME MODELS FOR THE EXCESS GIBBS ENERGY AND SUBSEQUENT ACTIVITY COEFFICIENTS FOR BINARY SYSTEMS

Name	ΔG^{EX}	Binary parameters	$\ln \gamma_1$ and $\ln \gamma_2$
UNIQUAC[d]	$\Delta G^{EX} = \Delta G^{EX}(\text{combinatorial}) + \Delta G^{EX}(\text{residual})$ $\frac{\Delta G^{EX}(\text{combinatorial})}{RT} = x_1 \ln \frac{\Phi_1}{x_1} + x_2 \ln \frac{\Phi_2}{x_2} + \frac{z}{2}\left(q_1 x_1 \ln \frac{\theta_1}{\Phi_1} + q_2 x_2 \ln \frac{\theta_2}{\Phi_2}\right)$ $\frac{\Delta G^{EX}(\text{residual})}{RT} = -q_1 x_1 \ln [\theta_1 + \theta_2 \tau_{21}] - q_2 x_2 \ln [\theta_2 + \theta_1 \tau_{12}]$ $\Phi_1 = \frac{x_1 r_1}{x_1 r_1 + x_2 r_2}, \quad \theta_1 = \frac{x_1 q_1}{x_1 q_1 + x_2 q_2}$ $\ln \tau_{21} = -\frac{\Delta u_{21}}{RT}, \quad \ln \tau_{12} = -\frac{\Delta u_{12}}{RT}$ r and q are pure-component parameters and coordination number $z = 10$	Δu_{12} and Δu_{21}[e]	$\ln \gamma_i = \ln \frac{\Phi_i}{x_i} + \frac{z}{2} q_i \ln \frac{\theta_i}{\Phi_i} + \Phi_j\left(\ell_i - \frac{r_i}{r_j}\ell_j\right) - q_i \ln (\theta_i + \theta_j \tau_{ji}) + \theta_j q_i \left(\frac{\tau_{ji}}{\theta_i + \theta_j \tau_{ji}} - \frac{\tau_{ij}}{\theta_j + \theta_i \tau_{ij}}\right)$ where $i = 1, j = 2$ or $i = 2, j = 1$ $\ell_i = \frac{z}{2}(r_i - q_i) - (r_i - 1)$ $\ell_j = \frac{z}{2}(r_j - q_j) - (r_j - 1)$

[a]Two-suffix signifies that the expansion for ΔG^{EX} is quadratic in mole fraction. Three-suffix signifies a third-order, and four-suffix signifies a fourth-order equation.

[b]NRTL, nonrandom two-liquid.

[c]$\Delta G_{12} = G_{12} - G_{22}$; $\Delta G_{21} = G_{21} - G_{11}$

[d]UNIQUAC, universal quasi-chemical.

[e]$\Delta u_{12} = u_{12} - u_{22}$; $\Delta u_{21} = u_{21} - u_{11}$.

Source: J. M. Prausnitz [*Molecular Thermodynamics of Fluid-Phase Equilibria* (Englewood Cliffs, N.J.: Prentice-Hall, 1969)] discusses the Margules, van Laar, Wilson, and NRTL equations. The UNIQUAC equation is discussed in *AIChE J.*, **21**, 116 (1975).

Example 10.4

At atmospheric pressure, the binary system of n-butanol–water exhibits a minimum-boiling azeotrope and partial miscibility in the liquid phase. x–y–T data at 1.02 bar are as follows:

Data for Example 10.4
Vapor and Liquid Mole Fractions of n-Butanol
in Equilibrium with Water at 1.022 Bar

T (°C)	y	x
100	0.0	0.0
95.8	0.150	0.008
95.4	0.161	0.009
92.8	0.237	0.019
92.8	0.240	0.020
92.7	0.246	0.098
92.7	0.246	0.099
92.7	0.246	0.247
93.0	0.250	0.454
93.0	0.247	0.450
96.3	0.334	0.697
94.0	0.276	0.583
96.6	0.340	0.709
100.8	0.444	0.819
106.4	0.598	0.903
106.8	0.612	0.908
110.9	0.747	0.950
117.5	1.000	1.000

Source: T. E. Smith and R. F. Bonner, "Vapor–Liquid Equilibrium Still for Partially Miscible Liquids," *Ind. Eng. Chem.*, **41**, 2867 (1949).

(a) Using only the azeotropic point, estimate the x–y–T equilibrium properties for this system. Use the van Laar activity coefficient correlation in Table 10.1.
(b) Estimate the compositions of the two liquid phases that occur in this immiscible system.

Solution

(a) From the table of data provided, the best estimate of the azeotropic point is $T \sim 92.7°\text{C} = 365.9$ K, $y(n\text{-butanol}) = x(n\text{-butanol}) = 0.246$.

At this low pressure, the vapor phase is essentially ideal ($\phi_i = 1.0$) and Eq. (10-98) may be used to estimate the activity coefficients. At 365.9 K, $P_{vp}(n\text{-butanol})$ is about 0.378 bar and $P_{vp}(\text{water}) = 0.774$ bar. Thus,

$$\gamma(n\text{-butanol}) = \frac{(1.023)(0.246)}{(0.378)(0.246)} = 2.71$$

$$\gamma(\text{water}) = \frac{(1.023)(0.754)}{(0.774)(0.754)} = 1.32$$

The van Laar activity coefficient expressions in Table 10.1 may be algebraically rearranged to solve for the constants A and B:

$$\frac{A}{R} = T \ln \gamma_1 \left(1 + \frac{x_2 \ln \gamma_2}{x_1 \ln \gamma_1}\right)^2$$

$$\frac{B}{R} = T \ln \gamma_2 \left(1 + \frac{x_1 \ln \gamma_1}{x_2 \ln \gamma_2}\right)^2$$

With n-butanol as component 1 and water as 2,

$$\frac{A}{R} = (365.9)(\ln 2.71)\left[1 + \frac{(0.754)(\ln 1.32)}{(0.246)(\ln 2.71)}\right]^2$$

$$= 1253 \text{ K}$$

Similarly,

$$\frac{B}{R} = 479 \text{ K}$$

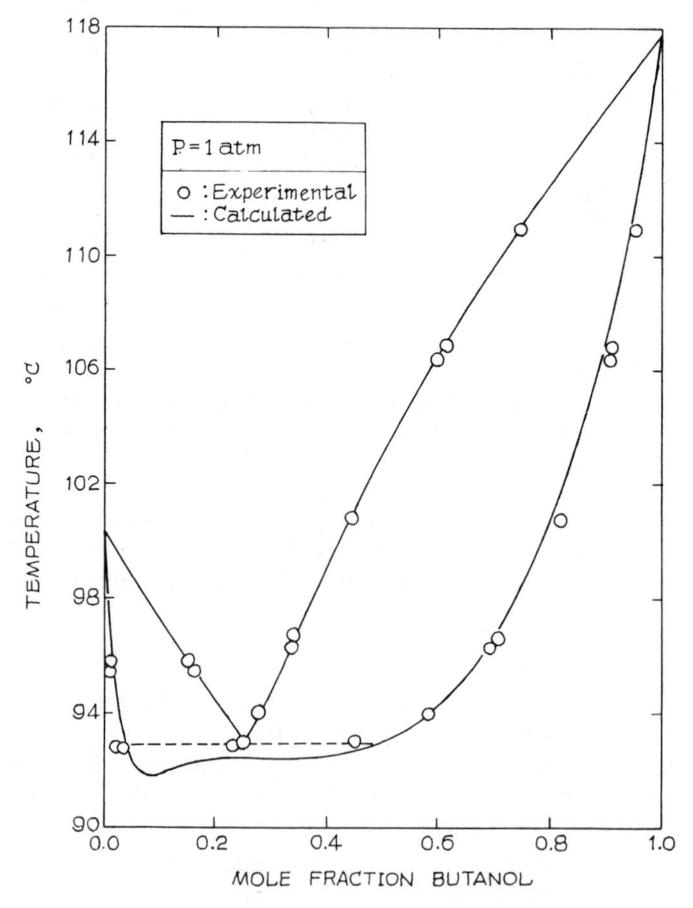

Figure 10.6 T-x-y diagram for n-butanol-water system.

With these parameters,

$$\ln \gamma_1 = \frac{1253/T}{[1 + 2.62(x_1/x_2)]^2}$$

$$\ln \gamma_2 = \frac{479/T}{[1 + 0.382(x_2/x_1)]^2}$$

If Eq. (10-98) is written for both n-butanol and water and added to eliminate vapor-phase compositions, with $\phi_i = 1.0$,

$$P = x_1 P_{vp_1}(T)\gamma_1(T, x_1) + x_2 P_{vp_2}(T)\gamma_2(T, x_1)$$

P is set at 1.02 bar. For a given value of x_1 and the activity coefficient correlations, one iterates to find the system temperature. Vapor compositions are then found from Eq. (10-98) with vapor pressures for each component at the boiling temperature. Accurate vapor pressures are necessary. Experimental data are compared with computed values in Figures 10.6 and 10.7. The agreement is quite good except for

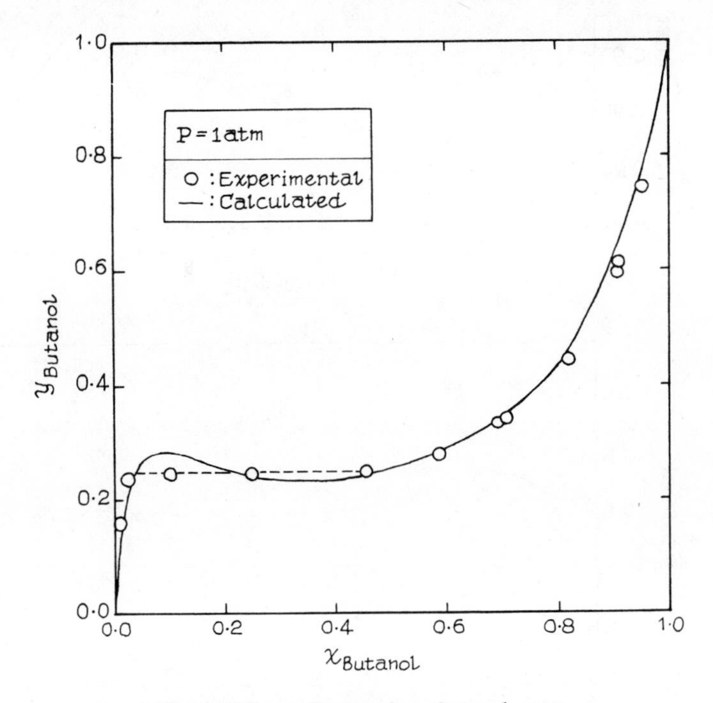

Figure 10.7 x-y diagram for n-butanol-water.

the liquid phase at low n-butanol concentrations. The temperature correction to the activity-coefficient correlations is very approximate ($T \ln \gamma =$ constant), but is probably satisfactory over the small temperature range involved.

(b) The computed maxima and minima in the x–y curve should alert one to the fact that a liquid-phase split has occurred in this region. For a binary system to be stable, as shown in Figure 9.7, the variation of the chemical potential with composition must be positive (under constant-temperature and constant-pressure constraints). Neglecting the small temperature variations, this criterion reduces to

$$\frac{d \ln (\gamma_i x_i)}{dx_i} > 0 \tag{10-102}$$

or

$$\frac{d \ln a_1}{dx_1} > 0, \qquad \frac{d \ln a_2}{dx_2} > 0 \tag{10-103}$$

where a_i is the activity.

Calculated values of a_1 and a_2 are shown in Figure 10.8. Since the stability criteria are violated on the spinodal curve, we know that the binodal curve lies outside the spinodal curve. Thus, by inspection of Figure 10.8, we see that the binodal (equi-

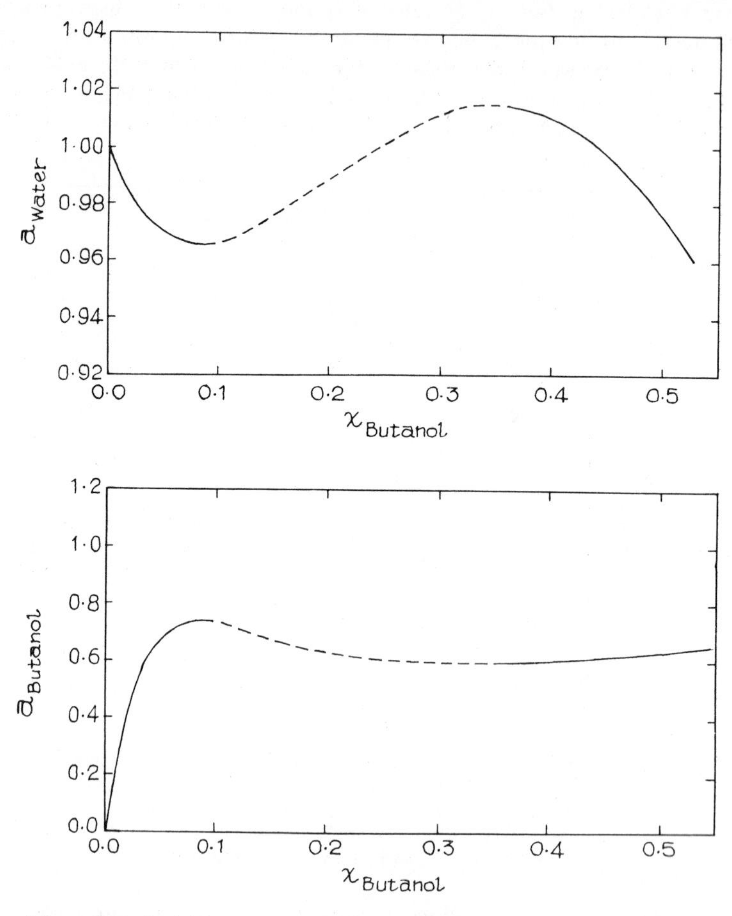

Figure 10.8 Activities in the *n*-butanol-water system at one atmosphere.

librium) mole fractions of *n*-butanol are *less* than about 0.09 and greater than about 0.34. To determine the concentrations in the two immiscible liquid phases, the equilibrium criterion, Eq. (10-100), is used for both components. A trial-and-error solution yields *n*-butanol mole fractions of about 0.04 and 0.50 in the two phases. Experimental values are about 0.02 and 0.34. The calculated compositions were used to draw the dashed lines on Figures 10.6 and 10.7.

In theory, if the activity-coefficient correlations contained an accurate temperature function, this procedure could be used to determine both liquid-phase compositions as a function of temperature—even to predict consolute temperatures. In general, however, the van Laar form with $T \ln \gamma$ = constant is not satisfactory for this purpose.

When the system is at high pressure (but still with the temperature below the critical temperature of any component), Eq. (10-97) is still applicable. The Poynting correction and v_i may not be negligible, and ϕ_i^v cannot ordinarily be set equal to unity. The more difficult problem in applying Eq. (10-97) lies in the activity-coefficient term. We saw in Table 10.1 that an approximate temperature correction was included, but no pressure correction. At low pressures, even for isothermal cases where pressure can vary, the activity coefficient is a weak function of pressure and such corrections are neglected. This may not be true at high pressures.

One way to treat this problem is to define a constant-pressure activity coefficient to be used in Eq. (10-97). Presumably, any of the activity coefficient correlations in Table 10.1 could then be used.

For illustration, let us still retain the reference state for the fugacity to be the system pressure (which can vary). Choose a *constant* reference pressure (RP) for the activity coefficient. Since, in Eq. (10-97) the activity coefficient is evaluated at P, then

$$\gamma_i(P) = \gamma_i(RP) \exp\left(\int_{RP}^{P} \frac{\overline{\Delta V}_i^\sigma}{RT}\, dP\right) \tag{10-104}$$

Substitution of Eq. (10-104) into Eq. (10-97) now leads to an activity coefficient that is only temperature and composition dependent. Of course, one has now to evaluate an exponential term involving partial volumes which themselves are a function of temperature, system pressure, and composition. Although correlations could (and have) been developed to estimate this exponential correction, none have been particularly successful.[12] It is worth noting, however, that when this pressure correction is made for γ_i, it is common to modify Eq. (10-94) and define the reference fugacity at the value of RP chosen for γ_i. While the exponential terms may then be combined, one loses the desirable feature of Eq. (10-94) that $f_{i,P_{vp}}^\sigma$ is easily related to the vapor pressure of pure i as in Eq. (10-96).

In summary, the discussion in this section has been concerned with the integral approach to express phase equilibrium relations. For nonpolar systems, especially at elevated pressures, more and more emphasis is being placed on equation-of-state approaches. Modern computers can readily solve the nonlinear equations [Eq. (10-88)] provided that one has available an equation of state which will correlate *mixture P–V–T* composition properties. Systems containing components whose critical temperature exceeds the system temperature may be treated.

[12] I. Wichterle [*Fluid Phase Eq.*, **2**, 143 (1978)] discusses many of the proposed values of RP, running from 0 to 1000 bar.

The alternative integral approach introduces the activity-coefficient and the reference-state fugacity for the condensed phase(s). At relatively low pressures and for systems with no supercritical components, this approach has been widely and successfully used even for very polar mixtures. Reference states are normally defined as the pure component in the same condensed state as the mixture. Many activity coefficient–composition correlations have been suggested (see Table 10.1). The temperature correction in these correlations is only approximate, and no pressure correction is present. These limitations are normally of little importance except in cases where the pressure is high and/or the temperature is near the mixture critical. Under the latter conditions, one can develop the thermodynamics to yield, for example, pressure-independent activity coefficients, but one must then include other terms involving volumetric properties of the condensed phase. Rarely does one have available the necessary data to determine such correction factors.

10.5 Equilibrium in Systems with Supercritical Components

We developed the integral approach for condensed phases using activity coefficients in the previous section. Therein, it was assumed that the system temperature did not exceed the critical temperature of any component. This allowed us to define reference fugacity states as *pure* components in the condensed state. (We could have selected reference fugacity states other than for the pure material, but there was no advantage to be gained by such a selection.)

When we have a condensed phase with one or more components that cannot exist in the same condensed state when pure and at the system temperature, this usually implies that the system temperature exceeds the critical temperature of the components. Carbon dioxide in a liquid hydrocarbon mixture at a temperature higher than 304 K (the critical temperature of CO_2) would be an example of a *supercritical* component. In Section 8.8 we considered such cases briefly for binary systems. We expand this treatment and extend it to multicomponent systems in this section.

Reviewing briefly the development in Section 8.8, refer to Figure 8.6. Component fugacities are plotted as a function of composition for a binary liquid system of solute (A) and solvent (B) at some temperature. Component fugacities are shown only over a limited concentration range since it was assumed that the solute could not exist as a liquid when in the pure state.

The reference fugacity for the solvent was defined (as in Section 10.4) as the pure liquid at the system temperature and pressure. For the solute, a fictitious reference fugacity state was defined by drawing a line tangent to the f_A curve as $x_A \longrightarrow 0$. The intersection of this tangent with the $x_A = 1.0$ ordinate yielded the reference fugacity state for the solute. This value, f_A^{**}, was called the infinite-dilution reference (or standard) state. When we introduce a reference fugacity state such as f_A^{**} (with the usual reference state for the solvent), we say that the system is *unsymmetrically normalized.* In Section 10.4 only symmetrically normalized systems were discussed. Activity coefficients for the unsymmetrical case are defined in Eqs. (8-168) and (8-169).

We wish to stress that f_A^{**} has no physical significance, whereas f_B is related to a well-defined liquid state (i.e., pure B at T, P). f_A^{**} must be obtained from experimental data or by an estimation method derived from some solution theory. An alternative type of reference state for the solute might also be proposed. Suppose that we "hypothetically" extrapolate the $\hat{f}_A$ versus x_A curve to $x_A = 1.0$. This would locate a reference state, f_A, which is also fictitious. Again, f_A is related to no physical state of the solute and must be determined from experimental data. But with its introduction, we now have a symmetrically normalized system with γ_A defined in the same way as γ_B [see Eq. (8-168)], that is,

$$\gamma_A = \frac{\hat{f}_A}{x_A f_A} \tag{10-105}$$

Since $\hat{f}_A$ and x_A do not depend on the choice of the reference state, we can divide Eq. (10-105) by Eq. (8-169) to obtain

$$\frac{\gamma_A}{\gamma_A^{**}} = \frac{f_A^{**}}{f_A} = \gamma_A^{\infty} = \text{constant} \tag{10-106}$$

Since γ_A^{**} approaches unity as $x_A \longrightarrow 0$, we can define the constant γ_A^{∞} as

$$\gamma_A^{\infty} = \lim_{x_A \to 0} \gamma_A \tag{10-107}$$

The infinite-dilution reference fugacity f_A^{**} can then be simply related to f_A and γ_A^{∞} as

$$f_A^{**} = f_A \gamma_A^{\infty} \tag{10-108}$$

or

$$\ln f_A^{**} = \ln f_A + \ln \gamma_A^{\infty} \tag{10-109}$$

and, with Eq. (10-106),

$$\ln \gamma_A - \ln \gamma_A^{**} = \ln \gamma_A^{\infty} \tag{10-110}$$

Example 10.5

Assume that we are interested in a liquid binary mixture of A and B with A as the supercritical component. Using the van Laar correlation in Table 10.1, show how one may relate activity coefficients of A to composition using both symmetrical and unsymmetrical normalization. How is f_A related to f_A^{**}?

Solution

In Table 10.1 we let component 1 be A and component 2 be B. Also, to avoid confusion, let the van Laar constants be C and D rather than A and B. Then, since Table 10.1 is developed for the symmetrically normalized case,

$$RT \ln \gamma_A = C\left(1 + \frac{Cx_A}{Dx_B}\right)^{-2}$$

$$RT \ln \gamma_B = D\left(1 + \frac{Dx_B}{Cx_A}\right)^{-2}$$

Note that even though component A is supercritical, the definition of γ_A is given by Eq. (10-105) (i.e., the reference fugacity f_A is implied). γ_B is given by Eq. (8-168). Then, with Eq. (10-107),

$$\ln \gamma_A^{\infty} = \frac{C}{RT}$$

so with Eqs. (10-109) and (10-107),

$$\ln f_A{}^* = \ln f_A + \frac{C}{RT} \tag{10-111}$$

$$\ln \gamma_A^{**} = \ln \gamma_A - \frac{C}{RT}$$

$$= \frac{C}{RT}\left[\left(1 + \frac{Cx_A}{Dx_B}\right)^{-2} - 1\right] \tag{10-112}$$

Clearly, $\gamma_A^{**} \longrightarrow 1$ as $x_A \longrightarrow 0$.

The object of Example 10.5 and the preceding development was to illustrate that, for binary liquid systems involving a supercritical component, one could employ either the symmetrical or unsymmetrical normalized activity coefficient. The two reference-state fugacities for the supercritical component are related by Eq. (10-109) and the resulting activity coefficients by Eq. (10-110). In both instances, the reference fugacities (f_A or f_A^{**}) are fictitious pure-component properties and numerical values must be found from experimental data. There is no advantage to be gained from working with one or the other for binary systems. For multicomponent cases, however, the treatment using f_A may have conceptual advantages.[13]

We consider a ternary liquid mixture containing A, B, and C with A the supercritical component. The reference-state fugacities are f_A, f_B, and f_C. The later two refer to the pure liquid states of B and C at T, P. The first is a fictitious reference state; for a binary system it can be defined in an unambiguous manner, but for multicomponent systems, it will be shown to vary with the *relative* composition of B and C. (Use of f_A^{**} would not improve the situation.)

Following the treatment of Van Ness and Abbott (see footnote 13), we make use of the *mixture* fugacity defined earlier in Eqs. (8-144) and (8-145), where we noted that the appropriate partial property for the mixture fugacity was

$$\frac{\partial}{\partial N_i}(N \ln f_m)_{T,P,N_j[i]} = \ln \frac{\hat{f}_i}{x_i} \tag{10-113}$$

For reasons that will become apparent later, we define the reference fugacity for the supercritical component, f_A, in terms of f_{AB} and f_{AC}, where these refer to the reference fugacities of A in A, B and A, C binaries with symmetrical normalization at the same T, P.

$$\ln f_A = x_B' \ln f_{AB} + x_C' \ln f_{AC} \tag{10-114}$$

[13] H. C. Van Ness and M. M. Abbott [*AIChE J.*, **25**, 645 (1979)] have presented a detailed analysis of both supercritical reference fugacities. Our treatment follows their development quite closely.

with

$$x'_B = \frac{x_B}{x_B + x_C}, \qquad x'_C = \frac{x_C}{x_B + x_C} \tag{10-115}$$

Therefore, the value for f_A in the ternary is related to binary properties and the relative ratios of components B and C.

We now define a function $(\Delta \ln f_m)$ as

$$\Delta \ln f_m = \ln f_m - x_A \ln f_A - x_B \ln f_B - x_C \ln f_C \tag{10-116}$$

The function $(\Delta \ln f_m)$ can be shown to be equivalent to $\Delta G^{EX}/RT$ for a mixture with the reference states as defined for f_A, f_B, and f_C. We multiply Eq. (10-116) by the total moles in the system, N, and then perform three partial differentiations with respect to N_A, N_B, and N_C. Using Eq. (10-113), one obtains:

$$\frac{\partial}{\partial N_A}(N \Delta \ln f_m)_{N_B, N_C} = \ln \frac{\hat{f}_A}{x_A} - \ln f_A \tag{10-117}$$

$$\frac{\partial}{\partial N_B}(N \Delta \ln f_m)_{N_A, N_C} = \ln \frac{\hat{f}_B}{x_B} - \ln f_B - N_A \frac{\partial}{\partial N_B}(\ln f_A) \tag{10-118}$$

$$\frac{\partial}{\partial N_C}(N \Delta \ln f_m)_{N_A, N_B} = \ln \frac{\hat{f}_C}{x_C} - \ln f_C - N_A \frac{\partial}{\partial N_C}(\ln f_A) \tag{10-119}$$

In the differentiation in Eq. (10-117), $\ln f_A$ is not a function of N_A at constant N_B and N_C, by virtue of Eq. (10-114).

With the definition of γ_i as

$$\gamma_i = \frac{\hat{f}_i}{x_i f_i} \tag{10-120}$$

we also define

$$N_A \frac{\partial}{\partial N_B}(\ln f_A) = \ln \gamma_B^R \tag{10-121}$$

$$N_A \frac{\partial}{\partial N_C}(\ln f_A) = \ln \gamma_C^R \tag{10-122}$$

and substitute Eqs. (10-120) through (10-122) into Eqs. (10-117) through (10-119):

$$\frac{\partial}{\partial N_A}(N \Delta \ln f_m)_{N_B, N_C} = \ln \gamma_A \tag{10-123}$$

$$\frac{\partial}{\partial N_B}(N \Delta \ln f_m)_{N_A, N_C} = \ln \frac{\gamma_B}{\gamma_B^R} \tag{10-124}$$

$$\frac{\partial}{\partial N_C}(N \Delta \ln f_m)_{N_A, N_B} = \ln \frac{\gamma_C}{\gamma_C^R} \tag{10-125}$$

Then, since $N \Delta \ln f_m$ has all the properties of a $\Delta \underline{B}$ function as discussed in Chapter 8, we multiply each term in Eqs. (10-123) through (10-125) by the appropriate mole fraction and sum. The result is

$$\Delta \ln f_m = x_A \ln \gamma_A + x_B \ln \frac{\gamma_B}{\gamma_B^R} + x_C \ln \frac{\gamma_C}{\gamma_C^R} \tag{10-126}$$

The terms γ_B^R and γ_C^R may be evaluated from their definitions [Eqs. (10-121) and (10-122)] with Eq. (10-114), the results are

$$\ln \gamma_B^R = \frac{x'_C x_A}{x_B + x_C} \ln \frac{f_{AB}}{f_{AC}} \tag{10-127}$$

$$\ln \gamma_C^R = \frac{x'_B x_A}{x_B + x_C} \ln \frac{f_{AC}}{f_{AB}} \tag{10-128}$$

where it can be shown that

$$x_B \ln \gamma_B^R + x_C \ln \gamma_C^R = 0 \tag{10-129}$$

So Eq. (10-126) becomes

$$\Delta \ln f_m = x_A \ln \gamma_A + x_B \ln \gamma_B + x_C \ln \gamma_C \tag{10-130}$$

with

$$\Delta \ln f_m = \frac{\Delta G^{\text{EX}}}{RT} \tag{10-131}$$

The partial properties of ($N \Delta \ln f_m$) or $\Delta \underline{G}^{\text{EX}}/RT$ are given by Eqs. (10-123) through (10-125).

As a simple example, assume that for the ternary mixture

$$\frac{\Delta G^{\text{EX}}}{RT} = \frac{\Delta G_{AB}^{\text{EX}}}{RT} + \frac{\Delta G_{AC}^{\text{EX}}}{RT} + \frac{\Delta G_{BC}^{\text{EX}}}{RT} \tag{10-132}$$

with

$$\frac{\Delta G_{ij}^{\text{EX}}}{RT} = C_{ij} x_i x_j \tag{10-133}$$

Then, for the ternary,

$$\frac{\Delta G^{\text{EX}}}{RT} = \Delta \ln f_m = C_{AB} x_A x_B + C_{AC} x_A x_C + C_{BC} x_B x_C \tag{10-134}$$

and

$$\frac{\partial}{\partial N_A}\left(\frac{\Delta \underline{G}^{\text{EX}}}{RT}\right)_{N_B, N_C} = C_{AB} x_B + C_{AC} x_C - \frac{\Delta G^{\text{EX}}}{RT} \tag{10-135}$$

$$\frac{\partial}{\partial N_B}\left(\frac{\Delta \underline{G}^{\text{EX}}}{RT}\right)_{N_A, N_C} = C_{BC} x_C + C_{AB} x_A - \frac{\Delta G^{\text{EX}}}{RT} \tag{10-136}$$

$$\frac{\partial}{\partial N_C}\left(\frac{\Delta \underline{G}^{\text{EX}}}{RT}\right)_{N_B, N_C} = C_{BC} x_B + C_{AC} x_A - \frac{\Delta G^{\text{EX}}}{RT} \tag{10-137}$$

Finally, with Eqs. (10-123) through (10-125) with Eq. (10-131),

$$\ln \gamma_A = C_{AB} x_B + C_{AC} x_C - \frac{\Delta G^{\text{EX}}}{RT} \tag{10-138}$$

$$\ln \gamma_B = C_{BC} x_C + C_{AB} x_A - \frac{\Delta G^{\text{EX}}}{RT} + \frac{x'_C x_A}{x_B + x_C} \ln \frac{f_{AB}}{f_{AC}} \tag{10-139}$$

$$\ln \gamma_C = C_{BC} x_B + C_{AC} x_A - \frac{\Delta G^{\text{EX}}}{RT} + \frac{x'_B x_A}{x_B + x_C} \ln \frac{f_{AC}}{f_{AB}} \tag{10-140}$$

Equations (10-138) through (10-140) allow one to calculate the ternary activity coefficients once the binary parameters C_{AB}, C_{AC}, C_{BC}, f_{AB}, and f_{AC} have been determined. The component fugacities of each component in the ternary are then expressed as

$$\hat{f}_i = x_i \gamma_i f_i \tag{10-141}$$

The extension of the ternary treatment to multicomponent systems is available (see footnote 13). No comment was made concerning the pressure correction of f_A, if such should be necessary. No satisfactory methods have yet been developed, and the case is similar to the subcritical cases discussed in Section 10-4.

PROBLEMS

10.1. If a small-diameter wire is passed over a block of ice and weights are attached to each end of the wire, the wire apparently cuts through the ice but leaves no trace of the path (Figure P10.1). This phenomenon is often called *regelation.*

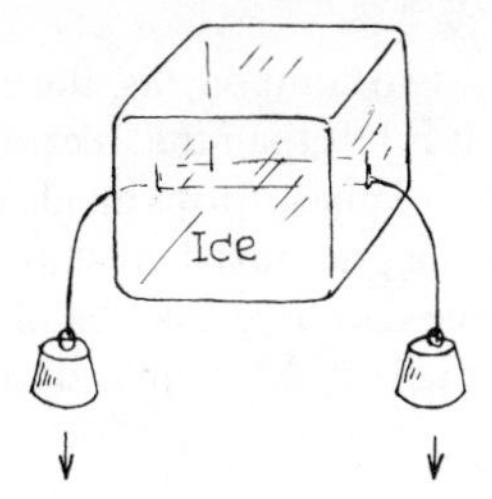

Figure P10.1

Many physics texts explain the phenomenon by the fact that the high pressure exerted by the wire lowers the ice freezing point and thus the wire "melts through." Similar reasoning is invoked to explain the occurrence of a water lubrication film beneath ice skates.

What is your opinion of this explanation?

10.2. For a pure component, a first-order transition is a change in which the first and all higher derivatives of the chemical potential, μ, have a discontinuity at the point of change. A second-order transition is one in which only the second and higher derivatives of μ with respect to P and T have a discontinuity. Solid methane, for instance, shows, at a particular temperature, a discontinuous C_p because of the onset of free rotation of the molecules.

In other words, in the first-order transition, the chemical potential is continuous at the point of change, but it has a discontinuous first derivative. In a second-order transition, it is the specific entropy and the specific volume that are continuous at the point of change and yet have discontinuous first derivatives.

Derive the analogs to the Clapeyron equation for a second-order phase transition.

10.3. A saturated solution of naphthalene in chlorobenzene is prepared at 293 K and 1 bar. The liquid phase is decanted and the liquid mole fraction of naphthalene is determined to be 25.6%. The solid-free solution is placed in a piston–cylinder and compressed isothermally to 500 bar.

Data: For solutions of naphthalene in chlorobenzene, the specific liquid volume, cm^3/mol, is independent of pressure and is given as

$$V = 102 + 26x, \qquad x = \text{mole fraction naphthalene}$$

Ideal liquid solution properties may be assumed.

	Naphthalene	Chlorobenzene
Melting point (K)	353.3	228.0
Molar volume (cm^3/mol)	115 (solid)	102 (liquid)
Latent heat of fusion (kJ/mol)	18.46	—

Estimate the mole fraction of naphthalene in the liquid under these conditions.

10.4. In a binary solution of two components, the eutectic point is the lowest freezing point of the mixture. It is less than the freezing point of either pure component.

Assume that the liquid phase forms an ideal solution and that all solid phases are pure components (i.e., no mixed crystals form). The vapor phase forms an ideal-gas mixture. Some data that may be of use are given below. Neglect any pressure effects. Assume that ΔH values do not vary with temperature.

	Nitrogen	Oxygen
Freezing point (K)	63.3	54.4
$\Delta H_{\text{vaporization}}$ (J/mol)	6000	7490
$\Delta H_{\text{sublimation}}$ (J/mol)	6720	7940
ΔH_{fusion} (J/mol)	721	447

Estimate the eutectic point (i.e., composition and temperature) for a liquid–air mixture (O_2 and N_2).

10.5. It is late fall and you suddenly remember that you have not put antifreeze in the radiator of your car. All service stations nearby are closed and it is predicted that the temperature will drop to -10°C tonight. In desperation you visit the chemical engineering laboratories to borrow some suitable chemical to mix with water to make a noncorrosive antifreeze. There are formamide, urea, methanol, ethanol, glycerol, sucrose, and other such chemicals: Which one to use, and how much? Then you dimly remember that in physical chemistry someone told you that in dilute solutions there was the same freezing-point depression per mole of solute for all such materials. Make any and all assumptions necessary to estimate rapidly the value of this constant ($\Delta T/N$). ΔT is the freezing-point depression in °C when N moles of solute are dissolved in 1 kg of water. $\Delta H_{f_{H_2O}} = 335$ kJ/kg.

The next day, when you have time to reflect on your action the preceding evening, you note that all commercial antifreezes are aqueous solutions. Why is this so? Do all aqueous *solutions* have a freezing point *below* that of pure water?

Is it possible to have an aqueous solution with a freezing point higher than pure water?

10.6. It is a fact that occasionally during rapid loading of liquid oxygen and hydrogen into missile tanks, one or the other of the tanks has imploded, with catastrophic results (see Figure P10.6). Such tanks are constructed to withstand an internal

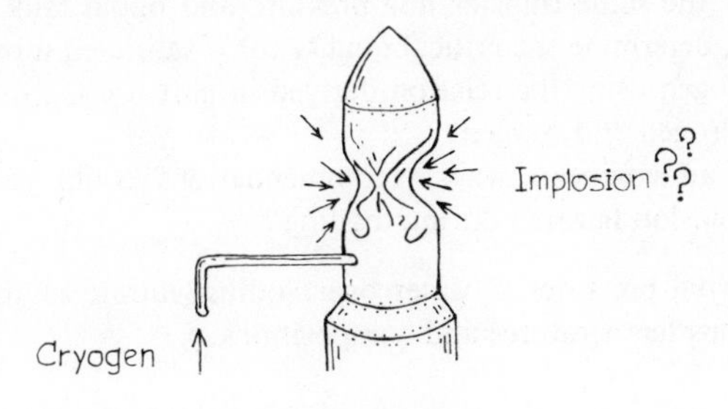

Figure P10.6

pressure somewhat above atmospheric but will collapse if the external pressure significantly exceeds the internal pressure.

During the initial part of the loading cycle, the cryogen may be all gas, a mixture of gas and liquid, or all liquid. The cryogen is pumped through a side port and has intimate contact with the gas already in the tank.

Data:

Tank volume: 300 m^3

	Vapor pressure (bar)	Saturation temperature (K)	Heat of vaporization (kJ/kg)
O_2	2.55	100	200
H_2	2.55	24	458

C_p(vapor) = 24.2(H_2), 29.3(O_2), J/mol K

Assume ideal gases.

(a) Demonstrate that a pressure decrease cannot occur if only gas flows into the tank.

(b) Derive a general relation to relate the fractional change in tank pressure and temperature with the fractional increase in the mass of gas in the tank for the case in which you think the pressure drops the maximum amount. The feed is all liquid, saturated at the pressure in the transfer line, and when this liquid enters, it is immediately vaporized by contact with gas present in the tank. Clearly state any assumptions made.

(c) If the feed were a mixture of liquid and gas, saturated at the transfer line pressure, derive a relation to determine the critical quality of the feed. The critical quality is defined as that fraction of vapor above which no pressure drop is possible irrespective of any rate process.

(d) Determine the fractional change in pressure for both hydrogen and oxygen for part (b) when the mass of gas in the tank has doubled. The liquid in the transfer line is saturated at 2.55 bar and the initial tank gas pressure and temperature are 1 bar and 278 K, respectively. Assume that the initial gas in the oxygen tank is oxygen and that the initial gas in the hydrogen tank is hydrogen.

(e) For the same transfer line pressure and initial tank conditions as in part (d), determine the critical quality for a saturated mixture of gas and liquid cryogen using the relation derived in part (c). Do the calculation for both hydrogen and oxygen.

(f) As an engineer, what recommendations could you make to minimize implosion hazards during loading?

10.7. The partial pressures of water over sodium nitrate solutions are shown below at various temperatures and compositions.

Concentration ($g\ NaNO_3/100\ g\ H_2O$)	Partial pressure of water (N/m^2)		
	0°C	25°C	50°C
0	610	3,168	12,340
10	589	3,057	11,890
20	571	2,952	11,480
30	553	2,852	11,080
40	539	2,758	10,610
50	524	2,672	10,330
60	511	2,589	9,990
70	497	2,510	9,670
73	493[a]		
80		2,438	9,370
90		3,369	9,080
92		2,356[a]	
100			8,810
110			8,560

[a]The solution is saturated.

In addition, the following data are available for saturated solutions of $NaNO_3$.

T (°C)	Partial pressure of water (N/m^2)	Concentration of $NaNO_3$ ($g/100\ g\ H_2O$)
0	493	73.0
20	1,760	88.2
25	2,356	92.0
40	5,210	104.8
60	13,150	124.0
80	28,810	148.0
100	56,260	176.0
120	99,730	210.6

Determine the enthalpy change for crystallization at 25°C.

$$\Delta H_{cry} = x_{H_2O}(H^L - \bar{H}^L)_{H_2O} + x_{NaNO_3}(H^S - \bar{H}^L)_{NaNO_3}$$

Also calculate $(H^S - \bar{H}^L)_{NaNO_3}$ at 25°C, saturated solution.

10.8. A calorimeter has recently been constructed for NASA to determine the thermal conductivities of various superinsulations. In essence, this calorimeter is a thick-walled copper sphere with a radius of 0.506 m; the superinsulation is placed on the outside of the sphere. The insulation is 1.27 cm thick and consists of many layers of 6-μm aluminized Mylar separated by thin spacers made of glass-fiber paper, silk netting, or other such materials. The tests are carried out in high-vacuum environmental chambers. The outer layer of insulation is maintained at 294 K. The calorimeter is filled with liquid hydrogen.

Heat transfer through the insulation is manifested by the boiling of the liquid hydrogen, and the mass flow of vapor is measured as a function of time. The vapor vent line may be considered adiabatic (actually, it has a separate liquid hydrogen guard to prevent axial conduction) and is of sufficient cross-sectional area that the pressure in the calorimeter is equal to the prevailing atmospheric pressure.

In any given test, the calorimeter is filled with liquid hydrogen and the liquid level held about constant until the calorimeter heat leak reaches a steady-state value. At this time the approximate head of liquid hydrogen is measured and the test started. During a test the flow rate of vented hydrogen vapor is measured at various times and the barometric pressure recorded. The assumption that the copper shell and both hydrogen phases are at the same temperature is believed to be quite good; also, the liquid and vapor phases are always in equilibrium with each other at the prevailing atmospheric pressure existing at the exit end of the vent line.

Some real test data are shown below.

TEST XX-3
MASS OF LIQUID HYDROGEN IN TANK
AT START OF TEST = 38.4 kg

Time (h)	Mass flow rate of vented hydrogen vapor [g/h ($\pm$0.2)]	Barometric pressure (bar)
0	40.8	1.016
2	41.3	1.011
4	42.2	1.007
6	44.5	0.989[a]
8	43.5	0.991
10	44.9	0.987
12	43.1	1.000
14	39.5	1.012
16	39.9	1.016
18	40.8	1.017
20	41.3	1.012

[a]A severe thunderstorm and squall occurred between the hours of about 6 and 10, hence the drop in barometric pressure.

Saturation Properties of Hydrogen

T (K)	P (bar)	Enthalpy (kJ/kg)	
		Liquid	Vapor
20.0	0.932	200	657
20.4	1.013	205	659
21	1.206	210	662

After an appropriate thermodynamic analysis, determine the effective thermal conductivity of the insulation.

10.9. The following data have been reported for the vapor–liquid equilibrium between ethyl alcohol and n-hexane at 1.013 bar.

T (°C)	Mole fraction alcohol in liquid	Mole fraction alcohol in vapor
76.0	0.990	0.905
73.2	0.980	0.807
67.4	0.940	0.635
65.9	0.920	0.580
61.8	0.848	0.468
59.4	0.755	0.395
58.7	0.667	0.370
58.35	0.548	0.360
58.1	0.412	0.350
58.0	0.330	0.340
58.25	0.275	0.330
58.45	0.235	0.325
59.15	0.102	0.290
60.2	0.045	0.255
63.5	0.010	0.160
66.7	0.006	0.065

The vapor pressures of ethyl alcohol and n-hexane are as follows:

T (°C)	P_{vp}(alcohol) (bar)
50.0	0.2933
52.0	0.3233
56.0	0.3891
58.0	0.4266
60.0	0.4670
62.0	0.5108
64.0	0.5578
66.0	0.6085
68.0	0.6629
70.0	0.7215
72.0	0.7844
74.0	0.8519
76.0	0.9241
78.4	1.0133

T (°C)	P_{vp}(n-hexane) (bar)
−53.9	1.33×10^{-3}
−34.5	6.66×10^{-3}
−25.0	1.33×10^{-2}
−14.1	2.67×10^{-2}
−2.3	5.33×10^{-2}
5.4	8.00×10^{-2}
15.8	1.33×10^{-1}
31.6	2.67×10^{-1}
49.6	5.33×10^{-1}
68.7	1.013
93.0	2.027
131.7	5.066
166.6	10.13

(a) What does thermodynamics tell us about the validity of these data?
(b) What can one conclude about the heat of solution and the entropy of solution?

10.10. (a) Suppose that one were interested in determining x–y–P relations for an *isothermal* system of components 1 and 2 at low pressures (where the vapor phase forms an ideal-gas mixture). In one type of experiment, no vapor composition measurements can be made directly but P–x data can be obtained over the entire range of liquid compositions. Show clearly how vapor compositions and liquid-phase activity coefficients may then be determined.
(b) Repeat part (a) assuming that pressure-vapor composition data were available and γ, x values were to be calculated.
(c) In a given binary vapor–liquid equilibrium system at constant temperature, the total pressure is graphed as a function of liquid mole fraction. A tangent is drawn to the total pressure curve as the mole fraction of one component becomes vanishingly small (Figure P10.10). What limits are placed on the slope of this tangent?

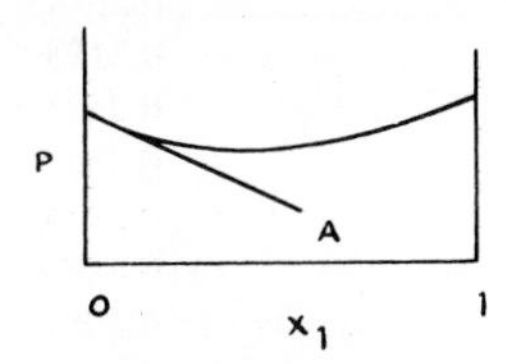

Figure P10.10

10.11. Vapor–liquid equilibrium data are shown below for the binary system isooctane–perfluoroheptane at 303, 323, and 343 K.

Data and allowable assumptions:

The vapor phase in equilibrium with the liquid may be considered an ideal-gas mixture.

Specific volumes of liquids are negligible compared to vapor specific volumes.

The enthalpies of vaporization of pure isooctane and perfluoroheptane at 323 K are 34.33 and 35.31 kJ/mol, respectively.

VAPOR–LIQUID EQUILIBRIUM DATA FOR ISOOCTANE–PERFLUOROHEPTANE ($T = 323$ K)

Mole fraction isooctane in liquid	Mole fraction isooctane in vapor	Pressure (bar)
0	0	0.3156
0.0560	0.1455	0.3632
0.1680	0.2940	0.4052
0.2555	0.3495	0.4176
0.2850	0.3615	0.4182
0.3760	0.3910	0.4204
0.4270	0.3997	0.4200
0.6450	0.4240	0.4146
0.7990	0.4540	0.4025
0.8258	0.4755	0.3915
0.8718	0.5015	0.3789
0.9085	0.5435	0.3612
0.9870	0.8790	0.2331
1	1	0.1955

VAPOR–LIQUID EQUILIBRIUM DATA FOR ISOOCTANE–PERFLUOROHEPTANE ($T = 303$ K)

Mole fraction isooctane in liquid	Mole fraction isooctane in vapor	Pressure (bar)
0	0	0.1308
0.0380	0.1345	0.1505
0.0712	0.2180	0.1620
0.0990	0.2614	0.1688
0.1514	0.3198	0.1765
0.2058	0.3595	0.1812
0.3060	0.3869	0.1845
0.3348	0.3935	0.1848
0.3910	0.4080	0.1848
0.4298	0.4081	0.1847
0.5505	0.4165	0.1842
0.8448	0.4555	0.1762
0.9195	0.5040	0.1612
0.9550	0.5705	0.1403
0.9881	0.6805	0.1007
1	1	0.0833

VAPOR–LIQUID EQUILIBRIUM DATA
FOR ISOOCTANE–PERFLUOROHEPTANE ($T = 343$ K)

Mole fraction isooctane in liquid	Mole fraction isooctane in vapor	Pressure (bar)
0	0	0.6715
0.0785	0.1540	0.7711
0.1600	0.2570	0.8190
0.3190	0.3565	0.8515
0.3992	0.3810	0.8532
0.3995	0.3825	0.8521
0.6938	0.4471	0.8242
0.8020	0.4870	0.7905
0.8815	0.5590	0.7300
0.9540	0.7155	0.5945
0.9825	0.8918	0.4789
1	1	0.4082

(a) A saturated liquid stream containing 50 mole% isooctane is flowing in a pipeline at 323 K. What is the minimum isothermal, isobaric work required to separate this stream into two liquid fractions that contain 90 and 5 mole% isooctane, respectively? Diagram your proposed process scheme.

(b) A batch vessel at 323 K contains 100 moles of liquid with 50 mole% isooctane. Devise a scheme to boil this liquid in such a manner that the pressure remains constant at the original value; the temperature is also to remain constant at 323 K. Calculate the heat interaction for the process ending with only vapor in the vessel.

10.12. With the current emphasis on the utilization of solar and geothermal energy, there has been interest in developing techniques to "store" this energy. One concept involves the melting of inorganic salts, as energy may then be recovered at a later time by allowing the fused salts to solidify. Most pure inorganic salts, however, melt at relatively high temperatures. Salt mixtures usually melt at temperatures below those of the pure components. The minimum melting temperature for a mixture is termed the *eutectic point*. (Assume that, in the case of interest, solid solutions do not form.)

We would like to estimate this eutectic point as accurately as possible to aid us in evaluating the concept of fused-salt thermal-energy storage.

(a) Consider the system LiCl–KCl. What would you estimate to be the atmospheric-pressure eutectic temperature and composition? Assume that the liquid phase of molten LiCl–KCl forms an ideal solution and liquid enthalpies are not a strong function of temperature. Some data for this system are shown on the following page.

Salt	Melting point (K)	Enthalpy of fusion (kJ/mol)	Entropy of fusion (J/mol K)
LiCl	883	19.93	22.57
KCl	1043	26.54	25.46

(b) The experimentally determined value of the eutectic point is $x_{\text{LiCl}} \sim 0.59$, $T \sim 618$ K. As can be seen from the ideal solution result from part (a), the prediction assuming ideal solutions is quite poor. What would be your estimate of the eutectic temperature for the LiCl–KCl system if regular solution behavior were assumed and the heat of mixing were expressed as

$$\Delta H = x_{\text{LiCl}} x_{\text{KCl}}(-17{,}580 - 377 x_{\text{LiCl}})$$

with ΔH in J/mol and x is the mole fraction? Assume that ΔH is not a strong function of temperature.

10.13. The *International Critical Tables*, Vol. III, p. 313, lists the boiling points for mixtures of acetaldehyde and ethyl alcohol at various pressures. The data are summarized below for mixtures containing 80 mole% ethyl alcohol in the liquid.

From these data determine the molar enthalpy of vaporization of ethyl alcohol at 320.7 K, from a liquid mixture containing 80% ethyl alcohol. [That is, what is $(\bar{H}^v - \bar{H}^L)$ alcohol?]

$$\bar{H}^v = \text{partial molar enthalpy of saturated vapor}$$

$$\bar{H}^L = \text{partial molar enthalpy of saturated liquid}$$

Data:
Mole fraction ethyl alcohol in liquid = 0.80

T (K)	P (N/m^2)	Mole fraction ethyl alcohol in vapor
331.3	9.319×10^4	0.318
320.7	5.306×10^4	0.385
299.9	1.027×10^4	0.330

Note: The vapor phase may be considered to be an ideal-gas mixture, but the liquid phase is a nonideal solution.

10.14. In preparing to make a simulation study of an evaporation process to remove water from concentrated sucrose solutions, we need to know, as accurately as possible, the enthalpy change in vaporizing water from these concentrated solutions. We do have some new, unpublished experimental data that show the boiling point of the solution at several pressures and compositions (see below).

BOILING TEMPERATURES FOR SUCROSE[a]–WATER SOLUTIONS

$P = 1.013$ bar		$P = 0.8781$ bar		$P = 0.7350$ bar	
Brix[b]	T (°C)	Brix[b]	T (°C)	Brix[b]	T (°C)
0	100	0	96.1	0	91.5
75.0	107.5	75.0	103.5	75.0	99.5
77.4	109	77.4	105.5	77.4	101.8
78.7	110	78.7	106		
80.0	110.5	80.0	107	80.0	103
81.3	112	81.3	108.5	81.3	104
82.7	113	82.7	109.8	82.7	105
84.2	114.5	84.2	111	84.2	106.5
85.8	116.5	85.8	113	85.8	108
87.2	118.5	87.2	115	87.2	110.5
88.8	122	88.8	118	88.8	112.5
90.5	127	90.5	122	90.5	116
92.3	130.5	92.3	125	92.3	121
94	138	94	131	94	128
		95	135	95	131

[a]Sucrose may be considered as pure (no invert) with the empirical formula $C_{12}H_{22}O_{11}$.
[b]Brix = weight percent sucrose.

Making *as few assumptions* as possible, calculate $H_W^V - \bar{H}_W^L$ for a 90 Brix solution at a pressure of 1.013 bar. Clearly show all steps in your calculations and justify any assumptions made. Discuss your result if you feel it is unreasonable.

10.15. We reproduce below a section on anesthesia from W. J. Moore, [*Physical Chemistry*, 4th ed. (Englewood Cliffs, N. J.: Prentice-Hall, 1972), p. 241].

Mechanism of Anesthesia

One of the most fascinating unsolved problems in medical physiology is the mechanism by which gases produce anesthesia and narcosis. Many anesthetics, such as krypton and xenon, are apparently inert chemically; in fact, it would appear that all gases produce an anesthetic effect at high enough pressures. Cousteau, in *The Silent World*, gives a memorable account of the nitrogen narcosis experienced at great depths, *l'ivresse des grandes profondeurs*, which has claimed the life of more than one diver.

[The accompanying table] summarizes some data on mice, which were tested by a criterion based on their righting reflex. Animals in a test chamber were rocked off their feet; if a mouse did not replace all four feet on the floor within 10 s, the anesthetic was awarded a "knockout." The results indicate that even helium produces narcosis at high pressures.

Early attempts to understand the causes of anesthesia were made by Meyer (1899) and Overton (1901), who found a good correlation between the solubility of a gas in a lipid (olive oil) and its narcotic efficacy. Since nerve cell membranes are composed mainly of lipids and proteins, it was suggested that the anesthetic molecules dissolve in the membranes and block the process of nerve conduction in some way as yet unknown. The activity of dissolved anesthetic necessary to produce anesthesia is in the range 0.02 to 0.05, so that the anesthetic can certainly alter the properties of the membrane to some considerable extent.

BEST ESTIMATES OF ANESTHETIC PRESSURES FOR MICE (RIGHTING REFLEX)

Key number	Gas	Pressure (atm)	Key number	Gas	Pressure (atm)
1	He	190	10	C_2H_4	1.1
2	Ne	>110	11	C_2H_2	0.85
3	Ar	24	12	Cyclo-C_3H_6	0.11
4	Kr	3.9	13	CF_4	19
5	Xe	1.1	14	SF_6	7.0
6	H_2	85	15	CF_2Cl_2	0.4
7	N_2	35	16	$CHCl_3$	0.008
8	N_2O	1.5	17	Halothane	0.017
9	CH_4	5.9	18	Ether	0.032

Estimate the anesthetic pressure for carbon tetrachloride. Clarify and justify (where possible) any assumptions made. Note that the value of X in the accompanying figure from Moore refers to the solubility of the gas in olive oil when the partial pressure of the gas is 1 bar.

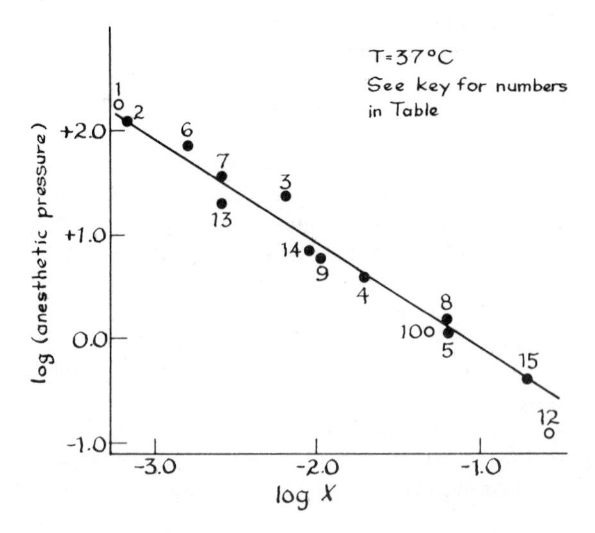

Property data for carbon tetrachloride and for the compounds listed in the table are as follows:

Gas	Critical pressure (bar)	Normal boiling point (K)	Critical temperature (K)
He	2.27	4.21	5.19
Ne	27.6	27.0	44.4
Ar	48.7	87.3	150.8
Kr	55.0	119.8	209.4
Xe	58.4	165.0	289.7
H_2	13.0	20.4	33.2
N_2	33.9	77.4	126.2
N_2O	72.4	184.7	309.6
CH_4	46.0	111.7	190.6
C_2H_4	50.4	169.4	282.4
C_2H_2	61.4	189.2	308.3
Cyclopropane	54.9	240.4	397.8
CF_4	37.4	145.2	227.6
SF_6	37.6	209.3	318.7
CF_2Cl_2	41.2	243.4	385.0
$CHCl_3$	54.7	334.3	536.4
Halothane[a]	—	—	—
Ether	36.4	307.7	466.7
CCl_4	45.6	349.7	556.4

[a] $C_2HBrClF_3$.

10.16. Some vapor–liquid equilibrium data for the system acetone–water are as follows:

$P = 1.013$ bar

T (°C)	Mole fraction acetone		Differential enthalpy of mixing for the saturated liquid (J/mol)	
	Liquid	Vapor	$\overline{\Delta H_A}$	$\overline{\Delta H_w}$
100	0	0	5296.3	0
84.75	0.02	0.4451	—	—
75.13	0.05	0.6340	3715.8	43.1
68.19	0.10	0.7384	2357.2	154.1
65.02	0.15	0.7813	—	—
63.39	0.20	0.8047	2357.2	481.5
61.45	0.30	0.8295	−564.4	817.7
60.39	0.40	0.8426	−990.2	1038.3
59.91	0.50	0.8518	−1030.0	1063.4
59.55	0.60	0.8634	−845.7	837.4
58.79	0.70	0.8791	−566.5	86.7
58.07	0.80	0.9017	−286.8	−548.5
57.07	0.90	0.9371	−78.7	−1737.5
54.14	1.00	1.0000	0	−3274.1

In addition to these data, the vapor pressures of the pure components are as follows:

T (°C)	P_{vp}, acetone (bar)	P_{vp}, water (bar)
15	0.1933	0.0171
20	0.2437	0.0233
30	0.3769	0.0424
40	0.5633	0.0737
50	0.8171	0.1233
60	1.154	0.1992
70	1.591	0.3116
80	2.148	0.4734
90	2.843	0.7010
100	3.697	1.013

(a) Assuming the vapor phase to be an ideal-gas mixture, determine the entropy change in mixing pure liquid water and liquid acetone, at 1.013 bar, to form a 40 mole % solution. The pure components and final solution are at 60.39°C.

(b) Calculate the total entropy change for a process in which 0.6 mole of liquid water and 0.4 mole of liquid acetone, both at 60.39°C, are separately vaporized, expanded reversibly and isothermally to a pressure so that they may be added to a vapor mixture at 1.013 bar (with 84.26 mole % acetone), reversibly across semipermeable membranes. One mole of this vapor is then condensed to a liquid with a composition of 40 mole % acetone at 60.39°C. Compare your result with the relation determined in part (a).

(c) Comment on the consistency of the data.

10.17. W. L. Jolley and Joel Hildebrand studied the solubility and entropy of solution of gases in nonpolar solvents. They have plotted the equilibrium solubility of various inert gases in benzene at 298 K when there is a partial pressure of 1.013 bar of the inert gas over the solution. x is the equilibrium solubility (mole fraction) of the gas in benzene, $\bar{S}$ is the partial molar entropy of the gas in solution, and S(gas) is the entropy of the pure gas at 298 K and 1.013 bar (see Figure P10.17).

(a) Which gases become more soluble with an increase in temperature at temperatures greater than 298 K? Prove your reasoning and state clearly any simplifying assumptions that you make in the proof.

(b) For some gases, there is an entropy *increase* when the pure gas dissolves in benzene to form a solution. How might you explain this fact?

(c) Estimate ΔH (vaporization) of pure benzene at 298 K and 1.013 bar. The vapor pressure of pure benzene at this temperature is 0.126 bar.

10.18. Use a liquid solution model or your own creative imagination to devise a two-constant arithmetic expression that you think will properly express ΔG^{EX} as a function of composition for a nonideal binary liquid mixture. Remember that $\Delta \underline{G}^{EX}$ is proportional to the total number of moles and should go to zero as $x_A, x_B = 0$.

Using this proposed equation, determine algebraic (constant temperature, constant pressure) expressions for the activity coefficients as a function of composition and estimate the x–y curve for ethanol and water at 1 atm (1.013 bar) using only the azeotropic point.

$$x_{\text{alcohol}} = y_{\text{alcohol}} = 0.8943 \text{ at } 78.17°\text{C}$$

The vapor pressures of the pure components as a function of temperature are

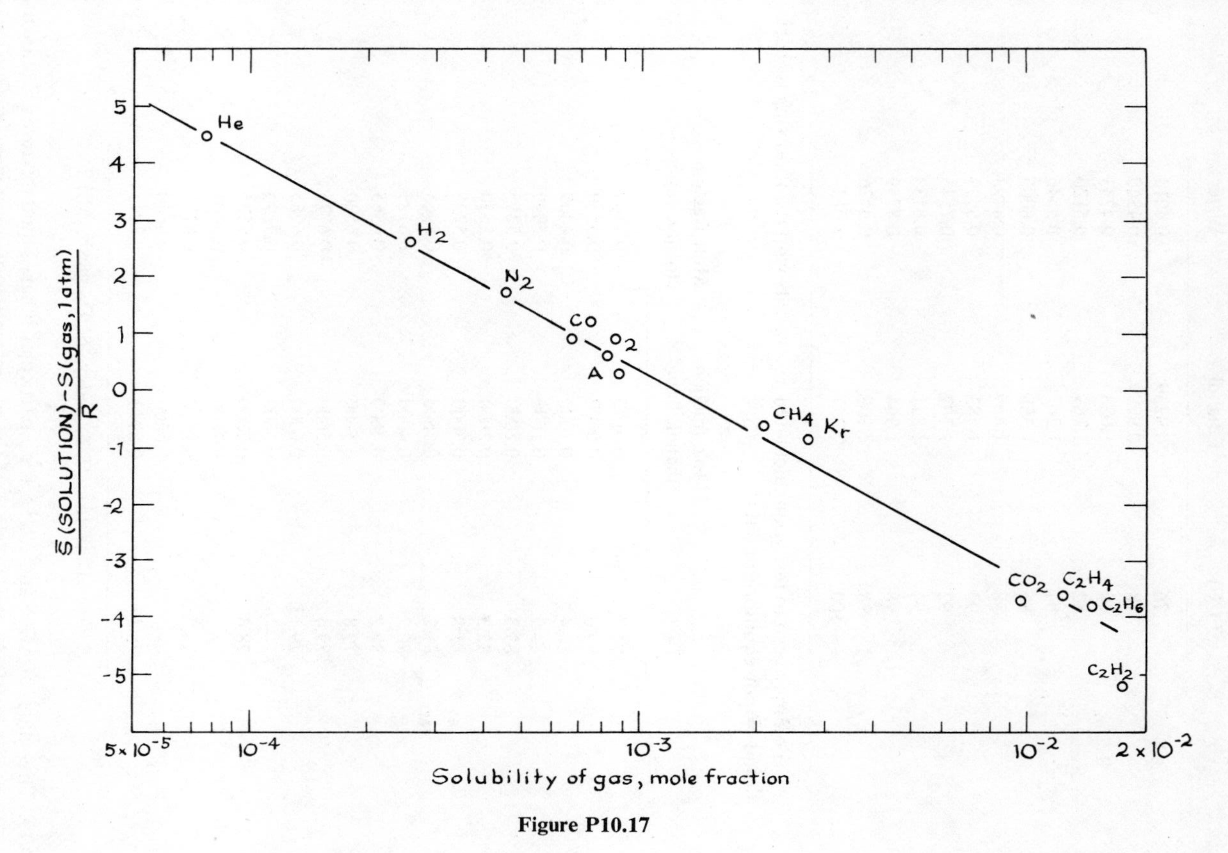

Figure P10.17

as follows:

T (°C)	Vapor pressure (bar)	
	Ethanol	Water
76	0.9239	0.4013
78	0.9999	0.4360
80	1.083	0.4733
82	1.169	0.5120
84	1.267	0.5546
86	1.368	0.6000
88	1.469	0.6493
90	1.583	0.6999
92	1.707	0.7546
94	1.831	0.8133
96	1.964	0.8759
98	2.108	0.9426
100	2.257	1.013

You may wish to test your calculated x–y results with the following isobaric liquid–vapor equilibrium data.

T (°C)	x Mole fraction ethanol, liquid	y Mole fraction ethanol, vapor
95.7	0.0190	0.1700
90.0	0.0600	0.3560
86.4	0.1000	0.4400
84.3	0.1600	0.5040
83.3	0.2000	0.5285
82.3	0.2600	0.5570
81.8	0.3000	0.5725
81.2	0.3600	0.5965
80.7	0.4000	0.6125
80.2	0.4600	0.6365
79.8	0.5000	0.6520
79.4	0.5600	0.6775
79.13	0.6000	0.6965
—	0.6600	0.7290
78.6	0.7000	0.7525
—	0.7600	0.7905
78.3	0.8000	0.8175
—	0.8600	0.8640
78.17	0.8943	0.8943

Source: Data from J. S. Carey, Sc.D. thesis, M.I.T., 1929.

10.19. The values of ΔG^{EX} and ΔH^{EX} for a particular binary liquid mixture are shown in Figure P10.19 as a function of the mole fraction of component A. One can describe these curves by: $\Delta G^{EX} = \eta\,\Delta H^{EX}$. It was also found for T, P constant that $\overline{\Delta G}_A^{EX} = \eta_A\,\overline{\Delta H}_A^{EX}$ and $\overline{\Delta G}_B^{EX} = \eta_B\,\overline{\Delta H}_B^{EX}$. The constants η, η_A, and η_B are not functions of temperature.

(a) Derive a relationship to show how the activity coefficient varies with temperature for component A if the pressure and composition were fixed.

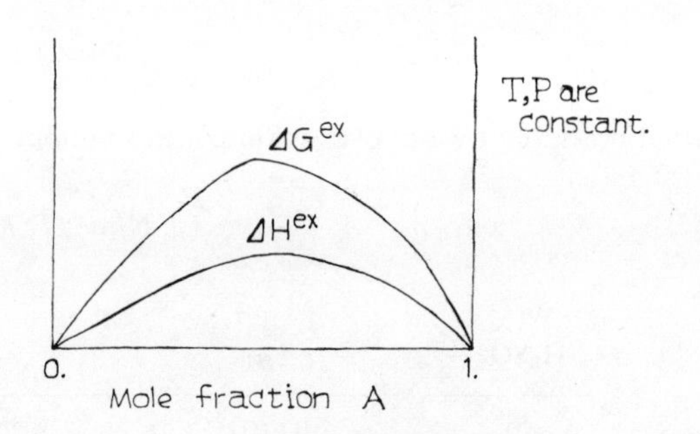

Figure P10.19

(b) Using the expression derived in part (a), keeping the pressure and composition constant, and supposing that $\gamma_A = 1.2$ at 300 K, what is γ_A at 400 K if $\eta = 2$?

(c) Discuss the cases $\eta_A > 1$, $\eta_A = 1$, and $0 < \eta_A < 1$ from the standpoint of $\overline{\Delta S}_A^{EX}$ and indicate what type of liquid solution one might expect in each of these cases. Assume that $\gamma_A > 1$.

10.20. It is claimed that a low-temperature liquid mixture may be made by mixing ice and pure sulfuric acid in a Dewar flask. If the ice and acid were originally at 0°C, what would be the final temperature and liquid composition when 1 kg of acid is poured over 4 kg of ice?

Data:

(1) The heat capacity of sulfuric acid solutions may be approximated by

$$C_p = 4.186 - 3.349x \text{ kJ/kg soln-K}\ .$$

where x is the weight fraction H_2SO_4. Assume that the heat capacity does not vary with temperature. The heat capacity of water ice is 2.09 kJ/kg K.

(2) Freezing point of sulfuric acid solutions

Wt% H_2SO_4	T (°C)
0	0
10	−5
20	−13
30	−35
40	−62
50	−35

(3) The enthalpy change in the dilution of 1 kg H_2SO_4 with water (at 0°C) to form a solution with x weight fraction H_2SO_4 is approximated as

$$\Delta H = 255.8x - 732.6 \quad \text{kJ/kg acid} \qquad (0.01 < x < 0.3)$$

(4) Vapor pressure of pure water ice

T (°C)	P (N/m²)
0	611
−10	260
−20	104
−30	38.1
−40	12.9

(5) Partial pressure of water over sulfuric acid solutions

$$\log_{10} p = A - \frac{B}{T} \qquad (p, \text{N/m}^2; T, K)$$

Wt% H_2SO_4	A	B
0	11.071	2260
10	11.050	2259
20	11.050	2268
30	10.989	2271
40	10.969	2299
50	10.957	2357

10.21. A well-insulated tank is vented to the atmosphere. Initially, it contains a hot aqueous solution of sodium hydroxide at 90°C, with a concentration of 45wt% caustic. Connected to the tank is a steam header with saturated steam at 1 bar, 100°C [Figure P10.21(a)].

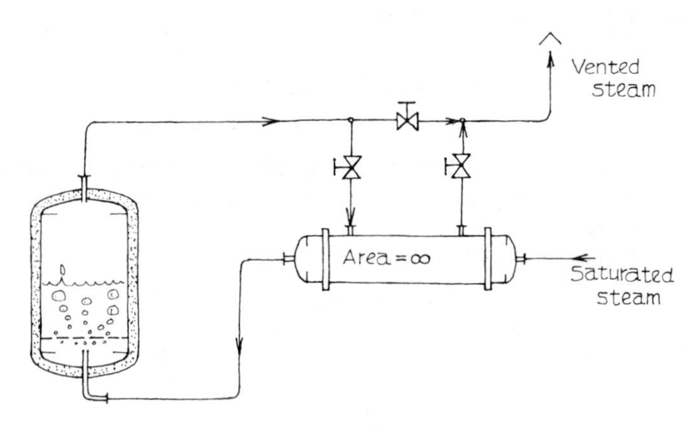

Figure P10.21(a)

Assume that the vented steam is in equilibrium with the solution at all times and neglect any sensible heat effects for the vessel walls or inert gases over the solution. The enthalpy of water vapor at 1 bar and 100°C is 2675 kJ/kg. The heat capacity of water vapor at 1 bar is about 1.91 kJ/kg K.

(a) If the steam line were opened and steam allowed to bubble slowly into the tank, plot the resulting solution temperature as a function of sodium hydroxide concentration.

(b) Plot a curve showing the kilograms of steam condensed and vented as a function of the sodium hydroxide concentration.

(c) If the vent line from the tank were arranged [as shown in Figure P10.21(a)] so that the vented gas was passed through an "infinite-area" heat exchanger countercurrent to the entering steam, how would the answers in parts (a) and (b) be affected? Data are given in Figures P10.21(b) and (c).

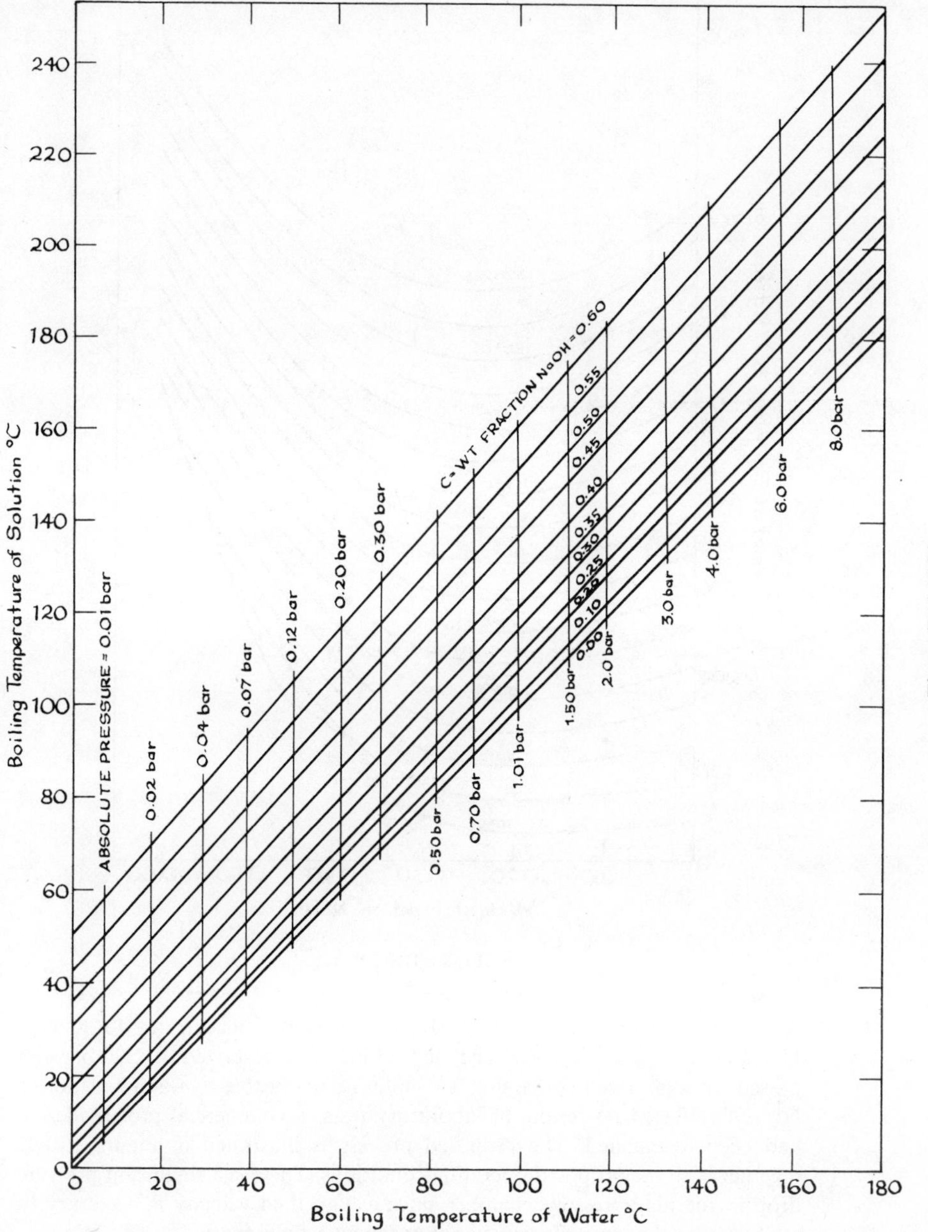

Figure P10.21(b)

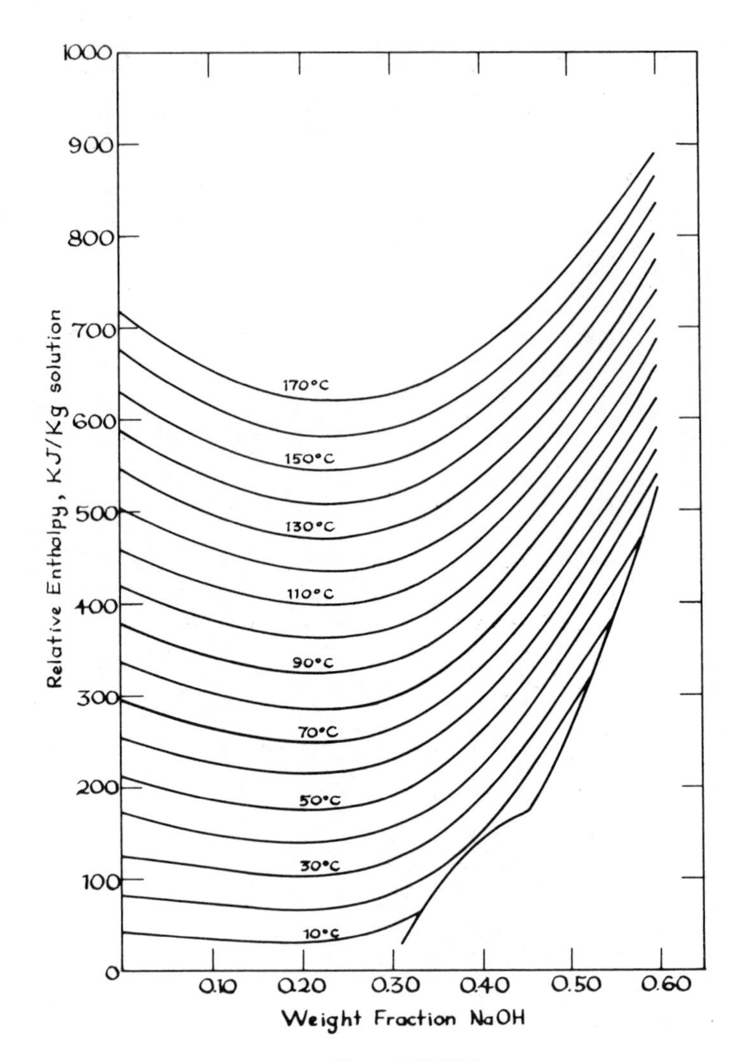

Figure P10.21(c)

10.22. An aqueous solution of 15 wt% NH_4NO_3 is to be concentrated to 30 wt% NH_4NO_3 by reverse osmosis. The feed solution is pressurized to 1520 bar and passed through a cell containing a membrane permeable to water but not to NH_4NO_3. Based on results of laboratory tests, a commercial process design had been formulated. The proposed process is illustrated in Figure P10.22 together with the proposed operating conditions. There is a significant pressure drop in the high-pressure chamber since a high fluid velocity is necessary to minimize the mass-transfer resistance in the bulk fluid phase.

Data: At 1 bar and 298 K the chemical potentials of both species have been determined as a function of composition. These data are shown in Figure 8.2. For the chemical potential, ammonium nitrate is referred to a saturated solution, while the reference for H_2O is pure water.

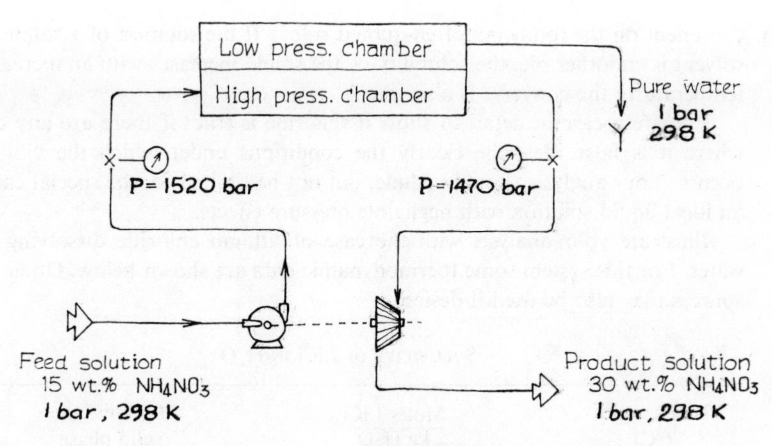

Figure P10.22

At 298 K and for pressures of 0.1 to 2000 bar, the following assumptions may be made:

V(pure water) = 1 cm³/g
$\bar{V}_W$(independent of NH_4NO_3 conc.) = 1 cm³/g
V(pure crystalline NH_4NO_3) = 0.58 cm³/g
$\bar{V}_{NH_4NO_3}$(15 wt% NH_4NO_3) = 0.622 cm³/g
$\bar{V}_{NH_4NO_3}$(30 wt% NH_4NO_3) = 0.632 cm³/g
Molecular weight of NH_4NO_3 = 80.04

Weight fraction NH_4NO_3	$T = 298$ K, $P = 1$ bar p/P_{vp} (water)	p/P_{vp} (NH_4NO_3)
0	1.0	0
0.10	0.9601	0.0483
0.15	0.9405	0.0750
0.20	0.9215	0.1120
0.25	0.8990	0.1565
0.30	0.8746	0.2115

At 298 K, $P_{vp(water)} = 0.0316$ bar, $P_{vp(NH_4NO_3)(pure)} = 10^{-5}$ bar. Assume ideal gases.

(a) Describe in detail a reversible process that will accomplish the same overall separation as that described (i.e., will accept the same feed solution and produce the same product solutions). You are free to use any kind of any finite number of pumps, membranes, heat exchangers, or any other device dictated by your creativity. Present a flow sheet summarizing your process; indicate the operating conditions (i.e., pressure and composition) at each point in the process.
(b) For your reversible process, calculate the work requirement.
(c) The net work requirement for the reverse osmosis process shown in Figure P10.22 is 86 J/g of feed solution. What is the overall efficiency of this process?

10.23. Comment on the following often-quoted rule: "If the solution of a solute in a solvent is endothermic, the solubility of the solute increases with an increase in temperature; the converse is also true."

Prepare a case in detail to show if this rule is true; if there are any cases where it is false, describe clearly the conditions under which the violation occurs. Your analysis should include, but not be limited to, the special case of an ideal liquid solution with negligible pressure effects.

Illustrate your analysis with the case of lithium chloride dissolving into water. For this system some thermodynamic data are shown below. Other data sources may also be used if desired.

SOLUBILITY OF LiCl IN H_2O

Temperature (°C)	$\frac{\text{Moles LiCl}}{\text{kg } H_2O}$	Equilibrium solid phase
0	15.3	$LiCl \cdot 2H_2O$
10	17.3	
12.5	17.8	$LiCl \cdot 2H_2O/LiCl \cdot H_2O$
20	18.8	$LiCl \cdot H_2O$
40	21.2	
60	24.0	
80	27.1	
98	30.7	$LiCl \cdot H_2O/LiCl$
100	30.8	LiCl

Source: International Critical Tables, Vol. IV (New York: McGraw-Hill, 1930), p. 233.

HEAT OF FORMATION AT 298.2 K

Formula	State	ΔH_f (kJ/mol)
LiCl	Crystal	−409.05
$LiCl \cdot H_2O$	Crystal	−713.05
$LiCl \cdot 2H_2O$	Crystal	−1013.6
LiCl (aq)	$3H_2O$	−429.65
	$4H_2O$	−434.30
	$5H_2O$	−437.232
	$8H_2O$	−440.912
	$10H_2O$	−441.933
	$15H_2O$	−443.164
	$20H_2O$	−443.746
	$25H_2O$	−444.069
	$50H_2O$	−444.726
	$100H_2O$	−445.120
	$200H_2O$	−445.392
	$400H_2O$	−445.601
	∞H_2O	−446.217
H_2O	liquid	−286.030

Source: U.S. Bureau of Standards Circular 500, p. 433.

10.24. A novel separation process involves the extraction of solid solutes into a gas which is somewhat above its critical temperature and pressure. The process is termed *supercritical* extraction.

(a) Assume that the solid solute is in equilibrium with the supercritical gas at some temperature T and pressure P. No gas dissolves in the solid solute, nor is the specific volume of the solid a strong function of pressure. Derive an expression to allow one to estimate the mole fraction of the solute in the supercritical gas phase. Assume that one has an equation of state to express the system pressure as a function of the volume, temperature, and composition of the vaporized solute–supercritical gas phase.

(b) Using the relationship found in part (a), write a computer program and use it to estimate the equilibrium mole fraction of phenanthrene in supercritical carbon dioxide at 313 K at pressures ranging from 1 to 300 bar, in 10-bar increments, Use the Peng–Robinson equation of state to describe the P–V–T–y relationship for the gas phase. This equation is given in Eqs. (8-55) through (8-59). Data for CO_2 and phenanthrene are as follows:

	Phenanthrene	CO_2
Molecular weight	178	44
T_c (K)	878	304.2
P_c (bar)	29.0	73.76
Acentric factor, ω	0.44	0.225
Solid volume (cm^3/mol)	181.9	
Normal boiling point, T_b (K)	612.6	
ΔH_v at T_b (kJ/mol)	55.68	
Interaction parameter, δ_{ij}	0.10	

(c) Using the thermodynamic relationship derived in part (a), at 313 K, what is the *minimum* solubility of phenanthrene in CO_2? Is there a *maximum* solubility?

(d) Derive an expression to allow an estimation of the heat of solution of solid phenanthrene into supercritical CO_2. What experimental data would you require to determine this heat of solution at various temperatures and pressures?

10.25. A 0.02832-m^3 chamber that is well insulated contains initially 13.6 kg of dry-ice flakes in equilibrium with CO_2 vapor. The pressure is maintained at 1.013 bar by a safety relief valve. A steady stream of propane at 1.013 bar and 273.2 K is injected into the bottom of the chamber and passes up through the dry ice bed.

Data: The acentric factors for CO_2 and propane are 0.225 and 0.152, respectively. Propane has a critical temperature of 369.8 K and a critical pressure of 42.4 bar. If you need vapor-phase properties of the superheated vapor phase, use the Peng–Robinson equation of state (see Chapters 7 and 8).

Assuming that any gases leaving the chamber are in equilibrium with any solid and/or liquid phase present, plot the temperature and mole fraction of propane in the vapor as a function of the amount of propane injected up until the time at which all dry-ice flakes disappear.

THERMODYNAMIC PROPERTIES OF SATURATED CARBON DIOXIDE[a]

T (K)	P (bar)	V (condensed phase) ($m^3/kg \times 10^4$)	V (vapor) ($m^3/kg \times 10^4$)	H (condensed phase) ($kJ/kg \times 10^{-2}$)	H (vapor) (kJ/kg)	S (condensed phase) (kJ/kg K)	S (vapor) (kJ/kg K)
177.6	0.219	6.293	15,180	−2.826	3.005	2.539	5.823
188.8	0.614	6.355	5,730	−2.698	3.070	2.609	5.709
199.9	1.53	6.442	2,375	−2.561	3.124	2.681	5.526
205.4	2.343	6.492	1,570	−2.482	3.142	2.721	5.457
211.0	3.506	6.542	1,060	−2.384	3.156	2.766	5.393
216.5	5.159	6.611	725	−2.279	3.161	2.815	5.328
216.6	5.178	6.611	722	−2.277	3.161	2.816	5.327
216.6	5.178	8.490	722	−0.319	3.161	3.720	5.327
222.2	6.529	8.640	579	−0.214	3.177	3.767	5.295
227.6	8.150	8.796	468	−0.109	3.191	3.814	5.264
233.2	10.05	8.971	382	0	3.205	3.859	5.235
238.8	12.26	9.152	314	0.105	3.215	3.904	5.207
244.3	14.82	9.351	260	0.218	3.222	3.948	5.180
249.9	17.74	9.564	217	0.323	3.226	3.991	5.151
255.4	21.06	9.801	181	0.437	3.231	4.034	5.128
261.0	24.83	10.08	152	0.558	3.226	4.080	5.103
266.5	29.08	10.38	128	0.684	3.217	4.127	5.077
272.1	33.84	10.73	108	0.823	3.205	4.177	5.052
277.6	39.15	11.16	90.1	0.970	3.180	4.225	5.022
283.2	45.06	11.66	75.2	1.126	3.140	4.278	4.989
288.8	51.61	12.30	62.1	1.291	3.073	4.335	4.951
294.3	58.84	13.18	50.2	1.482	2.966	4.396	4.909
299.9	66.79	14.79	37.9	1.719	2.761	4.477	4.838
304.2	73.82	21.45	21.45	2.256	2.256	4.647	4.647

[a]Properties at temperatures below 216.6 K refer to solid as the condensed phase.

THERMODYNAMIC PROPERTIES OF SATURATED PROPANE

T (K)	P (bar)	V (liquid) ($m^3/kg \times 10^3$)	V (vapor) (m^3/kg)	H (liquid) (kJ/kg)	H (vapor) (kJ/kg)	S (liquid) (kJ/kg K)	S (vapor) (kJ/kg K)
211.0	0.390	1.654	1.011	378.2	823.4	3.682	5.791
216.5	0.516	1.673	0.780	389.8	830.4	3.738	5.774
222.1	0.674	1.687	0.610	401.7	837.4	3.793	5.753
227.6	0.869	1.706	0.483	413.6	843.9	3.847	5.737
233.2	1.103	1.725	0.385	425.7	850.6	3.900	5.723
238.8	1.391	1.744	0.313	438.2	857.4	3.953	5.711
244.3	1.727	1.764	0.253	450.8	864.1	4.006	5.698
249.9	2.134	1.785	0.208	463.8	870.9	4.057	5.687
255.4	2.607	1.806	0.171	476.8	877.4	4.108	5.675
261.0	3.161	1.829	0.144	490.1	883.9	4.158	5.665
266.5	3.792	1.854	0.120	503.8	889.9	4.208	5.656
272.1	4.530	1.880	0.100	517.1	895.7	4.257	5.648
277.6	5.364	1.907	0.0830	530.1	901.3	4.305	5.641
283.2	6.309	1.936	0.0712	543.8	906.9	4.353	5.634
288.8	7.370	1.966	0.0614	557.3	912.3	4.401	5.627
294.3	8.570	2.003	0.0533	571.5	917.4	4.448	5.622
300.0	9.901	2.041	0.0465	585.9	922.0	4.495	5.616

Note: It may be assumed that, should propane and CO_2 form a liquid solution, the solution will exhibit ideal-solution behavior. State clearly and justify any additional assumptions that you make.

Chemical Equilibria 11

In Chapter 6 the concept of chemical equilibrium was considered briefly. We had chosen an *isolated* system with π phases and n components. No internal barriers were present. We allowed only a single chemical reaction to occur among i components ($i \leq n$) with specified stoichiometry. At equilibrium it was shown that the temperatures and pressures were equal in all phases, as were the chemical potentials of all components. The introduction of the single chemical reaction led to Eq. (6-55), which indicated that the sum of products of the chemical potential of each component times its stoichiometric multiplier in the assumed reaction was zero.

In the present chapter we analyze in greater depth the concepts behind chemical equilibria and employ the results to calculate compositions of equilibrium systems. There have been a number of published papers that relate to our objective. In most instances, however, the primary aim of the authors was to describe computational algorithms to obtain numerical results in complex systems. Although we note some of the algorithms, the primary intent of the chapter is on fundamental concepts, and the interested reader is referred to the literature for computing methods.

11.1 Formulation of the Problem

In essence we wish to obtain thermodynamic relationships applicable to a multicomponent system in chemical equilibrium. The system is defined as stated in Postulate I by two independently variable properties plus the masses of all components initially present. In the equilibrium state, we need also to specify two constraints so that the moles (or masses) of all chemical entities may be determined. In the development in Chapter 6, we chose the system to be isolated so that the constraints were constant *total* energy and volume (as well as mass). The equilibrium criterion was then given in Eq. (6-39). With the phase equilibrium relationships for T and P [Eqs. (6-44) and (6-45)] and the energy and volume restrictions [Eqs. (6-40) and (6-41)], Eq. (6-39) may be written as

$$\delta \underline{S} = \sum_{s=1}^{\pi} \sum_{j=1}^{n} S_{N_j}^{(s)} \, \delta N_j^{(s)} = 0 \tag{11-1}$$

With $S_{N_j}^{(s)} = \mu_j^{(s)}/T$, Eq. (11-1) becomes

$$\sum_{s=1}^{\pi} \sum_{j=1}^{n} \mu_j^{(s)} \, \delta N_j^{(s)} = 0 \tag{11-2}$$

Note that the mole balance given in Eq. (6-42) was not used since chemical reactions can occur; we will have to develop an alternative relation to allow the conservation of atoms rather than moles of specific components. Also, to this point the equality of chemical potentials in the different phases was not invoked since it will be convenient in our treatment to allow this fact to appear as a consequence of the analysis.

Let us stop briefly and examine Eq. (11-2). The variations $\delta N_j^{(s)}$ could arise due to a transfer of δN_j between phases or as a result of one or more chemical reactions that involve N_j. Irrespective of the origin of the variations, within the limitation of constant total $\underline{U}$ and $\underline{V}$, the sum in Eq. (11-2) must be zero for an equilibrium state. If we had not chosen a system at constant total $\underline{U}$ and $\underline{V}$, but, instead, a system at constant T and P, then the equilibrium criterion would have been Eq. (6-32) (i.e., the Gibbs energy is minimized). Expanding $\underline{G}(T, P, N_j^{(s)})$ with T, P constant, one again obtains Eq. (11-2). Thus the same equilibrium criterion is obtained and virtual variations of $\delta N_j^{(s)}$ must be now be conceived bearing the constant T, P restrictions in mind. Equation (11-2) would also have been obtained if the system were held at constant volume and temperature [here the Helmholtz energy would be minimized and Eq. (6-29) would be used]. Other constraints could be used, but as long as no restrictions are placed on any $\mu_j^{(s)}$ or $\delta N_j^{(s)}$ (as by, for example, a semipermeable membrane), Eq. (11-2) is obtained as the basic equilibrium criterion.

There is one additional important comment relevant to the understanding and use of Eq. (11-2). The component summation is designated as $j = 1, \ldots, n$. The choice of the components present in the equilibrium mixture must be made by the investigator on some rational basis. Clearly, the components in the j vector are not limited to those that might have been used to synthesize the system initially. Usually, but not always, the j vector will contain the original components in addition to others which may appear as a result of some chemical transformation.

Consider some simple examples. If H_2 and O_2 were the only species charged to the original system, what should we include in our j vector at equilibrium? Obviously, no component could be specified that contained elements other than

H or O (i.e., no nuclear transformations are considered). H_2O would be an obvious choice of another component if evidence were available indicating that it could form under the actual state of the system. Should the H radical be included—or H_2O_2—or other components? If water is to be present, is it in the vapor phase, the solid, the liquid, all three? Questions such as these are vital to the solution of any chemical equilibrium problem, but they are often not easily answered. Generally, one can include all the possible components desired—in as many phases as desired—and then force the computational scheme to show that actual equilibrium compositions of most are negligibly small (or zero). This general inclusion principle also applies to selecting components to exist in more than one phase. The disadvantage of including many components lies in greater computational complexity. But if a possible component is not included, thermodynamics provides no mechanism for its introduction.

In summary, to treat systems in chemical equilibrium, we specify the overall system by the total number of atoms (or the mass) of each element present, and we place two additional independent restraints (e.g., T and P) as required by Postulate I. We then can use Eq. (11-2) to compute the moles of all *selected* components in all phases.

This is only one type of *problem formulation.* In Section 11.7 we consider alternative types.

11.2 Conservation of Atoms

It was stressed above that there is a selection process to specify the components present in the equilibrium system. These components are energetically favorable aggregates of atoms which comprise the system. Let us assume that the total number of elements present is ℓ and that each component present can be represented as W_j. Following Smith[1] we consider identical chemical species in different phases as distinct components. Thus $H_2O(g)$ and $H_2O(l)$ would be represented by two different W_j's (the formula vector).

A matrix $\mathfrak{D}$ is formed with the columns as formula vectors, W_j ($j = 1, \ldots, n'$)[2] and rows as elements ($i = 1, \ldots, \ell$). The element d_{ij} then represents the number of atoms i in component j. If b_i is the total number of atoms of i (divided by Avogadro's number) present in the system, the conservation of atoms leads to

$$\sum_{j=1}^{n'} d_{ij} N_j - b_i = 0, \qquad i = 1, \ldots, \ell \tag{11-3}$$[3]

[1]W. R. Smith, *Ind. Eng. Chem. Fundam.*, **19**, 1 (1980).

[2]Since we are treating the same chemical species in different phases as distinct entities, the value of n' is set as $n \leq n' \leq n\pi$, where n is the number of distinct chemical entities [as in Eq. (11-2)] and π is the number of phases.

[3]As Zeleznik and Gordon [F. J. Zeleznik and S. Gordon, *Ind. Eng. Chem.*, **60**(6), 27 (1968)] point out, if ionization occurs, Eq. (11-3) should also include a conservation of charges. This is done by assuming charge is the $(\ell + 1)$st element, letting $b_{\ell+1} = 0$ and then $d_{\ell+1,j}$ is the charge on the j component.

In addition to Eq. (11-3), there is a nonnegativity constraint,

$$N_j \geq 0 \tag{11-4}$$

11.3 Nonstoichiometric Formulation

In this section we combine Eqs. (11-2) and (11-3) to provide a set of equations from which N_j may be calculated. The technique is termed *nonstoichiometric* because no specific chemical reactions are introduced.

We first write Eq. (11-2) with the chosen formula vector set $j = 1, \ldots, n'$ as

$$\sum_{j=1}^{n'} \mu_j \, \delta N_j = 0 \tag{11-5}$$

We then multiply each equation in Eq. (11-3) by a Lagrange multiplier η_i and, after taking variations, we combine with Eq. (11-5). The result is

$$\sum_{j=1}^{n'} \left(\mu_j + \sum_{i=1}^{\ell} \eta_i d_{ij}\right) \delta N_j + \sum_{i=1}^{\ell} \left(\sum_{j=1}^{n'} d_{ij} N_j - b_i\right) \delta \eta_i = 0 \tag{11-6}$$

Treating the variations of N_j and η_i as independent, we recover Eq. (11-3) and, also,

$$\mu_j + \sum_{i=1}^{\ell} \eta_i d_{ij} = 0, \qquad j = 1, \ldots, n' \tag{11-7}$$

Equations (11-3) and (11-7) comprise the necessary relations to solve for the n' values of N_j and the ℓ values of η_i. That is, the unknowns total $n' + \ell$ while the equations in (11-3) and (11-7) also total $n' + \ell$. Also, since η_i and d_{ij} are independent of the phase(s) that contain components with the same structural atomic formula, Eq. (11-7) implies that for these chemically identical components, μ_j is the same in all phases.

Example 11.1

Assume that we have a given system in chemical equilibrium at some specified T and P. The elements comprising the system are C, H, and O. We have assumed that the only components present in the equilibrium mixture are CO_2, H_2, H_2O, and C. We also have reason to believe that water is present only in the gas phase, but inconsequential amounts of carbon appear in the gas. We then have four components: (1) $CO_2(g)$; (2) $H_2(g)$; (3) $H_2O(g)$; (4) C(s). Develop the nonstoichiometric relations to allow calculation of N_{CO_2}, N_{H_2}, N_{H_2O}, and N_C at equilibrium, assuming that b_C, b_H, and b_O are, respectively, the total quantities of C, O, and H present.

Solution

We first develop the matrix $\mathfrak{D}$.

$$\begin{array}{c|cccc|l} & CO_2 & H_2 & C & H_2O & \\ \hline C & 1 & 0 & 1 & 0 & \\ H & 0 & 2 & 0 & 2 & = \mathfrak{D} \\ O & 2 & 0 & 0 & 1 & \end{array}$$

Note that the columns denote the components ($j = 1, 2, 3, 4$), whereas the rows include the elements ($i = 1, 2, 3$). A matrix element, say d_{24}, would then represent the atoms of H in a molecule of H_2O. With Eqs. (11-7) and (11-3),

$$\mu_{CO_2} + \eta_C + 2\eta_O = 0 \quad \text{(A)}$$

$$\mu_{H_2} + 2\eta_H = 0 \quad \text{(B)}$$

$$\mu_C + \eta_C = 0 \quad \text{(C)}$$

$$\mu_{H_2O} + 2\eta_H + \eta_O = 0 \quad \text{(D)}$$

$$N_{CO_2} + N_C = b_C \quad (\alpha)$$

$$2N_{H_2} + 2N_{H_2O} = b_H \quad (\beta)$$

$$2N_{CO_2} + N_{H_2O} = b_O \quad (\gamma)$$

We can express the chemical potential of any component j as

$$\mu_j = G_j^\circ + RT \ln \frac{\hat{f}_j}{f_j^\circ} \tag{11-8}$$

where Eq. (8-116) was used twice, once to specify μ_j in the reacting mixture and once for μ_j in some arbitrarily defined standard state[4] where

$$\mu_j^\circ = G_j^\circ$$

and the fugacity of j in this standard state is f_j°.

For a gas phase, Eq. (8-132) may be used to express $\hat{f}_j$ in terms of pressure, mole fraction, and nonideality, for example,

$$\hat{f}_j = \phi_j(y_j, T, P)Py_j \tag{11-9}$$

For a liquid solution, Eq. (11-9) may be used if ϕ_j(liquid) can be determined, or Eq. (8-149) can be introduced to employ activities. If the reference state for j is the pure material, Eq. (8-155) is applicable:

$$\hat{f}_j = \gamma_j(x_j, T, P) f_j(T, P) x_j \tag{11-10}$$

With either Eq. (11-9) or (11-10) substituted into Eq. (11-8), μ_j becomes a function of mole fractions (as well as T and P and the reference-state conditions). Thus, the set of seven equations (A) through (D), (α) through (γ) for N_{CO_2}, N_{H_2}, N_C, N_{H_2O}, η_C, η_H, η_O may be solved.

The astute reader of Example 11.1 will probably have noted that the seven equations developed can be greatly simplified and the solution expedited if (B) is multiplied by 2, (C) by -1, and (D) by -2, and then (A) through (D) added. Then all η_i terms cancel and the set of equations is immediately reduced from seven to four, with the unknowns being the mole numbers of CO_2, H_2, C, and H_2O.

Suppose, however, that the problem discussed in Example 11.1 is made more realistic by adding CO as a component in the equilibrium mixture. Equations (A) through (D) would be unaffected, but the $\mathfrak{D}$ matrix would have an additional column of CO with the column elements 1, 0, 1. Thus, a new relation is necessary:

$$\mu_{CO} + \eta_C + \eta_O = 0 \quad \text{(E)}$$

[4]The standard state is, however, always at the temperature of the reacting system.

and (α) and (γ) would have to be modified to include the moles of CO. There are now eight equations in eight unknowns. One still suspects that there is some convenient scheme to collapse the set of equations (A) through (E) to eliminate η_C, η_H, and η_O. The systematic method that will accomplish this elimination is, in fact, embodied in the *stoichiometric* formulation discussed in Section 11.4.

Example 11.2

Suppose that in Example 11.1 we had reason to believe that water existed in the equilibrium mixture both in the gas and liquid states. Also, we suspect that significant quantities of CO_2 dissolve in the water, but we neglect any solubility of H_2 (and carbon) in the aqueous phase. How do the nonstoichiometric relations change?

Solution

Following the format of Example 11.1, since $n' = 6$, the $\mathfrak{D}$ matrix is expanded to include two columns for CO_2 [$CO_2(g)$ and $CO_2(l)$] and two columns for water [$H_2O(g)$ and $H_2O(l)$]. The elements in each of these column sets are, of course, identical. When Eq. (11-7) is applied, we find that we have two equations for CO_2 [for $\mu_{CO_2}(g)$ and $\mu_{CO_2}(l)$] and, similarly for H_2O. Combining these, we obtain

$$\mu_{CO_2}(g) = \mu_{CO_2}(l)$$

$$\mu_{H_2O}(g) = \mu_{H_2O}(l)$$

There are three equations from Eq. (11-3), containing as variables $N_{CO_2}(g)$, $N_{CO_2}(l)$, N_C, $N_{H_2O}(g)$, $N_{H_2O}(l)$, and N_{H_2}. Those with the three Lagrange multipliers, η_C, η_H, and η_O, total nine. The solution is obtained from the set of four equations from Eq. (11-7), three equations from Eq. (11-3), and the two equalities of chemical potential of CO_2 and H_2O since $\mu = f(T, P, \text{composition})$.

11.4 Stoichiometric Formulation

Before developing this formulation, we should return briefly to Eq. (11-3) and modify slightly to make it more general. The index on i was set to include $i = 1, \ldots, \ell$. Actually, ℓ should be replaced by the integer that yields the rank of the $\mathfrak{D}$ matrix. In most cases this rank will equal the number of chemical elements, ℓ, but as shown in Example 11.3, there are exceptions.

Example 11.3

Assume that an equilibrium mixture contains CO, H_2 and methanol. Write the nonstoichiometric relations.

Solution

$$\begin{array}{c|ccc|l} & CO & H_2 & CH_3OH & \\ \hline C & 1 & 0 & 1 & \\ H & 0 & 2 & 4 & = \mathfrak{D} \\ O & 1 & 0 & 1 & \end{array}$$

The three equations from Eq. (11-7) are

$$\begin{aligned} \mu_{CO} \quad + \eta_C \quad\quad\quad + \eta_O &= 0 \\ \mu_{H_2} \quad\quad\quad + 2\eta_H \quad\quad &= 0 \\ \mu_{CH_3OH} + \eta_C + 4\eta_H + \eta_O &= 0 \end{aligned}$$

but, with Eq. (11-3)

$$\begin{aligned} N_{CO} \quad\quad + N_{CH_3OH} &= b_C \\ 2N_{H_2} + 4N_{CH_3OH} &= b_H \\ N_{CO} \quad\quad + N_{CH_3OH} &= b_O \end{aligned}$$

We see that the C and O balances are identical. Thus, the moles of CO and CH_3OH are not independent. The rank of $\mathfrak{D}$ is 2, whereas $\ell = 3$. Therefore, in Eq. (11-3), i should equal 1, 2.

With this minor, but important change in Eq. (11-3), we now introduce a variation of the forumlation which effectively eliminates the Lagrange multipliers η_i. As Zeleznik and Gordon[5] point out, these undetermined multipliers originated since we desired to treat each δN_j as independently variable. Another approach is to use the concept of *independent reactions*, in which the δN_j for each reaction is not independent. If we had no atom restrictions, the number of conceivable reactions one could formulate would be n'. But with the restrictions imposed by Eq. (11-3), this number is reduced to $n' - C$, where C is the rank of the $\mathfrak{D}$ matrix. We write each of these independent reactions as

$$\sum_{j=1}^{n'} \nu_{jr} W_j = 0, \qquad r = 1, \ldots, n' - C \tag{11-11}$$

where W_j is the formula vector for component j. The term ν_{jr} is termed the *stoichiometric coefficient* for component j in reaction r. To determine the (column) vector ν_j for any reaction r, the conservation of elements requires that

$$\sum_{j=1}^{n'} d_{ij} \nu_{jr} = 0, \qquad r = 1, \ldots, n' - C \tag{11-12}$$

or

$$\mathfrak{D}\nu = 0 \tag{11-13}$$

Example 11.4

For the CO_2, H_2, C, H_2O, CO example described in Example 11.1 and in the following text, specify the number of independent reactions and suggest appropriate stoichiometric coefficient vectors.

Solution

Rewriting the $\mathfrak{D}$ matrix for this case,

	CO_2	H_2	C	H_2O	CO
C	1	0	1	0	1
H	0	2	0	2	0
O	2	0	0	1	1

$= \mathfrak{D}$

[5] *Op. cit.*

When $\mathfrak{D}$ is reduced to eschelon form by Gauss–Jordan elimination, it can be shown the rank C of the $\mathfrak{D}$ matrix is 3. With $n' = 5$, the number of independent reactions is $5 - 3 = 2$. Equation (11-11) or (11-12) states that

$$\begin{vmatrix} d_{11} & d_{12} & d_{13} & d_{14} & d_{15} \\ d_{21} & d_{22} & d_{23} & d_{24} & d_{25} \\ d_{31} & d_{32} & d_{33} & d_{34} & d_{35} \end{vmatrix} \times \begin{vmatrix} \nu_1 \\ \nu_2 \\ \nu_3 \\ \nu_4 \\ \nu_5 \end{vmatrix} = 0$$

With the given elements in the $\mathfrak{D}$ matrix,

$$\begin{aligned} \nu_1 \quad\quad + \nu_3 \quad\quad + \nu_5 &= 0 \\ + 2\nu_2 \quad\quad + 2\nu_4 \quad\quad &= 0 \\ 2\nu_1 \quad\quad\quad\quad + \nu_4 + \nu_5 &= 0 \end{aligned}$$

Eliminating, for example, ν_3, ν_4, and ν_5,

$$\nu_3 = \nu_1 - \nu_2$$

$$\nu_4 = -\nu_2$$

$$\nu_5 = \nu_2 - 2\nu_1$$

If $\nu_1 = 1$, $\nu_2 = 0$, then $\nu_3 = 1$, $\nu_4 = 0$, $\nu_5 = -2$ and one independent reaction would be

$$CO_2 + C - 2CO = 0$$

For the second reaction, let $\nu_1 = 0$ and $\nu_2 = 1$; then $\nu_3 = -1$, $\nu_4 = -1$, and $\nu_5 = 1$. Then

$$H_2 - C - H_2O + CO = 0$$

Thus, there are two stoichiometric column vectors, {1, 0, 1, 0, −2} and {0, 1, −1, −1, 1} for the two reactions.

These are two possible independent reactions. Other methods of eliminating ν_j values or in choosing different values of ν_1, ν_2 may lead to different reactions, but these can always be obtained by linear combinations of the two shown above.

Equation (11-12) [or (11-13)] is a very important relation. We can use it directly to provide an equilibrium formulation with no Lagrange multipliers as follows. In Eq. (11-7), for a particular reaction r, multiply by ν_{jr} and sum

$$\sum_{j=1}^{n'} \nu_{jr}\mu_j + \sum_{j=1}^{n'} \nu_{jr} \sum_{i=1}^{\ell} \eta_i d_{ij} = 0 \tag{11-14}$$

Inverting the two summations in the second term, and using Eq. (11-12), we obtain

$$\sum_{j=1}^{n'} \nu_{jr}\mu_j = 0, \qquad r = 1, \ldots, n' - C \tag{11-15}$$

In this stoichiometric formulation we have n' variables (N_j). There are $n' - C$ equations of the form of Eq. (11-15), and C atom constraint balances from Eq. (11-3). (As noted earlier, in the majority of all situations one may replace C by ℓ, the number of chemical elements.) Whereas in this method we have eliminated the Lagrange multipliers, it is now necessary to specify the appropriate sets of stoichiometric multipliers for all independent reactions. In complex systems, this can sometimes lead to difficulties.

It is emphasized that the independent reactions introduced in the stoichiometric formulation would not normally have any connection with the actual chemical reaction path. They were developed as an artifice to relate δN, for each reaction, to changes in a single variable by the stoichiometric coefficient vector. That is, for each independent reaction,

$$\delta N_{jr} = \nu_{jr}\, d\xi_r, \qquad r = 1, \ldots, n' - C \tag{11-16}$$

The variable ξ_r is called the *extent of reaction* for reaction r.

$$\delta N_j = \sum_{r=1}^{n'-C} \delta N_{jr} = \sum_{r=1}^{n'-C} \nu_{jr}\, d\xi_r \tag{11-17}$$

Integrating Eq. (11-17) yields

$$N_j = N_j(\text{initial}) + \sum_{r=1}^{n'-C} \nu_{jr}\xi_r, \qquad j = 1, \ldots, n' \tag{11-18}$$

N_j represents the moles of j in the equilibrium mixture and equals the moles of N_j present initially as modified by chemical reactions. Since μ_j is a function of N_j at a given P, T, then Eq. (11-18) may be used with Eq. (11-15) to provide $n' - C$ relations between the $n' - C$ extents of reaction. For example, assume that we have only a gas phase which behaves as an ideal gas mixture; then Eq. (11-8) becomes

$$\mu_j = G_j^\circ + RT \ln y_j \tag{11-19}$$

$$= G_j^\circ + RT \ln \frac{N_j}{N} \tag{11-20}$$

In arriving at Eq. (11-19) from Eq. (11-8), we made several assumptions: (1) the fugacity of j in the standard state is given as P since an ideal gas is chosen; and (2) the fugacity coefficient of j in the mixture is unity—this is a consequence of the ideal-gas assumption.

In Eq. (11-20) we substituted for the mole fraction of j the mole ratio of N_j to the total moles,[6]

$$N = \sum_{j=1}^{n'} N_j \tag{11-21}$$

If we now combine Eqs. (11-15), (11-18), and (11-20), we obtain

$$\sum_{j=1}^{n'} \nu_{jr}\left[G_j^\circ + RT \ln \frac{N_j(\text{initial}) + \sum_{r=1}^{n'-C} \nu_{jr}\xi_r}{N}\right] = 0, \qquad r = 1, \ldots, n' - C \tag{11-22}$$

Solutions of Eqs. (11-22) are described in review papers.[7] For cases wherein there are many phases or many components or several independent reactions, numerical computation is necessary. In simple cases, where hand calculation

[6]In Eq. (11-21), the total moles N would include all reactants and products *and* any inert components (where $\nu_{jr} = 0$). If desired, the inert components may be separated from the sum in Eq. (11-21) and carried along as an additive constant.

[7]Smith, *op. cit.*; R. Gautam and W. R. Seider, *AIChE J.*, **25**, 991 (1979).

is possible, the approach may be modified. Return to Eq. (11-15) and use Eq. (11-8) for μ_j. Assume for the moment that there is only a single independent reaction. Then

$$\sum_{j=1}^{n'} \nu_j\left(G_j^\circ + RT \ln \frac{\hat{f}_j}{f_j^\circ}\right) = 0$$

or

$$\frac{-\left(\sum_{j=1}^{n'} \nu_j G_j^\circ\right)}{RT} = \ln\left[\prod_{j=1}^{n'}\left(\frac{\hat{f}_j}{f_j^\circ}\right)^{\nu_j}\right] \tag{11-23}$$

The *standard Gibbs energy for the reaction* is written

$$\Delta G^\circ = \sum_{j=1}^{n'} \nu_j G_j^\circ \tag{11-24}$$

and the *equilibrium constant* for the reaction as

$$\prod_{j=1}^{n'}\left(\frac{\hat{f}_j}{f_j^\circ}\right)^{\nu_j} = K_a \tag{11-25}$$

Therefore, Eq. (11-23) becomes

$$K_a = \exp\left(\frac{-\Delta G^\circ}{RT}\right) \tag{11-26}$$

It is clear that the same procedure could be followed if there were more than one independent reaction. Equations (11-23) through (11-26) would then be written for each reaction. The appropriate stoichiometric coefficient vector ν_{jr} could be found as before. Then

$$\Delta G_r^\circ = \sum_{j=1}^{n'} \nu_{jr} G_j^\circ$$

$$\prod_{j=1}^{n'}\left(\frac{\hat{f}_j}{f_j^\circ}\right)^{\nu_{jr}} = K_{a_r}, \qquad r = 1, \ldots, n' - C \tag{11-27}$$

$$K_{a_r} = \exp\left(\frac{-\Delta G_r^\circ}{RT}\right)$$

Note that G_j°, f_j°, and $\hat{f}_j$ are independent of which reaction is considered.

The introduction of equilibrium constants into the stoichiometric formulation is usually only of value when there are one or two independent reactions. For more complex systems, the general methods described earlier are used. However, it is of value to explore further the concept of a single independent reaction employing Eqs. (11-24) through (11-26), as the concepts developed are applicable in the general case even when no equilibrium constant is formally used.

11.5 Equilibrium Constants

As noted above, we will only treat the case of a system in chemical equilibrium with a single independent reaction. Also, we assume that the stoichiometric vector has been determined. Thus, of the n' components in the π phases, the stoichiometric vector element of some may be zero; these are termed *inert components*, as they do not enter into the equilibrium calculations.

When computing ΔG° in Eq. (11-24) or μ_j in Eq. (11-8), values of the standard Gibbs energies for the reacting components must be specified. This standard state must be chosen at the temperature of the equilibrium system, but the pressure, composition, and state (gas, liquid, solid) may be selected for convenience. It is clear from Eq, (11-26) that K_a depends only on the properties of the reacting components in their assumed standard states. Within the expression for K_a [see Eq. (11-25)], there also appear the fugacities of the components in the same standard state.

Thus, for any numerical value of K_a, there is an implied set of standard states chosen for the components. One normally denotes this by writing the independent reaction with a shorthand notation to specify the standard states. For example, if the reaction were written

$$\nu_1 W_1(\mathrm{g}) + \nu_2 W_2(l) + \nu_3 W_3(\mathrm{s}) = 0 \tag{11-28}$$

the designation in the parentheses following the formula vector W_j indicates the standard state chosen. The symbol (g) after W_1 means that the standard state of component 1 is a pure gas, in an ideal-gas state at 1 bar (often abbreviated as "unit fugacity"), at the system temperature. The symbol (l) following W_2 normally means that the standard state for component 2 is a pure liquid at its vapor pressure at the system temperature. In some instances, however, the same symbol would indicate that the standard state for 2 is a pure liquid at the system pressure (or at 1 bar) at the system temperature. Since Gibbs energies or fugacities are relatively insensitive to pressure—except near the critical point—the selection of a standard state pressure is often not important. In a similar way, (s) indicates a solid in its most stable state at T. The defining pressure is usually, but not always, the vapor pressure.

To convert Eq. (11-25) to a function of concentrations, the mixture fugacity, $\hat{f}_j$, is expanded as a function of concentration by using Eq. (11-9) or (11-10) if j is present in a vapor phase or in a condensed phase, respectively. If a component is present in two or more phases, either fugacity expansion may be used since $\hat{f}_j$ is equal in all phases at equilibrium.

Should all species taking part in the reaction be present as vapors, substitution of Eq. (11-9) into Eq. (11-25) yields

$$K_a = \left(\prod_{j=1}^{n'} \phi_j^{\nu_j}\right)\left(\prod_{j=1}^{n'} y_j^{\nu_j}\right) P^{\nu} = K_\phi K_y P^{\nu} \tag{11-29}$$

where $\nu = \sum_{j=1}^{n} \nu_j$, and the standard-state fugacities, f_j°, have been taken as unit fugacity. If the vapor mixture forms an ideal soluiton, then (see Section 8.5)

$$\phi_j = \frac{f_j}{P} \qquad \text{(ideal solution)} \tag{11-30}$$

where f_j is the fugacity of pure vapor j at the temperature and pressure of the mixture. Substituting Eq. (11-30) into Eq. (11-29) yields

$$K_a = K_{f/P} K_y P^{\nu} \qquad \text{(ideal vapor solution)} \tag{11-31}$$

For liquid or solid mixtures, with Eq. (11-10),

$$\frac{\hat{f}_j}{f_j^\circ} = \frac{f_j}{f_j^\circ}\gamma_j x_j \tag{11-32}$$

If the standard-state pressure is taken as P_{vp_j} or 1 bar and the system pressure is P, then

$$RT \ln \frac{f_j}{f_j^\circ} = \int_{(P_{vp_j} \text{ or } 1 \text{ bar})}^{P} V_j \, dP \tag{11-33}$$

Unless P is very large, the integral in Eq. (11-33) is normally small; if it were neglected, then substitution of Eq. (11-32) into Eq. (11-25) yields

$$K_a = \left(\prod_{j=1}^{n'} \gamma_j^{\nu_j}\right)\left(\prod_{j=1}^{n'} x_j^{\nu_j}\right) = K_\gamma K_x, \tag{11-34}$$

(condensed phase; pressure correction neglected)

We recall that K_a is dependent on the standard state. The selection of the standard state is made in a pragmatic sense; that is, it is usually determined by the available data. For a few simple reactions, K_a has been expressed as a function of temperature, as shown later in Figure 11.1. When such data are unavailable, K_a can be determined from ΔG° by using Eq. (11-24) or from ΔH° and ΔS° by Eq. (11-35).

$$\Delta G^\circ = \Delta H^\circ - T\,\Delta S^\circ \tag{11-35}$$

Since it is impractical to list ΔG° (or ΔH°, ΔS°) for every reaction, tables are available for a large number of compounds showing the Gibbs energy and enthalpy of formation of the species from the elements. In these tables the function ΔG° becomes, for each species, ΔG_f° and similarly, ΔH° becomes ΔH_f°. To obtain ΔG° and ΔH°, at the temperature of interest,

$$\Delta G^\circ = \sum_{j=1}^{n'} \nu_j \, \Delta G_{f_j}^\circ \tag{11-36}$$

and

$$\Delta H^\circ = \sum_{j=1}^{n'} \nu_j \, \Delta H_{f_j}^\circ \tag{11-37}$$

where, again, $\nu_j = 0$ for those species that do not participate in the reaction. For elements, by convention,[8]

$$\Delta H_f^\circ = \Delta G_f^\circ = 0 \qquad \text{at all temperatures} \tag{11-38}$$

Several tabulations[9] form an excellent reference source for values of ΔH_f° and ΔG_f° over a wide range of temperatures. These compilations also show values of C_p°, the heat capacity in the ideal-gas standard state.

[8]For elements such as oxygen, nitrogen, etc., there is no problem with this convention. For elements that are solids at the temperature of interest, a clear statement of the crystal form is also necessary. For example, the standard state for carbon is based on graphite. Should other forms of carbon be present in a system, the ΔG_f° and ΔH_f° for such forms are not zero.

[9]D. R. Stull, E. F. Westrum, Jr., and G. C. Sinke, *The Chemical Thermodynamics of Organic Compounds* (New York: Wiley, 1969); D. R. Stull and H. Prophet, *JANAF Thermochemical Tables*, 2nd ed., NSRDS-NBS 37, June 1971.

If ΔG° and ΔH° are available at one temperature, ΔG° can be found at any other temperature as follows:

$$\frac{d(\Delta G^\circ/T)}{dT} = \frac{1}{T}\frac{d\,\Delta G^\circ}{dT} - \frac{\Delta G^\circ}{T^2} = -\frac{\Delta S^\circ}{T} - \frac{\Delta G^\circ}{T^2} = -\frac{\Delta H^\circ}{T^2} \tag{11-39}$$

Equation (11-39) may be integrated if ΔH° is known at one temperature T_1 since

$$\Delta H_T^\circ = \Delta H_{T_1}^\circ + \int_{T_1}^{T} \Delta C_p^\circ \, dT \tag{11-40}$$

where

$$\Delta C_p^\circ = \sum_{j=1}^{n'} \nu_j C_{p_j}^\circ \tag{11-41}$$

Before completing this discussion, it is appropriate to describe briefly the Third Law of Thermodynamics. Experimental evidence indicates that the entropy change in a chemical reaction becomes negligible as the absolute temperature approaches zero, that is,

$$\Delta S_0^\circ = 0 \tag{11-42}$$

With the additional stipulation that the entropy state of elements is zero at $T = 0$, it follows that $\Delta S_0^\circ = 0$ for all materials; or, as more often stated, $S_{0_j}^\circ = 0$. With this base, it is possible to refer to an absolute entropy that can be calculated by integrating with actual heat capacity data through the solid phase (at very low temperatures), the liquid phase, and into the vapor phase. Entropy changes that result from phase transitions are also included. The determination of such absolute entropies obviously requires considerable data[10] and will not be considered further in this text.

In many instances the equilibrium constant of a reaction is known at one temperature and an extrapolation to other temperatures is necessary. From Eqs. (11-25) and (11-39),

$$\frac{d \ln K_a}{dT} = -\frac{1}{R}\frac{d(\Delta G^\circ/T)}{dT} = \frac{\Delta H^\circ}{RT^2} \tag{11-43}$$

Often, ΔH° does not vary appreciably with temperature and, therefore, $\ln K_a$ is nearly linear in $1/T$. A number of reaction equilibrium constants have been plotted in this fashion on Figure 11.1. From the slope one can find a temperature-average value of ΔH°, that is,

$$-\Delta H^\circ = \frac{R \ln (K_{T_2}/K_{T_1})}{1/T_2 - 1/T_1} \tag{11-44}$$

Positive slopes on Figure 11.1 then correspond to reaction with $\Delta H^\circ < 0$ (i.e., exothermic reactions).

[10]For example, since as $T \longrightarrow 0$, $C_p(s) \longrightarrow 0$, special caution must be used to integrate at low temperatures. All phase transformations (first and higher order) must also be included.

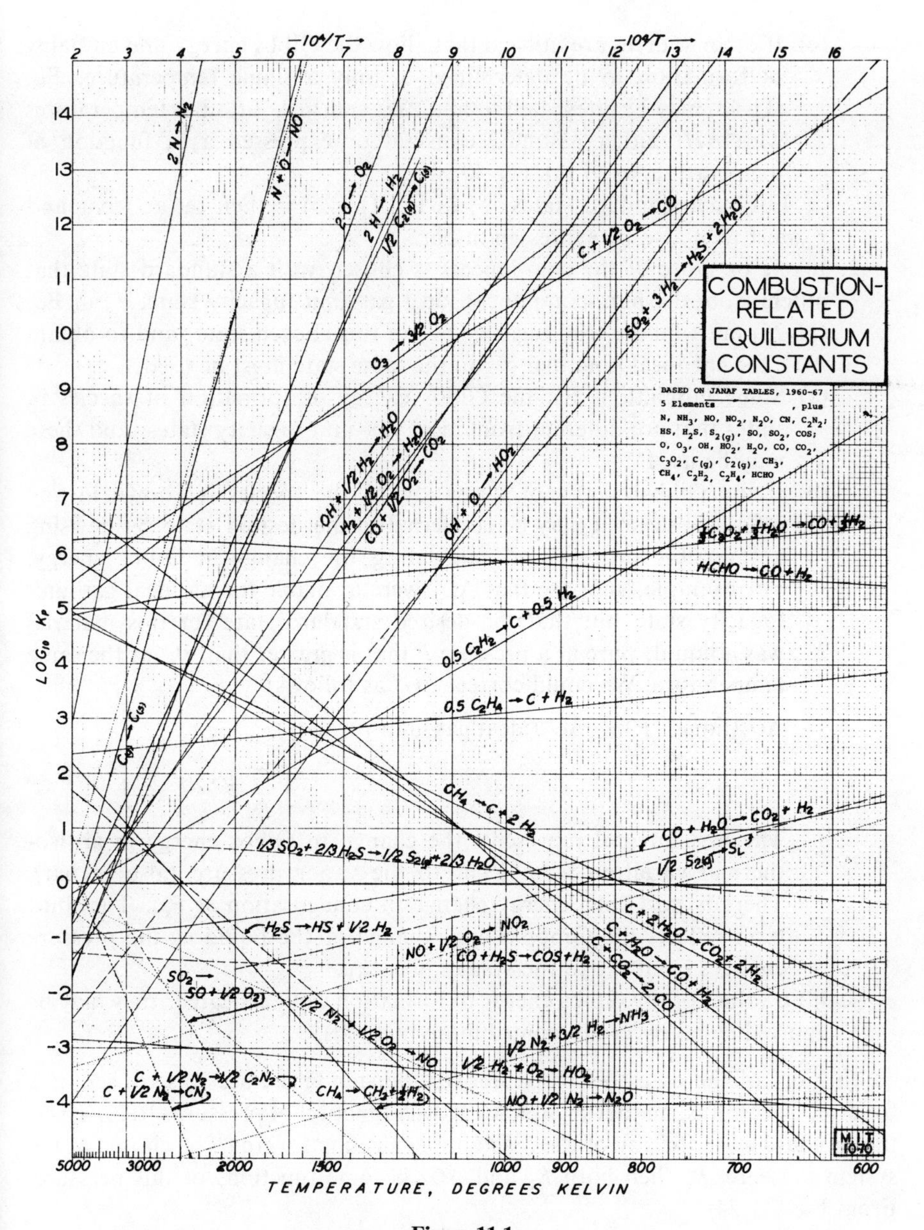

Figure 11.1

In summary, to determine the equilibrium constant of a reaction:

1. If the standard states of all components are chosen as pure materials in an ideal-gas unit fugacity state:
   (a) Use Figure 11.1, if applicable.
   (b) Determine the enthalpy and Gibbs energy of formation for all components (see footnote 7) and use Eqs. (11-36) and (11-37).

(c) If other sources are utilized that allow the Gibbs energy and enthalpy of formation to be determined at only a single temperature, Eq. (11-43) may be integrated to find ΔG° and ln K_a at other temperatures with ΔH° either assumed constant or expressed as a function of temperature, as in Eq. (11-40).

For such standard states, f_j° in Eq. (11-25) is then set at 1 bar and $\hat{f}_j$ must then also be expressed in bar.

2. If one or more of the components is chosen with a standard state that differs from the pure-component ideal gas–unit fugacity state, f_j° in Eq. (11-25) must be the true fugacity of j in this chosen state, and to obtain ΔG°, the chemical potential of j in the same state must be used.
   (a) Some tabulations list the Gibbs energy of formation of species in states that are different from ideal gas–unit fugacity states, and these may be used.
   (b) If the Gibbs energy of formation of the species is known in any reference state, it may be converted to the desired state by devising a reversible process and calculating the change in Gibbs energy. For example, suppose that ΔG_f° were available for an ideal gas–unit fugacity state but that the desired standard state for this material was a liquid, pure at a pressure P that is greater than P_{vp} at the same temperature. We could correct ΔG_f° as follows:

$$\Delta G_f^\circ(\text{liquid}, P) = \Delta G_f^\circ(\text{ideal gas–unit fugacity}) + \int_{1\ \text{bar}}^{P_{vp}} V^v\, dP + \int_{P_{vp}}^{P} V^L\, dP$$

   The first integral represents the change in Gibbs energy in an isothermal variation from 1 bar to the vapor pressure (also in bar). There is no Gibbs energy change in condensation at P_{vp}. The third term reflects the Gibbs energy change of the liquid as the pressure changes from P_{vp} to the system pressure.
3. If no values of the Gibbs energy of formation can be located, they may be approximated by group-contribution methods.[11]

In most of the cases discussed above, the standard state was chosen at a fixed pressure; thus, neither K_a nor ΔG° was a function of pressure. If, however, the standard state for any of the reactants or products were to be chosen at the system pressure, P, then both K_a and ΔG° become functions of this pressure. From Eq. (11-24),

$$\left(\frac{\partial \Delta G^\circ}{\partial P}\right)_T = \sum_{j=1}^{n'} \nu_j \left(\frac{\partial G_j^\circ}{\partial P}\right)_T = -RT\left(\frac{\partial \ln K_a}{\partial P}\right)_T \tag{11-45}$$

If G_j° is not a function of P, this derivative vanishes. If G_j° is a function of P, then $(\partial G_j^\circ/\partial P)_T = V_j^\circ$, the molar volume of j in the chosen standard state. It is

[11] R. C. Reid, J. M. Prausnitz, and T. K. Sherwood, *Properties of Gases and Liquids*, 3rd ed. (New York: McGraw-Hill, 1977), Chap. 7.

usually more convenient to use a ΔG° (or K_a) that is independent of pressure. In this regard, referring to the liquid-phase example shown in 2(b) above, we see that if the standard state were chosen as the pure liquid at its vapor pressure, ΔG_f° would be pressure independent.

Example 11.5

As noted in Section 11.1, a closed system at constant temperature and pressure is in equilibrium when it attains a minimum Gibbs energy. Consider a system initially charged with 1 mole of pure I_2, which is maintained at 800°C and 1 bar, in which the following dissociation reaction occurs:

$$I_2(g) = 2I(g)$$
$$\Delta H^\circ = 156.6 \text{ kJ/mol}$$
$$\Delta S^\circ = 108.4 \text{ J/mol K}$$

The standard states are pure vapors at unit fugacity.

Determine the equilibrium composition by first calculating the Gibbs energy of the mixture as a function of the moles of I_2 dissociated (denoted by x) and then determining the minimum in $\underline{G}$ with respect to x. Assume that the gas mixture is ideal.

Solution

Let us form a mixture containing $(1 - x)$ moles of I_2 and $2x$ moles of I by a two-step process. First select x moles of I_2 in its standard state and assume that it decomposes to $2x$ moles of I in its standard state. Then let us mix the remaining $(1 - x)$ moles I_2 with the newly formed $2x$ moles of I. The changes in enthalpy and entropy are:

$$\Delta \underline{H}_1 = x\,\Delta H^\circ$$
$$\Delta \underline{S}_1 = x\,\Delta S^\circ$$
$$\Delta \underline{H}_2 = \Delta \underline{H}_{\text{mix}} = 0 \qquad \text{(ideal-gas mixture)}$$
$$\Delta \underline{S}_2 = \Delta \underline{S}_{\text{mix}} = \Delta \underline{S}^{\text{ID}} = -R[(1-x)\ln y_{I_2} + (2x)\ln y_I]$$
$$= -R\left[(1-x)\ln\frac{1-x}{1+x} + (2x)\ln\frac{2x}{1+x}\right]$$

Since

$$\Delta \underline{G} = \Delta \underline{H} - T\,\Delta \underline{S}$$
$$\Delta \underline{G} = x\,\Delta H^\circ - Tx\,\Delta S^\circ + RT\left[(1-x)\ln\frac{1-x}{1+x} + (2x)\ln\frac{2x}{1+x}\right]$$

Values for the enthalpy and entropy changes are shown in Figures 11.2 and 11.3.

$\Delta \underline{G}$ attains a minimum value when x is about 0.052 (i.e., when some 5.2% of the original iodine has decomposed). Returning to the $\Delta \underline{G}$ expression derived above, to find the minimum value of $\Delta \underline{G}$ as x is varied, we can also differentiate with respect to x and set the result equal to zero; that is,

$$0 = (\Delta H^\circ - T\,\Delta S^\circ) + RT\ln\frac{4x^2}{1-x^2} = \Delta G^\circ + RT\ln\frac{y_I^2}{y_{I_2}}$$

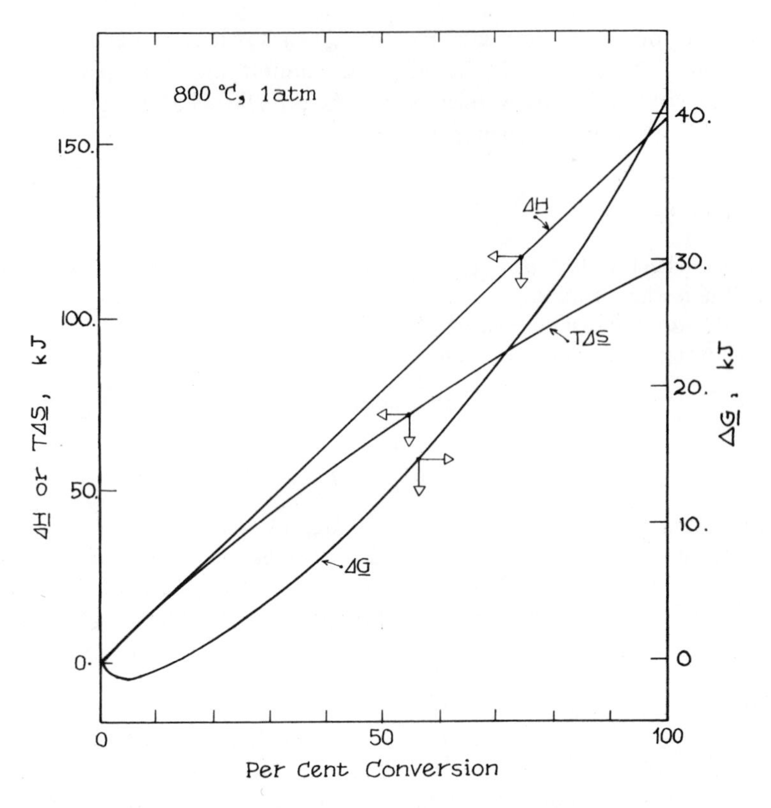

Figure 11.2 ΔH, $T\Delta S$, and ΔG for the reaction I2(g) $\longrightarrow$ 2I(g).

This is, of course, the same result that we would obtain from Eqs. (11-26) and (11-29) using unit-fugacity standard states and assuming ideal gases and 1 bar pressure (i.e., $K_\phi = 1$, $P^\nu = 1$), with $\Delta G^\circ = \Delta H^\circ - T\,\Delta S^\circ = 156{,}000 - (1073)(108.4) = 40.3$ kJ/mol I_2.

The important point of this simple example is to note that it was the entropy change in mixing that led to the minimization of $\Delta \underline{G}$. Without it, in this case, the equilibrium mixture would have contained only pure iodine as I_2.

With only a slight increase in complexity, gas-phase nonidealities may be included. For such a case,

$$\Delta \underline{H} = x\,\Delta H^\circ + \Delta \underline{H}^{EX}$$

$$\Delta \underline{S} = x\,\Delta S^\circ + \Delta \underline{S}^{ID} + \Delta \underline{S}^{EX}$$

$$\Delta \underline{H}^{EX} - T\,\Delta \underline{S}^{EX} = \Delta \underline{G}^{EX} = 2x(\bar{G}_I - \bar{G}_I^{ID}) + (1-x)(\bar{G}_{I_2} - \bar{G}_{I_2}^{ID})$$

$$= 2xRT\ln\gamma_I + (1-x)RT\ln\gamma_{I_2}$$

Then

$$\frac{d\,\Delta G}{dx} = 0 = \Delta G^\circ + RT\ln\frac{\gamma_I^2}{\gamma_{I_2}} + RT\ln\frac{y_I^2}{y_{I_2}}$$

or

$$\Delta G^\circ = -RT\ln\frac{(y_I\gamma_I)^2}{y_{I_2}\gamma_{I_2}} = -RT\ln K_a$$

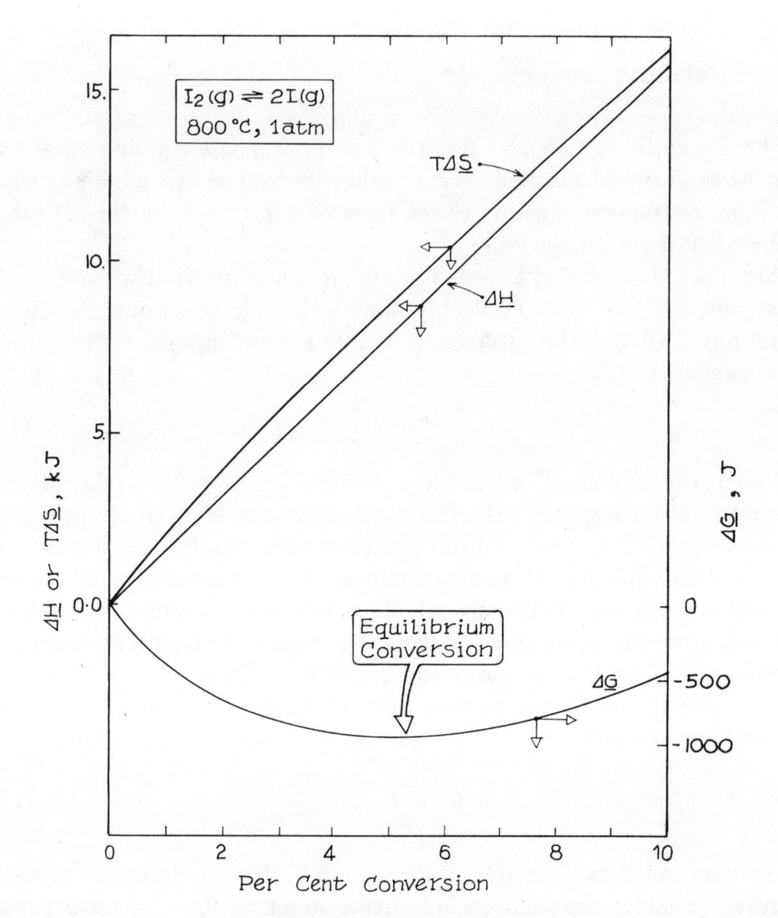

Figure 11.3 Expanded view of Fig. 11.2.

since $y_j \gamma_j = \hat{f}_j$. To proceed further in this case requires data for the variation of γ_{I} and γ_{I_2} (or $\Delta\underline{H}^{\text{EX}}$, $\Delta\underline{S}^{\text{EX}}$) with composition.

Although ΔH° is commonly referred to as the enthalpy of reaction, it is, strictly speaking, the enthalpy change between products and reactants in their respective standard states. ΔH° is not the enthalpy change when the reactants and products are present in an equilibrium mixture. For an equilibrium mixture,

$$\Delta H_r = \sum_{j=1}^{n'} \nu_j \bar{H}_j \tag{11-46}$$

where $\bar{H}_j$ is the partial molar enthalpy of component j at the system temperature, pressure, and in the equilibrium mixture. For ideal-gas mixtures, there are no enthalpy changes on mixing or pressure effects on enthalpies, so that, in this case, $\Delta H^\circ = \Delta H_r$.

11.6 The Phase Rule for Chemically Reacting Systems

We have seen in Section 11.3 that the criteria of phase equilibrium are valid even when chemical reactions occur within the system. In addition, we find one more restraining equation of the form of Eq. (11-15) or (11-25) for each independent chemical reaction.

Since the criteria of phase equilibrium were used previously to develop the phase rule, Eq. (10-7), we need only modify the rule to account for chemical equilibrium. That is, the variance should be reduced by one for each indepedent chemical reaction:

$$\mathcal{F} = n + 2 - \pi - r \tag{11-47}$$

where r is the number of independent chemical reactions and n the number of chemically distinct species. The phase rule does not imply that any set of $\mathcal{F}$ intensive variables can be used to describe completely all other intensive variables. It does imply that there are certain sets of $\mathcal{F}$ intensive variables that can be used to describe completely all other intensive variables. Consider, for example, a two-phase system containing five species in which one isomerization reaction occurs between species 1 and 2. For this system, $\mathcal{F} = 5 + 2 - 2 - 1 = 4$. If one of the phases present were an ideal vapor phase, the chemical equilibrium expression,

$$K_a = K_y = \frac{y_1}{y_2} \tag{11-48}$$

can be interpreted as a function of y_1, y_2, and T. Thus, only two of these three intensive variables are independent, hence a set of four intensive variables necessary to satisfy the phase rule must contain at least two variables other than y_1, y_2, and T.[12]

Since Postulate I is valid for systems in which chemical reactions occur, $n + 2$ independently variable properties are still required to specify the intensity and extent of the system. Thus, it follows that a minimum of $\pi + r$ extensive variables are necessary in addition to the maximum of $n + 2 - \pi - r$ intensive variables. As illustrated in the following problem, there are special cases in which some intensive variables are set by specifying certain extensive variables. In such cases, the dependent intensive variables cannot be used in the set of $n + 2$ that are necessary to describe completely the system.

[12]If the phase containing the reactant species were not ideal, Eq. (11-48) would contain a K_ϕ term which, in theory, is a function of all y_j, and hence the restriction on y_1, y_2, and T would appear to be removed. If, however, the nonideality were not large, we might find that astronomical pressures are required to satisfy a set of y_1, y_2, T which was far removed from that necessary to satisfy Eq. (11-48).

Example 11.6

For each of the following cases, determine a set of $n + 2$ independent variables which include $n + 2 - \pi - r$ intensive variables. Consider also the special case in which only the reactants are initially charged and these are fed in the ratio of their stoichiometric coefficients. Assume that the vapor phase is ideal in each case.

(a) A(g) + B(g) = C(s) (A and B not soluble in C)
(b) A(g) + B(g) = C(g) + D(g)

Solution

(a) Since $n = 3$, $\pi = 2$, and $r = 1$, $\mathscr{f} = n + 2 - \pi - r = 3 + 2 - 2 - 1 = 2$. Possible intensive variables are T, P, y_A (species C is present as a pure solid). The equilibrium relationship is

$$K_a = \frac{\hat{f}_C^S/f_C^\circ}{(\hat{f}_A^V/f_A^\circ)(\hat{f}_B^V/f_B^\circ)} = g(T) \tag{11-49}$$

Choose f_C° as the fugacity of pure solid C at T and the vapor pressure of C(s) at T and neglect any effect of pressure on the fugacity of the pure solid. Then, $\hat{f}_C^S/f_C^\circ = 1$. Choosing the standard states of A and B as unit fugacity, $\hat{f}_A^V/f_A^\circ = y_A P$ and $\hat{f}_B^V/f_B^\circ = y_B P$. Thus, Eq. (11-49) becomes

$$K_a = (y_A y_B)^{-1} P^{-2} \tag{11-50}$$

or, simply,

$$g(y_A, T, P) = 0 \tag{11-51}$$

Thus, any two of the three intensive variables can be used in conjunction with Eq. (11-51) to determine the third intensive variable.

To determine the extent of the system, we must specify $\pi + r = 3$ extensive variables. Let us choose these as N_{A_0}, N_{B_0}, and N_{C_0}. Defining ξ as the extent of reaction [see Eq. (11-18)], Eq. (11-50) can be written as

$$y_A y_B = \frac{N_{A_0} - \xi}{N_{A_0} + N_{B_0} - 2\xi}\left(\frac{N_{B_0} - \xi}{N_{A_0} + N_{B_0} - 2\xi}\right) = K_a^{-1} P^{-2} \tag{11-52}$$

If T and P were the two intensive variables specified, Eq. (11-52) could be used to determine ξ and, thence, y_A and y_B. Thus, T, P, N_{A_0}, N_{B_0}, and N_{C_0} would be an independent set of $n + 2$ variables.

For the special case in which A and B are fed in the stoichiometric ratio of 1 : 1, $N_{A_0} = N_{B_0}$ and Eq. (11-52) reduces to

$$\tfrac{1}{4} = K_a^{-1} P^{-2} \tag{11-53}$$

Clearly, both T and P could not be considered independent. Only one can be selected as independent since Eq. (11-53) can then be used to calculate the other. Since $\nu_A = \nu_B$, y_A and y_B are fixed at 0.5 by the stoichiometric feed condition. Thus, the mole fraction is an intensive variable which is set by the stoichiometric feed condition. Only one other intensive variable can be set independently for this special case, and that intensive variable must be either T or P. A set of $n + 2$ variables would then have to include three extensive variables in addition to T or P (e.g., T, y_A, N_{B_0}, N_{C_0}, $\underline{V}$).

(b) Since $n = 4$, $\pi = 1$, $r = 1$, $\mathcal{f} = n + 2 - \pi - r = 4 + 2 - 1 - 1 = 4$. One set of intensive variables is T, P, y_A, y_B, y_C. Choosing the standard-state fugacities as unity, the equilibrium relationship becomes

$$K_a = \frac{y_C y_D}{y_A y_B} \tag{11-54}$$

or, simply,

$$g(T, y_A, y_B, y_C) = 0 \tag{11-55}$$

Thus, any three variables from the set T, y_A, y_B, y_C are independent, and three of these, in addition to P, can be used to specify the intensity of the system. Let us assume that we are given T, P, y_A, y_B. To define the extent of the system, we require $\pi + r = 2$ extensive variables. Let us choose these as N_{A_0} and N_{B_0}.

Given $T, P, y_A, y_B, N_{A_0}, N_{B_0}$, we could first calculate K_a from T and then solve for y_C using (11-54) in the form

$$\frac{y_C(1 - y_A - y_B - y_C)}{y_A y_B} = K_a \tag{11-56}$$

Knowing y_C, we can solve next for y_D by difference. The total moles initially fed, N_0, and the extent of reaction at equilibrium, ξ, can be determined by solving simultaneously the two equations for A and B:

$$y_A = \frac{N_{A_0} - \xi}{N_0} \tag{11-57}$$

$$y_B = \frac{N_{B_0} - \xi}{N_0} \tag{11-58}$$

The initial moles of C and D can then be found from the following equations:

$$y_C = \frac{N_{C_0} + \xi}{N_0} \tag{11-59}$$

or

$$N_{C_0} = y_C N_0 - \xi \tag{11-60}$$

and

$$N_{D_0} = N_0 - N_{A_0} - N_{B_0} - N_{C_0} \tag{11-61}$$

For the special case in which A and B are fed in the stoichiometric ratio of 1 : 1, $N_{A_0} = N_{B_0}$, and $N_{C_0} = N_{D_0} = 0$, and Eq. (11-56) reduces to

$$\frac{(0.5 - y_A)^2}{y_A^2} = K_a \tag{11-62}$$

Thus, y_A and T cannot be considered as independent. We are free to choose y_A and P or T and P, but not y_A, T, and P. A set of $n + 2$ variables would then have to include four extensive variables in addition to y_A, P or T, P. Such a set could be $N_{A_0}, N_{B_0}, N_{C_0}, N_{D_0}, y_A, P$.

In Example 11.6, the special cases of stoichiometric feeds are sometimes cited as examples of cases that do not obey the phase rule, as stated in Eq. (11-47). If we recall that the phase rule states that there are certain sets of $\mathcal{f}$ intensive variables that describe the intensity of the system, the special cases do not disobey the rule. For example, in case (a) of Example 11.6, when we specify stoichiometric feed conditions, y_A is set and must be included as one of the two degrees of freedom allowed by the phase rule. Similarly, in case (b), the two variables y_B and y_C are set by specifying y_A, P, and the stoichiometric feed condition, and hence y_B and y_C must be included as two of the four degrees allowed by the phase rule.

It should be noted that for a single-phase system, specifying the moles charged plus T and P is always sufficient to determine $\mathscr{f}$ intensive variables. In this case, we need only determine the r values of ξ_r (one for each independent reaction) to calculate the $(n - 1)$ equilibrium compositions. Given T and P, there is one equilibrium relationship for each reaction, or r equations in the r unknown values of ξ.

A final word of caution is in order. When the phase rule is applied to a given system, it is assumed that we know the number of species and phases present and the number of reactions that occur. We can, however, always charge materials to a "black box" and control the temperature and pressure; nature will then decide for us which reactions occur and which species and phases are present at equilibrium. If we try to specify T and P and also n, π, and r, we run the risk of overspecifying the system. In such cases, one or more phases and species will disappear by one or more reactions going to completion. For example, if we charged $CaCO_3$(s), CaO(s), and CO_2(g) to a vessel and set T and P, it is unlikely that the three phases will coexist in chemical equilibrium. In this case, $\mathscr{f} = n + 2 - \pi - r = 3 + 2 - 3 - 1 = 1$. Thus, we can control the temperature or the pressure, but not both, if we are to have three phases present at equilibrium. If we set the temperature at 898°C, the equilibrium pressure of CO_2 is 1 atm. If we set the pressure higher than 1 atm, all the CO_2 will disappear by reaction with CaO; if we set the pressure lower than 1 atm, all the $CaCO_3$ will decompose to CO_2 and CaO. In either case, $\mathscr{f} = 2 + 2 - 2 - 0 = 2$, which is in agreement with the fact that we have set T and P independently.[13]

Example 11.7

Barium sulfide is made by reducing barium sulfate-rich ore with coke in a rotary kiln. It has been proposed that the reduction be carried out by contacting the ore with CO in a fluidized bed. At the design temperature, the equilibrium constant for the reduction reaction,

$$BaSO_4(s) + 4CO(g) = BaS(s) + 4CO_2(g) \tag{1}$$

was determined experimentally by equilibrating a mixture of CO and CO_2 with a mixture of BaS and $BaSO_4$ solids. The applicability of the experimental results to the industrial process has been questioned for the following reason: the barium sulfate ore contains appreciable amounts of Fe_2O_3 and Fe_3O_4. Since the reaction

$$3Fe_2O_3(s) + CO(g) = 2Fe_3O_4(s) + CO_2(g) \tag{2}$$

is known to proceed at the temperature under consideration, it is suggested that the gas phase in equilibrium with $BaSO_4$–BaS will be different from that in equilibrium with the ore ($BaSO_4$–BaS–Fe_2O_3–Fe_3O_4).

[13]Note that whereas reactions in a heterogeneous system can go to completion, reactions in a homogeneous phase can never be complete.

(a) How many intensive variables can be set independently for an equilibrium system containing $BaSO_4$(s), BaS(s), Fe_2O_3(s), Fe_3O_4(s), CO(g), CO_2(g)? Assume that no solid solutions are formed.

(b) If the following initial system was maintained at a fixed temperature and 1 bar pressure, what would be the final equilibrium composition?

1 mole ore (70 mole% $BaSO_4$, 5 mole% Fe_2O_3, 25 mole% Fe_3O_4)

0.5 mole CO (no CO_2 initially present)

Assume that at this temperature $K_{a_1} = 10^8$, $K_{a_2} = 20$, and the vapor phase is an ideal-gas mixture.

Solution

(a) There are six species, five phases, and two independent reactions; thus, $\mathcal{f} = 1$. In this particular case, however, the two equilibria cannot be satisfied simultaneously because

$$\frac{y_{CO_2}}{y_{CO}} = (K_{a_1})^{1/4} = K_{a_2}$$

Since $(K_{a_1})^{1/4}$ is 100 and K_{a_2} is 20, at equilibrium, one of the solids must disappear, leaving five species, four phases, and only one reaction. Thus, $\mathcal{f} = 2$.

(b) Since the temperature and pressure are set, we must determine which solid phase disappears. The initial (y_{CO_2}/y_{CO}) ratio is zero; as Fe_2O_3 and $BaSO_4$ are reduced, (y_{CO_2}/y_{CO}) increases. When this ratio reaches 20, reaction (2) is satisfied, but reaction (1) is not. Thus, the ratio will lie between 20 and 100 while Fe_3O_4 is oxidized to Fe_2O_3 and $BaSO_4$ is reduced to BaS. The cycle will continue until Fe_3O_4 or $BaSO_4$ is depleted. The first to vanish will be the limiting reactant for the overall reaction

$$BaSO_4(s) + 8Fe_3O_4(s) = BaS(s) + 12Fe_2O_3(s)$$

Since N_{j_0}/ν_i is 0.7 for $BaSO_4$ and 0.031 for Fe_3O_4, the limiting reactant is Fe_3O_4 and it will disappear. Thus, at equilibrium, we will have $0.05 + (0.25)(3)/(2) = 0.425$ mole of Fe_2O_3. From K_{a_1},

$$\frac{y_{CO_2}}{y_{CO}} = \frac{1 - y_{CO}}{y_{CO}} = 100$$

or $y_{CO} = 0.01$, $y_{CO_2} = 0.99$. Thus, $N_{CO} = (0.01)(0.5) = 0.005$ and $N_{CO_2} = 0.495$. Originally, by an oxygen balance,

$$4N_{BaSO_4} + 3N_{Fe_2O_3} + 4N_{Fe_3O_4} + N_{CO}$$
$$= (4)(0.7) + (3)(0.05) + (4)(0.25) + 0.5 = 4.45$$

At equilibrium,

$$4.45 = 4N_{BaSO_4} + 3N_{Fe_2O_3} + N_{CO} + 2N_{CO_2}$$
$$= 4N_{BaSO_4} + (3)(0.425) + 0.005 + (2)(0.495)$$

or

$$N_{BaSO_4} = 0.545 \text{ mole} \quad \text{and} \quad N_{BaS} = 0.7 - 0.545 = 0.155 \text{ mole}$$

Since we have found that reactions (1) and (2) cannot be satisfied simultaneously, we might question if they are truly independent. If we apply the criteria of Section 11.4 to the system of $BaSO_4$(s), BaS(s), Fe_2O_3(s), Fe_3O_4(s), CO(g), and CO_2(g), we would indeed find that there are two independent reactions. Applying the phase rule to this system, we find that it is univariant. Thus, if we specify the pressure, we can determine the temperature for which $(K_{a_1})^{1/4} = K_{a_2}$. Alternatively, if we specify only the temperature, we should be able to find a pressure and gas-phase composition such that

$$K_{a_1} = \frac{a_{BaS}(a_{CO_2})^4}{a_{BaSO_4}(a_{CO})^4}$$

and

$$K_{a_2} = \frac{(a_{Fe_3O_4})^2 a_{CO_2}}{(a_{Fe_2O_3})^3 a_{CO}}$$

Since activities of the solids are weak functions of pressure, it may be necessary to go to extremely high or low pressures before these expressions are simultaneously satisfied.

11.7 Effect of Chemical Equilibrium on Thermodynamic Properties

In the nonstoichiometric and stoichiometric formulations for systems in chemical equilibrium, we implicitly assumed that the system was at a known temperature and pressure. In many real problems, different properties may be specified to define the equilibrium state. For example, suppose that we had a tubular, adiabatic reactor into which reactants are passed at a steady flow. For the exit mixture, the total enthalpy must be the same as the entering material, that is,

$$\underline{H}(T, P, N_1, \ldots, N_n)_{\text{leaving}} = \underline{H}(T, P, N_1, \ldots, N_n)_{\text{entering}}$$

By specifying chemical equilibrium and one other variable at the reactor exit (e.g., the pressure), the problem is then defined.

In other cases there may be a relationship between the initial (or entering) entropy and that in the final state, as, for example, in an expansion turbine.

Often, but not always, it is convenient to estimate the final temperature and pressure and then iterate to obtain appropriate values to conform to given enthalpy or entropy balances. This technique is illustrated for a very simple case in Example 11.8.

Example 11.8

The system NO_2–N_2O_4 is a particularly good one to study the effects of chemical reactions on thermodynamic properties. The stoichiometry is simple, the enthalpy of dimerization of the NO_2 is large, and the kinetics are so rapid that the system may always be assumed to be in chemical equilibrium.

If the reaction is written as

$$N_2O_4(g) = 2NO_2(g)$$

with unit fugacity standard states, then

$$\Delta G^\circ = 5.733 \times 10^4 - 176.7T, \qquad \text{J/mol } N_2O_4$$

The heat capacity of the mixture of N_2O_4 and NO_2, at constant composition (i.e., with no chemical reaction) is essentially independent of composition and is given as

$$C_p(\text{no reaction}) = 4.18 \times 10^{-4}T + 0.7264 \text{ J/g K}$$

Suppose that we choose a base of 1 mole (92 g) of N_2O_4 initially. Then, with Eq. (11-18), $N_{N_2O_4} = 1 - \xi$, $N_{NO_2} = 2\xi$, $N = 1 + \xi$. Let us assume that NO_2–N_2O_4 systems form an ideal-gas mixture. Then $\hat{f}_{N_2O_4} = y_{N_2O_4}P = (1 - \xi)P/(1 + \xi)$, $\hat{f}_{NO_2} = y_{NO_2}P = 2\xi P/(1 + \xi)$. Since $f^\circ_{N_2O_4} = f^\circ_{NO_2} = 1$ bar, then, with Eqs. (11-25) and (11-26),

$$K_a = \frac{4\xi^2 P}{1 - \xi^2} = \exp\left(\frac{-\Delta G^\circ}{RT}\right)$$

$$\xi = \left(\frac{K_a}{K_a + 4P}\right)^{1/2}$$

For example, if $T = 400$ K, and $P = 10$ bar, $\Delta G^\circ = -1.335 \times 10^4$ J/mol N_2O_4, $K_a = \exp[-(-1.335 \times 10^4)/8.314 \times 400] = 55.39$ and $\xi = [55.39/(55.39 + 4 \times 10)]^{1/2} = 0.762$. Under these conditions, the mole fraction $NO_2 = (2)(0.762)/(1 + 0.762) = 0.865$.

(a) Derive expressions to calculate $\Delta \underline{H}$ and $\Delta \underline{S}$ for a mixture of NO_2 and N_2O_4 between T_1, P_1 and T_2, P_2. Assume a base of 1 mole of N_2O_4 (or 92.0 g of mixture).

(b) Suppose that one had a N_2O_4–NO_2 mixture at 400 K and 10 bar, and this mixture was expanded across an insulated valve to 1 bar. Estimate the exit temperature.

(c) A N_2O_4–NO_2 mixture at 400 K and 10 bar is expanded across a reversible, adiabatic turbine to 1 bar. How much work is produced per mole of NO_2(92 g of mixture)?

(d) Calculate the effective heat capacity $(\partial \underline{H}/\partial T)_P$ of a gas mixture of N_2O_4 and NO_2 in chemical equilibrium between 290 and 370 K at 1 bar.

Solution

(a) At a given T and P, the total enthalpy of the N_2O_4–NO_2 gas mixture may be written as

$$\underline{H} = \bar{H}_{N_2O_4} N_{N_2O_4} + \bar{H}_{NO_2} N_{NO_2}$$

The partial molar enthalpy of, say, $\bar{H}_{N_2O_4}$ can be expanded as follows:

$$\bar{H}_{N_2O_4}(T, P, y) = \Delta H^{EX}_{N_2O_4}(T, P, y) + \overline{\Delta H}^{ID}_{N_2O_4}(T, P, y) + \int_{P^\circ}^{P} \left(\frac{\partial H}{\partial P}\right)_T dP \Bigg|_{N_2O_4} + H^\circ_{N_2O_4}(T, P^\circ)$$

where P° is the standard-state pressure, 1 bar. This expansion can be greatly simplified since $\overline{\Delta H}^{ID}_{N_2O_4} = 0$ and, if the mixture is ideal, $\overline{\Delta H}^{EX}_{N_2O_4} = 0$. Also, if the effect of pressure on the enthalpy of N_2O_4 between the system pressure P and the standard state pressure (1 bar) is negligible, then

$$\bar{H}_{N_2O_4} = H^\circ_{N_2O_4}(T, P^\circ)$$

$\bar{H}_{NO_2}$ can be obtained in a similar manner. With $N_{N_2O_4} = 1 - \xi$ and $N_{NO_2} = 2\xi$,

$$\underline{H} = \xi(2H^\circ_{NO_2} - H^\circ_{N_2O_4}) + H^\circ_{N_2O_4}$$

Then, since $(2H^\circ_{NO_2} - H^\circ_{N_2O_4}) = \Delta H^\circ_r$,

$$\underline{H} = \xi \Delta H^\circ_r + H^\circ_{N_2O_4}$$

The enthalpy difference between two states T_1, P_1 (with $\xi = \xi_1$) and T_2, P_2 ($\xi = \xi_2$), is, assuming that $\Delta H_r^\circ \neq f(T)$,

$$\underline{H}_2 - \underline{H}_1 = \Delta H_r^\circ(\xi_2 - \xi_1) + \int_{T_1}^{T_2} C_{p_{N_2O_4}}^\circ \, dT \tag{A}$$

A similar formulation may be obtained for $\underline{S}$. Here, however, even for an ideal-gas mixture, $\overline{\Delta S}_{N_2O_4}^{ID}$ is not zero but

$$\overline{\Delta S}_{N_2O_4}^{ID} = -R \ln y_{N_2O_4} = -R \ln \frac{1-\xi}{1+\xi}$$

Also, for N_2O_4 treated as an ideal gas,

$$\int_{P^\circ}^{P} \left(\frac{\partial S}{\partial P}\right)_T dP = -R \ln \frac{P}{P^\circ}$$

Then,

$$\underline{S} = \bar{S}_{N_2O_4} N_{N_2O_4} + \bar{S}_{NO_2} N_{NO_2} = \bar{S}_{N_2O_4}(1-\xi) + \bar{S}_{NO_2}(2\xi)$$

$$= \left[\overline{\Delta S}_{N_2O_4}^{EX} - R \ln \frac{1-\xi}{1+\xi} - R \ln P + S_{N_2O_4}^\circ(T, P^\circ)\right](1-\xi)$$

$$+ \left[\overline{\Delta S}_{NO_2}^{EX} - R \ln \frac{2\xi}{1+\xi} - R \ln P + S_{NO_2}^\circ(T, P^\circ)\right]2\xi$$

Assuming that $\overline{\Delta S}_{N_2O_4}^{EX} = \overline{\Delta S}_{NO_2}^{EX} = 0$,

$$\underline{S} = -R\xi\left(\ln \frac{4\xi^2 P}{1-\xi^2}\right) + \xi \, \Delta S_r^\circ - R \ln \frac{1-\xi}{1+\xi} - R \ln P + S_{N_2O_4}^\circ(T, P^\circ)$$

The first two terms may be simplified to give

$$\underline{S} = \frac{\xi \, \Delta H_r^\circ}{T} - R \ln \left[\frac{P(1-\xi)}{1+\xi}\right] + S_{N_2O_4}^\circ(T, P^\circ)$$

where ΔH_r is the standard enthalpy change of reaction. Then,

$$\underline{S}_2 - \underline{S}_1 = \Delta H_r^\circ\left(\frac{\xi_2}{T_2} - \frac{\xi_1}{T_1}\right) - R \ln \left[\frac{P_2}{P_1}\left(\frac{1-\xi_2}{1+\xi_2}\right)\frac{1+\xi_1}{1-\xi_1}\right] + \int_{T_1}^{T_2} C_{p_{N_2O_4}}^\circ \frac{dT}{T} \tag{B}$$

If no chemical reaction occurred, then $\xi_2 = \xi_1$ and

$$\underline{S}_2 - \underline{S}_1 = \int_{T_1}^{T_2} C_{p_{N_2O_4}}^\circ \frac{dT}{T} - R \ln \frac{P_2}{P_1}$$

as expected.

(b) We use Eq. (A). ΔH_r° is estimated from $\ln K_a = -\Delta G^\circ/RT = -(5.733 \times 10^4/RT) + (176.7/R)$ and $[d \ln K_a/d(1/T)] = -\Delta H_r^\circ/R = -5.733 \times 10^4/R$ or $\Delta H_r^\circ = 5.733 \times 10^4$ J/mol N_2O_4 reacted. Also, at the start of this example, ξ was calculated at 400 K, 10 bar to be 0.762. Then, across a cracked valve, $\Delta \underline{H} = 0$:

$$0 = 5.733 \times 10^4(\xi_2 - 0.762) + 92\left[\int_{400}^{T_2} (4.18 \times 10^{-4}T + 0.7264) \, dT\right]$$

Since $\xi_2 = f(T_2, P_2)$, then, by iteration,

$$T_2 = 360 \text{ K}, \qquad P_2 = 1 \text{ bar}, \qquad \xi_2 = 0.819$$

In passing through the valve, N_2O_4 reacted to form more NO_2, and the temperature dropped since the reaction is endothermic. If no reaction had occurred, the exit temperature would have been the same as that at the inlet since ideal-gas behavior was assumed.

(c) Across the turbine $\Delta \underline{S} = 0$ and we will use Eq. (B) to determine T_2 ($P_2 = 1$ bar). With $T_1 = 400$ K, $P_1 = 10$ bar, $\xi_1 = 0.762$, and $\Delta H_r^\circ = 5.733 \times 10^4$ J/mol N_2O_4,

$$0 = 5.733 \times 10^4\left(\frac{\xi_2}{T_2} - \frac{0.762}{400}\right) - 8.314 \ln\left[\frac{1}{10}\left(\frac{1-\xi_2}{1+\xi_2}\right)\frac{1.762}{0.238}\right] + 92\int_{400}^{T_2} (4.18 \times 10^{-4}T + 0.7264)\frac{dT}{T}$$

$$T_2 = 340.5 \text{ K}, \qquad P_2 = 1 \text{ bar}, \qquad \xi_2 = 0.636$$

In this case, while passing through the turbine, NO_2 reacted to form more N_2O_4 even as the system pressure dropped. With these $T_1, P_1, \xi_1, T_2, P_2, \xi_2$ values and Eq. (A),

$$\Delta \underline{H} = 5.733 \times 10^4(0.636 - 0.762) + 92\int_{400}^{340.5} (4.18 \times 10^{-4}T + 0.7264)\, dT$$
$$= -1.12 \times 10^4 \text{ J/mol } N_2O_4$$

Thus for each 92 g of mixture fed to the reversible turbine, 1.12×10^4 J of work could be obtained.

If no reaction had occurred, $T_2 = 315.5$ K and $\Delta \underline{H} = -6.8 \times 10^3$ J/mol N_2O_4. Thus, even though the outlet temperature is smaller in the "no-reaction" case, less work is obtained per mole (92 g) of N_2O_4–NO_2 mixture than in the chemical equilibrium case.

(d) In this case we can use Eq. (A); differentiating with respect to T, we obtain

$$\left(\frac{\partial \underline{H}}{\partial T}\right)_P = \Delta H_r^\circ\left(\frac{d\xi}{dT}\right)_P + C_{p_{N_2O_4}}^\circ$$

But since $K_a = 4\xi^2P/(1-\xi^2)$ and $d \ln K_a/d(1/T) = -\Delta H_r^\circ/R$, then

$$\left(\frac{d\xi}{dT}\right)_P = \frac{\xi(1-\xi^2)\,\Delta H_r^\circ}{2RT^2}$$

so

$$\left(\frac{\partial \underline{H}}{\partial T}\right)_P = \frac{\xi(1-\xi^2)(\Delta H_r^\circ)^2}{2RT^2} + C_{p_{N_2O_4}}^\circ$$

Basis: 1 mole N_2O_4 or 92 g of Mixture

T (K)	$C_{p_{N_2O_4}}^\circ$ (J/mol N_2O_4 K)	$\frac{\xi(1-\xi^2)(\Delta H_r^\circ)^2}{2RT^2}$ (J/mol N_2O_4 K)	$\left(\frac{\partial \underline{H}}{\partial T}\right)_P$ (J/mol N_2O_4 K)
290	78.0	158	236
300	78.4	433	511
310	78.8	549	628
320	79.1	645	724
330	79.5	686	766
340	79.9	649	729
350	80.3	545	625
360	80.7	411	492
370	81.1	287	368

It can be seen from this example that the heat capacity of a reacting gas mixture can be significantly affected by the chemical reaction in those temperature regions where ξ is a strong function of temperature.

In the more general case, where there may be many independent reactions, the simple techniques developed in Example 11.8 would not be practicable, and a sensitivity matrix is developed in conjunction with the computer algorithm.[14] Such a matrix is a convenient method to determine the effects on the equilibrium solution by changing some input parameter—or the specified path. Le Châtelier's principle, discussed in Section 11.8, addresses the same question of the system sensitivity, although usually only single reaction systems are treated.

11.8 Le Châtelier's Principle in Chemical Equilibria

If a system in thermal, mechanical, and chemical equilibrium were perturbed from the equilibrium state, changes would usually occur within the system to reestablish a new equilibrium state. For example, if the system temperature were suddenly increased at constant pressure and extent of reaction, the system would no longer be in chemical equilibrium. If the system were then isolated from the environment, reaction would occur in such a way as to decrease the temperature (i.e., the system would adjust itself to reduce the effect of the initial perturbation). This simple example can easily be proved from Eq. (11-43), irrespective of whether the reaction is exothermic or endothermic. Almost all equilibrium systems behave in this manner for all kinds of perturbations, and the general rule that defines such behavior is called the Le Châtelier principle. One of the simplest statements of this principle is given by Maass and Steacie:[15] "If an attempt is made to change the pressure, temperature, or concentration of a system in equilibrium, then the equilibrium will shift, if possible, in such a manner as to diminish the magnitude of the alteration in the factor that was varied."

A system that follows Le Châtelier's principle is said to moderate with respect to the perturbation. It is interesting to examine the behavior of a chemically reacting system in light of this principle. The simple method proposed by Katz[16] will be used first. For those interested, other treatments are available in Prigogine and Defay and in Callen.[17]

[14] W. R. Smith, *Can. J. Chem. Eng.*, **47**, 95 (1969).

[15] Otto Maass and E. W. R. Steacie, *Introduction to the Principles of Physical Chemistry* New York: Wiley, 1939).

[16] Lewis Katz, "A Systematic Way to Avoid Le Châtelier's Principle in Chemical Reactions," *J. Chem. Ed.*, **38**, 375 (1961).

[17] I. Prigogine and R. Defay, *Chemical Thermodynamics*, trans., D. H. Everett (New York: Longmans, Green, 1954), Chap. 17; H. B. Callen, *Thermodynamics* (New York: Wiley, 1960), Chap. 8.

To develop the Katz technique, the mathematics are considerably simplified if only ideal-gas mixtures are considered and standard states are limited to those corresponding to a pure ideal gas at 1 bar. A brief discussion is presented later to indicate the extension to other cases.

Consider the general reaction

$$aA(g) + bB(g) = cC(g) + dD(g)$$

Define the following ratios:

$$Q_p \equiv \frac{p_C^c p_D^d}{p_A^a p_B^b} \tag{11-63}$$

$$Q_N \equiv \frac{N_C^c N_D^d}{N_A^a N_B^b} = \frac{Q_p N^\nu}{P^\nu} \tag{11-64}$$

$$Q_y \equiv \frac{y_C^c y_D^d}{y_A^a y_B^b} = \frac{Q_N}{N^\nu} = \frac{Q_p}{P^\nu} \tag{11-65}$$

where $\nu = c + d - a - b$. At equilibrium, $K_p = Q_p$, $K_y = Q_y$, etc., but the Q definitions are applicable whether or not equilibrium has been established.

If the reaction proceeds to the right, Q_N increases. We can easily prove this by taking the logarithm of Eq. (11-64) and differentiating with respect to the extent of reaction, ξ. The result is

$$\frac{dQ_N}{d\xi} = Q_N\left(\frac{c^2}{N_C} + \frac{d^2}{N_D} + \frac{a^2}{N_A} + \frac{b^2}{N_B}\right) \tag{11-66}$$

Since $dQ_N/d\xi$ is clearly positive, Q_N increases as the reaction proceeds to form more C and D. One may also show that Q_y increases with ξ by a similar technique.

The key to the entire treatment is to consider the system in chemical equilibrium; if a perturbation should occur, the equilibrium constants are unaffected (except for temperature changes), but the values of Q may change. The subsequent increase or decrease in Q_N or Q_y indicates the direction of the shift in the reaction. Several examples are treated below. All cases are isothermal.

Addition of an inert gas at constant pressure

$$K_p = \text{constant} = \frac{Q_N P^\nu}{N^\nu}$$

P is a constant and as N increases, Q_N will then increase if $\nu > 0$ and decrease if $\nu < 0$. Thus, reaction occurs to form more C and D if $\nu > 0$ and less C and D if $\nu < 0$.

Addition of an inert gas at constant volume

$$K_p = \text{constant} = \frac{Q_N P^\nu}{N^\nu} = Q_N\left(\frac{RT}{\underline{V}}\right)^\nu$$

Since T and $\underline{V}$ are constant, Q_N does not change and there is no change in the moles of reactants and products present at equilibrium.

Variation of system pressure

$$K_p = \text{constant} = Q_y P^\nu$$

For an increase in P, the reaction shifts to the right if $\nu < 0$ and to the left if $\nu > 0$. Similar, but opposite, conclusions hold if P is decreased.

Addition of a reactant or product at constant volume

$$K_p = \text{constant} = Q_N\left(\frac{RT}{V}\right)^\nu$$

If C or D were added, Q_N increases initially, but since all other terms are constant, reaction must take place to the left to decrease Q_N to its original value. Similarly, reaction occurs to the right if A or B were to be added.

Addition of a reactant or product at constant pressure

This case requires a little more consideration. Since $Q_p = Q_y P^\nu$, and P is constant, a change in Q_p indicates a corresponding change in Q_y.

If the addition of one of the components leads to an increase in the ratio denoted by Q_p, then reaction must occur in such a way to lower Q_p to its original value since, at equilibrium, $K_p = Q_p$. Now

$$Q_p = \frac{P^\nu Q_N}{N^\nu}$$

If component j is added, with all the moles of other components constant:

$$\left(\frac{\partial Q_p}{\partial N_j}\right)_{T,P,N_i[j]} = \frac{P^\nu}{N^{2\nu}}\left[N^\nu\left(\frac{\partial Q_N}{\partial N_j}\right)_{T,P,N_i[j]} - (Q_N)\left(\frac{\partial N^\nu}{\partial N_j}\right)_{T,P,N_i[j]}\right]$$

But

$$Q_N = \prod_{j=1}^{k} N_j^{\nu_j}; \quad \left(\frac{\partial Q_N}{\partial N_j}\right)_{T,P,N_i[j]} = \frac{\nu_j Q_N}{N_j}; \quad \text{and} \quad \left(\frac{\partial N^\nu}{\partial N_j}\right)_{T,P,N_i[j]} = \nu N^{(\nu-1)}$$

Substituting, we obtain

$$\left(\frac{\partial \ln Q_p}{\partial N_j}\right)_{T,P,N_i[j]} = \frac{1}{N_j}(\nu_j - y_j \nu) \tag{11-67}$$

From Eq. (11-67), Q_p increases with the addition of N_j provided that $(\nu_j - y_j\nu) > 0$. In this case, *reaction* then occurs to decrease Q_p to its original value. For example, consider the reaction,

$$2NH_3 = N_2 + 3H_2, \qquad \nu = 2$$

For ammonia, the term $[-2 - y_{NH_3}(2)]$ is certainly never greater than zero and Q_p decreases with addition of NH_3. Reaction then occurs to increase Q_p (i.e., the amounts of N_2 and H_2 will increase because of ammonia decomposition). For hydrogen addition, it is easily shown that reaction will occur to form more ammonia.

For nitrogen, $\nu_{N_2} - y_{N_2}\nu = 1 - 2y_{N_2}$. This term is greater than zero if $y_{N_2} < 0.5$, but less than zero if $y_{N_2} > 0.5$. For the case $y_{N_2} < 0.5$, Q_p increases with the addition of nitrogen, so reaction will occur to reduce Q_p (i.e., N_2 and H_2 react to form more NH_3). If, however, $y_{N_2} > 0.5$, Q_p decreases with nitrogen addition and additional NH_3 must decompose to reestablish Q_p at its original value. Such behavior as noted in the last case is often referred to as a case in which the system does not "moderate" with respect to the variation. It is interesting to note that, although the ammonia system may not moderate with respect to the addition of nitrogen (i.e., react in such a way that the effect of the nitrogen addition is decreased), the final concentration of nitrogen, expressed, say, as mole/liter, is still less than the original nitrogen concentration before addition. Thus, to this extent, the system moderates even with respect to nitrogen.

Another example involves the methanol synthesis reaction:

$$CO(g) + 2H_2(g) = CH_3OH(g)$$

We can define the following mole numbers:

	Initial moles	Moles in an equilibrium state
CO	A	$A - x$
H_2	B	$B - 2x$
CH_3OH	0	x
		$A + B - 2x$

$$K_p = \frac{p_{CH_3OH}}{p_{CO}p_{H_2}^2} = \frac{y_{CH_3OH}}{P^2 y_{CO} y_{H_2}^2}$$

$$= \frac{(x)(A + B - 2x)^2}{(A - x)(B - 2x)^2 P^2}$$

At about 500°C, $K_p \sim 10^{-6}$; unless the pressure is high, the amount of methanol formed is small. To illustrate the effect of adding H_2 or CO, assume a low pressure of 1 bar. The equilibrium constant expression may then be approximated as

$$\frac{x}{K_p} \sim \frac{AB^2}{(A + B)^2}$$

Consider two cases; in the first, start with 1 mole of CO and allow chemical equilibrium to be attained with various amounts of hydrogen. As shown in Figure 11.4, the amount of CH_3OH present at equilibrium continually increases with an increase in hydrogen. This is what is predicted, since

$$\nu_{H_2} - y_{H_2}\nu = -2 - (-2)y_{H_2} < 0$$

for all values of y_{H_2} up to unity. From Eq. (11-67), Q_p decreases and reaction occurs to raise Q_p to its original value. For the second case, start with 1 mole

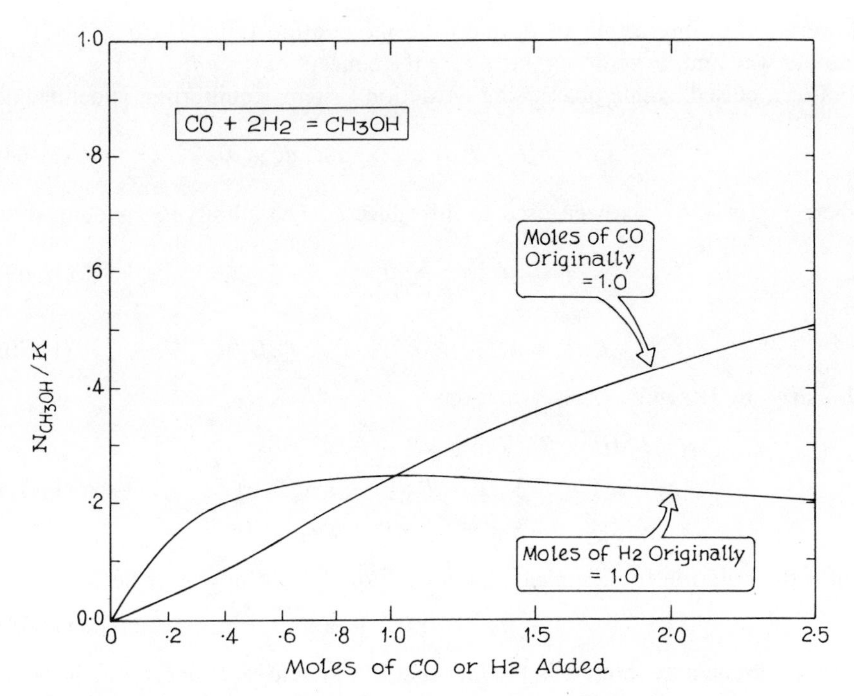

Figure 11.4 Variation of methanol produced with the addition of CO or H_2.

of H_2 and add different quantities of CO. The value of x/K_p increases as CO is first added, but then decreases if more than 1 mole is added. Again we note that

$$\nu_{CO} - y_{CO}\nu = -1 - (-2)y_{CO} = 2y_{CO} - 1$$

If $y_{CO} < 0.5$, from Eq. (11-67), Q_p decreases with CO addition and subsequent reaction occurs to form more CH_3OH and bring Q_p to its original level. If $y_{CO} > 0.5$ the opposite effect occurs, and the addition of CO leads to a decrease in the yield of CH_3OH. The maximum yield of CH_3OH occurs when 1 mole of CO has been added to 1 mole of H_2.

In the more general sense, moderation may be examined from the point of view of starting with a system in equilibrium. The system is perturbed by changing one of the variables. We are interested in following the behavior of the system back to an equilibrium state (which may be different from the original state). If, in the change from the nonequilibrium state to the equilibrium state, there is a corresponding change in the perturbing variable opposite to that which originally caused the upset, we speak of the system as moderating with respect to that variable. We have already seen, for example, in the ammonia–hydrogen–nitrogen equilibria, if the system were perturbed at constant temperature and pressure by adding hydrogen, the system will react such that more ammonia is formed. This change reduces the amount of hydrogen and the system moderates. Many other examples could be cited. Instead of considering many in detail, however, it is interesting to develop the moderation criteria, Eq. (11-67), in another way and, in so doing, introduce the concept of the *affinity*, Y.

For a closed, single-phase, single-reaction system, equilibrium is defined as

$$T\,d\underline{S} = d\underline{U} + P\,d\underline{V} - \sum_{j=1}^{n} \mu_j \nu_j\,d\xi \leq 0 \tag{11-68}$$

where Eq. (11-16) has been used to introduce ξ. The affinity, Y, is defined as

$$Y \equiv -\sum_{j=1}^{n} \mu_j \nu_j \tag{11-69}$$

Thus,

$$T\,d\underline{S} = d\underline{U} + P\,d\underline{V} + Y\,d\xi \leq 0 \tag{11-70}$$

Similarly, by Legendre transformations,

$$\begin{aligned} d\underline{H} &= T\,d\underline{S} + \underline{V}\,dP - Y\,d\xi \geq 0 \\ d\underline{A} &= -\underline{S}\,dT - P\,d\underline{V} - Y\,d\xi \geq 0 \\ d\underline{G} &= -\underline{S}\,dT + \underline{V}\,dP - Y\,d\xi \geq 0 \end{aligned} \tag{11-71}$$

Thus, the criterion for chemical equilibria, Eq. (11-15), now may be written as

$$Y = 0 \tag{11-72}$$

Of more immediate interest is the fact that if a system were *not* in equilibrium, then, regardless of what sets of variables (P, T), $(\underline{U}, \underline{V})$, $(\underline{S}, \underline{V})$, or $(T, \underline{V})$ are kept constant, reaction must occur so that

$$Y\,d\xi > 0 \qquad \text{(real process)} \tag{11-73}$$

or, for simultaneous reactions,

$$\sum_{m=1}^{r} Y^{(m)}\,d\xi^{(m)} > 0 \tag{11-74}$$

Presumably, some reaction could take place, so that viewed singly, Eq. (11-73) is violated, but for the entire system Eq. (11-74) still holds. Considering only a single reaction, a generalized reaction rate R may be defined as

$$R = \frac{d\xi}{dt} \tag{11-75}$$

so that for any actual chemical reaction,

$$YR > 0 \tag{11-76}$$

Returning to the original problem, we start with a system in chemical equilibria and perturb it by addition of a small amount of one of the reactive components. We are interested in the behavior of the perturbed system as it approaches the new equilibrium state. We hold the pressure and temperature constant. Let i be the added component. The change in affinity is

$$Y \sim Y_0 + \left(\frac{\partial Y}{\partial N_i}\right)_{T,P,N_j[i]} \delta N_i \tag{11-77}$$

for small values of δN_i. Equation (11-72) requires that $Y_0 = 0$ because the derivative is evaluated at the original equilibrium state. Then from Eq. (11-76) and (11-77),

$$\left[\left(\frac{\partial Y}{\partial N_i}\right)_{T,P,N_j[i]} \delta N_i\right]\left(\frac{1}{\nu_i}\frac{dN_i}{dt}\right) > 0 \tag{11-78}$$

where δN_i is the amount of i added (positive in this case) and dN_i/dt is the response of the system to the addition. If $dN_i/dt < 0$, the system moderates with respect to an addition of N_i. For dN_i/dt to be negative, however, and still satisfy the inequality (11-78),

$$\frac{1}{\nu_i}\left(\frac{\partial Y}{\partial N_i}\right)_{T,P,N_j[i]} < 0 \qquad \text{(for moderation)} \tag{11-79}$$

To explore the consequences of Eq. (11-79), from Eq. (11-69)

$$\begin{aligned}\frac{1}{\nu_i}\left(\frac{\partial Y}{\partial N_i}\right)_{T,P,N_j[i]} &= -\frac{1}{\nu_i}\sum_{j=1}^{n}\nu_j\left(\frac{\partial \mu_j}{\partial N_i}\right)_{T,P,N_j[i]} \\ &= -\frac{1}{\nu_i}\sum_{j\neq i}\nu_j\left(\frac{\partial \mu_j}{\partial N_i}\right)_{T,P,N_j[i]} - \left(\frac{\partial \mu_i}{\partial N_i}\right)_{T,P,N_j[i]}\end{aligned}$$

But

$$\left(\frac{\partial \mu_i}{\partial N_i}\right)_{T,P,N_j[i]} = -\frac{1}{N_i}\sum_{j\neq i}N_j\left(\frac{\partial \mu_j}{\partial N_i}\right)_{T,P,N_j[i]}$$

so

$$\frac{1}{\nu_i}\left(\frac{\partial Y}{\partial N_i}\right)_{T,P,N_j[i]} = -\frac{1}{\nu_i}\sum_{j=1}^{n}N_j\left(\frac{\nu_j}{N_j} - \frac{\nu_i}{N_i}\right)\left(\frac{\partial \mu_j}{\partial N_i}\right)_{T,P,N_j[i]} < 0 \qquad \text{(for moderation)} \tag{11-80}$$

The right-hand side of Eq. (11-80) must be < 0 for the system to moderate with additional N_i. Note that this expression may be greatly simplified for the case of ideal gases. Here

$$\left(\frac{\partial \mu_j}{\partial N_i}\right)_{T,P,N_j[i]} = RT\left(\frac{\partial \ln y_j}{\partial N_i}\right)_{T,P,N_j[i]} \tag{11-81}$$

Equation (11-80) then simplifies to

$$\frac{1}{\nu_i}\left(\frac{\partial Y}{\partial N_i}\right)_{T,P,N_j[i]} = \frac{RT}{N_i}\left(\frac{y_i\nu}{\nu_i} - 1\right) \tag{11-82}$$

Thus, for moderation,

$$\frac{y_i\nu}{\nu_i} - 1 < 0$$

or

$$\frac{y_i\nu - \nu_i}{\nu_i} < 0 \qquad \text{(to moderate)} \tag{11-83}$$

Equation (11-83) is a more compact form than Eq. (11-67), which requires subsequent steps to determine changes in Q_P.

This technique, in which the affinity concept is used to study the effects of system upsets, is very powerful and not limited to ideal gases, although the final results may be of such a form as to hinder immediate physical visualization of the consequences [i.e., Eq. (11-80)].

Example 11.9

Apply Eq. (11-83) to the decomposition reaction

$$NH_2COONH_4(s) = 2NH_3(g) + CO_2(g)$$

Solution

The amount of solid present is unimportant. Thus,

$$\nu_{CO_2} = 1, \qquad \nu_{NH_3} = 2, \qquad \nu = 3$$

For CO_2 addition, to have the reaction proceed in order to form solid,

$$\frac{(3)(y_{CO_2}) - 1}{1} < 0, \qquad y_{CO_2} < \tfrac{1}{3}$$

Similarly, to form more solid by ammonia addition,

$$\frac{(3)(y_{NH_3}) - 2}{2} < 0, \qquad y_{NH_3} < \tfrac{2}{3}$$

PROBLEMS

11.1. A reaction occurs between A and B to form C as follows:

$$A(l) + 3B(g) \rightleftharpoons C(l)$$

where the gas standard state denotes unit fugacity, and the liquid standard states are pure liquid at the vapor pressure corresponding to the reaction temperature.

(a) Set up an expression for the chemical equilibrium constant in terms of the compositions, fugacities of pure components, and activity coefficients. Assume that there are two phases (liquid and gas) at equilibrium and that both phases are ideal solutions. Clearly define all quantities and describe briefly how they may be determined from theory or experiment.

(b) If the chemical reaction equilibrium constant were written so that all standard states were gas at unit fugacity, how would this constant be related to the K calculated in part (a)?

(c) Which of the values of K calculated above would vary more if the system pressure were doubled? Why?

(d) From the phase rule, how many independent variants are there for this system in chemical and phase equilibrium?

11.2. We are interested in studying the reduction of carbon dioxide with hydrogen to form carbon and water over an iron catalyst. This reaction may be of value in future long-range manned NASA missions.

Although the feed gases are pure CO_2 and H_2, the reaction mixture is known to contain, in addition, CO, CH_4, H_2O, and C(solid). The catalyst, although elemental iron initially, may be partially oxidized to FeO or reacted with carbon to form FeC at the reaction conditions.

Data:

	ΔG_f°(298 K) (kJ/mol)	ΔH_f°(298 K) (kJ/mol)	Vapor pressure (298 K)	
CO_2	−394.6	−393.8	[a]	
H_2(g)	0	0	[b]	
C(graphite)	0	0	Not known	
H_2O(g)	−228.7	−242.0	3160 N/m^2	(liquid)
C(diamond)	2.87	2.02×10^{-3}	Not known	

[a]Sublimes.
[b]Above critical temperature.

(a) The temperature and pressure are set and the mole fraction of CO_2 in the $CO_2 \cdot H_2$ feed is varied to study the equilibrium yield. How many independent reactions are there present for this system?

(b) One of the suspected reactions is

$$CO_2(g) + 2H_2(g) = C(s) + 2H_2O(g)$$

What is the value of K for this reaction at 298 K? Feel free to use some or all of the data given above.

(c) Should the reaction in part (b) have been written so that the standard state for water, instead of being a vapor at unit fugacity, was pure liquid at its vapor pressure at 298 K, how would the equilibrium constant be affected? (That is, calculate the new value.) Would the yield of water increase or decrease at equilibrium?

11.3. Two engineers are discussing a particular reaction A = B + 2D and there seems to be a dispute as to whether the yield of products would increase or decrease with temperature. Both agree that an increase in temperature produces less favorable yields if the reaction were exothermic.

Johnny O. has written the reaction as

$$A(g) = B(g) + 2D(g)$$

whereas Lita Q. indicates

$$A(g) = B(l) + 2D(g)$$

At 373 K, heats of formation are as follows:

	ΔH_f°(kJ/mol at 373 K)
A(g)	−146.5
B(g)	−41.9
D(g)	−46.1

At 373 K, the enthalpy of vaporization of B is 29.3 kJ/mol.

Assuming that these heats of formation are relatively insensitive to temperature and that the gas phase is ideal, would an increase in temperature give a larger or smaller yield in the reactor for a feed of 1 mole of A? Discuss.

11.4. We are in the process of designing a sulfuric acid concentrator, and we need accurate vapor–liquid equilibrium data for the H_2SO_4–H_2O system. We have located an article which reports the partial pressures of H_2SO_4 and H_2O as a function of liquid composition at constant temperature. Since the article is rather old, one of our engineers suggested that we determine the consistency of the data by applying the Gibbs–Duhem equation. He used the following equation:

$$N_A d \ln p_A + N_w d \ln p_w \frac{\underline{V}}{RT} dP \tag{A}$$

where N_A and N_w are the total moles (liquid plus vapor) of H_2SO_4 and H_2O, respectively; p is partial pressure, P is total pressure, and $\underline{V}$ is the total volume (liquid plus vapor).

The calculations from the data do not provide consistency, as determined by Eq. (A). However, another one of our engineers claims that Eq. (A) is not applicable because sulfuric acid decomposes in the vapor phase according to the reaction

$$H_2SO_4(g) = H_2O(g) + SO_3(g)$$

She claims that the proper consistency test is

$$N_A d \ln p_A + N_w d \ln p_w + N_T d \ln p_T = \frac{\underline{V}}{RT} dP \tag{B}$$

where subscripts A, w, and T refer to H_2SO_4, H_2O, and SO_3, respectively.

(a) Given that the decomposition reaction occurs, is either one or both of Eqs. (A) and (B) a valid test of consistency? Clearly explain your reasoning.

(b) At the temperature in question, we have calculated a value of K_p $(= p_w p_T / p_A)$ from Gibbs energies of formation. Although the article does not clearly state how the partial pressures were obtained, we assume that they measured the total pressures and the mole fraction of sulfur-containing species in the vapor phase (i.e., $y_A + y_T$). Assuming that we know p_A $[= (y_A + y_T)P]$ and P as functions of liquid composition at constant T, and assuming that K_p is known, derive an expression to test the consistency of the data.

11.5. Zinc oxide is reduced by roasting it with carbon in a closed retort in which the gaseous and liquid products of reaction may be continuously removed. Air is carefully excluded from the retort so that the reaction is free to proceed under its own equilibrium pressure. The actual process involves two reactions:

$$ZnO(s) + CO(g) \rightleftharpoons Zn(g) + CO_2(g) \tag{1}$$

$$C(s) + CO_2(g) \rightleftharpoons 2CO(g) \tag{2}$$

Zinc produced in the reaction may be removed from the retort as either a gas or a liquid or both, depending on the operating temperature and pressure.

Data:

Temperature (K)	$\Delta G^\circ_{1,T}$ (kJ/mol)	$\Delta G^\circ_{2,T}$ (kJ/mol)
1100	58.54	−24.99
1200	47.01	−42.53
1300	35.22	−59.94
1400	23.76	−77.26
1500	12.27	−94.47

Material	Melting point (K)	Boiling point (K)	Density (g/cm³)	Molecular weight
Zn	692	1180	7.2	65.38
ZnO	2248	Unknown	5.6	81.37
C	3823	5100	2.3	12.01

(a) What does the phase rule indicate about the variance of this system under conditions where liquid zinc is produced? Under conditions where it is not produced? *Explain carefully.*
(b) Calculate the equilibrium reaction pressure as a function of the roasting temperature.
(c) Determine the temperature at which the operation must be conducted in order that the products may be withdrawn at a pressure of 1.0 bar.

11.6. Assume that the equations shown below represent chemical reactions. Compute the value of ΔG°_{298} for each. The standard states are as follows: I_2 gas, unit fugacity; I_2 liquid, pure liquid, 1 bar; I_2 solid, pure solid, 1 bar. The densities of liquid and solid iodine may be taken to be 1.03 g/cm³ and the gas iodine to be an ideal gas.
(a) $I_2(s) \rightleftharpoons I_2(g)$
(b) $I_2(s) \rightleftharpoons I_2(l)$
(c) $I_2(l) \rightleftharpoons I_2(g)$
Some vapor-pressure data for iodine are as follows:

T (K)	P_{vp} (N/m²)	T (K)	P_{vp} (N/m²)
223	4.9×10^{-3}	333	5.75×10^{2}
233	2.5×10^{-2}	343	1.096×10^{3}
243	1.07×10^{-1}	353	2.01×10^{3}
253	4.0×10^{-1}	363	3.57×10^{3}
263	1.32	373	6.07×10^{3}
273	3.99	387.4[a]	1.20×10^{4}
283	1.077×10^{1}	393	1.48×10^{4}
293	2.69×10^{1}	403	2.09×10^{4}
303	6.28×10^{1}	423	3.92×10^{4}
313	1.37×10^{2}	433	5.25×10^{4}
323	2.88×10^{2}	457.6[b]	1.013×10^{5}

[a]Melting point.
[b]Normal boiling point.

11.7. An inventor claims to be able to produce diamonds from β-graphite at room temperature by a process involving the application of 37 kbar pressure. In view of the data shown below, are his claims to be taken seriously?

Specific gravity, β-graphite = 2.26

Specific gravity, diamond = 3.51

$$C_{\beta\text{-graphite}} \rightleftharpoons C_{\text{diamond}}, \qquad \Delta G^\circ_{298} = 2870 \text{ J/g-atom}$$

Both solids are incompressible and no solid solutions are formed.

11.8. Gilliland et al.[18] investigated the hydration of ethylene by placing ethylene, water, and sulfuric acid in an agitated pressure vessel kept at constant temperature by a condensing vapor. After analyzing the liquid phase for ethanol until constant composition indicated that equilibrium had been attained, a sample of the vapor phase was withdrawn and analyzed. One run showed the following data:

$$T = 527 \text{ K}, \qquad P = 267.7 \text{ bar}$$

	Mole fraction	
	Liquid phase	Gas phase
Ethyl alcohol	0.084	0.075
Ethylene	0.0197	0.250
H_2O	0.881	0.675
H_2SO_4	0.015	—

(a) Estimate ΔG° at 527 K for this reaction.

(b) The values reported in the literature for absolute entropies, S°, and heats of formation are as follows:

	S°(298 K) (J/mol K)	ΔH_f°(298 K) (kJ/mol)
$C_2H_5OH(g)$	278.0	−237.0
$C_2H_4(g)$	220.6	+52.3
$H_2O(g)$	188.8	−241.14

Compute ΔG°_{527} form these data.

11.9. In a laboratory investigation a high-pressure gas reaction $A \rightleftharpoons 2B$ is being studied in a flow reactor at 473 K and 100 bar. At the end of the reactor the gases are in chemical equilibrium and their composition is desired.

Unfortunately, to make any analytical measurements on this system it is necessary to bleed off a small side stream through a low-pressure conductivity cell operating at 1 bar. It is found that when the side stream passes through the sampling valve, the temperature drops to 373 K, and the conductivity cell gives compositions of $y_A = 0.5$ and $y_B = 0.5$.

Data and allowable assumptions:

Heat of the reaction, $\Delta H = 29.3$ kJ/mol of A reacting, independent of temperature.

Heat capacity: 29.3 J/mol K for B independent of temperature. The gas mixture is ideal at all pressures, temperatures, and compositions. Assume no heat transfer in the sampling line or across the sampling valve.

(a) Calculate the composition of the gas stream before the sampling valve.

(b) Are the gases in chemical equilibrium after the sampling valve? (Show definite proof for your answer.)

[18] E. R. Gilliland, R. C. Gunness, and V. O. Bowles, "Free Energy of Ethylene Hydration," *Ind. Eng. Chem.*, **28**, 370 (1936).

11.10. Brewer, Bromley, Gilles, and Lofgren (*The Transuranic Elements*, NNES, Div. IV, Volume 14B, McGraw-Hill, 1949, p. 861) discuss some interesting experiments with plutonium chloride ($PuCl_3$).

Small vials of solid $PuCl_3$ crystals were heated to 973 K and a mixture of argon and chlorine gas passed through them at 8.0×10^4 Pa. The exit gases were cooled to 298 K in a trap that contained a fine filter. Solid $PuCl_3$ was found on the filter after an experiment.

Four experiments were made. In each, 2000 cm^3 of an argon–chlorine gas mixture was passed through the $PuCl_3$. (The 2000 cm^3 was measured at 8.00×10^4 Pa and 298 K.) The reported data are as follows:

Mole fraction in the sweep gas		Quantity of $PuCl_3$ found in trap (mg)
Chlorine	Argon	
0	1.0	0
0.0625	0.9375	32
0.25	0.75	73
1.0	0	143

There seems to be a disagreement concerning the mechanism of transport of the $PuCl_3$. One group simply states that although $PuCl_3$ is not particularly volatile, it is vaporized in the sweep gas at 973 K and trapped at the low temperature outside the vial. The other group suggests that a chemical reaction is involved and proposes that

$$PuCl_3(s) + \tfrac{1}{2}Cl_2(g) = PuCl_4(g)$$

occurs at 973 K and the $PuCl_4$ vapor is carried into the trap, where it decomposes to $PuCl_3$ and Cl_2. Skepticism has greeted the latter explanation since $PuCl_4$ has never been observed. The atomic weight of plutonium is 239; for chlorine, 35.5, and for argon, 40.

What opinions do you have concerning these two theories? Support your statement with quantitative calculations.

11.11. We do not seem to be able to locate any data for the Gibbs energy of reaction, nor ΔH° for a reaction shown as

$$D(g) = A(g) + B(g)$$

However, one of our research students has conducted a few experiments as described below. A constant-volume pressure vessel was immersed in a constant-temperature bath. The vessel was first evacuated and then filled with pure gaseous D at 298 K and 1 bar. After heating the closed vessel to 473 K, reaction occurred to form A and B. When no further pressure change was noted, the gauge read 3 bar. Further heating to 523 K caused a pressure increase to 3.30 bar.

With these data, can one calculate ΔG°_{523} and ΔH°? If so, provide numerical values. If not, describe what additional data are required. You may assume that the vapor phase forms an ideal-gas mixture, and that ΔH° is not a function of temperature.

11.12. A very efficient plant finds that it has two gas streams, the properties of which are tabulated below. The management would like to extract the maximum work possible from the streams before discarding the products in a collection vessel at 300 K and 1 bar. Rocky Jones, has suggested a black box, which he claims will serve the purpose (see Figure P11.12).

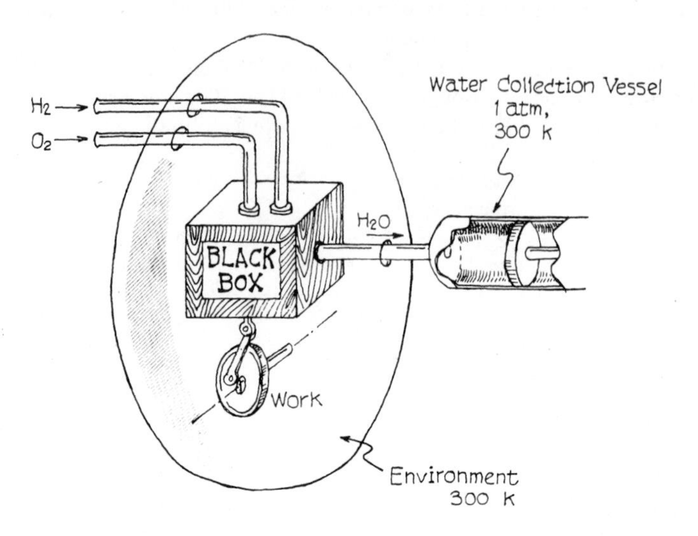

Figure P11.12

	Pure oxygen	Pure hydrogen
Temperature (K)	300	600
Pressure (bar)	2	3
Flow rate (mole/s)	1	2
C_p (J/mol K)	29.3	29.3

Data: $\Delta G_f^\circ(H_2O) = -228.8$ kJ/mol; $\Delta H_f^\circ(H_2O) = -241.9$ kJ/mol at 300 K with products and reactants in the standard states of pure gases, unit fugacity. The vapor pressure of liquid water at 300 K is 0.036 bar.

What is the maximum power obtainable from Rocky's box?

11.13. In our sulfuric acid plant we have a process stream of pure SO_2 gas at 2 bar and 600 K flowing at a rate of 6.4 kg/s. This is to be mixed with a pure oxygen stream at 293 K and 1 bar. The oxygen flow rate is 1.6 kg/s. The SO_2 and O_2 are reacted to form SO_3, which is then separated from any unreacted SO_2 and O_2 (which are recycled) and cooled to 293 K before flowing to the absorbers.

One of our new employees has indicated that we could start with the same reactants and end with the same product SO_3 but, at the same time, obtain significant "free" work from the process. His ideas are a bit sketchy, but I illustrate them in Figure P11.13. He indicates that all heat sinks and sources in the environment are at 293 K.

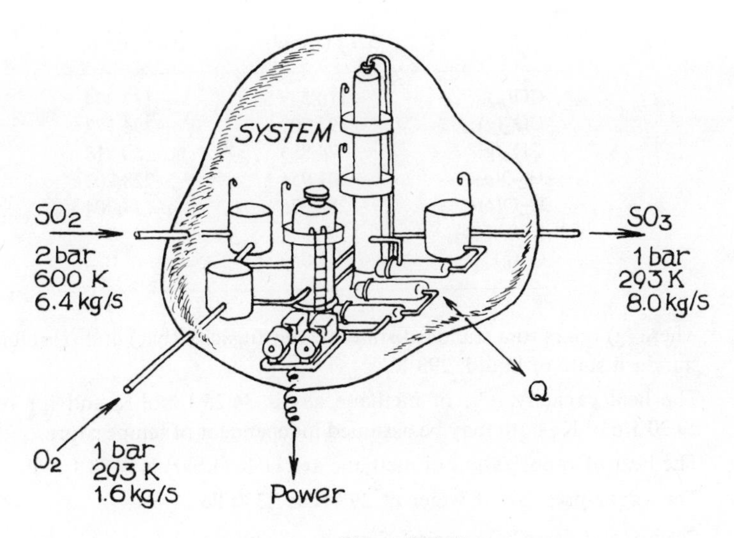

Figure P11.13

Some Gibbs energy and heats of formation are shown below at 293 and 600 K. Also assume that $C_p(SO_2 \text{ and } SO_3) = 41.9$ J/mol K and $C_p(O_2) = 29.3$ J/mol K.

	$\Delta G^\circ_{f,293}$ (kJ/mol)	$\Delta H^\circ_{f,293}$ (kJ/mol)	$\Delta G^\circ_{f,600}$ (kJ/mol)	$\Delta H^\circ_{f,600}$ (kJ/mol)
SO_2	−300.2	−297.1	−302.2	−301.0
O_2	0	0	0	0
SO_3	−412.4	−436.7	−428.3	−483.2

Could you tell me how much power we could produce from his process? Make a clear sketch of your proposed scheme.

11.14. Suppose that you have at your disposal 1 mole of liquefied natural gas (LNG) and a cylinder containing 4 moles of high-pressure oxygen gas.

Data:

Assume that the LNG is pure methane, saturated liquid, at 1 bar pressure. The saturation temperature is 111 K.

The oxygen cylinder is at 150 bar and 298 K.

The environment is the atmosphere at constant temperature and pressure of 298 K and 1 bar. The composition of the atmosphere, on a dry basis, in mole fraction, is

O_2: 0.20
CO_2: 0.00025

with the remainder as nitrogen and inert gases.

The relative humidity of the atmosphere is 100%.

The Gibbs energies of formation and enthalpies of formation of a few compounds are as follows: ($T = 298$ K)

	ΔH_f° (J/mol)	ΔG_f° (J/mol)
CO(g)	−110,577	−137,334
CO_2(g)	−393,702	−394,572
CH_4(g)	−74,883	−50,818
H_2O(g)	−241,951	−228,702
H_2O(*l*)	−285,975	−237,304
C(s)	0	0
O_2(g)	0	0

where (g) refers to a standard state of unit fugacity (bar) and (*l*) refers to a standard state of liquid, 298 K.

The heat capacity, C_p, of methane gas is 34.25 J/mol K and for oxygen 29.30 J/mol K; both may be assumed independent of temperature.

The heat of vaporization of methane at 111 K (1 bar) is 8183 J/mol.

The vapor pressure of water at 298 K is 2330 Pa.

Devise (and describe) a series of processes and calculate the *maximum* work you could obtain from these starting materials.

11.15. It has been suggested that liquid air can serve as the ultimate pollution-free automotive fuel. The liquid would be used as the working fluid in an open-cycle Rankine engine that only has heat interactions with the atmosphere.

The cycle is shown schematically in Figure P11.15. Liquid air in a vented, insulated fuel-storage tank (a) is pumped into a high-pressure vaporizer (b), where it is heated by atmospheric air to ambient temperature (298.2 K). The resulting high-pressure gas is then fed to a multistage expansion engine (c) that drives the vehicle. The engine obviously operates without a condenser and exhausts to the atmosphere (d). The air would probably be reheated between expansion stages by the use of suitable reheat coils (e).

You are invited to *comment* on the theoretical, as well as practical feasibility of the scheme. A useful result would be to estimate the energy cost per kilojoule delivered to the wheels of a small "city"-type automobile, by a liquid-air-fueled engine, and compare this cost with those for gasoline and lead–acid battery operation. The essential question of whether this engine, in drawing heat only from the atmosphere, violates the Second Law, should also be discussed. Other thermodynamic data for liquid air and gasoline can be obtained from standard handbooks.

Efficiency: The efficiency relative to a reversible adiabatic process for high-pressure air motors that are currently commercially available is about 85%.

On the same basis, the efficiency of an automobile (new) gasoline engine is about 12%.

Costs:

Liquid air: ≈ \$0.22/kg

Gasoline: ≈ \$400/m^3

Lead–acid battery: Stores 22.05 W-h/kg

Replacement cost = \$6.6/kg

Lifetime = 300 charge cycles

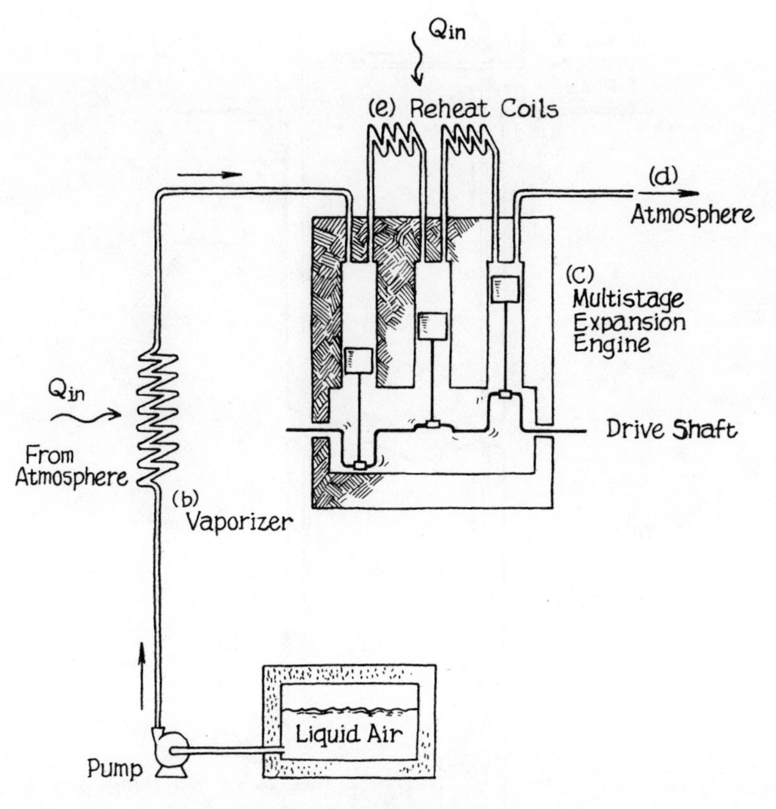

Figure P11.15

Charge efficiency: 1.33 W-h charge to store 1.0 W-h

Cost of charging: 0.27/kW-h

11.16. Whillier (*J. So. African Inst. Min. Met.*, April 1977, p. 183) discusses the problem of supplying cooling water to deep gold mines.

As normally carried out, a water-chilling refrigeration plant is located on the surface to cool water before it is circulated down into the mine (see Figure P11.16). Water leaves this plant at point 1 and flows down to control valve 2. Assume that the flow is adiabatic and frictionless.

At 2, the water is reduced in pressure and flows to an underground storage tank 3, from which it is circulated through the mine and back to the surface. The bottom of the mine where the workers are located and where tank 3 is placed may be assumed to be at the same pressure as at the mine surface—1 bar.

The deepest gold mine in South Africa is about 3.4 km. For this mine, if the water temperature at point 1 is 10°C, what is the temperature in tank 3?

Whillier has suggested that control valve 2 be replaced by a power-producing turbine. If such a turbine were installed and operated at 70% efficiency, what would the temperature in tank 3 be for this case? (Assume that the turbine operates in an adiabatic manner.)

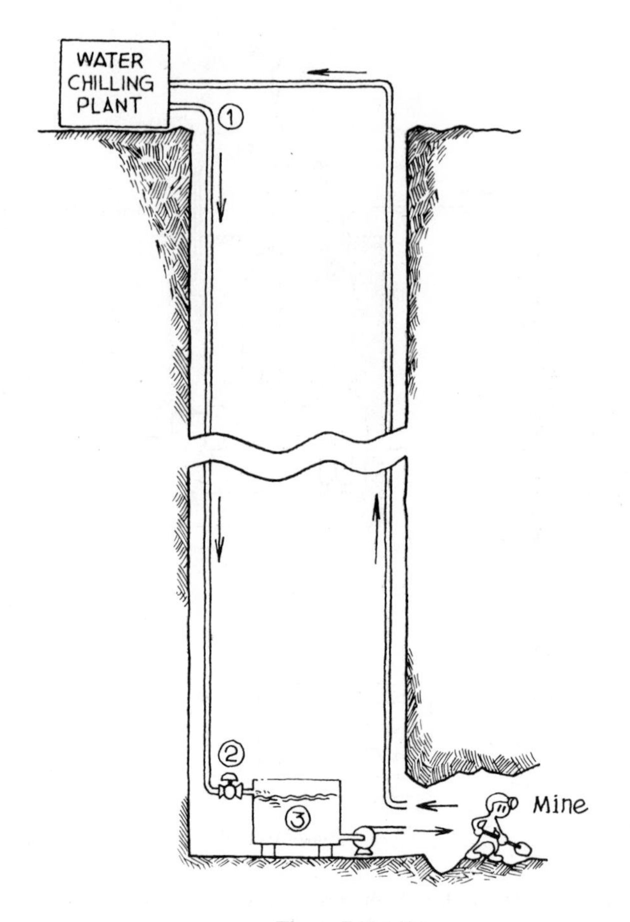

Figure P11.16(a)

Data:

$g = 9.8\ \text{m/s}^2$

C_p(water, liquid) $= 4.19$ kJ/kg K

ρ(water, liquid) $= 1000\ \text{kg/m}^3$ (assume constant)

Neglect heat leaks to piping and tanks

At the time I was writing this problem, I discussed it with one of my closest friends, Rocky Jones. He was only mildly interested in such a prosaic situation, but he did return in a short while with a startling idea related to the problem. It seems, as usual, that he has found the solution to our energy problems. What he proposes is really quite simple. He will dig a deep hole in the earth, and at the bottom he would place a power-producing turbine. He would allow water to flow down a pipe to the bottom of the hole and pass through the turbine. Then he would use the power from the turbine to drive a generator to produce electric power that would be used to electrolyze the water that passed through the turbine. The hydrogen and oxygen produced would fill the hole (except for the water pipe) and, at the surface, these gases would be burned to produce power and form liquid water to be recirculated.

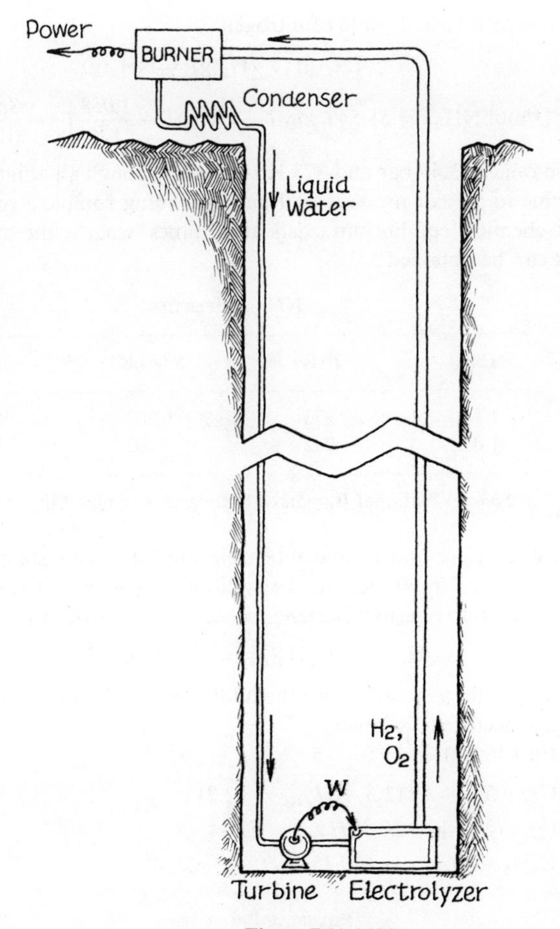

Figure P11.16(b)

Could you let me have your valued judgment on the feasibility of this process?

(*Note:* For water

$$\Delta H^\circ_{f_{298}} = -1.59 \times 10^7 \text{ J/kg}$$

$$\Delta G^\circ_{f_{298}} = -1.32 \times 10^7 \text{ J/kg})$$

11.17. (a) For a general reaction at a given temperature and pressure,

$$aA + bB = cC + dD$$

Suppose that one began initially with pure A and B and desired to vary the initial mole ratio (A/B) to maximize the concentration of C and D in the equilibrium mixture. Show for the case of an ideal-gas mixture that the desired initial ratio $(N_{A_0}/N_{B_0}) = a/b$.

(b) What is the maximum conversion of N_2 to ammonia obtainable in a Haber ammonia unit operating at 300 bar and 773 K when stoichiometric quantities of hydrogen and nitrogen are used?

Use as a basis 1 mole of nitrogen:

$$\tfrac{1}{2}N_2(g) + \tfrac{3}{2}H_2(g) = NH_3(g)$$

$$\Delta G^\circ(\text{J/mol } NH_3) = 51.54T\log T - 51.50T + \frac{1.058}{10^3}T^2 - \frac{3.54}{10^6}T^3 - 38{,}390$$

(c) Ammonia at 3.45 bar and 373 K is passed through an adiabatic reversible turbine to an exit pressure of 1 bar. Assuming complete reversibility and that chemical equilibrium exists at all times, what is the maximum work that can be obtained?

NH_3 PROPERTIES

P (bar)	T (K)	S (J/gK)	H (J/g)
3.45	373	6.202	1686.4
1.01	282	6.202	1493.5

Source: National Bureau of Standards Circular 142.

11.18. One mole of ethylene and 1 mole of benzene are fed to a constant-volume batch reactor and heated to 600 K. On the addition of a Friedel–Crafts catalyst, an equilibrium mixture of ethylbenezene, benzene, and ethylene is formed:

$$C_6H_6(g) + C_2H_4(g) \rightleftharpoons C_6H_5C_2H_5(g)$$

The pressure in the reactor before the addition of the catalyst (i.e., before any reaction has occurred) is 2 bar.

Data (in kJ/mol):

$C_6H_6(l)$: $\Delta G^\circ_{f_{298}} = 113.5$, $\Delta H^\circ_{f_{298}} = 52.21$

$C_2H_4(g)$: $G^\circ_{f_{298}} = 68.20$, $\Delta H^\circ_{f_{298}} = 52.34$

$C_6H_5C_2H_5(g)$: $G^\circ_{f_{298}} = 130.75$, $\Delta H^\circ_{f_{298}} = 29.81$

PHYSICAL PROPERTIES

	Boiling point (°C)	ΔH_{vap} at 1 bar (kJ/mol)	Average C_p (J/mol K)
C_6H_6	80.1	30.77	113
C_2H_4	−103.71	13.52	54
$C_6H_5C_2H_5$	136.19	36.01	167

Calculate the total heat removed by the cooling fluid in a heat exhanger used to maintain the reactor temperature constant at 600 K as the reaction proceeds to equilibrium.

11.19. Hydrogen molecules exist in ortho and para forms. The difference results from the fact the nuclear spins of the individual atoms in a molecule may be either parallel or antiparallel. The equilibrium ratio between the forms is determined almost completely by the temperature. At room temperature and above, the ratio is constant at 25% para and 75% ortho; at liquid hydrogen temperatures (20 K) the ortho form is present in negligible amounts. Equilibrium ratios at other temperatures are shown in the accompanying table. The actual ratios for any hydrogen-gas system are, however, strongly influenced by the rates of transformation. For example, at room temperature and above, the homogeneous rates are reasonably large, but a catalyst is necessary to achieve any appreciable transformation at 20 K.

From an engineering point of view, gaseous hydrogen when cooled and liquefied converts to the lower-energy form (para) in an exothermic reaction, and as cooling and liquefaction rates are fast compared to the conversion rate, freshly liquefied hydrogen differs only slightly from the 25%-*p*, 75%-*o* composition of the feed gas. Transformation to the para form then occurs at low temperatures, and this requires additional low-temperature refrigeration.

A prospective client is proposing to use stored liquid hydrogen (which is available and fully converted to the para form) to cool a stream of hot helium. The hydrogen gas is to be exhausted around 420 K and fed to another part of the system. Although the size of the heat exchanger is not particularly important, one would like to cool the helium stream as much as possible with a given hydrogen flow rate. The client feels that the residence time of the hydrogen in the system is so short that the exit will still exist in a pure para form. It is felt, however, that if a good catalyst could be developed, the endothermic transformation from $p \longrightarrow o$ could be carried out in the heat exchanger and this additional heat effect used to cool the helium to a lower temperature.

At present he wants us to share in the cost to develop a catalyst that will yield an equilibrium gas mixture [of $H_2(o)$ and $H_2(p)$] at all positions in the heat exchanger.

Write a concise memorandum indicating your opinion of the scheme and recommending our course of action.

Temperature (K)	Equilibrium fraction *para*-hydrogen
< 10	1.0
10	0.999999
20	0.99821
30	0.97021
40	0.88727
50	0.77054
60	0.65569
70	0.55991
80	0.48537
90	0.42882
100	0.38620
120	0.32959
150	0.28603
200	0.25974
250	0.25264
300	0.25072
> 300	0.25

11.20. A rigid, well-insulated gas storage tank of 0.3-m^3 capacity is filled originally with hydrogen gas at 1 bar and 30 K. Connected to the tank is a large high-pressure manifold containing hydrogen gas at 100 bar and 40 K. The valve is opened and hydrogen is allowed to flow into the tank until the gas inside attains a temperature of 50 K. Assume ideal gases and that the ortho–para equilibrium shift is infinitely rapid (see Problem 11.19 for data on equilibrium composition). Heat-capacity data for pure *para*-hydrogen indicate that over the temperature range of interest here, $C_p/R \simeq 2.50$. The tank walls and hardware are assumed to have zero heat capacity. Assume that the contents of the tank are well mixed at all times.

(a) What is the pressure when the temperature of the gas inside the tank reaches 50 K?

(b) What would the temperature be if the manifold and tank pressure were increased without limit?

(c) What is the entropy change of the universe for this process?

11.21. Freshly liquefied hydrogen that has not been catalyzed consists of a 3 : 1 ortho–para mixture. On standing, there is a slow shift of the mixture toward the equilibrium concentration, a fact that complicates the problem of storing the liquid for any length of time.

Assume that 1 mole of hydrogen at room temperature and 1 bar is quickly liquefied (in order not to affect the ortho/para ratio) and then allowed to reach an equilibrium state. From the data in Problem 11.19 showing the equilibrium fraction of *para*-hydrogen as a function of temperature, this should occur when the concentration of para is about 99.8%. During this change the system is chosen to be adiabatic and always vented at 1 bar.

Data: Heat of vaporization of hydrogen is assumed independent of ortho/para ratio and equals 454 kJ/kg at 20.4 K, the boiling temperature at 1 bar. There is no appreciable difference between the vapor- and liquid-phase concentration of the two components (i.e., the relative volatility of *ortho*- to *para*-hydrogen is unity over the entire liquid range).

How many moles of liquid hydrogen remain after the final equilibrium state is reached?

11.22. Gaseous nitrogen peroxide consists of a mixture of NO_2 and N_2O_4, and chemical equilibrium between these components is rapidly established. It has been suggested that this gas mixture be employed as a heat transfer medium. To evaluate this proposal, the heat capacity of the equilibrium mixture must be determined as a function of temperature.

Ideal gases may be assumed for this temperature range at 1 bar. $\Delta G^\circ = 57{,}330 - 176.72T(\text{K})$ for the reaction $N_2O_4 = 2NO_2$. Unit fugacities are assumed for the standard states, and ΔG° is in J/mol. C_p for the frozen equilibrium mixture may be approximated as:

$$C_p(\text{frozen equilibrium}) = 4.19 \times 10^{-4}T(\text{K}) + 0.726 \text{ J/gK}$$

Calculate and plot the effective heat capacity $(\partial H/\partial T)_P$ for the equilibrium mixture between 290 and 370 K at 1 bar.

11.23. The system N_2O_4–NO_2 remains in chemical equilibrium under essentially all conditions (i.e., if a step change in pressure or temperature is imposed, it takes only about 0.1 to 0.2 μs for the system to react and attain equilibrium under the new conditions).

In a portion of a plant design, we are faced with estimating the change in temperature of a high-pressure N_2O_4–NO_2 gas mixture as it is throttled across a valve. The valve is well insulated and the flow rate is steady. The upstream temperatures and pressures are 353 K and 5 bar. The upstream mole fractions are $y_{NO_2} = 0.637$ and $y_{N_2O_4} = 0.363$. The downstream pressure is 1 bar.

Available data: The standard Gibbs energy of the reaction and the *frozen* heat capacity are given in Problem 11.22. The molecular weights of N_2O_4 and NO_2 are 92 and 46, respectively. Although not strictly true, the gas mixture may be assumed to be ideal (i.e., the relation $P\underline{V} = NRT$ is applicable and there is no effect of pressure on the enthalpy of the pure components). The fraction N_2O_4 dissociated to NO_2 is shown in Figure P11.23.

What is the downstream temperature and composition?

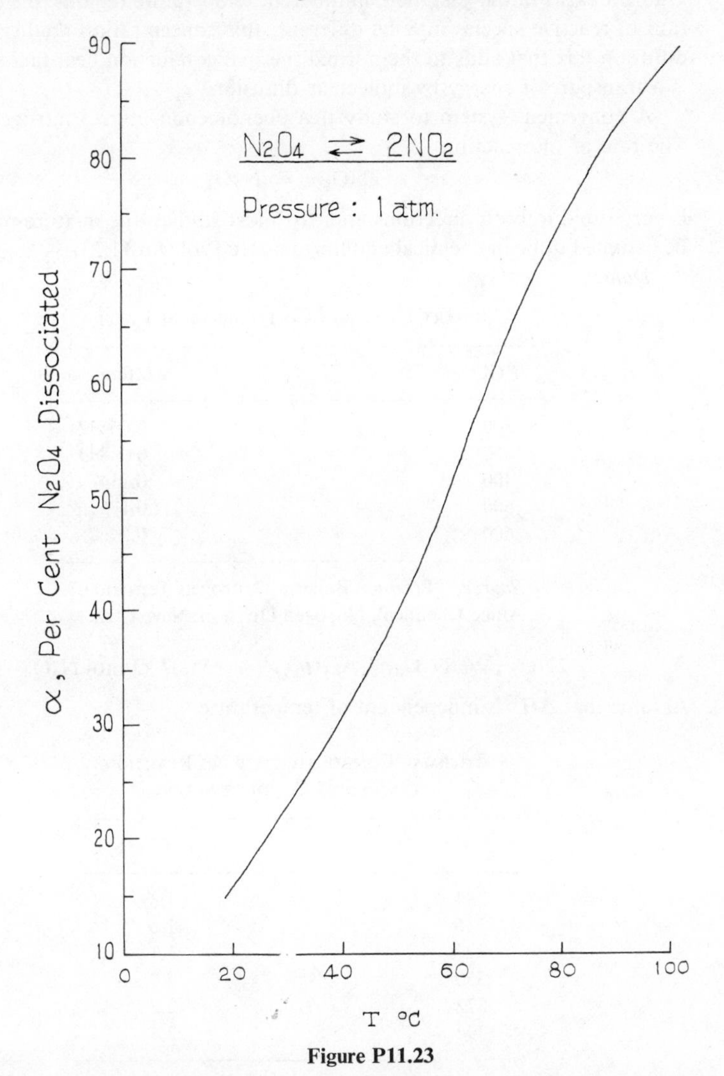

Figure P11.23

11.24. As noted in Problem 11.23, the chemical kinetics are such that gas mixtures of NO_2–N_2O_4 are always in chemical equilibrium.

Suppose that a mixture of these gases is expanded from a pressure and temperature, P_1, T_1 to a lower pressure P_2 in an adiabatic, reversible turbine.

(a) Write a differential equation that expresses how the pressure, temperature, and fraction N_2O_4 dissociated vary during the expansion. (Assume a basis of 1 mole N_2O_4 and let α be the fraction dissociated.)
(b) Show clearly how the adiabatic, reversible work may be calculated.
(c) Repeat parts (a) and (b) if the turbine were isothermal (at T_1) and reversible. Assume ideal gas mixtures.

11.25. The thermal conductivity of reacting gas mixtures is often found to be much larger than would be expected from molecular considerations. If a temperature gradient exists in the gas, then, in different temperature regions, the concentration of reactive species may be different; this concentration gradient causes a diffusion flux that adds to the normal thermal conduction heat flux since there is a transport of energy by molecular diffusion.

A convenient system to study this phenomenon utilizes nitrogen dioxide. The rate of the reaction

$$2NO_2 \rightleftharpoons N_2O_4$$

is very rapid in both directions and, for most studies, the mixture may always be assumed to be in chemical equilibrium (see Problem 11.23).

Data:

BINARY DIFFUSION COEFFICIENTS (at 1 bar)

T (K)	D (cm^2/s)
300	0.06532
350	0.08843
400	0.1147
500	0.1757
600	0.2467

Source: "Product Bulletin, Nitrogen Tetroxide," Allied Chemical, Nitrogen Division, New York.

$$2NO_2 \rightleftharpoons N_2O_4, \qquad \Delta H^\circ_{298} = -58.07 \text{ kJ/mol } N_2O_4$$

Assume that ΔH° is independent of temperature.

THERMAL CONDUCTIVITY IF NO REACTION ("FROZEN" CONDUCTIVITY)

T (K)	W/m K $\times 10^2$
294	1.3
316	1.6
327	1.7
350	2.0
372	2.2
394	2.4

Equilibrium data are shown in Figure P11.23.

Derive an expression that shows the additional contribution to thermal conductivity because of the diffusion flux for this system. Plot this pifference as a function of temperature. Also plot the temperature gradient between two plates 1 cm apart if the top plate is at 400 K and the bottom plate at 300 K and there is a mixture of NO_2 and N_2O_4 between the plates. The pressure is 1 bar. What is the heat flux between the plates?

11.26. One mole of H_2 and 1 mole of I_2 are placed in a constant-volume container at 473 K, the total pressure being 1 bar. Since a catalyst is present, the following reaction is always in equilibrium:

$$\tfrac{1}{2}H_2(g) + \tfrac{1}{2}I_2(s) = HI(g)$$

$$\Delta G^\circ_{298} = 1323 \text{ J}$$

$$\Delta H^\circ_{298} = 24.74 \text{ kJ, independent of temperature}$$

The vapor pressure of HI from 222.4 K (melting point) to 237.7 K (normal boiling point) is

$$\log_{10} P_{vp} = 9.755 - \frac{1129}{T}$$

where P_{vp} is in N/m^2 and T is in K. The vapor pressure of solid HI is 80.40 N/m^2 at 184.3 K. The vapor pressures of solid and liquid iodine are given in Problem 11.6.

Make a plot of the total pressure in the container as a function of the temperature as the temperature in the box is reduced from 473 K to 223 K. Assume that I_2 is insoluble in liquid HI. Also plot the partial pressure of each component as a function of temperature.

11.27. When one is studying the properties of a "pure" substance in vapor–liquid equilibria, many simple equations, such as the Clausius–Clapeyron equation, may be derived. A little thought about this kind of analysis reveals, however, that the situation is more complex. For example, take liquid HF. It undoubtedly exists in rather complex equilibria between monomeric HF and polymers of the form $(HF)_x$; also vapor HF is known to consist of a chemical equilibrium mixture of HF and $(HF)_6$ with traces of other polymers. Other examples might include acetic acid, alcohols, or the classic N_2O_4–NO_2 case wherein the liquid and vapor phases contain both N_2O_4 and NO_2 in phase and chemical equilibrium. Finally, one might even cite the case of water; in the liquid phase we are reasonably certain that the molecules are not completely monomeric in nature; in the vapor, there have been theories advanced to allow for the presence of $(H_2O)_x$. With these thoughts in mind, the question is again raised about results obtained from analyses that consider only the monomer.

To demonstrate your ability to handle problems of this sort, consider a situation in which there is an equilibrium of the form

$$nA \rightleftharpoons B$$

This chemical equilibrium relation holds for both liquid and vapor phases; also, phase equilibria is attained. Examples of this relation might be:

B	A	n
N_2O_4	NO_2	2
$(HF)_6$	HF	6
$(CH_3COOH)_2$	CH_3COOH	2

Choose a base of 1 gram-formula weight of A in each phase to simplify the analysis. Let α^v be the number of moles of monomer reacting in the vapor phase to form α^v/n moles of polymer. Thus, in the vapor phase there are $(1 - \alpha^v)$ moles of monomer and α^v/n moles of polymer. Similarly, let α^L be the comparable parameter in the liquid phase.

(a) Suppose that you could measure experimentally the vapor pressure of this two-phase A–B system as a function of temperature. What would the slope $(dP/dT)_{\text{saturation}}$ represent? Make a rigorous analysis and express your results in the usual thermodynamic nomenclature, such as partial molar quantities, α^v, α^L, etc. Also describe in words what your answer means.

(b) If the vapor pressure data were plotted as $\ln P$ against $1/T$, the "apparent latent heat" can be calculated as

$$\Delta H(\text{apparent}) = -R\frac{d\ln P}{d(1/T)}$$

How would ΔH(apparent) be related to the enthalpy term found in part (a)? In these calculations assume that liquid volumes are negligible compared to vapor volumes and that the vapor phase behaves as an ideal-gas mixture.

(c) For the case $6HF \rightleftharpoons (HF)_6$, vapor-pressure data, plotted as $\ln P$ versus $1/T$ gives an apparent ΔH of 24.70 kJ/20.1 g HF at 273.2 K [R. L. Jarry and Wallace Davis, Jr., "The Vapor Pressure, Association, and Heat of Vaporization of Hydrogen Fluoride," *J. Phys. Chem.*, **57**, 600 (1953)]. Also, a calorimetric enthalpy of vaporization gives 7.03 kJ/20.1 g HF at the same temperature [Karl Fredenhagen, "Physikalischchemische Messungen am Fluorwasserstoff," *Z. Anorg. Chem.*, **210**, 210 (1933)]. Estimate the mole fractions of HF and $(HF)_6$ in the vapor phase at 273.2 K.

11.28. It is well known that vapors of very polar gases are often quite nonideal even at low pressures. One reason for this behavior is that polymerization undoubtedly occurs and the effective number of moles is less than the apparent value based on the mass of gas and the molecular weight of the monomer.

Suppose that we attempt to account for the "apparent" nonideality by allowing a monomer W to undergo the following dimerization:

$$2W \rightleftharpoons W_2$$

We further assume that:

1. No higher polymerization reactions occur.
2. W and W_2 behave as ideal gases in the usual sense and their mixture is an ideal gas mixture.
3. The mixture is always in chemical equilibrium with

$$K \equiv \frac{p_{w_2}}{p_w^2}$$

Let α be the fraction W which associates.

(a) Derive an equation to relate the apparent compressibility factor Z to α, where $Z \equiv P\underline{V}/NRT$. N is the mass of W in volume $\underline{V}$ divided by the molecular weight of monomer.

(b) Derive an equation relating the apparent compressibility factor to the system temperature and pressure if $\Delta G^\circ = \Delta H^\circ - T\Delta S^\circ$ and unit fugacity standard states are employed. (Assume that ΔH° and ΔS° are not functions of temperature.)

(c) Let W be monomeric acetic acid. For acetic acid:

$$\Delta H^\circ = -58.62 \text{ kJ/mol of dimer}$$

$$\Delta S^\circ = -138.2 \text{ J/mol K}$$

At 391 K and 1 bar, what fraction of the acetic acid would you predict would be in the form of dimers? What would be the apparent compressibility factor?

11.29. Some recent experiments have been carried out to determine the equilibrium partial pressure of oxygen over molten potassium oxides. The data were taken in the following way. Samples of pure KO_2 were placed in a MgO boat in an evacuated tube. The tube was inserted into an oven at a sufficiently high temperature that the oxide melted. The pressure of the evolved oxygen was measured after equilibrium was attained. From this pressure measurement, the tube volume, and the oven temperature, the moles of oxygen evolved could be ascertained. From this value and the original sample weight, the atomic O/K ratio of the oxide liquid could be calculated. Next, a known amount of oxygen was bled out of the system, and the system allowed to come to equilibrium.

Again the oxygen pressure was measured, and the liquid O/K ratio calculated. The data indicated that at any given temperature level the oxygen partial pressure depends only on the liquid O/K ratio. The data are as follows:

	Partial pressure of oxygen (bar)		
O/K ratio	773 K	873 K	923 K
1.0	0.018	0.071	0.13
1.1	0.10	0.13	0.21
1.2	0.26	0.26	0.31
1.3	0.44	0.41	0.44
1.4	0.62	0.56	0.61

(a) Are these data consistent with the phase rule? Demonstrate. There is essentially nothing known about the structure of the liquid phase. It is black, probably has a high electrical conductivity, and contains some or all of the following species: K^+, O_2^-, K_2O, K_2O_2, KO_2.

(b) Estimate accurately the heat evolved or absorbed if a reaction occurs so that the liquid absorbs oxygen isothermally at 873 K as it changes from an O/K ratio of 1.0 to 1.4. Express your answer on a basis of 1 g-atom of potassium in the liquid.

(c) A "simple" picture of the liquid shows it to be an "ideal" mixture of liquid K_2O, K_2O_2, and KO_2. Demonstrate how you would calculate equilibrium constants for the reactions given below, using only the p_{O_2} and O/K values measured experimentally.

$$2KO_2(l) \longrightarrow K_2O_2(l) + O_2(g) \qquad (1)$$

$$2KO_2(l) \longrightarrow K_2O(l) + \tfrac{3}{2}O_2(g) \qquad (2)$$

(d) How could you test the hypothesis in part (c) to see if it were reasonable?

(e) Many thermodynamicists would describe the system as follows:

$$\tfrac{1}{2}O_2(g) \rightleftharpoons [O(l)]$$

where [O(*l*)] represents the oxygen in the liquid phase.

$$K = \frac{[O(l)]}{p_{O_2}^{1/2}}$$

By defining an activity coefficient such that

$$[O(l)] = (\gamma)(\text{conc. } O_2 \text{ in liquid})$$

with $\gamma \longrightarrow 1.0$ as concentration $\longrightarrow 0$. Show how the activity coefficient and equilibrium constant may be determined from the data at any given temperature and composition. Assume for simplicity that you have at your disposal the partial pressures of oxygen over the entire range of O/K ratios from 0 to 2.

11.30. In recovery furnaces where kraft paper liquor is burned, evidence has been accumulating that the sulfur emission is far less than might be expected from the sulfur present initially in the liquor. This result, of course, is very desirable, but to optimize furnace operation, some ideas regarding the mechanism of such a sulfur trapping would be helpful.

The liquor is a very complex mixture of carbonaceous material and a number of inorganic salts (both solids and liquids) through which air is blown to evaporate water and to burn the carbon (and other combustibles). Sulfur in the original liquor ends up as various salts (e.g., Na_2S), a few ppm are found in the flue gas as SO_2 and SO_3. Surprisingly, a significant amount of S also appears as Na_2SO_4 in the stack electrostatic precipitator.

Several of the current theories hypothesize a general mechanism which allows a decomposition of sodium carbonate in the furnace. Ideas differ as to the reaction products, but the net result seems to be the formation of reactive sodium species in the furnace gases. These sodium species then can trap SO_2/SO_3 as sulfites or sulfates and end as precipitates in the flue.

Prepare a brief memorandum (with suitable appendices for documentation) to indicate your opinion of the sodium carbonate decomposition mechanism. To allow a more closely defined and simpler case, let us assume that we have, initially, an evacuated, cold bed of Na_2CO_3 and carbon. We plan to heat this bed to temperatures typical of those reported for kraft liquor furnaces (i.e., between 1000 and 1400 K). Then we will measure the pressure and gas composition. To aid in the design of our experiment, estimate what pressures and gas compositions you would expect in this temperature range. The JANAF thermochemical tables are a good source of data.

11.31. (a) Following are some interesting articles clipped from *The New York Times*.[19] Please read them carefully and prepare a well-written, brief memorandum which we can submit to our Director of Research. She needs this analysis to advise our Board of Directors whether to purchase an interest in the invention.

(b) Mr. Rocky Jones has also approached us to suggest methods to produce hydrogen from water. As usual, he is very indefinite, but I was able to obtain at least one specific proposal that is described in Figure P11.31(a).

In our West Bacon plant, we have a situation where we are now venting (wastefully!) 3 kg of steam each and every second. This steam is piped from the Deleted, proprietary data reactor at 1000 K and 39.2 bar. Mr. Jones proposes to run this steam into his unit and vent saturated steam at 1 bar. A portion of

[19]©1977 by The New York Times Company. Reprinted by permission.

Experts Dispute Contention That Device Can Extract Cheap Energy From Water

By ROBERT LINDSEY
Special to The New York Times

LOS ANGELES, March 28—A machine that its developers contend can extract cheap energy from water has been tested here and has touched off a flurry of speculation on Wall Street, a Federal investigation of possible stock manipulation, and virtually unanimous skepticism from scientists who say the machine cannot do what its inventors say it can.

Its inventors say the device separates tap water, in a continuous, self-sustaining reaction with virtually no outside energy, into its two component parts—oxygen and hydrogen, a fuel that can be used to heat homes or power automobiles without pollution.

Scientists who were questioned say such a machine is theoretically impossible, but developers insist recent tests have shown its feasibility. Rumors of the invention have caused wild trading in the stock of one company associated with the device, as well as an investigation by the Securities and Exchange Commission.

Last week, officials of two commercial testing laboratories here said that, in preliminary reports, they had examined and tested the device and that, as the inventors contend, it produced combustible hydrogen as well as oxygen from water, for periods of 19 and 30 minutes.

Yesterday, a reporter watched as water, apparently from the municipal supply system of a suburban community here, was piped into the prototype device, a stainless steel box somewhat larger than a big trunk. A few knobs were turned, and a jet of flaming gas shot out of a tube and kept burning until the machine was turned off about 15 minutes later.

It was impossible to determine if flammable gas or other energy sources had been concealed in the box, or whether the demonstration was a hoax in any other way. However, Sam Leach, the machine's 61-year-old inventor, said that, except for electricity necessary to initiate the process, the energy from the reaction had come from the water itself.

Specialists Skeptical

"The water is being split into hydrogen and oxygen," he asserted. "The reaction is self-sustaining."

When the few details known about the process were explained to a half dozen of the country's leading specialists in hydrogen energy technology, everyone said the device could not operate as the inventors contended. Several called it a fraud.

The dream of liberating the energy of hydrogen in every drop of water has excited scientists for at least a century and has accelerated in the last three years. A discovery of a cheap hydrogen-oxygen separation system would obviously have enormous impact on world economics, industry, and the balance of power.

It has long been known that hydrogen and oxygen can be split by applying great amounts of energy. High school chemistry students observed the process in electrolysis, in which an electric current passes through water containing a salt or alkali and separates oxygen and hydrogen.

Nuclear reactors have also been used to do the job. Every method requires the use of far more energy to obtain hydrogen than that derived from it.

Mr. Leach, a well-to-do, reclusive Southern California inventor who says he has more than 70 patents, accused other researchers of "tunnel vision" in failing to apply generally known principles to the problem that he says he has solved.

"As described to me, the system violates the principles of thermodynamics," said Dr. Bernard M. Abraham of the Argonne National Laboratory, Argonne, Ill. "You can't get something for nothing; there's no way they can do it."

"You can't get anymore energy out of it than you put into it," said Dr. R. H. Wentorf Jr., a physical chemist in the General Electric Company's research and development center at Schenectady, N.Y. "I'd be very cautious about it; some trick is being used."

Scoffers Were Expected

"Sure, they've got a magic machine—they put something in it called 'dreamium,'" scoffed a scientist at the California Institute of Technology.

"We expect them to say we can't do it," said Morris Mirkin, who founded and then sold the national Budget Rent-a-Car Company, and who with his family owns rights to the technology for all applications except housing. "But they're wrong, and we'll prove it."

The researchers who were interviewed said there were several ways to stage a demonstration that seemed to prove water was being separated into a combustible fuel—for instance, concealing a hidden source of energy within the machine, perhaps a hidden battery or electric lines, or using one of several kinds of metals that could be liberated of inherent hydrogen and oxygen in a relatively short time until exhausted.

However, a spokesman for the two laboratories that checked the machine, the Smith-Emery Company of Los Angeles and the Approved Engineering Test Laboratories of Encino, Calif., said the machine had been disassembled and reassembled in front of their observers and no source of hidden energy had been found.

"I could find no evidence of hanky-panky," said Gordon Walker, who directed the test for Smith-Emery.

Rumors that a machine had been invented that might provide an almost inexhaustible amount of cheap energy have circulated in the financial community since last summer. They have caused extensive speculation and volatile ups and downs in the price of stock of the Presley Companies, a Newport Beach, Calif., homebuilding concern that has been hit hard by the national slump in construction.

The value of its stock has soared from about $2 last summer to 20¾ on the American Stock Exchange, until the Securities and Exchange Commission suspended trading on Thursday until April 3.

According to government informants, the commission is investigating dealing in this stock, alleged failure to disclose certain information about the project, and other possible violations of Federal securities laws. An S.E.C. investigator told a reporter today, "You'll be hearing more about this from us."

According to Mr. Leach, hydrogen and oxygen are separated in the machine by a process that uses two stainless steel chambers. Each contains a granular metal called a "reactant," not otherwise indentified. The reactant acts much like a sponge to absorb oxygen.

Started by Electricity

To start the process, a small electric charge is said to be applied to heat the reactant and vaporize a flow of water into steam. The steam is said to pass over the heated reactant in one chamber, and the reactant grabs, or "sequesters," the oxygen while letting hydrogen pass through for any intended use.

At some point, the reactant "sponge" becomes filled with oxygen, Mr. Leach said, and it has to be emptied for the next cycle.

This is done, he continued, by heating the chamber in such a way that the oxygen is removed, and the reactant is regenerated. The hydrogen-producing reaction is said to take place at about 850 to 900 degrees Fahrenheit, and the oxygen-releasing reaction at 1,000 to 1,200 degrees.

According to Mr. Leach, the energy used to remove the oxygen in the second step is heat generated during the oxygen-steam reaction in the first step in the adjacent chamber. This heat is also used to provide additional steam, he said.

The two-step process—first removing the oxygen, then exhausting it from the saturated reactor—is said to continue indefinitely, as long as more water is added, according to Mr. Leach.

He said he had run his prototype machine, the fourth since the development began, a maximum of two hours, and contended that the limit was not in the fundamental process but in the equipment. Despite the doubts of other scientists, who say they do not see where he could obtain energy for the oxygen-purging stage, Mr. Leach said no additional energy was needed, and that the metal reactant was not consumed, eroded or damaged in the process.

Details not Explained

Mr. Leach will not disclose the specific developments he learned in order to make the reaction work. However, an aide said that patents on the procedure and equipment had been applied for in 86 countries and that at least one would probably disclose details about May 1.

"It's an exothermic reaction that is self-sustaining," Mr. Leach said. "The scientists [who doubt the development] are well aware of it. I did the mathematical modeling, and after two years, before I had worked on any piece of equipment, I knew the inventing was over.

"These are all commonly known; what I've done is bring some things together that other people hadn't thought about doing," he said. "The important thing is we provided an environment for these reactions to occur, the timing and so forth. I know why they're saying it can't work; I can't fault them for that, but they're missing something; thermodynamicists follow certain things blindly, like tunnel vision."

"There'll be additional tests," Mr. Leach said, "to certify that we are in fact splitting water, but I have no doubt at all that we have done it."

Split of Water Elements for Fuel Draws Interest, Along With Scorn

By ROBERT LINDSEY
Special to The New York Times

LOS ANGELES, May 5—Despite continuing skepticism over its effectiveness, a controversial process whose inventor claims can separate water into its component parts—hydrogen and oxygen—at low cost has drawn the interest of at least two major corporations.

The Bechtel Corporation of San Francisco began negotiating with Sam Leslie Leach, the inventor of the process, late last winter, and has indicated that it was interested in purchasing a worldwide exclusive license for the process. Today, the Aluminum Company of America in Pittsburgh acknowledged that it had expressed interest in the process, although no contract negotiations nor any testing of the technique have taken place.

An invention capable of separating the hydrogen from water at low cost, theoretically at least, would have substantial importance in helping the world meet energy needs. Pure hydrogen can be used as a fuel much as gasoline or natural gas.

Inventors have tried to split water for more than a century, much as alchemists sought to turn stone into gold. But every process tested in the past has required more energy to separate the hydrogen and oxygen than could be derived from the process itself. Scientists have derided assertions, including those of Mr. Leach, saying that what he claims to have invented amounts to a "perpetual motion machine" that violates certain basic laws of physics.

Assertion First Made Last Year

Mr. Leach, a well-to-do 62-year-old Californian who has a number of patents in optics and other fields, first made the assertion that he had developed an economic water-splitting process last year. Since then, he said he had repeatedly asked the United States Energy Research and Development Administration to evaluate his process, but he says the Government has not taken him seriously.

No information has been made public to indicate that Mr. Leach has made the technical breakthrough that he claims. Neither Bechtel nor Alcoa has tested the Leach machine, but the inventor has discussed some elements of the scientific principles involved with research specialists from both companies and they apparently have taken his presentations seriously enough to enter into detailed consideration of the process.

Bechtel is a large, diversified construction and engineering concern heavily involved in a wide variety of energy projects. Alcoa is the nation's largest producer in the energy-intensive aluminum industry.

A Bechtel spokesman said today the company would have no comment on the company's interest in the machine.

Technical Experts Sent by Others

A number of other major corporations and representatives of at least two foreign governments, Israel and Brazil, have sent technical experts to a suburban laboratory near here to witness demonstrations of the prototype machine in which tap water is introduced at one end and oxygen and hydrogen—the hydrogen in the form of a flame—is exhausted.

Although Mr. Leach's claims have brought the disapproval and derision of scientists, they have perplexed industry, the Securities and Exchange Commission and Wall Street. After word leaked out about the machine last year, shares of the Presley Companies, a California home builder that has a license for one application of the process, skyrocketed.

Subsequently, the S.E.C. investigated and concluded that, based on his patent applications, there was no evidence to support Mr. Leach's assertions of the breakthrough.

Mr. Leach's response was that in his early 10 applications he had outlined only portions of the technology and that he has since submitted additional ground.

He has hinted obliquely that the process is related to what he says is the discovery of an overlooked source of energy in certain materials that, when processed in a certain way, provides the energy needed to sustain the water-splitting operation.

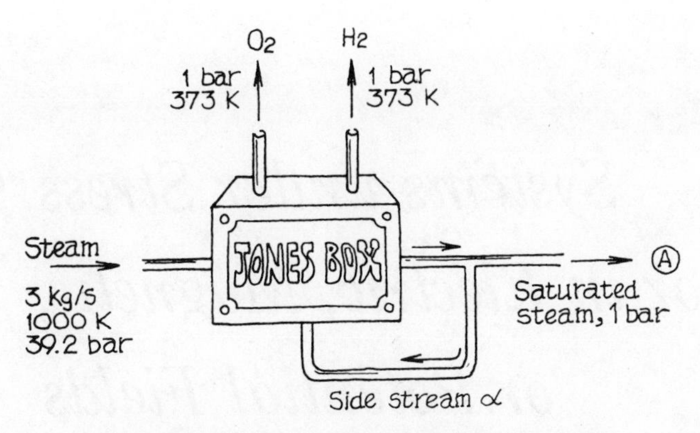

Figure P11.31(a) Box appears to be adiabatic and rigid.

this saturated steam effluent is recycled to the unit and hydrogen and oxygen gases are (somehow) produced in separate streams.

Please analyze this "Jones box" and let me have your considered opinion as to its operability and the amount of hydrogen we might hope to obtain each hour.

Data: Assume ideal gases:

$$C_p(\text{water vapor}) \sim 35 \text{ J/mol K}$$

$$\Delta H^\circ_{f_{373}}(\text{water vapor}) \sim -2.42 \times 10^5 \text{ J/mol}$$

$$\Delta G^\circ_{f_{373}}(\text{water vapor}) \sim -2.254 \times 10^5 \text{ J/mol}$$

$$\Delta H_v(373 \text{ K}) \sim 40{,}000 \text{ J/mol}$$

(c) You know I have a high opinion of Rocky, but I wonder if he hasn't neglected a real opportunity in his simple Jones box. It would seem to me that we could go even further! Could we take his exit steam (saturated vapor, 1 bar) and condense it in another "box" to saturated liquid water at 1 bar and produce even more H_2?! For example, starting from stream *A* in Figure P11.31(a), we would have the result shown in Figure P11.31(b). Please comment on my extension.

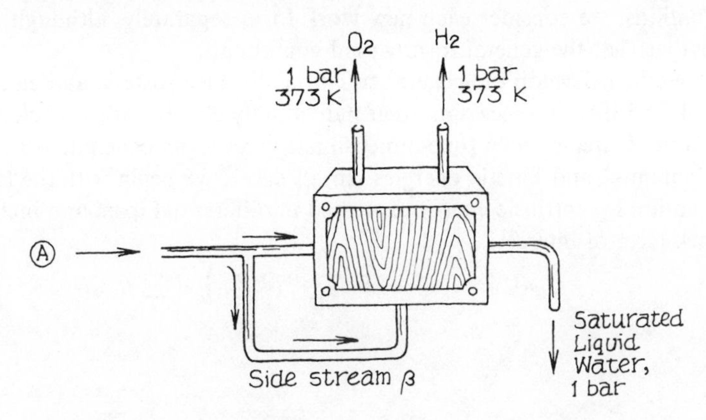

Figure P11.31(b) Again the box (?) appears to be adiabatic and rigid.

Systems under Stress or in Electric, Magnetic, or Potential Fields 12

To this point we have dealt almost entirely with simple systems in which work done on or by the system could be associated with a change in system boundaries (i.e., $P\,d\underline{V}$). With these limitations we expressed $\underline{U}$ as a function of $(\underline{S}, \underline{V}, N_1, \ldots, N_n)$ in a Fundamental Equation, and from this starting point developed useful Legendre transforms, various equilibrium relationships, etc.

There exist other types of work that may also be done on or by a system and it is occasionally necessary to take these into account. For example, if our system were in an electric field, then, by changing the field or by moving the system from one point in the field to another, there is a *work interaction* between the system and the environment. In this chapter we establish Fundamental Equations that include these new variables. To minimize the number of terms in the derived equations, we consider each new work form separately, although it should be obvious that the general form would contain all.

We first develop the general equations for electrostatic and electromagnetic work and then in succession treat individually electrostatic and electromagnetic systems, systems under (one-dimensional) stress, and conclude with a treatment of potential and kinetic energies. In all cases, we begin with the Fundamental Equation for intrinsic energy expressed in differential form and include the new work term of interest:

$$d\underline{U} = T\,d\underline{S} - P\,d\underline{V} - [dW_{\text{rev}}] + \sum_j \mu_j\,dN_j \tag{12-1}$$

12.1 Electrostatic and Electromagnetic Work

In the theory of electrodynamics, Maxwell's equations occupy the same preeminent position reserved for Newton's laws in classical mechanics (i.e., a vast amount of empirical evidence accumulated over the past century has led scientists to believe that all macroscopic electromagnetic phenomena are governed by these equations). They are adopted here as the basis of an expression for electromagnetic work. Following the treatment of Stratton,[1] the final result is a relation for the work *done on the system by external sources to establish the field*:

$$W = \int_{\underline{V}_s} \left[\int_0^{\mathfrak{D}} \boldsymbol{\mathcal{E}} \cdot d\mathfrak{D} + \int_0^{\mathcal{B}} \mathcal{H} \cdot d\mathcal{B} \right] d\underline{V}_s \tag{12-2}$$

where:[2] $\mathcal{E}$ = electric field strength, V/m

$\mathfrak{D}$ = electric flux density, A s/m²

$\mathcal{H}$ = magnetic field strength, A/m

$\mathcal{B}$ = magnetic flux density, V s/m²

The integral is taken over the entire volume subject to the field, $\underline{V}_s$, and includes the free space as well as any volume occupied by the system. $\underline{V}_s$ is a constant, independent of the temperature and pressure. Also, for the work to be given by Eq. (12-2), the electric and magnetic fields must vanish at the boundaries of $\underline{V}_s$.

The electric vectors $\mathcal{E}$ and $\mathfrak{D}$ are related by the permittivity, ξ:

$$\mathfrak{D} = \xi \mathcal{E} \tag{12-3}$$

where in free space $\xi = \xi_0 = (8.854)(10^{-12})$ F/m.[3] For real substances ξ is a function of the material, temperature, pressure, and electric field strength.

In a similar manner $\mathcal{B}$ is related to $\mathcal{H}$ by the permeability, μ:

$$\mathcal{B} = \mu \mathcal{H} \tag{12-4}$$

where in free space $\mu = \mu_0 = (4\pi)(10^{-7})$H/m.[4] For real substances, the permeability, μ, is a function of the material, temperature, pressure, and magnetic field strength.

The constants ξ_0 and μ_0 are related to the velocity of electromagnetic waves in free space by Eq. (12-5):

$$c = (\xi_0 \mu_0)^{-1/2} = (2.9979)(10^8) \text{ m/s} \tag{12-5}$$

In general, ξ and μ are symmetric tensors of rank 2,[5] but as used here, we consider only the cases in which $\mathfrak{D}$ and $\mathcal{E}$ are parallel, as are $\mathcal{B}$ and $\mathcal{H}$. Thus, ξ and μ can be visualized as scalar multipliers. This simplification limits the treat-

[1] J. A. Stratten, *Electromagnetic Theory* (New York: McGraw-Hill, 1941).

[2] Rational mks units are used. Note that in these definitions, A s ≡ coulomb and V s/m² ≡ tesla = weber/m².

[3] F = farad = A s/V.

[4] H = henry = V s/A.

[5] For example,

$$\mathfrak{D}_x = \xi_{xx}\mathcal{E}_x + \xi_{xy}\mathcal{E}_y + \xi_{xz}\mathcal{E}_z, \text{ etc.}$$

ment of electric systems to simple geometries (e.g., to parallel-plate condensers or long uniformly wound solenoids), but the principal concepts and results are not greatly affected.

Several other parameters are commonly used to describe electric systems. For example, the dielectric constant, ϵ, and magnetic susceptibility, χ, are defined as:

$$\epsilon \equiv \frac{\xi}{\xi_0} \tag{12-6}$$

$$\chi \equiv \frac{\mu}{\mu_0} - 1 \tag{12-7}$$

It is also convenient to define two other quantities. We shall refer to $\mathcal{P}$ as the *electric polarization* and $\mathfrak{M}$ as the *magnetization* and define them as:

$$\mathcal{P} \equiv \mathfrak{D} - \xi_0 \mathcal{E} = \xi_0(\epsilon - 1)\mathcal{E} \tag{12-8}$$

$$\mathfrak{M} \equiv \frac{\mathcal{B}}{\mu_0} - \mathcal{H} = \chi \mathcal{H} \tag{12-9}$$

It is obvious that, as defined, $\mathcal{P}$ and $\mathfrak{M}$ vanish in free space.

Since we are limiting our treatment to isotropic, one-dimensional systems, the vectorial notation may be dropped. Also, since we are interested primarily in systems in which the electric and magnetic fields do not vary over the volume ($\underline{V}_s$) or if an average field can be ascertained, Eq. (12-2) can be integrated over $\underline{V}_s$. Finally, let us replace $\mathfrak{D}$ and $\mathcal{B}$ by using Eqs. (12-8) and (12-9). The resulting work is

$$W = \underline{V}_s\left(\int_0^{\mathcal{P}} \mathcal{E}\, d\mathcal{P} + \xi_0 \int_0^{\mathcal{E}} \mathcal{E}\, d\mathcal{E} + \mu_0 \int_0^{\mathfrak{M}} \mathcal{H}\, d\mathfrak{M} + \mu_0 \int_0^{\mathcal{H}} \mathcal{H}\, d\mathcal{H}\right) \tag{12-10}$$

The first and third terms represent the reversible work required to polarize or magnetize a material system in $\underline{V}_s$. The second and fourth terms represent the reversible work required to establish an electric and magnetic field in $\underline{V}_s$. These latter terms are independent of the material in the field. This convenient separation of work terms was one of the principal reasons for defining $\mathcal{P}$ and $\mathfrak{M}$.

12.2 Electrostatic Systems

To write the Fundamental Equation we use Eq. (12-1) with (dW_{rev}) as the *negative* of the first two terms in Eq. (12-10). The sign change is necessary since we are interested in the work done *by* the system. Thus,

$$d\underline{U} = T\, d\underline{S} - P\, d\underline{V} + \underline{V}_s\left(\mathcal{E}\, d\mathcal{P} + \frac{\xi_0\, d\mathcal{E}^2}{2}\right) + \sum_j \mu_j\, dN_j \tag{12-11}$$

As noted previously, the term $(\underline{V}_s \xi_0 \mathcal{E}^2/2)$ does not depend on the material within the system. Consequently, it is usually incorporated as a constant in the definition of the internal energy. Following this procedure,

$$\underline{U}' = \underline{U} - \frac{\underline{V}_s \xi_0 \mathcal{E}^2}{2} \tag{12-12}$$

Then

$$d\underline{U}' = T\, d\underline{S} - P\, d\underline{V} + \underline{V}_s \mathcal{E}\, d\mathcal{P} + \sum_j \mu_j\, dN_j \tag{12-13}$$

Again we emphasize that $\mathcal{P}$ is the component of the electric polarization vector parallel to the field $\mathcal{E}$.

Equation (12-13) is the Fundamental Equation that we were seeking. It is immediately obvious that a number of Legendre transforms may be written in order to obtain independent variable sets other than $\underline{U}'(\underline{S}, \underline{V}, \mathcal{P}, N_1, \ldots, N_n)$. Also, since the Fundamental Equation is homogeneous, the Euler form is

$$\underline{U}' = T\underline{S} - P\underline{V} + \underline{V}_s \mathcal{E}\mathcal{P} + \sum_j \mu_j N_j \tag{12-14}$$

and the Gibbs–Duhem relation becomes

$$\underline{S}\, dT - \underline{V}\, dP + \underline{V}_s \mathcal{P}\, d\mathcal{E} + \sum_j N_j\, d\mu_j = 0 \tag{12-15}$$

These equations are identical to those derived earlier with the exception of the new electrostatic terms. The chemical potential now, however, becomes

$$\mu_k = \bar{U}'_k - T\bar{S}_k + P\bar{V}_k - \underline{V}_s \mathcal{E} \left(\frac{\partial \mathcal{P}}{\partial N_k}\right)_{T,P,N_j[k]} \tag{12-16}$$

$$= \bar{U}'_k - T\bar{S}_k + P\bar{V}_k - \underline{V}_s \mathcal{E} \bar{\mathcal{P}}_k \tag{12-17}$$

To maintain our previous contention that μ_k is a partial molar Gibbs energy, we see that we should define the Gibbs energy as the transform

$$\underline{G}(T, P, \mathcal{E}, N_j) = \underline{U}' - T\underline{S} + P\underline{V} - \underline{V}_s \mathcal{E}\mathcal{P} \tag{12-18}$$

It is important to recognize that this function is no different from the Gibbs energy used previously in this book. The designation of the variable $\mathcal{E}$ was not noted earlier since we implicitly assumed that our systems were either not dielectrics ($\epsilon = 1.0$), the field was zero, or there was no interaction between an imposed field and our system. This same reasoning holds for all other transforms and, in addition, to the equilibrium and stability criteria developed earlier.

For dielectric materials in an electric field, the concept of the pressure is somewhat different from the same material in the absence of a field. Pressures within a body are modified by the presence of the field and to apply the Fundamental Equation, one should consider that the pressure term refers to that value acting on the boundaries of the system. Solids, however, often present a complex problem in the theory of elasticity. A detailed discussion of this case is given by Landau and Lifshitz.[6]

Several examples are presented below to illustrate the application of the Fundamental Equation including an electrostatic term.

[6]L. P. Landau and E. M. Lifshitz, "Electrodynamics of Continuous Media," Vol. 8, *Course of Theoretical Physics*, translated by G. B. Sykes and G. S. Bell (New York: Pergamon Press, 1960).

Example 12.1

For a dielectric material present in an electrostatic field, the variation in the system volume with changes in field strength at constant temperature, pressure, and mole numbers is called *electrostriction*. Estimate the fractional change in volume of hydrogen gas at 293 K and 20.2 MN/m² when the electric field on the system is increased from 0 to 10^6 V/m. Experimental data show that for hydrogen under these conditions a plot of ln P versus the dielectric constant yields essentially a straight line with a slope of 24.8; that is, at 293 K and for fields between 0 and 10^6 V/m,

$$\left(\frac{\partial \ln P}{\partial \epsilon}\right)_{T,\mathcal{E},N} = 24.8$$

Solution

We first wish to obtain the partial derivative $(\partial \underline{V}/\partial \mathcal{E})_{T,P,N}$. The third Legendre transform of Eq. (12-14) is

$$y^{(3)} = f(T, P, \mathcal{E}, N) = \underline{U}' - T\underline{S} + P\underline{V} - \underline{V}_s \mathcal{E}\mathcal{P}$$

$$dy^{(3)} = -\underline{S}\, dT + \underline{V}\, dP - \underline{V}_s \mathcal{P}\, d\mathcal{E} + \sum_j \mu_j\, dN_j$$

Assuming that the field is constant and taking $\mathcal{P}$ as the component of $\mathcal{P}$ parallel to $\mathcal{E}$, Eq. (12-8) leads to

$$dy^{(3)} = -\underline{S}\, dT + \underline{V}\, dP - \underline{V}_s \xi_0(\epsilon - 1)\mathcal{E}\, d\mathcal{E} + \sum_j \mu_j\, dN_j$$

and therefore

$$\left(\frac{\partial \underline{V}}{\partial \mathcal{E}}\right)_{T,P,N} = -\frac{\partial}{\partial P}[\underline{V}_s \xi_0(\epsilon - 1)\mathcal{E}]_{T,\mathcal{E},N} = -\underline{V}_s \xi_0 \mathcal{E}\left(\frac{\partial \epsilon}{\partial P}\right)_{T,\mathcal{E},N}$$

Separating variables and integrating, we obtain

$$\frac{\underline{V}_{\mathcal{E}} - \underline{V}_{\mathcal{E}=0}}{\underline{V}_s} = -\frac{\xi_0}{P(24.8)(2)}\mathcal{E}^2$$

ξ_0 is a constant equal to $(8.85)(10^{-12})$ A s/V m. P is 20.2 MN/m², and $\mathcal{E} = 10^6$ V/m.

$$\frac{\underline{V}_{\mathcal{E}} - \underline{V}_{\mathcal{E}=0}}{\underline{V}_s} = \frac{-(8.85)(10^{-12})(10^6)^2}{(20.2)(10^6)(2)(24.8)} \sim -(9)(10^{-9})$$

In this case, the contraction in volume is extremely small. Normally, for macroscopic systems, electrostriction produces negligible changes in volume, although for microscopic systems in which local fields can be very large, there can be significant effects. For example, applied to ions in solution, electrostriction is often invoked to account qualitatively for the decrease in total volume when ionic solutions are mixed.

Example 12.2

Indicate how the entropy of a dielectric changes as an electric field is applied at constant temperature, pressure, and mole numbers.

Solution

If we use the same Legendre transform developed in Example 12.1,

$$\left(\frac{\partial \underline{S}}{\partial \mathcal{E}}\right)_{T,P,N} = \frac{\partial}{\partial T}[\underline{V}_s \xi_0(\epsilon - 1)\mathcal{E}]_{P,\mathcal{E},N} = \xi_0 \mathcal{E}\underline{V}_s\left(\frac{\partial \epsilon}{\partial T}\right)_{P,\mathcal{E},N}$$

For most materials, ξ, and therefore ϵ, decrease with an increase in temperature; thus, the entropy decreases as the material is polarized. Perhaps this result could have been anticipated from a consideration of the fact that, in an electric field, the molecules would tend to align their dipoles with the field and the overall randomness of the system would decrease. At constant $\underline{V}$ and N, during polarization there also would be a heat interaction of $T\,d\underline{S}$ and, since $d\underline{S} < 0$, heat would be evolved.

Example 12.3

Estimate the work required to establish a field when a dielectric is present within a parallel-plate capacitor and the field is increased from 0 to $\mathcal{E}$.

Solution

This work is given by the first term in Eq. (12-10). With Eq. (12-8),

$$W = \frac{\underline{V}_s \xi_0 \epsilon \mathcal{E}^2}{2}$$

To estimate W, choose a basis of 1 kg-mol. For a field of 10^6 V/m, selecting as examples, air, methyl alcohol, and fused silica:

Material	Conditions	ϵ	Volume, $\underline{V}_s$ (m³/kg-mol)	Work (J/kg-mol)	Approximate breakdown (V/m)
Air	273 K, 1 bar	1.0006	22.4	100	$(3)(10^6)$
Methyl alcohol (liquid)	293 K	31.2	0.0402	5.5	—
Fused silica	293 K	4	0.027	0.48	10^7

Compared to the value of RT at 300 K [$\sim(2.5)(10^6)$ J/kg-mol], these work terms are indeed insignificant; more energy could be stored at higher field strengths, but the value of 10^6 V/m used in the example is close to the breakdown strength of most materials.

Example 12.4

A flat plate capacitor is placed inside a system containing a pure dielectric material. As the capacitor is charged, show how the chemical potential and concentration vary with field strength.

Solution

Let us first find the derivative

$$\left(\frac{\partial \mu_i}{\partial \mathcal{E}}\right)_{T,\underline{V},N}$$

For this derivative we desire the Legendre transform,

$$y^{(2)} = f(T, \mathcal{E}, \underline{V}, N)$$

$$y^{(2)} = \underline{U}' - T\underline{S} - \underline{V}_s \mathcal{E} \mathcal{P}$$

$$dy^{(2)} = -\underline{S}\,dT - \underline{V}_s \mathcal{P}\,d\mathcal{E} - P\,d\underline{V} + \sum_j \mu_j\,dN_j$$

and

$$\left(\frac{\partial \mu_i}{\partial \mathcal{E}}\right)_{T,\underline{V},N} = -\underline{V}_s\left(\frac{\partial \mathcal{P}}{\partial N_i}\right)_{T,\underline{V},\mathcal{E},N_j[i]} = -\underline{V}_s\xi_0\mathcal{E}\left(\frac{\partial \epsilon}{\partial N_i}\right)_{T,\underline{V},\mathcal{E},N_j[i]}$$

where Eq. (12-8) has been used. Define the variation of ϵ with N_i as ϵ' and assume that it is independent of field strength; then, integrating, we get

$$\mu_i(\mathcal{E}) - \mu_i(\mathcal{E} = 0) = -\frac{\underline{V}_s\xi_0\mathcal{E}^2\epsilon'}{2}$$

This equation indicates the variation of μ_i with $\mathcal{E}$. As suggested by Guggenheim,[7] the value of μ_i ($\mathcal{E} = 0$) can be related to a standard-state value at the same temperature by

$$\mu_i = \mu_i^\circ + RT \ln \frac{\hat{f}_i}{\hat{f}_i^\circ} \qquad (\mathcal{E} = 0)$$

where the superscript represents some arbitrarily chosen standard state at T. Also, for simplicity, let us assume that the material is an ideal gas:

$$\hat{f}_i = y_i P = P = \frac{N_i RT}{\underline{V}}$$

Substituting yields

$$\mu_i(\mathcal{E}) = \mu_i^\circ + RT \ln \frac{N_i RT}{\hat{f}_i^\circ \underline{V}} - \frac{\underline{V}_s\xi_0\mathcal{E}^2\epsilon'}{2}$$

The dielectric constant can be expressed by the Debye equation as

$$\frac{\epsilon - 1}{\epsilon + 2} = \frac{NL}{3\underline{V}}\left(\alpha + \frac{\mu^{+2}}{3\xi_0 kT}\right)$$

ϵ is the dielectric constant, α the molecular polarizability, μ^+ the dipole moment, and L is Avogadro's number. For ideal gases $\epsilon \sim 1$; thus (for a pure gas),

$$\epsilon' = \left(\frac{\partial \epsilon}{\partial N}\right)_{T,\underline{V}} = \frac{L}{\underline{V}}\left(\alpha + \frac{\mu^{+2}}{3\xi_0 kT}\right)$$

Then, assuming that $\underline{V}_s \sim \underline{V}$,

$$\mu_i(\mathcal{E}) = \mu_i^\circ + RT \ln \frac{CRT}{\hat{f}_i^\circ} - \frac{\xi_0 L\mathcal{E}^2(\alpha + \mu^{+2}/3\xi_0 kT)}{2}$$

where C is the concentration.

Assume now that this gas in the field $\mathcal{E}$ is in equilibrium with more pure gas outside the field at temperature T. For this external gas, $\mathcal{E} = 0$, we could again write a similar expression for μ_i:

$$\mu_i(\text{outside}) = \mu_i^\circ + RT \ln \frac{C^{\text{outside}} RT}{\hat{f}_i^\circ}$$

At equilibrium, since $\mu_i(\text{outside}) = \mu_i(\text{inside})$,

$$RT \ln \frac{C^{\text{inside}}}{C^{\text{outside}}} = \frac{\xi_0 L\mathcal{E}^2(\alpha + \mu^{+2}/3\xi_0 kT)}{2}$$

[7]E. A. Guggenheim, *Thermodynamics* (Amsterdam: North-Holland, 1959), Chap. 11.

Suppose that our system consisted of HCL. $\alpha \sim 2.6 \times 10^{-30}$ m³/molecule and $\mu^+ \sim 1$ debye $= [1/(3)(10^{29})]C$ m/molecule. k is $(1.38)(10^{-23})$ J/molecule K and let $T = 300$ K. For the other terms $\xi_0 = (8.854)(10^{-12})$A s/V m, $L = (6.032)(10^{23})$ molecules/mol, and assume that $\mathcal{E} = 10^7$ V/m. Then

$$RT \ln \frac{C^{\text{inside}}}{C^{\text{outside}}} = (8.854)(10^{-12})(6.023)(10^{23})(10^{14})$$

$$\times \{2.6 \times 10^{-30} + [(3)^2(10^{58})(8.854)(10^{-12})(1.38)(10^{-23})(300)(3)]^{-1}\}0.5$$

$$= (52.3)(10^{25})(2.6 + 101)(10^{-30})(0.5) = 2.8 \times 10^{-2} \text{ J/mol}$$

$$\frac{C^{\text{inside}}}{C^{\text{outside}}} = \exp\left[\frac{(2.8)(10^{-2})}{(8.314)(300)}\right] = \exp(1.1 \times 10^{-5})$$

The enhancement is obviously very small. Only at much higher field strengths would one be able to show an appreciable difference between the concentrations inside and outside the field.

It is perhaps clear from the examples shown above that in most cases electric fields affect the thermodynamic properties of a system very little. To obtain significant effects, extremely large field strengths are required, but before they may be attained, breakdown normally occurs. Nevertheless, in the immediate vicinity of ions in a solution, very high field strengths do exist and significantly affect the microscopic properties of matter.

12.3 Electromagnetic Systems

We can develop the thermodynamics of systems in electromagnetic fields in much the same way as we did for systems in electrostatic fields (see Section 12.2). In this case, however, we must deal with the magnetic flux density, $\mathcal{B}$, the magnetic field strength, $\mathcal{H}$, and the magnetization (or magnetic moment per unit volume or magnetic polarization vector), $\mathcal{M}$. They are related as shown in Eq. (12-9). $\mathcal{M}$ is zero unless there is material in the field.

As before, the work done *by* the system is given by the negative of the last two terms in Eq. (12-10), and the Fundamental Equation then becomes

$$d\underline{U} = T\,d\underline{S} - P\,d\underline{V} + \underline{V}_s[\mu_0\mathcal{H}\,d\mathcal{M} + \mu_0\mathcal{H}\,d\mathcal{H}] + \sum_j \mu_j\,dN_j \qquad (12\text{-}19)$$

The second term in the electromagnetic work is normally separated and combined with $\underline{U}$ in a manner analogous to Eq. (12-12) to define or modify an internal energy $\underline{U}'$:

$$\underline{U}' = \underline{U} - \frac{\underline{V}_s\mu_0\mathcal{H}^2}{2} \qquad (12\text{-}20)$$

so that

$$d\underline{U}' = T\,d\underline{S} - P\,d\underline{V} + \underline{V}_s\mu_0\mathcal{H}\,d\mathcal{M} + \sum_j \mu_j\,dN_j \qquad (12\text{-}21)$$

This Fundamental Equation is analogous to Eq. (12-13) for electrostatic systems[8] and as noted there the Euler form, the Gibbs–Duhem, and various Legendre transforms are readily obtained. To illustrate the use of such forms, several examples are given below.

Example 12.5

Holding the temperature, pressure, and mole numbers constant, how is the entropy of a material affected by changes in the magnetic field strength?

Solution

We shall want to find a Legendre transform for the variables $(T, P, \mathcal{H}, N)$. Thus,

$$y^{(3)} = \underline{U}' - T\underline{S} + P\underline{V} - \mu_0 \underline{V}_s \mathcal{H}\mathcal{M}$$

$$dy^{(3)} = -\underline{S}\, dT + \underline{V}\, dP - \mu_0 \underline{V}_s \mathcal{M}\, d\mathcal{H} + \sum_j \mu_j\, dN_j$$

and

$$\left(\frac{\partial \underline{S}}{\partial \mathcal{H}}\right)_{T,P,N} = \mu_0 \underline{V}_s \left(\frac{\partial \mathcal{M}}{\partial T}\right)_{P,\mathcal{H},N} = \mu_0 \underline{V}_s \mathcal{M} \delta_I$$

where δ_I is equal to $(\partial \ln \mathcal{M}/\partial T)$ and is called the thermal magnetization coefficient. Values of δ_I are not often known. For water, Camp and Johnson[9] report that $\delta_I \sim (5)(10^{-4})\ \text{K}^{-1}$. For water, $\mu \sim (1.3)(10^{-6})$ V s/m A and $\chi = (-9.06)(10^{-6})$. Then, for a value of $\mathcal{B} = 1000$ gauss[10] $= 0.1\ \text{Wb/m}^2 = 0.1\ \text{V s/m}^2$, from Eq. (12-4),

$$\mathcal{H} = \frac{\mathcal{B}}{\mu} = \frac{0.1}{(1.3)(10^{-6})} = (7.7)(10^4)\ \text{A/m}$$

$$\mathcal{M} = \chi\mathcal{H} = -(9.06)(10^{-6})(7.7)(10^4) \sim -0.7\ \text{A/m}$$

Let $\underline{V} = 1\ \text{m}^3$, so that

$$\mathcal{M}\underline{V}_s = -0.7\ \text{A m}^2$$

Then

$$\left(\frac{\partial \underline{S}}{\partial \mathcal{H}}\right)_{T,P,N} = (4\pi)(10^{-7})(-0.7)(5)(10^{-4}) \sim -3 \times 10^{-10}\ \text{J/(A/m) K}$$

For 1 m^3 of water, with an increase in $\mathcal{B}$ from 0 to 1000 gauss, the entropy decreases, but very slightly.

[8]See, however, F. W. Camp and E. F. Johnson, "The Effect of Strong Magnetic Fields on Chemical Engineering Systems," MATT-67, Plasma Physics Laboratory, Princeton University, Princeton, N.J. It is argued in this report that there should be an additional term in Eq. (12-21) to account for the expansion of the system at constant $\mathcal{M}$. This term is $\mu_0 \mathcal{H}\mathcal{M}\, d\underline{V}_s$ and should be included with the $P\, d\underline{V}$ term to give $-(P - \mu_0 \mathcal{H}\mathcal{M})\, d\underline{V}$. If this is true, then to Eq. (12-13) one should correct the $-P\, d\underline{V}$ term to $-(P - \mathcal{E}\mathcal{P})\, d\underline{V}$ to allow the expansion of a system in an electrostatic field at constant P. In our treatment we have considered $\underline{V}_s$ to be a constant and equal to the total volume affected by electric or magnetic fields, whereas $\underline{V}$ is taken as the volume of the material system under consideration.

[9]*Ibid.*

[10]In SI units, this would be abbreviated as 0.1 tesla (T).

Example 12.6

For a pure component, at constant pressure, show how the boiling point varies with the magnetic intensity, $\mathcal{H}$.

Solution

For phase equilibrium, the chemical potential is equal in the vapor and liquid. From Eq. (12-21), a total Legendre transform for a pure component yields the Gibbs–Duhem equation:

$$0 = -\underline{S}\, dT + \underline{V}\, dP - \mu_0 \underline{V}_s \mathcal{M}\, d\mathcal{H} - N\, d\mu$$

If we divide by the number of moles, N, and solve for $d\mu$:

$$d\mu = -S\, dT + V\, dP - \mu_0 V_s \mathcal{M}\, d\mathcal{H}$$

Holding the pressure constant and using

$$\mu^V = \mu^L$$

then

$$\left(\frac{\partial T}{\partial \mathcal{H}}\right)_{[L-V],P} = -\mu_0 \mathcal{M} \frac{V^V - V^L}{S^V - S^L}$$

Since entropy and volume are weak functions of $\mathcal{H}$, with $\mathcal{M} = \chi\mathcal{H}$,

$$\Delta T = T_{\mathcal{H}} - T_{\mathcal{H}=0} = -\mu_0 \chi \frac{(V^V - V^L)\mathcal{H}^2}{2(S^V - S^L)}$$

To illustrate the use of this relation, consider water at 100°C[11] and determine ΔT for the case in which $\mathcal{H}$ increases from zero to a field of $(8)(10^5)$ A/m. This corresponds to 10,000 gauss. Also,

$$\mu_0 = (4\pi)(10^{-7}) \text{ V s/A m}$$

$$\chi = (-9.06)(10^{-6})$$

$$V^V - V^L = 30.1 \text{ m}^3\text{/kg-mol}$$

$$S^V - S^L = (1.09)(10^5) \text{J/kg-mol K}$$

$$\Delta T = -\frac{(4\pi)(10^{-7})(-9.06)(10^{-6})(30.1)(8)^2(10^5)^2}{(1.09)(10^5)(2)} = 10^{-3} \text{ K}$$

The increase is inconsequential.[12]

It is clear from the examples shown that electromagnetic fields affect thermodynamic properties only slightly. The phenomenon of superconductivity at low temperatures is, however, one important exception. Another is the magneto-

[11]Camp and Johnson, *op. cit.*

[12]See also A. J. Meachin and M. W. Biddulph (*Cryogenics*, Jan. 1978, p. 29), who show theoretically and experimentally that there is essentially no effect of magnetic fields (up to about 10^7 A/m) on the vapor pressures of oxygen, nitrogen, or argon. A similar relation can be derived for estimating the change in boiling temperature for materials in an electrostatic field [R. K. Lyon, *Nature*, **192**, 1285 (1961); R. C. Sharma, *J. Appl. Phys.*, **42**, 1234 (1971)].

caloric effect that is utilized for cooling below 1 K.[13] Referring to Example 12.5, we showed that

$$\left(\frac{\partial \underline{S}}{\partial \mathcal{H}}\right)_{T,P,N} = \mu_0 \underline{V}_s \left(\frac{\partial \mathcal{M}}{\partial T}\right)_{P,\mathcal{H},N} \tag{12-22}$$

From Eq. (12-9), $\mathcal{M} = \chi \mathcal{H}$ and at constant N, Eq. (12-22) becomes

$$\left(\frac{\partial \underline{S}}{\partial \mathcal{H}^2}\right)_{T,P,N} = \mu_0 \frac{\underline{V}_s}{2} \left(\frac{\partial \chi}{\partial T}\right)_{P,\mathcal{H},N}$$

For many paramagnetic materials at low temperatures the relation between χ and T is given by Curie's Law:

$$\chi = \frac{C}{T}$$

Therefore,

$$\left(\frac{\partial \underline{S}}{\partial \mathcal{H}^2}\right)_{T,P,N} = -\frac{\mu_0 C \underline{V}_s}{2T^2} \tag{12-23}$$

At low temperatures, the entropy will decrease considerably when the material is isothermally magnetized. The next step is to isolate thermally the sample and remove it from the field. A temperature drop occurs on demagnetization as the interested reader can verify by determining the sign of $(\partial T/\partial \mathcal{H})_{\underline{S},P,N}$.

12.4 Thermodynamics of Systems under Stress

It is rare for chemical engineers to become involved with thermodynamic analyses of systems that are in tension or compression. Significant problems arising in this area usually fall heir to applied mechanicists, who are interested in the behavior of materials under static or dynamic loads. The materials involved are solid instead of fluid since, for the latter, the consequences of the imposition of a force can be treated by pressure–volume terms introduced earlier in the book. In fact, solids are almost always considered to be elastic (i.e., although they may be stressed and deformed, the total volume is invariant). In the thermodynamics of such systems, the system volume is then assumed constant and no $P\underline{V}$ terms are used.

In the discussion of stressed systems, we refer to the *strain* resulting from an applied *stress*. These terms are readily visualized in a qualitative way since it is common experience to apply a force to, for example, rubber and expect a change in the dimensions of the specimen. A careful definition of terms is, however, necessary before any quantitative relations can be meaningful. Such definitions

[13] W. F. Giauque, "Paramagnetism and the Third Law of Thermodynamics. Interpretation of the Low-Temperature Magnetic Susceptibility of Gadolinium Sulfate," *J. Am. Chem. Soc.*, **49**, 1870 (1927); P. Debye, "Einige Bemerkungen zur Magnetisierung bei tiefer Temperatur," *Ann. Phys.*, **81**, 1154 (1926); W. F. Giauque and D. P. MacDougall, "Attainment of Temperatures Below 1° Absolute by Demagnetization of $Gd_2(SO_4)_3 \cdot 8H_2O$," *Phys. Rev.*, **43**, 768 (1933).

will be our first task, although the ultimate objective is to formulate a Fundamental Equation for stressed systems.

When we use the term "stress," we obviously imply a force/area. Stress is, therefore, a vectorial quantity with both a magnitude and direction. Similarly, for strain we describe the movement of an element of the system in a particular direction. Consider Figure 12.1. A bar of an elastic material is fixed to a rigid

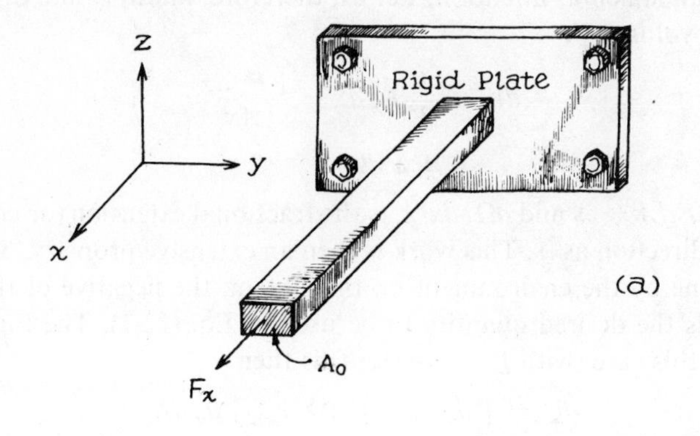

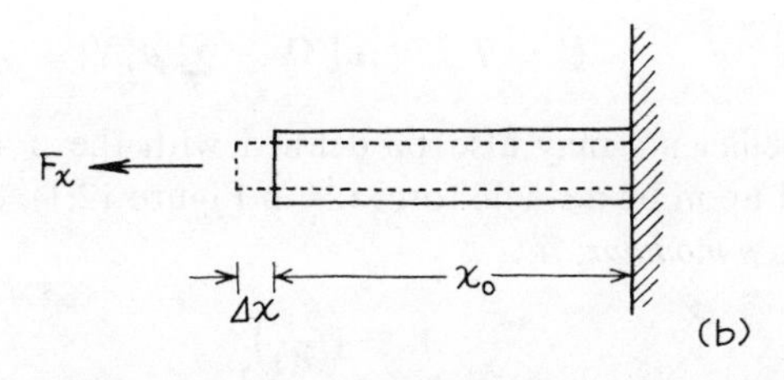

Figure 12.1 Extension under stress.

plate and subjected to a tensile force, F_x. The sides are unconstrained. Because of F_x, the bar is elongated by Δx and is in tension. The stress in this case, F_x/A, produces a strain, Δx. Also, as shown, there are induced strains Δy and Δz as well as induced stresses in the y and z directions. Thus, the situation is complicated even in this relatively simple example. For any body under stress, we must consider the strain in the three coordinate directions. Also, one should allow for *shear* (i.e., a stress so directed that there is a rotation or a twisting of the system).

It is not within the scope of this book to develop in detail three-dimensional elastic theory as applications rarely occur in chemical engineering. The interested reader is referred to the excellent presentation of Callen.[14] We will limit our treatment to a one-dimensional case in which there is but one applied stress or

[14]H. B. Callen, *Thermodynamics* (New York: Wiley, 1960), Chap. 13.

force. Thus we return to Figure 12.1. Here the work done on the system is $F_x \Delta x$. The fact that there is a finite Δy or Δz is immaterial since we are only interested in force–distance interactions operating over the system boundaries and F_x is the only force to fit this criterion.[15]

As used in a thermodynamics study, the designation of work as simply $F_x\, dx$ is inconvenient since the terms do not blend neatly with other extensive properties in a Fundamental Equation. Let us, therefore, multiply and divide by the total system volume $\underline{V}$,

$$
\begin{aligned}
dW &= \frac{\underline{V}}{\underline{V}} F_x\, dx = \frac{\underline{V} F_x\, dx}{Ax} \\
&= \underline{V} \mathscr{S}\, d\Omega
\end{aligned}
\tag{12-24}
$$

The stress is $F_x/A = \mathscr{S}$ and $d\Omega$, dx/x, is the fractional extension (or contraction) in the same direction as $\mathscr{S}$. This work is then an extensive property. Since this is the work done by the environment on the system, the negative of the work in Eq. (12-24) is the desired quantity to be used in Eq. (12-1). The Fundamental Equation in this case (with $\underline{V}$ = constant) is then:

$$
d\underline{U} = T\, d\underline{S} + \mathscr{S} \underline{V}\, d\Omega + \sum_j \mu_j\, dN_j \tag{12-25}
$$

Once we have reached this point, the Euler relation, the Gibbs–Duhem equation, and many Legendre transforms can be written immediately. The Euler form is, for example,

$$
\underline{U} = T\underline{S} + \mathscr{S}\underline{V}\Omega + \sum_j \mu_j N_j \tag{12-26}
$$

Various coefficients may also be defined with the $\mathscr{S}$–Ω notation. When a system is acted upon by a single force (as in Figure 12.1), the stress–strain ratio is called *Young's modulus*, Y:

$$
Y \equiv \left(\frac{\partial \mathscr{S}}{\partial \Omega}\right)_T \tag{12-27}
$$

If, however, the sides are constrained from expanding or contracting in the other directions, then the derivative

$$
\left(\frac{\partial \mathscr{S}}{\partial \Omega}\right)_{T,\,\text{restrained}} \equiv \textit{isothermal elastic stiffness coefficient} \tag{12-28}
$$

Many other solid properties can be introduced, including heat capacities at constant length (or stress), etc. We illustrate one application of thermodynamics to such systems in the example below and provide several other problems at the end of the chapter.

[15]In the general case with both linear movement and rotation we should have to consider that the work done on the system consisted of six terms. Three terms would involve the linear movement in the three rectangular coordinates due to stresses in these directions. There would, however, be three other terms to account for motion in, for example, the y and z direction from an x-directed stress. There are not nine terms because of the conservation of angular momentum (i.e., the x-movement due to a y-stress is equal to the y-movement due to an x-stress). See Callen, *ibid.*

Example 12.7

An adiabatic, elastic deformation of a body is normally accompanied by a change in temperature. This phenomenon is known as the *thermoelastic effect*. Estimate the magnitude of this effect for a specimen of Armco iron that is stressed from 0 to 103 MPa at 773 K. Under these conditions, the coefficient of linear thermal expansion is

$$\alpha_{\mathscr{s}} = \frac{1}{L_0}\left(\frac{\partial L_0}{\partial T}\right)_{\mathscr{s}} = (16.8)(10^{-6})\ \text{K}^{-1}$$

and the heat capacity at constant stress is

$$C_{\mathscr{s}} = T\left(\frac{\partial S}{\partial T}\right)_{\mathscr{s}} \sim 9.1\ \text{cal/g-mol K}$$

Solution

We desire the partial derivative $(\partial T/\partial \mathscr{s})_{\underline{S},N}$. Thus, we employ a Legendre transform, $y^{(1)} = f(\mathscr{s}, \underline{S}, N)$. From Eq. (12-25),

$$dy^{(1)} = -\Omega \underline{V}\, d\mathscr{s} + T\, d\underline{S} + \sum_j \mu_j\, dN_j$$

$$\left(\frac{\partial T}{\partial \mathscr{s}}\right)_{\underline{S},N} = -\underline{V}\left(\frac{\partial \Omega}{\partial \underline{S}}\right)_{\mathscr{s},N}$$

$$= \frac{\underline{V}(\partial\Omega/\partial T)_{\mathscr{s},N}}{(\partial \underline{S}/\partial T)_{\mathscr{s},N}}$$

From the definition of Ω,

$$\alpha_{\mathscr{s}} = \left(\frac{\partial \Omega}{\partial T}\right)_{\mathscr{s},N}$$

Thus,

$$\int_{T_1}^{T_2} \frac{dT}{T} = -\int_{\mathscr{s}_1}^{\mathscr{s}_2} \frac{\alpha_{\mathscr{s}}}{C_{\mathscr{s}}N/\underline{V}}\, d\mathscr{s}$$

The term $(C_{\mathscr{s}}N/\underline{V})$ is simply the heat capacity at constant stress, on a unit volume basis. Assuming that both it and $\alpha_{\mathscr{s}}$ are not strong functions of $\mathscr{s}$ (see below), then

$$\ln\frac{T_2}{T_1} = -\frac{\alpha_{\mathscr{s}}(\Delta\mathscr{s})}{C_{\mathscr{s}}N/\underline{V}}$$

For small differences between T_2 and T_1,

$$\ln\frac{T_2}{T_1} \sim \frac{T_2 - T_1}{T} = \frac{\Delta T}{T}$$

and

$$\Delta T = -\frac{\alpha_{\mathscr{s}} T(\Delta\mathscr{s})}{C_{\mathscr{s}}N/\underline{V}}$$

$\alpha_{\mathscr{s}}$ has been given as $(16.8)(10^{-6})\ \text{K}^{-1}$.

$$\frac{C_{\mathscr{s}}N}{\underline{V}} = \frac{(9.1)(4.19)}{(55.8)}(7.6)(10^6) = 5.18\ \text{MJ/m}^3$$

where the atomic weight was chosen as 55.8 and the density as 7.6 g/cm³. Thus,

$$\Delta T = -\frac{(16.8)(10^{-6})(103)(773)}{5.18} = -0.26\ \text{K}$$

The expected temperature drop would be about 0.26 K for this adiabatic stress. This value is in excellent agreement with the value of −0.25 K reported by Rocca and Bever.[16] These same authors discuss the variation of $\alpha_{\mathscr{s}}/(C_{\mathscr{s}}N/\underline{V})$ with $\mathscr{s}$ and show that for stresses of the order of 100 MPa, this group varies only 2 to 3%.

It is also interesting to note that a temperature *drop must occur upon stretching* if $\alpha > 0$. For some substances in which there is a contraction upon heating, the *thermoelastic effect* will be positive (i.e., the material will heat when adiabatically stretched).

12.5 Systems in Body-Force Fields or under Acceleration Forces

We have now treated several systems in which there were work terms other than of the $P\,d\underline{V}$ type. Once we expressed this work (in an extensive manner), we simply included it in the general Fundamental Equation and from there proceeded to derive the desired Legendre transforms and partial derivatives. Implicit in this treatment was the supposition that the internal energy $\underline{U}$ was a function of variables other than $\underline{S}, \underline{V}, N_1, \ldots, N_n$ (e.g., for systems in electrostatic fields $\underline{U} = \underline{U}(\underline{S}, \underline{V}, \mathcal{P}, N_1, \ldots, N_n)$, where $\mathcal{P}$ is the component of the electric polarization parallel to the electric field $\mathcal{E}$).

It was then a simple task to develop any desired partial derivative from the appropriate Fundamental Equation. We also indicated in Section 12.3 that it was logical to expand the definitions of $\underline{H}$, $\underline{A}$, $\underline{G}$, etc., to include new system variables. Some authors, however, retain the earlier definitions for these well-known properties and define new properties if electrostatic, electromagnetic, etc., terms are to be included. Either approach is quite satisfactory, although the duality is often confusing. It is preferable to avoid definitions if possible and only consider the necessary independent variables. A Legendre transform is then employed to obtain the desired new Fundamental Equation.

Except for learning new notations introduced in Sections 12.2 through 12.4, there are no new principles introduced. When, however, one wishes to study systems in potential (body-force) fields or as acted upon to produce acceleration (or deceleration), then a somewhat different approach is taken. As we will show below, the work can readily be expressed in terms appropriate for these systems, but it is not immediately obvious how to use these terms to develop the Fundamental Equation. We must carefully consider the question whether the internal energy of a system depends on the position of the system in a potential field or on the velocity. Although we could, if we desired, extend the definition of $\underline{U}$ to include these variables, the results would run counter to our intuitive desire to preserve the character of $\underline{U}$ to be independent of position and velocity as much as possible. True, if we have a nonhomogeneous electromagnetic field, $\underline{U}$ is already a function of position, but we shall overlook such inconsistencies.

[16]Robert Rocca and M. B. Bever, "The Thermoelastic Effect," *Trans. AIME*, **88** (Feb. 1950), *Journal of Metals*, p. 327.

Although this is a text on classical thermodynamics, and little or no note has been taken of the constituent molecules in a system, we must admit that the desire to eliminate potential and acceleration fields from the definition of $\underline{U}$ relates to the fact that we would like to feel that those variables that affect $\underline{U}$ also modify some molecular properties. Perhaps the molecular velocity or spacing, or rotational frequency, etc., might be affected. These properties cannot, however, be related neatly to the overall system velocity or, for example, to the position in a potential field.

The net result of these admittedly intuitive feelings is to arrive at the decision to express the total energy of a system as a sum of the familiar internal energy and other energies that may be associated with potential or acceleration fields:

system energy = (internal + potential + kinetic) energy

$$\underline{E} = \underline{U} + \underline{PE} + \underline{KE} \tag{12-29}$$

The $\underline{PE}$ term is found by considering the work effects necessary to move a system in a force field. If a system is moved a distance $d\mathbf{L}$ against or with a colinear force $\mathbf{F}$, then

$$d(\underline{PE}) = -dW = -\mathbf{F} \cdot d\mathbf{L} = -M\mathbf{a} \cdot d\mathbf{L} = M\,d\phi \tag{12-30}$$

where the potential ϕ is defined as

$$d\phi \equiv -\mathbf{a} \cdot d\mathbf{L} \tag{12-31}$$

The potential energy of the system is then

$$\underline{PE} = M\phi \tag{12-32}$$

since the mass M was considered constant.

To illustrate, consider a system in a gravitational field with an acceleration $\mathbf{g}$. Measure $\mathbf{L}$ in the *up* direction and $\mathbf{g}$ in the *down* direction. Then, since $\mathbf{L}$ and $\mathbf{g}$ are colinear vectors but with opposite directions,

$$d\phi = g\,dL \tag{12-33}$$

$$\underline{PE} = M \int d\phi = M\phi = M \int g\,dL \tag{12-34}$$

No limits have been placed on the integration since one may arbitrarily define the potential energy to be zero at some reference value of L and then relate all other values to this base state.

As another example, consider a solid-body rotation. Let $\mathbf{L} = \mathbf{r}$ = the radius measured *outward* from the center of rotation. Here the centrifugal acceleration is $\boldsymbol{\omega}^2\mathbf{r}$ and is also directed radially *outward*. Then

$$d\phi = -\boldsymbol{\omega}^2\mathbf{r} \cdot d\mathbf{r} \tag{12-35}$$

$$\underline{PE} = M \int d\phi = M\phi = -M \int \omega^2 r\,dr \tag{12-36}$$

Example 12.8

What is the potential of a synchronous satellite at a distance r from the center of the earth and at an altitude where $g = g_r$?

Solution

From the definitions above,

$$d\phi = (g_r - \omega^2 r)\, dr$$

For synchronous operation, $d\phi = 0$ or

$$g_r = \omega^2 r$$

The kinetic energy of a system is, of course,

$$\underline{\mathrm{KE}} = \tfrac{1}{2}Mv^2 \tag{12-37}$$

and thus

$$\underline{E} = \underline{U} + \underline{\mathrm{PE}} + \underline{\mathrm{KE}} = \underline{U} + M\phi + \tfrac{1}{2}Mv^2 \tag{12-38}$$

To express the differential form for $\underline{E}$, and still employ the customary expansion for $\underline{U}$, we see immediately that there must be terms on the right-hand side involving M, ϕ, and v. It we use Eq. (12-1) with no electrical, etc., work terms, for simplicity, with the molecular weight of j being m_j,

$$\begin{aligned} d\underline{E} &= d\underline{U} + d(\underline{\mathrm{PE}}) + d(\underline{\mathrm{KE}}) \\ &= T\, d\underline{S} - P\, d\underline{V} + M\, d\phi + M\, d\frac{v^2}{2} \\ &\quad + \sum_j \left(\mu_j + m_j\phi + \frac{m_j v^2}{2}\right) dN_j \end{aligned} \tag{12-39}$$

Let us now examine these terms more carefully. The $M\, d\phi$ and $M\, d(v^2/2)$ terms indicate the negative value of the work the system *does on the environment* in undergoing a change in ϕ or v. For example, let the process be the fall of the system from a higher to a lower elevation in a gravitational field and, in so falling, a weight is raised in the environment. Now $d\phi = g\, dL < 0$ (since L is decreasing) and thus the work done by the system on the environment is positive as indicated by the rise of the weight.

Within the summation we also find terms involving ϕ and v. The interpretation is simple. As the mole numbers change because of mass addition or loss, there is introduced or removed potential and kinetic energy from the system. Of course, since Eq. (12-39) represents a quasi-static process, all such interchange must be such that the specific potential and kinetic energies of the mass entering and leaving the system must be equal to that present in the system. This is rather obvious since only one ϕ or v is noted. The analogy is the Fundamental Equation $\underline{U} = \underline{U}(\underline{S}, \underline{V}, N)$ compared to the open-system First Law, Eq. (3-60). We could write the comparable open-system relation for this case by inspection:

$$\begin{aligned} d\underline{E} &= d\left(\underline{U} + \phi M + \frac{Mv^2}{2}\right) \\ &= đQ - đW + \left[\sum_{j=1}^{n} \left(\bar{h}_j + \phi m_j + \frac{m_j v^2}{2}\right) dn_j\right]_e \\ &\quad - \left[\sum_{j=1}^{n} \left(\bar{h}_j + \phi m_j + \frac{m_j v^2}{2}\right) dn_j\right]_\ell \end{aligned} \tag{12-40}$$

Then it is obvious that without the potential and kinetic energy terms, Eq. (12-40) reduces to the general open-system First Law expression, Eq. (3-60). The show the connection between Eqs. (12-39) and (12-40) in another way, assume that the interchange of mass, heat, and work with the environment is reversible,

$$\bar{h}_{je} = \bar{h}_{j\ell} = \bar{H}_j \text{ (system)}$$

$$\phi_e = \phi_\ell = \phi \text{ (system)}$$

$$v_e = v_\ell = v \text{ (system)}$$

Then

$$d\underline{S} = \frac{đQ}{T} + \left[\sum_{j=1}^{n} \bar{S}_j \, dn_j\right]_e - \left[\sum_{j=1}^{n} \bar{S}_j \, dn_j\right]_\ell \tag{12-41}$$

and

$$đW = P\,d\underline{V} - M\,d\phi - Mv\,dv \tag{12-42}$$

$$d\underline{E} = T\,d\underline{S} - P\,d\underline{V} + M\,d\phi + Mv\,dv + \left[\sum_{j=1}^{n}\left(\mu_j + m_j\phi + \frac{m_j v^2}{2}\right) dn_j\right]_e - \left[\sum_{j=1}^{n}\left(\mu_j + m_j\phi + \frac{m_j v^2}{2}\right) dn_j\right]_\ell \tag{12-43}$$

or, since $(dn_j)_e - (dn_j)_\ell = dN_j$, by substitution, Eq. (12-39) is obtained.

Applying Euler's theorm to Eq. (12-39) and noting that ϕ and v are invariant with changes in mass, we obtain

$$\begin{aligned} \underline{E} &= T\underline{S} - P\underline{V} + \sum_{j=1}^{n}\left(\mu_j + m_j\phi + \frac{m_j v^2}{2}\right)N_j \\ &= \underline{U} + \sum_{j=1}^{n}\left(m_j\phi + \frac{m_j v^2}{2}\right)N_j \end{aligned} \tag{12-44}$$

Since $\underline{E} = \underline{E}(\underline{S}, \underline{V}, \phi, v, N_j)$, it is now possible to employ Legendre transforms as developed in Chapter 5 to obtain other fundamental representations and relationships between partial derivatives.

Example 12.9

Obtain the transforms (a) $f(T, P, \phi, v, N_1, \ldots, N_n)$ and (b) $f(T, P, M, v, N_1, \ldots, N_n)$ and show how the pressure varies with potential at constant T and velocity in a closed system.

Solution

(a) $f(T, P, \phi, v, N_1, \ldots, N_n)$

$$y^{(2)} = \underline{E} - T\underline{S} + P\underline{V}$$

$$dy^{(2)} = -\underline{S}\,dT + \underline{V}\,dP + M\,d\phi + Mv\,dv + \sum_{j=1}^{n}\left(\mu_j + m_j\phi + \frac{m_j v^2}{2}\right) dN_j$$

(b) $f(T, P, M, v, N_1, \ldots, N_n)$

$$y^{(3)} = \underline{E} - T\underline{S} + P\underline{V} - M\phi$$

$$dy^{(3)} = -\underline{S}\,dT + \underline{V}\,dP - \phi\,dM + Mv\,dv + \sum_{j=1}^{n}\left(\mu_j + m_j\phi + \frac{m_j v^2}{2}\right) dN_j$$

From this last relation,

$$\left(\frac{\partial P}{\partial \phi}\right)_{T,v,N} = -\left(\frac{\partial M}{\partial \underline{V}}\right)_{T,P,v,N} = -\rho$$

where ρ is the mass density. This relation is, of course, the usual equation to determine a hydrostatic pressure.

The final point of interest in this section involves the criteria of equilibrium in body-force fields. We develop such criteria only for a system in a potential energy field because no useful criteria are found for systems with kinetic energy (i.e., it is readily shown that systems with the lowest velocities are most stable).

For a system in a potential energy field, one criterion often given is that

$$dP = -\rho \, d\phi \tag{12-45}$$

but this was already derived in Example 12.9. Furthermore, it may easily be shown that, at equilibrium, there can be no gradients of temperature for a system in a potential field. The more interesting criteria result from an examination of the variation of chemical potential with ϕ.

Consider Figure 12.2. We show a system at equilibrium in a potential field. That is, there is some difference in ϕ between points A and B. (One might consider the mixture to be in a vertical pipe with A and B representing two

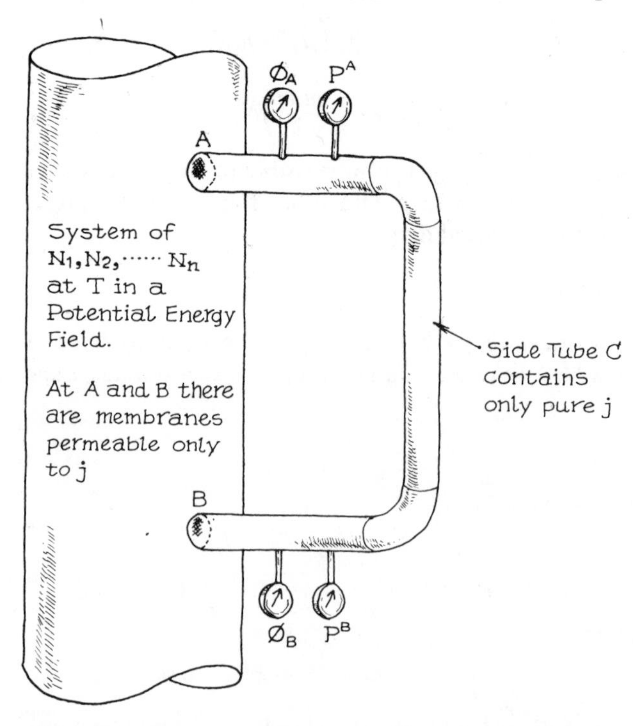

Figure 12.2 The effect of potential fields on chemical potentials.

heights.) The temperature of the system is uniform although the pressure varies with ϕ [see Eq. (12-45)]. At A and B we have semipermeable membranes that allow component j to pass freely into side tube C. At A the chemical potential of j is equal on both sides of the membrane; a similar statement may be made at B. Thus,

$$\mu^A_{j,\text{mix}} = \mu^A_{j,\text{pure}}$$
$$\mu^B_{j,\text{mix}} = \mu^B_{j,\text{pure}}$$

Subtracting yields

$$\mu^A_{j,\text{mix}} - \mu^B_{j,\text{mix}} = \mu^A_{j,\text{pure}} - \mu^B_{j,\text{pure}} \tag{12-46}$$

But in the side tube, where only pure j is present,

$$\mu^A_{j,\text{pure}} - \mu^B_{j,\text{pure}} = \int_{P^B}^{P^A} V_j\, dP = \int_{\phi^B}^{\phi^A} (-V_j\rho_j)\, d\phi = -m_j(\phi^A - \phi^B) \tag{12-47}$$

Combining Eqs. (12-46) and (12-47) and noting that A and B could have been chosen at random,

$$\mu^A_{j,\text{mix}} + m_j\phi^A = \mu^B_{j,\text{mix}} + m_j\phi^B = \text{constant} \tag{12-48}$$

Equation (12-48) is the desired result. It shows that the chemical potential is not a constant in a potential energy field but rather the sum of chemical potential *and* the product of the molecular weight times the potential is constant.

Example 12.10

A deep, small well hole has been capped and left undisturbed for years. It is believed that the temperature of the gas does not vary significantly with depth and that convection currents are negligible. The pressure at the well top is 2 bar, the temperature 300 K, and the gas composition is 70 mole% helium and 30 mole% methane. The hole is 2 km deep. What is the helium mole fraction at the bottom? Assume ideal gases.

Solution

With Z measured down from the surface and $\phi = 0$ on the surface, $\phi = -gZ$. Then from Eq. (12-47)

$$\mu_{\text{He},b} - \mu_{\text{He},t} = -m_{\text{He}}(\phi_b - \phi_t) = m_{\text{He}} g Z_b$$

where the subscripts b and t represent bottom and top conditions. $m_{\text{He}} = 4$ kg/kg-mol, $g = 9.81\ \text{m/s}^2$, and $Z_b = 2000$ m. Also, for ideal gases, with $T = 300$ K, $R = 8314$ J/kg-mol K,

$$\mu_{\text{He},b} - \mu_{\text{He},t} = RT \ln \frac{\hat{f}_{\text{He},b}}{\hat{f}_{\text{He},t}} = RT \ln \frac{p_{\text{He},b}}{p_{\text{He},t}}$$

Thus,

$$p_{\text{He},b} = p_{\text{He},t} \exp\left[\frac{(4)(9.81)(2000)}{(8314)(300)}\right] = (2)(0.7)(1.032) = 1.45 \text{ bar}$$

In a similar manner, $p_{\text{CH}_4,b} = 0.68$ bar, so the fraction helium at the bottom is $1.45/(1.45 + 0.68) = 0.68$.

PROBLEMS

12.1. We have a system α confined between the plates of a condenser in which a strong electric field is present. The material in α is a pure substance. At the boundaries of the condenser, the α system is contiguous with a system β made of the same pure material; the electric field in β is zero. The system β is itself in contact with a reservoir R_P through an adiabatic, frictionless piston; reservoir R_P is large and the pressure within is constant. System β is also in contact with a large reservoir R_T through a rigid, diathermal gate. The temperature in R_T is constant. What are the equilibrium criteria for the α–β system?

12.2. For a mixture of reacting gases, show how the chemical equilibrium constant varies with the electric field strength.

12.3. Present a similar analysis as requested in Problem 12.2 but for the magnetic field strength.

12.4. What is the work required to establish a magnetic field of 10,000 gauss in a long solenoid that contains 1 kg-mol of air at standard conditions? Assume that $\mu_{\text{air}} = \mu_0$.

12.5. A student is carrying out an experiment in which he desires to hang a balance pan a definite distance above a tabletop (see Figure P12.5). To accomplish

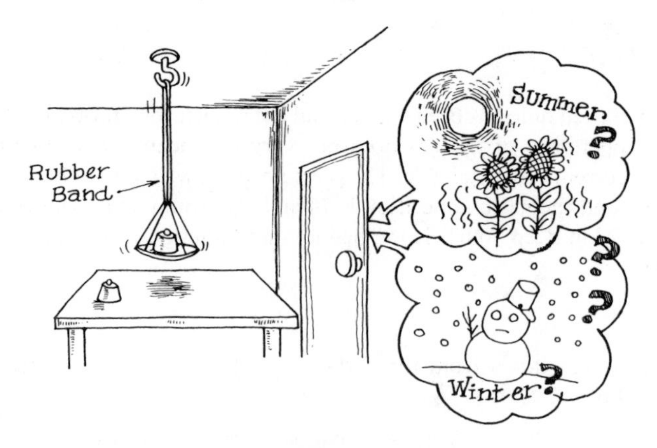

Figure P12.5

this, he decides to use a rubber band that he has conveniently found in his pocket. He readily obtains the desired separation distance by adding mass to the pan in order to stretch the rubber band the correct amount. The student has, however, just come in from the outside and when used, the rubber band is at the temperature outside the laboratory. He works rapidly and when the pan and weights are hung on the rubber band, it quickly stretches to a length of 30 cm.

Continuing the experiment, as the rubber band approaches room temperature, the student notices with irritation that the pan is rising. To counteract this, he gradually adds weights to the pan to maintain the same stretched length (30 cm). A total of 14 g has to be added until the rubber band is at room temperature and no more movement occurs.

Is it summer or winter outside?

Assume that the original extension of the rubber band is adiabatic and reversible. Initially, the rubber band was 7.5 cm long and had an (unstretched) cross section measuring 1.5 × 1.5 mm. It does not crystallize when stretched and data for this particular rubber band show that $(\partial \mathscr{s}/\partial T)_L = 4410\,\text{N/m}^2\,\text{K}$, and $T(\partial S/\partial T)_L = 1900$ J/kg K. The rubber band has a specific gravity of 0.95 and the room temperature is 298 K.

12.6. Crackpot Inventions, Inc. is currently designing an elastic-rod Carnot engine (see Figure P12.6). In this device, work will be produced by heat transfer from

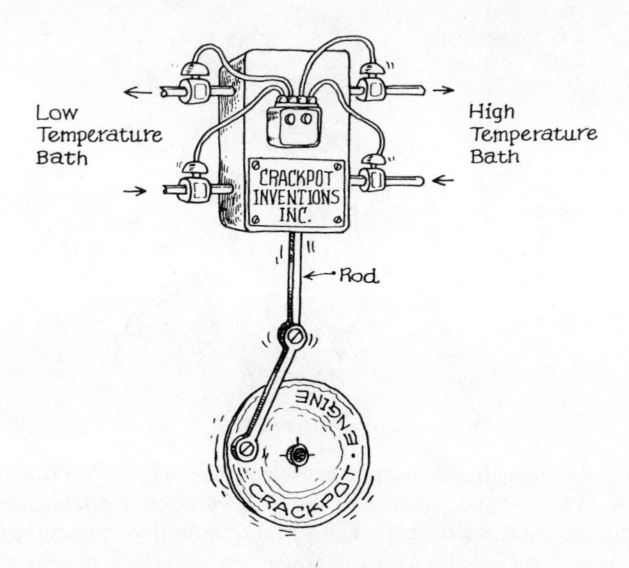

Figure P12.6

a high-temperature (T_A) to a low-temperature (T_B) bath by the extension and relaxation of an elastic rod. This device will operate in a four-step cycle:

(1) Isothermally stretching the elastic rod from L_1 to L_2 while transferring Q_B to the low-temperature bath.
(2) Adiabatically and reversibly stretching the rod from L_2, T_B to L_3, T_A.
(3) Isothermally relaxing the rod from L_3 to L_4, while absorbing Q_A from the high-temperature bath.
(4) Adiabatically and reversibly relaxing the rod to its initial length.

As it now stands, the device is fitted for a 25.4-cm rod that is capable of stretching to 30.5 cm. Crackpot wants to know what temperature they should use in their high-temperature bath so that their device can produce at least 34 J of work per cycle while using the room-temperature environment as the cold bath.

Their elastic rod has the following characteristics:

Size: 25.4 by 0.25 by 0.25 cm (unstretched)

$(\partial \mathscr{s}/\partial T)_L = 7.0$ MPa/K

$C_L = T(\partial S/\partial T)_L = 3.2$ J/g K

Specific gravity: 1.3

Room temperature is 297 K. What must be the temperature of the high-temperature bath? What is the efficiency of the engine?

12.7. A manned spacecraft is to be the payload of a large, multistage rocket. At liftoff, the rocket will accelerate vertically with a constant acceleration. Only later will a trajectory be programmed (Figure P12.7).

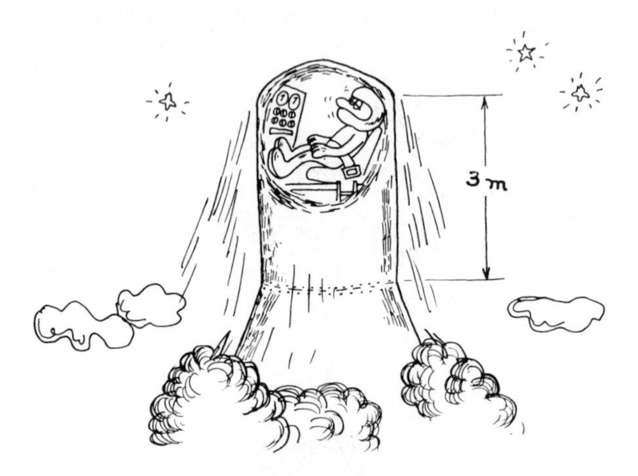

Figure P12.7

Prior to launch, the manned capsule is filled with 20% O_2 and 80% He at 290 K and 1 bar and is sealed until the high-acceleration launch phase is completed. During this same period, any heat transfer to or within the manned capsule may be neglected (i.e., it will remain isothermal), and assume that there is no composition variation due to respiration.

It is necessary to ensure that during the launch acceleration period the oxygen partial pressure never drops below 0.17 bar. Assuming that the capsule is 3 m high and of constant cross section, what is the total pressure at the leading edge (top) of the capsule at the maximum tolerable acceleration?

12.8. We have just received a proposal to support a project designed to desalinate water for submerged submarines. Attached to the skin of the submarine would be a membrane that is semipermeable to water. When the submarine is submerged to the correct depth, water would pass into the submarine (Figure P12.8).

We assume that the seawater density and temperature remain constant with depth and are 1024 kg/m³ and 290 K, respectively. The submarine interior is at 1 bar and has a humidity equivalent to saturation at 290 K (1767 Pa). The partial pressure over the ocean at the surface is 1671 Pa.

At what depth should the submarine cruise to make the system work?

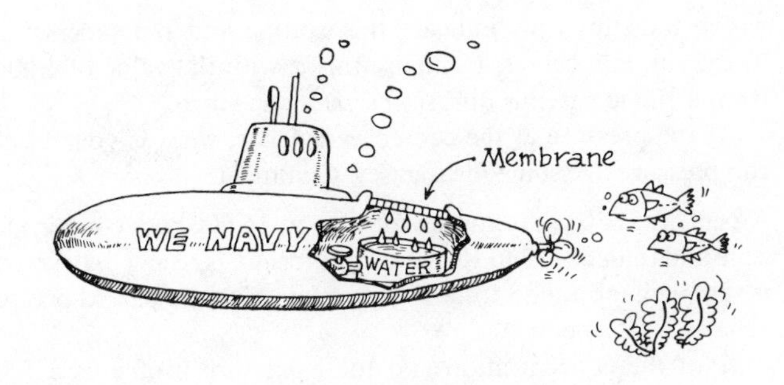

Figure P12.8

12.9. An equimolar mixture of He^3 and He^4 is pumped into a small-diameter, 30.5-m-long, well-insulated vertical tube. The initial pressure is 1 bar and the temperature 5 K. The gas mixture is ideal and no convection currents are present.

Derive a relation to express the equibrium composition as a function of height and determine the fraction of He^4 and the total pressure at the bottom of the tube. How would your results change if the tube length were allowed to increase without limit?

12.10. Preliminary reports indicate that the atmosphere of Saturn consists primarily of methane and hydrogen (see Figure P12.10). At the outer boundary, skimpy

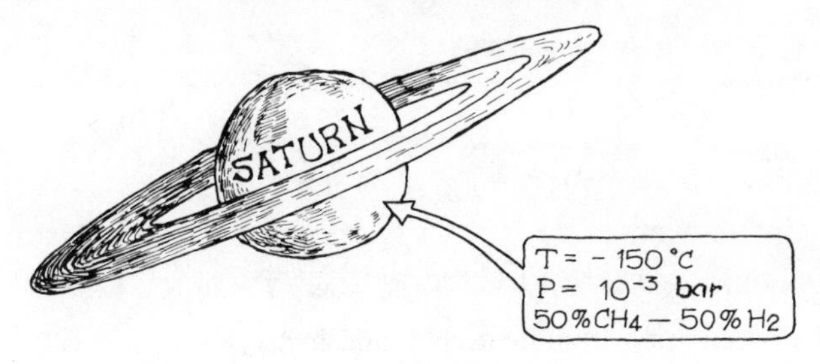

Figure P12.10

data show:

$T \sim -150°C$

$P \sim 10^{-3}$ bar

50% CH_4 and 50% H_2

From other observations, it is believed that the depth of the atmosphere is about 25,000 km. With a constant gravitational acceleration equal to 1.1 g_{earth}, describe the atmospheric concentration and pressure variations with altitude and point out any interesting conclusions.

12.11. To separate the uranium isotope $^{235}UF_6$ from $^{238}UF_6$ by centrifugation, one is limited by the mechanical strength of the material on the periphery. Recent reports, however, have indicated that some new alloys will withstand tangential velocities of 400 m/s at 400 K. Estimate a separation factor α_{8-5}:

$$\alpha_{8-5} = \frac{y_8^r y_5^0}{y_5^r y_8^0}$$

where the subscripts indicate the isotope and the superscripts r and 0 refer to the rim and center. Compare this α with the value of 1.0043 as determined from a single gaseous diffusion separation stage.

If the pressure at the center were 1 bar, what is your best estimate of the rim pressure? Assume ideal gases. Comment.

12.12. A centrifuge 30 cm in radius rotating at 15,000 rpm contains a mixture of benzene and toluene at 340 K. After equilibrium is reached, the mixture at the center is sampled and found to be a vapor with 6700 Pa partial pressures of both benzene and toluene.

Plot the concentration and total pressure profile as a function of radius. Assume that the gas is ideal and that any liquid phase present forms an ideal solution. At 340 K, the vapor pressures of the benzene and toluene are 0.624 and 0.234 bar, respectively.

12.13. In some of the new high-thrust rockets, extreme acceleration fields are expected. Combustion experts are already analyzing problems expected to be encountered in such an environment. You are asked to aid in the program to estimate the equilibrium constant for the reaction

$$CO + \tfrac{1}{2}O_2 = CO_2$$

at 1000 K and 1000 times normal gravity. At 1 g, $\log_{10} K_{1000} = 10.3$. Other data are shown below:

	$\Delta H_{f,298}$ (J/mol)	A	B $\times 10^3$	C $\times 10^7$
O_2	0	26.2	11.50	-32.2
CO_2	-3.93×10^5	28.7	35.73	-103.6
CO	-1.11×10^5	26.2	8.76	-19.2

C_p (J/mol K) $= A + BT/2 + CT^2/3$, where T is in °C.

12.14. Tritium separation from deuterium and hydrogen is not easily accomplished. Yet there are now suggestions that it can be more readily carried out in ultracentrifuges. Estimate what increase in tritium concentration you might expect at the rim draw-off cock if the feed is at atmospheric pressure and is comprised of 5% T_2, 50% D_2, and 45% H_2. The unit is to operate isothermally at 43 K and with a rim velocity of 400 m/s. Do you visualize any practical problems?

12.15. A graduate student adept in the application of thermodynamics was overheard disputing the reasoning of a pragmatic meteorologist. They were discussing the pressure and composition profile in the earth's atmosphere. Both agreed that as a basis, they would choose the earth's surface, where

$N_2/O_2 = 4.0$

Air is saturated with water (= 840 Pa at 278 K)

$T = 278$ K

$P = 1$ bar

The student would prefer to model the atmosphere as quiescent and isothermal in an equilibrium state. Using this model, estimate the N_2/O_2 ratio, the partial pressure of water, and the total pressure at an altitude of 1 km. (Neglect CO_2.)

The meteorlogist claims that this model is sheer nonsense. His own model involves visualizing a ùnit mass of air at sea level expanding polytropically as it is moved up above the surface of the earth. By polytropic expansion, he means that the air expands by a relation PV^n = constant. He has selected n to be 1.2. No compositional changes are allowed. Again calculate the pressure, the water partial pressure, and the temperature at an altitude of 1 km. Compare and contrast both models.

12.16. We present a variation on a scheme that has surfaced several times in the last few years. (See, for example, *Scientific American*, Dec. 1971, p. 100 and Apr. 1972, p. 110, as well as *Science*, June 2, 1972, p. 1011 and Dec. 15, 1972, p. 1199.)

We suggest in Figure P12.16 a simple scheme to reclaim fresh water from the

Figure P12.16

ocean. A long pipe is lowered into the ocean and on the bottom of this pipe we install a membrane permeable to water. As we all know, the density of seawater exceeds that of fresh water; therefore, at any depth, the pressure in the ocean would exceed that which would be exerted by a comparable column of fresh water. If we make the pipe of sufficient length, the pressure difference at the bottom would exceed the osmotic pressure and water from the sea would flow into the pipe.

Comment on the feasibility of this process. Assume that the ocean is an equilibrium system.

How would your evaluation change if we took a more realistic ocean and assume that currents mix the ocean to such a degree that there is neither concentration nor temperature variation with depth?

Data:

Assume that the ocean temperature = 283 K.

P_{vp} for water at 283 K = 1230 Pa.

The partial pressure of water over the ocean, at the surface, is 1207 Pa (this value assumes a 3.5% NaCl solution).

Density of seawater = 1.0255 g/cm³.

Seawater and fresh water are incompressible.

The partial molar volume of H_2O in seawater is equal to the pure molar volume of H_2O.

Thermodynamics of Surfaces 13

In the absence of gravity or other body forces, free liquids assume a spherical shape. Intermolecular attractive forces operate to pull the molecules together; this attraction results in the formation of spheres. A convenient way to characterize this behavior is to assume that the surface is in tension (i.e., each element of the surface layer experiences tensile forces from neighboring elements); the net result is a surface somewhat similar to that in an elastic balloon.

The analogy to the balloon may be carried further. We know that there exists a pressure difference between the inside and outside of a balloon because of these tensile forces. Similarly, in a bubble or drop of liquid, we will show that the internal pressure also exceeds that outside the drop. To demonstrate this, we will again use the concept of surface tension forces.

In this chapter we first examine surfaces from the standpoint of work interactions (i.e., we derive an expression to allow us to determine the work necessary to vary both the *volume and area* of a surface layer). Following this, we explore the effects of curvature on small liquid drops and crystal nuclei. In the remainder of the chapter we develop the important thermodynamic expressions for surface layers and also apply the criteria of equilibrium to the nucleation of new phases. The extensive subject matter of surfaces is examined only from a classical thermodynamic point of view; interesting and very important phenomena such as wetting, surface activity, and adhesion must be left for the reader to enjoy in specialized texts.

13.1 Surface Tension

To illustrate more clearly the mechanical analog of surface forces, consider a plane interface between phases α and β. The element shown in Figure 13.1 measures x units wide, y units deep, and has a thickness τ such that the top and bottom of the element are located in the homogeneous phases α and β, respectively. Each side of this parallelpiped is subjected to a normal, compressive

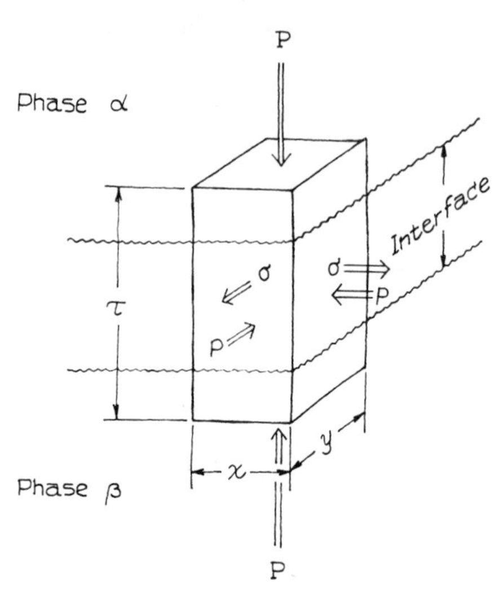

Figure 13.1 Element in plane interface.

pressure P that is equal at all points. This equality results from the criteria of phase equilibria derived for *plane* surfaces in Chapter 6. Also, in the two directions perpendicular to the thickness, τ, there is the tension force σ *per unit length* (i.e., the tensile force on the front and back faces would be σx; on each side face, σy).

If we choose the parallelpiped as our system, the work done by the system when the volume is increased by $dx\, dy\, d\tau$ is given as

$$\begin{aligned} đW &= (Pxy)\, d\tau + (P\tau y - \sigma y)\, dx + (P\tau x - \sigma x)\, dy \\ &= P(xy\, d\tau + \tau y\, dx + \tau x\, dy) - \sigma(y\, dx + x\, dy) \\ &= P\, d\underline{V} - \sigma\, dA \end{aligned} \tag{13-1}$$

In our later development of the thermodynamics of surface layers, the work term must be that given by Eq. (13-1).

13.2 Equilibrium Considerations

Consider the equilibrium system shown in Figure 13.2. It is composed of two distinct parts: a bulk phase α and a small fragment of a different phase β. Phase α is maintained at constant temperature and pressure by contact with the

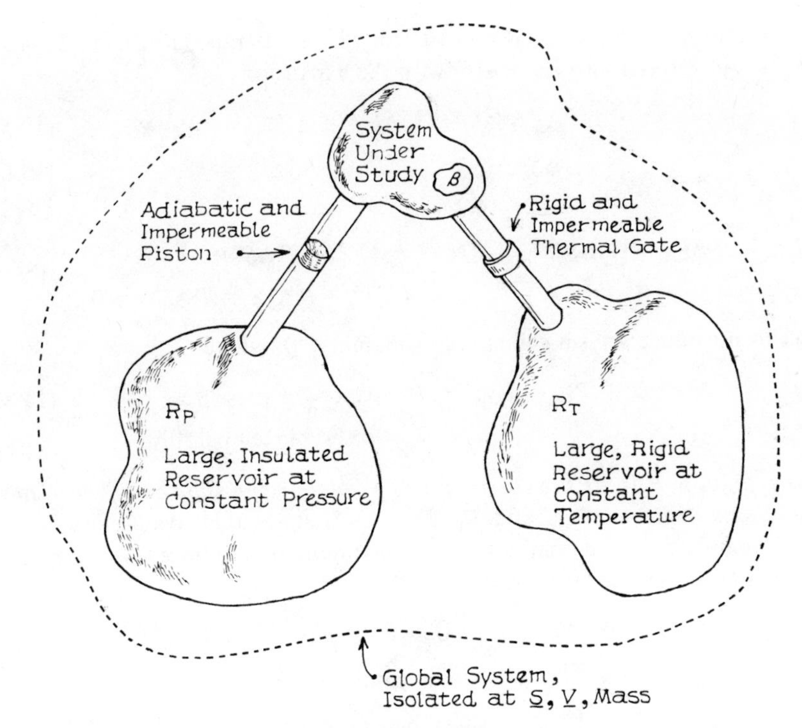

Figure 13.2 System interactions with constant pressure or temperature reservoirs.

large isobaric and isothermal reservoirs R_P and R_T. Except for the presence of the β phase, the situation is identical to that shown in Figure 6.4 and as described as case (c) in Section 6.3. There, with no β phase, it was shown that the system attained an equilibrium state when the total Gibbs energy was a minimum.

The present case is, however, somewhat different because the small element of phase β does not necessarily have the same intensive properties as phase α even though both phases are in equilibrium. In addition, we must consider energy terms for the surface phase between α and β.

If the same nomenclature as in Section 6.3 is used, the criterion for *stable* equilibrium states is

$$\Delta \underline{U}^{\Sigma} = \Delta \underline{U} + \Delta \underline{U}^{R_P} + \Delta \underline{U}^{R_T} > 0 \tag{6-16}$$

where the superscript $\sum$ represents the global system and terms that are not superscripted refer to the α–β system. As shown before, for variations in volume or entropy,

$$\Delta \underline{U}^{R_T} = -P\,\Delta \underline{V}^{R_P} = P\,\Delta \underline{V} \tag{13-2}$$

$$\Delta \underline{U}^{R_T} = T\,\Delta \underline{S}^{R_T} = -T\,\Delta \underline{S} \tag{13-3}$$

where the global conservation of total volume and entropy has been employed. The P and T values in Eqs. (13-2) and (13-3) are those that characterize phase α but not necessarily phase β or σ. Also,

$$\Delta \underline{U} = \Delta \underline{U}^{\alpha} + \Delta \underline{U}^{\beta} + \Delta \underline{U}^{\sigma} \tag{13-4}$$

where β and σ represent the β and surface phases. If Eqs. (13-2), (13-3), and (13-4) are substituted into Eq. (6-16) with the expansions

$$\Delta \underline{V} = \Delta \underline{V}^{\alpha} + \Delta \underline{V}^{\beta} + \Delta \underline{V}^{\sigma} \tag{13-5}$$

$$\Delta \underline{S} = \Delta \underline{S}^{\alpha} + \Delta \underline{S}^{\beta} + \Delta \underline{S}^{\sigma} \tag{13-6}$$

then

$$(\Delta \underline{U}^{\alpha} + P\,\Delta \underline{V}^{\alpha} - T\,\Delta \underline{S}^{\alpha}) + (\Delta \underline{U}^{\beta} + P\,\Delta \underline{V}^{\beta} - T\,\Delta \underline{S}^{\beta}) + (\Delta \underline{U}^{\sigma} + P\,\Delta \underline{V}^{\sigma} - T\,\Delta \underline{S}^{\sigma}) > 0 \tag{13-7}$$

Let us introduce a term called the availability, $\underline{\mathcal{V}}$, where

$$\underline{\mathcal{V}} \equiv \underline{U} + P\underline{V} - T\underline{S}; \qquad P = \text{reservoir pressure} \quad T = \text{reservoir temperature} \tag{13-8}$$

In this definition, P and T *are constants and equal to the pressure and temperature in the external reservoirs* R_P and R_T. Eq. (13-7) states that availability is the thermodynamic function that attains a maximum or minimum value in the α–β–σ system; then, at equilibrium,

$$d\underline{\mathcal{V}} = 0 \tag{13-9}$$

$$d^m\underline{\mathcal{V}} > 0 \quad \text{(stable equilibrium)}$$
$$< 0 \quad \text{(unstable equilibrium)} \tag{13-10}$$

where $d^m\underline{\mathcal{V}}$ is the lowest order, nonvanishing derivative. We discuss the stability criterion in Section 13.7.

At equilibrium, with Eqs. (13-7), (13-8), and (13-9),

$$d\underline{\mathcal{V}}^{\alpha} + d\underline{\mathcal{V}}^{\beta} + d\underline{\mathcal{V}}^{\sigma} = 0 \tag{13-11}$$

To determine the specific equilibrium criteria for this case, $\underline{U}^{\alpha}$, $\underline{U}^{\beta}$, and $\underline{U}^{\sigma}$ are expanded using the Fundamental Equations in energy representation,

$$d\underline{U}^{\alpha} = T\,d\underline{S}^{\alpha} - P\,d\underline{V}^{\alpha} + \sum_{j=1}^{n} \mu_j^{\alpha}\,dN_j^{\alpha} \tag{13-12}$$

$$d\underline{U}^{\beta} = T^{\beta}\,d\underline{S}^{\beta} - P^{\beta}\,d\underline{V}^{\beta} + \sum_{j=1}^{n} \mu_j^{\beta}\,dN_j^{\beta} \tag{13-13}$$

$$d\underline{U}^{\sigma} = T^{\sigma}\,d\underline{S}^{\sigma} - P^{\sigma}\,d\underline{V}^{\sigma} + \sigma\,dA + \sum_{j=1}^{n} \mu_j^{\sigma}\,dN_j^{\sigma} \tag{13-14}$$

With Eqs. (13-11) through (13-14) and with the understanding that $T = T^{\alpha}$ and $P = P^{\alpha}$ are constants and are determined by the reservoir temperatures and pressures,

$$(T^{\beta} - T)\,d\underline{S}^{\beta} + (T^{\sigma} - T)\,d\underline{S}^{\sigma} + \sum_{j=1}^{n} \mu_j^{\alpha}\,dN_j^{\alpha} + \sum_{j=1}^{n} \mu_j^{\beta}\,dN_j^{\beta} + \sum_{j=1}^{n} \mu_j^{\sigma}\,dN_j^{\sigma} - (P^{\beta} - P)\,d\underline{V}^{\beta} - (P^{\sigma} - P)\,d\underline{V}^{\sigma} + \sigma\,dA = 0 \tag{13-15}$$

Before examining Eq. (13-15), there is one additional restraint equation that must be used (i.e., there is conservation of mass in the system). For each component,

$$dN_j = dN_j^\alpha + dN_j^\beta + dN_j^\sigma = 0 \qquad (13\text{-}16)$$

If Eq. (13-16) is multiplied by $-\mu_j^\alpha$ (chosen as a Lagrange multiplier) and added to Eq. (13-15), each variation is independent and the coefficients must be zero. Therefore,

$$T^\beta = T = T^\alpha \qquad (13\text{-}17)$$

$$T^\sigma = T = T^\alpha \qquad (13\text{-}18)$$

$$\mu_j^\beta = \mu_j^\alpha \qquad (13\text{-}19)$$

$$\mu_j^\sigma = \mu_j^\alpha \qquad (13\text{-}20)$$

and

$$(P^\beta - P)\, d\underline{V}^\beta + (P^\sigma - P)\, d\underline{V}^\sigma - \sigma\, dA = 0 \qquad (13\text{-}21)$$

The first four equalities are as we expected: the temperatures and chemical potentials of each component are equal throughout the system. In this regard, our results do not differ from what we found earlier for phase equilibrium criteria. Equation (13-21) is, however, a different result.[1]

First, let us consider the term $(P^\sigma - P)\, d\underline{V}^\sigma$. We have some latitude in defining the extent of the surface phase and, depending on our choice, we could have different values of P^σ and $d\underline{V}^\sigma$. If we were to define our surface phase boundary so that the surface had properties similar to phase α, then $P^\sigma \sim P^\alpha = P$ and $(P^\sigma - P)\, d\underline{V}^\sigma$ is essentially zero.

Usually, however, the surface layer is associated with the discontinuous phase (in this case, β), so the $(P^\sigma - P)\, d\underline{V}^\sigma$ term is incorporated with the $(P^\beta - P)\, d\underline{V}^\beta$ term. Equation (13-21) then reduces to

$$P^\beta - P = P^\beta - P^\alpha = \sigma \frac{dA}{d\underline{V}^\beta} \qquad (13\text{-}22)$$

This final relation expresses the pressure difference across a curved interface in an equilibrium system.

Example 13.1

(a) If phase β were spherical in shape, what would be the pressure difference $(P^\beta - P^\alpha)$?

(b) Suppose that phase β were assumed to be a solid with the shape of a pancake of radius r and some fixed height h. (h is some multiple of the lattice spacing.) This case is often used in two-dimensional nucleation theories involving crystallization on a flat surface. For such a situation, how would $(P^\beta - P^\alpha)$ be related to the radius r?

(c) Repeat part (a) but assume that the phase boundary between the α and β phases is characterized by two radii of curvature C_1 and C_2.

[1]The case where β is in contact with two discrete α phases is noted in Section 13.7.

Solution

(a) Let the radius of the sphere of phase β be r. Then, $A = 4\pi r^2$, $\underline{V}^\beta = (4/3)\pi r^3$ and then, from Eq. (13-22),

$$P^\beta - P^\alpha = \sigma \frac{dA}{d\underline{V}^\beta} = \frac{2\sigma}{r} \tag{13-23}$$

(b) For the assumed pancake shape, the area of the curved surface, $A = 2\pi rh$ and $\underline{V}^\beta = \pi r^2 h$.

$$P^\beta - P^\alpha = \sigma \frac{dA}{d\underline{V}^\beta} = \frac{\sigma}{r} \tag{13-24}$$

(c) In Figure 13.3 a small element of the surface is shown. The area is xy. If we

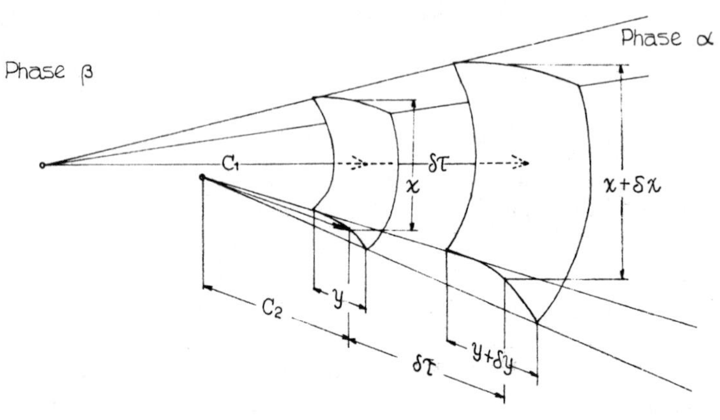

Figure 13.3

extend the element by $\delta\tau$ and thereby increase C_1 to $C_1 + \delta\tau$ and C_2 to $C_2 + \delta\tau$, then the increase in area and volume of phase β becomes

$$\delta A = (y + \delta y)(x + \delta x) - xy = y\,\delta x + x\,\delta y$$

$$\delta \underline{V}^\beta = xy\,\delta\tau$$

But, by similar triangles,

$$\frac{x + \delta x}{C_1 + \delta\tau} = \frac{x}{C_1}$$

$$\frac{y + \delta y}{C_2 + \delta\tau} = \frac{y}{C_2}$$

Then

$$\delta x = \frac{x}{C_1}\,\delta\tau$$

$$\delta y = \frac{y}{C_2}\,\delta\tau$$

Substituting the δx and δy terms into the expression for δA and using Eq. (13-22),

$$P^\beta - P^\alpha = \sigma\left(\frac{dA}{d\underline{V}^\beta}\right) = \frac{(xy\,\delta\tau/C_1 + xy\,\delta\tau/C_2)\sigma}{xy\,\delta\tau}$$

$$P^\beta - P^\alpha = \sigma\left(\frac{1}{C_1} + \frac{1}{C_2}\right) \tag{13-25}$$

For a sphere, $C_1 = C_2 = r$ and Eq. (13-25) reduces to Eq. (13-23); for a plane surface, $C_1 = C_2 = \infty$ and $P^\beta = P^\alpha$. The latter case yields the pressure equality we found earlier when we studied phase equilibrium for planar surfaces.

13.3 Effects of Pressure Differences across Curved Interfaces

For our equilibrium system containing a large bulk phase α and fragments of a different phase β, we have shown that the pressure in phase β exceeds that in α. If β is spherical, the pressure difference is given by Eq. (13-23). If β has a different curvature, we can always revert to the general case as shown by Eq. (13-25). These relations show how the pressure difference depends on the size and shape of the fragment of phase β.

To estimate the pressure inside the small fragments of phase β, it is convenient to relate P^β to the true equilibrium pressure in the system when a phase with the same *composition* and *temperature* as phase α is in equilibrium with a *planar* surface of phase β. Several examples are given below to illustrate the methodology.

Phase β: small liquid drops; Phase α: vapor

Let us begin with the case where a *pure* component vapor at P^α and T^α is in equilibrium with small liquid drops with radii r. The pressure inside the drop is P^β but, by Eq. (13-17), the drop temperature $T^\beta = T^\alpha$. The chemical potential equality between the drop and vapor phases is expressed in terms of fugacity (see Section 8.7):

$$f^\beta(P^\beta) = f^\alpha(P^\alpha) \tag{13-26}$$

The fugacity of the vapor (α) phase may be written in terms of a fugacity coefficient ν:

$$f^\alpha(P^\alpha) = \nu(P^\alpha)P^\alpha \tag{13-27}$$

For the fugacity of the liquid drop (β) phase, we can relate the fugacity at P^β to the planar equilibrium pressure P_{vp} by the Poynting correction [see Eq. (8-121)],

$$f^\beta(P^\beta) = f^\beta(P_{vp}) \exp\left[\left(\frac{1}{RT}\right)\int_{P_{vp}}^{P^\beta} V^\beta \, dP\right] \tag{13-28}$$

The fugacity of the liquid drop at P_{vp} is equal to the fugacity of the vapor at P_{vp}, so, analogous to Eq. (13-27),

$$f^\beta(P_{vp}) = \nu(P_{vp})P_{vp} \tag{13-29}$$

Combining Eqs. (13-26) through (13-29), we obtain

$$\nu(P_{vp})P_{vp} \exp\left[\left(\frac{1}{RT}\right)\int_{P_{vp}}^{P^\beta} V^\beta \, dP\right] = \nu(P^\alpha)P^\alpha \tag{13-30}$$

Equation (13-30) relates the pressures P^β and P^α to P_{vp} and other properties of the system. If we further assume that the pure vapor behaves as an ideal gas at P_{vp} and P^α, both fugacity coefficients are unity. Also, assume that the liquid molar volume, V^β, is not a strong function of pressure. Then Eq. (13-30) simplifies to

$$P^\alpha = P_{vp} \exp\left[\frac{V^\beta(P^\beta - P_{vp})}{RT}\right] \tag{13-31}$$

From Example 13.1, $P^\beta = P^\alpha + 2\sigma/r$, so

$$P^\alpha = P_{vp} \exp\left[\frac{V^\beta(P^\alpha - P_{vp} + 2\,\sigma/r)}{RT}\right] \tag{13-32}$$

In most cases $(P^\alpha - P_{vp}) \ll 2\,\sigma/r$, so Eq. (13-32) becomes

$$P^\alpha = P_{vp} \exp\left(\frac{2\sigma V^\beta}{r\,RT}\right) \tag{13-33}$$

To illustrate the use of Eq. (13-33), suppose that our α phase were pure water vapor at P^α and 313 K. At this temperature, the vapor pressure of water is 7370 Pa. That is, if the water vapor were in equilibrium with a *planar liquid* water surface, the pressure would be 7370 Pa. However, if the water vapor were in equilibrium with small spherical water droplets, the pressure would exceed 7370 Pa. For instance, if the droplet radii were 10 nm, then $P^\alpha = 8070$ Pa. If droplet radii could be made as small as 1 nm, $P^\alpha \sim 20{,}000$ Pa. In using Eq. (13-33) to make such estimates, we have assumed that the surface tension of water is independent of drop size. This assumption is questionable for very small drops.

Example 13.2

If spherical drops of water (of the same size) were in equilibrium with water vapor at 1.1 bar and 373 K, estimate the droplet radii and the pressure within the drops. At 373 K, the surface tension of water is 0.0589 N/m and the molar volume is 1.87×10^{-5} m^3/mol. The vapor pressure of water at 373 is 1.013 bar.

Solution

With Eq. (13-33),

$$P^\alpha = 1.1 = 1.013 \exp\left[\frac{(2)(0.0589)(1.87 \times 10^{-5})}{(8.314)(373)r}\right]$$

$$r = 8.6 \times 10^{-9}\ \text{m} = 8.6\ \text{nm}$$

$$P^\beta - P^\alpha = \frac{2\sigma}{r} = \frac{(2)(0.0589)}{8.6 \times 10^{-9}}$$

$$= 1.37 \times 10^7\ \text{Pa}$$

$$P^\beta = 1.1 \times 10^5 + 1.37 \times 10^7 = 1.38 \times 10^7\ \text{Pa}$$

Within such drops, the *calculated* pressure is very large. It is this high pressure that increases the fugacity of the liquid water so that a higher vapor phase fugacity results at equilibrium.

A similar treatment is possible for the case where we have a vapor-phase *mixture* at T^α and P^α in equilibrium with small liquid drops. Let x_j^β represent liquid mole fractions within the droplet and y_j^α be the vapor mole fractions in phase α. The analogs for Eqs. (13-26) through (13-30) are

$$\hat{f}_j^\beta(P^\beta, x_j^\beta) = \hat{f}_j^\alpha(P^\alpha, y_j^\alpha) \tag{13-34}$$

$$\hat{f}_j^\alpha(P^\alpha, y_j^\alpha) = \phi_j(P^\alpha, y_j^\alpha)\, P^\alpha y_j^\alpha \tag{13-35}$$

$$\hat{f}_j^\beta(P^\beta, x_j^\beta) = \nu_j(P_{vp_j})P_{vp_j}\gamma_j^\beta x_j^\beta \exp\left[\left(\frac{1}{RT}\right)\int_{P_{vp_j}}^{P^\beta} V_j^\beta\, dP\right] \tag{13-36}$$

In Eq. (13-35), ϕ_j is the fugacity coefficient of component j in the vapor. γ_j^β is the activity coefficient of j in the liquid drop phase and depends on P^β and x_j^β. $\nu_j(P_{vp_j})$ is the fugacity coefficient of pure vapor j at T^α and P_{vp_j}. Substituting Eqs. (13-35) and (13-36) into (13-34) leads to an unwieldy expression. Let us assume the vapor phase is an ideal-gas mixture ($\phi_j = 1.0$), the liquid drop is an ideal solution ($\gamma_j^\beta = 1.0$), the pure vapor of j at P_{vp_j} is ideal ($\nu_j = 1.0$), and that $V_j^\beta \neq f(P)$. Then

$$P^\alpha y_j^\alpha = P_{vp_j} x_j^\beta \exp\left[\frac{V_j^\beta(P^\beta - P_{vp_j})}{RT}\right] \tag{13-37}$$

This equation may be compared to Eq. (13-31) for a pure component under similar conditions.

There are n relations of the form of Eq. (13-37) and, as shown in Example 13.1, one equation showing how P^α and P^β depend on drop size (r): These are sufficient to solve for the $n - 1$ unknown mole fractions in the β phase as well as for P^α and P^β. Note that the exponential term in Eq. (13-37) can be considered as a factor that enhances the vapor pressure of each component in the liquid. Since molar volumes and vapor pressures differ from one component to another, the equilibrium liquid drop composition will differ from that of a planar liquid surface in equilibrium with the same vapor.

Phase β: vapor bubble; Phase α: liquid

Let us again start with a pure-component case. The α phase is now liquid at P^α, while the β phase consists of vapor bubbles of radii r. (The temperatures in both phases are equal.) Following the same technique illustrated above,

$$f^\beta(P^\beta) = f^\alpha(P^\alpha) \tag{13-38}$$

$$\nu(P^\beta)P^\beta = \nu(P_{vp})P_{vp} \exp\left[\left(\frac{1}{RT}\right)\int_{P_{vp}}^{P^\alpha} V^\alpha\, dP\right] \tag{13-39}$$

Assuming ideal gases and $V^\alpha \neq f(P)$,

$$P^\beta = P_{vp} \exp\left[\frac{V^\alpha(P^\alpha - P_{vp})}{RT}\right] \tag{13-40}$$

Equation (13-40) is identical to Eq. (13-31) with the α–β superscripts interchanged. The interpretation is, however, different, as illustrated in Example 13.3.

Example 13.3

We have a bulk liquid phase of pure ethane at 1 bar pressure and at a temperature of 270 K. This liquid is in equilibrium with small vapor bubbles of radii r. Calculate r and, also, the pressure within the bubbles.

At 270 K the vapor pressure of ethane is 22.1 bar, the liquid molar volume is 7.38×10^{-5} m^3/mol, and the surface tension is 3.5×10^{-3} N/m.

Solution

In this case we have liquid ethane, which is highly superheated. (From Problem 9.3 we see that the ethane is in a state very close to the limit of stability from a thermodynamic point of view.) If the liquid ethane were in equilibrium with a planar vapor interface, the pressure would be 22.1 bar.

With Eq. (13-40)

$$P^{\beta} = 22.1 \exp\left[\frac{(7.38 \times 10^{-5})(10^5 - 22.1 \times 10^5)}{(8.314)(270)}\right]$$

$$= 20.6 \times 10^5 \text{ Pa} = 20.6 \text{ bar}$$

$$P^{\beta} - P^{\alpha} = \frac{2\sigma}{r} = (20.6 \times 10^5 - 10^5) = \frac{(2)(3.5 \times 10^{-3})}{r}$$

$$r = 3.6 \times 10^{-9} \text{ m} = 3.6 \text{ nm}$$

As we noted in the liquid drop–vapor case, the pressure inside the small fragments of the β phase is significantly higher than the bulk α phase.

As we shall see later, calculation of the equilibrium bubble (or drop) size is an important step in formulating a theory of nucleation of phase β from a bulk α phase.

The equilibrium bubble pressure and size for a superheated liquid mixture can be determined by a similar approach.

Phase β: solid particles; Phase α: liquid

For this situation, consider a pure liquid in equilibrium with finely divided solid of the same material. If the solid phase were large and planar, the temperature would correspond to the freezing point of the liquid. The fact that the solid is composed of small fragments leads to a prediction that the freezing point will be depressed. If we assume that the solid fragments are spherical in shape with radii r,

$$\Delta T = T_r - T_m \sim \frac{-2V^s \sigma T_m}{r\,\Delta H_f} \tag{13-41}$$

where T_r = freezing point of the liquid in equilibrium with the finely divided solid

T_m = freezing point of the liquid in equilibrium with a planar solid

V^s = molar volume of the solid at T_m and at the liquid pressure

σ = solid–liquid interfacial tension

ΔH_f = enthalpy of fusion

The derivation of Eq. (13-41) is left as a problem at the end of the chapter. Actual calculations of ΔT are difficult as few values of σ are known. In fact, the measurement of ΔT for a known value of r has been used to infer values of σ.

Phase β: solid particles; Phase α: liquid solution

In this case one is interested in exploring whether the solubility of a solute in a solution is modified if the solution is in equilibrium with small solid fragments rather than with large (planar) solute crystals. To formulate the problem, suppose that we have a binary solution of 1 and 2 at some pressure and temperature P^α, T^α. This solution is in equilibrium with small fragments of a pure solute phase (component 1) at T^α and P^β. Then

$$f_1^\beta(P^\beta) = \hat{f}_1(P^\alpha, x_1^\alpha) \tag{13-42}$$

Next, assume that a liquid solution of 1 and 2 at the same temperature and pressure (P^α) were in equilibrium with a planar solute surface. In this case the solute concentration in the liquid is the "equilibrium" solubility $x_{1,\text{eq}}^\alpha$:

$$f_1^\beta(P^\alpha) = \hat{f}_1^\alpha(P^\alpha, x_{1,\text{eq}}^\alpha) \tag{13-43}$$

Note that in the planar case, the solid pressure is P^α.

Taking logarithms of Eqs. (13-42) and (13-43) and subtracting, then, with Eq. (8-149), we obtain

$$\ln \frac{a_1}{a_{1,\text{eq}}} = \frac{V^s}{RT}(P^\beta - P^\alpha) \tag{13-44}$$

a_1 is the activity of the solute in the solution that is in contact with the small solid solute particles. Approximating the ratio of activities by the ratio of concentrations and using $P^\beta - P^\alpha = 2\sigma/r$ yields

$$C_1 = C_{1,\text{eq}} \exp\left(\frac{V^s 2\sigma}{rRT}\right) \tag{13-45}$$

Since the exponential term is greater than unity, the solubility C_1 exceeds the equilibrium solubility $C_{1,\text{eq}}$. If the solid ionizes during dissolution, the exponential argument should be divided by i, the increase in the number of particles during dissolution and electrolytic dissociation.

Equation (13-45) may be criticized from many points of view. Small particles are probably not spherical, and a departure from this simple shape can considerably affect C/C_{eq} values.[2] Also, the correspondence between concentration and activity may not be acceptable for highly soluble compounds. Finally, and most important, it is probably not reasonable to associate σ with a solid–liquid interfacial tension, since, for small aggregates of only a few molecules, this interfacial tension loses any meaning.[3] Harbury suggests that a fictitious interfacial tension, σ', be used where, from the few data available, σ' is much less than σ.

[2] W. J. Jones, "Über die Beziehung zwischen geometrischer Form und Dampfdruck, Löslichkeit, und Formenstabilität," *Ann. Phys.*, **41**, 441 (1913).

[3] Lawrence Harbury, "Solubility and Melting Point as Functions of Particle Size," *J. Phys. Chem.*, **50**, 190 (1946).

A modification of Eq. (13-45) has been used in the development of two-dimensional nucleation growth theories,[4] where it is assumed that crystal growth is controlled by the rate of formation of small pancake-shaped nuclei on a large planar crystallite surface. For such a solid shape, as shown in Example 13.1, $P^\beta - P^\alpha = \sigma/r$, so, when Eq. (13-45) is derived, the factor of 2 has been dropped from the exponential term. Then, if the solution supersaturation for solute 1 is defined as

$$S = \frac{C_1 - C_{1,\text{eq}}}{C_{1,\text{eq}}} \tag{13-46}$$

the equilibrium pancake radius is

$$r = \frac{V^s}{RT \ln (1 + S)} \tag{13-47}$$

A definite size of two-dimensional nucleus is then associated with a given supersaturation of the solution.

13.4 Pure-Component Relations

The Fundamental Equations for any two phases α and β, separated by a surface phase σ, have already been given in Eqs. (13-12) through (13-14). Let us assume that the interface between α and β is not highly curved so that $P^\alpha = P^\beta = P^\sigma$ and begin our development by considering a closed system in which there is but a single nonreacting component present. We have also shown that, at equilibrium, $T^\alpha = T^\beta = T^\sigma = T$ and $\mu_j^\alpha = \mu_j^\beta = \mu_j^\sigma = \mu_j$. The total Legendre transform of Eqs. (13-12) through (13-14) yields the Gibbs–Duhem equations [see Eq. (5-45)]. For the α and β phases,

$$-S^\alpha \, dT + V^\alpha \, dP - d\mu = 0 \tag{13-48}$$

$$-S^\beta \, dT + V^\beta \, dP - d\mu = 0 \tag{13-49}$$

For the surface phase, in extensive form,

$$-\underline{S}^\sigma \, dT + \underline{V}^\sigma \, dP - A \, d\sigma - N^\sigma \, d\mu = 0 \tag{13-50}$$

and we divide Eq. (13-50) by A to obtain

$$-S^\sigma \, dT + \tau \, dP - d\sigma - \Gamma \, d\mu = 0 \tag{13-51}$$

In Eq. (13-51), $S^\sigma = \underline{S}^\sigma/A$ is the entropy per unit area. Similarly, $\tau = \underline{V}^\sigma/A$ is the thickness of the surface and $\Gamma = N^\sigma/A$ is a surface concentration, moles per unit area.

Eliminating dP and $d\mu$ from Eqs. (13-48), (13-49), and (13-51),

$$-\frac{d\sigma}{dT} = (S^\sigma - \Gamma S^\beta) - \frac{(\tau - \Gamma V^\beta)(S^\alpha - S^\beta)}{V^\alpha - V^\beta} \tag{13-52}$$

[4] M. Ohara and R. C. Reid, *Modelling Crystal Growth Rates from Solution*, (Englewood Cliffs, N.J.: Prentice-Hall, 1973), Chap. 2.

Equation (13-52) is symmetrical in α and β and terms containing these superscripts may be interchanged. The Euler form of Eq. (13-14) may be written as

$$\sigma = H^{\sigma} - TS^{\sigma} - \Gamma\mu \tag{13-53}$$

If Eq. (13-52) is multiplied by T and the result added to Eq. (13-53) and then the Euler form of Eq. (13-13), as multiplied by Γ, is used to eliminate the $\Gamma\mu$ term, there results

$$\sigma - T\frac{d\sigma}{dT} = (H^{\sigma} - \Gamma H^{\beta}) - \frac{(\tau - \Gamma V^{\beta})(H^{\alpha} - H^{\beta})}{V^{\alpha} - V^{\beta}} \tag{13-54}$$

where $H^{\sigma} \equiv \underline{H}^{\sigma}/A = (\underline{U}^{\sigma} + P\underline{V}^{\sigma})/A$.

Equations (13-52) and (13-54) relate the surface tension and surface tension-temperature gradient to thermodynamic properties such as entropy, enthalpy, and volume. It is not immediately obvious, however, that these relations are invariant with respect to the variations in thickness, τ, of the surface. Suppose for instance, that τ is extended by a slight amount into phase β, then

$$\frac{dS^{\sigma}}{d\tau} = \frac{S^{\beta}}{V^{\beta}}$$

$$\frac{d\Gamma}{d\tau} = \frac{1}{V^{\beta}}$$

Differentiating say, Eq. (13-52) with respect to a variation in τ into phase β, the right-hand side becomes zero. Thus, the relation is not dependent on this particular variation in τ. Similar results are obtained if τ is increased into phase α.

This invariance leads, of course, to an indefiniteness about terms such as S^{σ}, H^{σ}, τ, Γ. As noted above, for a given τ, there is a given S^{σ}. If τ is increased in phase β, per unit area, the entropy will increase by $S^{\beta}(d\tau/V^{\beta})$. This new S^{σ} will depend on the value of S^{β} to a degree proportionate to the increase in τ. We are accustomed to visualizing the properties of the surface layer to be more representative of the less mobile phase or of the discontinuous phase in a two-liquid system. For example, in a liquid–gas system, we often imagine the surface phase to possess nearly liquid-like properties. (Note that this is always possible by choosing the position of the surface layer in a manner to exclude any appreciable gas phase.) By choices of this nature, we may simplify Eqs. (13-52) and (13-54) to more common forms. For example, in a gas–liquid system, if phase β were liquid and phase α gas, then τ/Γ is comparable to $V^{\beta} = V^{L}$, which is much less than $(V^{\alpha} - V^{L}) = (V^{V} - V^{L})$. Equations (13-52) and (13-54) then become

$$-\frac{d\sigma}{dT} = S^{\sigma} - \Gamma S^{L} \tag{13-55}$$

$$\sigma - T\frac{d\sigma}{dT} = H^{\sigma} - \Gamma H^{L} \tag{13-56}$$

so

$$\sigma = G^{\sigma} - \Gamma G^{L} \tag{13-57}$$

Even in these simplified forms, S^σ, H^σ, and Γ are still indefinite and depend on τ, although neither $(S^\sigma - \Gamma S^L)$ nor $(H^\sigma - \Gamma H^L)$ depend on τ.

Surface tension decreases with temperature, and a convenient estimation equation for nonpolar materials is given by Brock and Bird:[5]

$$\sigma = P_c^{2/3} T_c^{1/3} Q (1 - T_r)^{11/9} \tag{13-58}$$

where P_C is in bar, T_c in kelvins, and σ in N/m.

$$Q = 1.207 \times 10^{-4} \left(1 + \frac{T_{b_r} \ln P_c}{1 - T_{b_r}}\right) - 2.81 \times 10^{-4} \tag{13-59}$$

and $T_{b_r} = T_b/T_c$.

Other estimation techniques have been suggested,[6] and Scriven and Davis with their colleagues have employed mean field theory to calculate surface tensions.[7]

Since $d\sigma/dT$ is negative, $(S^\sigma - \Gamma S^L)$ is positive. This quantity represents the difference between the entropy (per unit area) in the surface film and the entropy the same surface would have if it had the properties of the bulk liquid phase. In a similar manner, $\sigma - T(d\sigma/dT)$ may be visualized as the enthalpy difference between the surface film and bulk liquid. With Eq. (13-57), the surface tension itself may be considered to be the difference in Gibbs surface energy and the bulk liquid Gibbs energy expressed on a unit area basis. This isothermal Gibbs energy change is often defined to be the reversible work required to bring material from the bulk liquid to form a unit area of surface.

Before leaving this section on pure component surface thermodynamics, it is instructive to review the developments presented thus far. In the initial sections of this chapter, from mechanical considerations only, the concept of surface tension was developed. These same arguments led to some interesting conclusions such as the pressure difference existing across curved interfaces, the enhanced vapor pressure over small convex liquid or solid surfaces, etc. The application of thermodynamics in this development was minimal although it was proved that in any equilibrium-multiphase system, regardless of curvature, the temperature and chemical potentials were equal in all phases.

From these concepts, the surface tension and its temperature gradient were related to fundamental thermodynamic properties of the surface. The utility of this latter development may be questioned. The surface layer emphasized in the foregoing paragraphs is held by some to be completely fictitious and devoid of true physical significance; its depth is indefinite, it is a composite of the properties of the bounding phases, and the absolute values of S^σ, H^σ, etc., are dictated solely by the exact position and thickness of the layer—as defined by the user—not by nature.

[5] J. R. Brock and R. B. Bird, *AIChE J.*, **1**, 174 (1955).

[6] R.C. Reid, J. M. Prausnitz, and T. K. Sherwood, *Properties of Gases and Liquids*, 3rd ed. (New York: McGraw-Hill, 1977).

[7] B. S. Carey, L. E. Scriven, and H. T. Davis, *AIChE J.*, **24**, 1076 (1978); V. Bongiorno, L. E. Scriven, and H. T. Davis, *J. Coll. Int. Sci.*, **57**, 462 (1976).

13.5 Multicomponent Relations

The thermodynamics of surface layers in multicomponent systems is a straightforward extension from those developed in Section 13.4 for single components, although the algebra is considerably more complex. The relations employed are shown below; as before, the surface-layer equations are expressed in terms of a unit area of surface (i.e., $S^\sigma = \underline{S}^\sigma/A$; $\tau = \underline{V}^\sigma/A$; $\Gamma_j = N_j^\sigma/A$). The Gibbs–Duhem relation, Eq. (13-51), becomes

$$-d\sigma = S^\sigma\, dT - \tau\, dP + \sum_{k=1}^{n} \Gamma_k\, d\mu_k \tag{13-60}$$

For both phases α and β one may express the chemical potential of any component j as, for example, in phase β,

$$d\mu_j = -\bar{S}_j^\beta\, dT + \bar{V}_j^\beta\, dP + \sum_{k \neq i} \left(\frac{\partial \mu_j}{\partial x_k^\beta}\right)_{T, P, x[i, k]} dx_k^\beta \tag{13-61}$$

At *constant composition*, in the β and α phases, with Eqs. (13-60) and (13-61),

$$-d\sigma = \left(S^\sigma - \sum_{k=1}^{n} \Gamma_k \bar{S}_k^\beta\right) dT - \left(\tau - \sum_{k=1}^{n} \Gamma_k \bar{V}_k^\beta\right) dP \tag{13-62}$$

and from equations of the form of Eq. (13-61) for any component k, with all $dx = 0$,

$$dP = \frac{\bar{S}_k^\alpha - \bar{S}_k^\beta}{\bar{V}_k^\alpha - \bar{V}_k^\beta}\, dT, \qquad \text{all } x^\beta \text{ constant} \tag{13-63}$$

Equation (13-62) may be substituted into Eq. (13-63) to eliminate all pressure terms. Applying Eq. (13-62) to a vapor–liquid system where the vapor phase is α and the liquid phase is β, $\tau/\sum\Gamma_k$ is comparable to the liquid-phase volume $V^L = \sum_{k=1}^n x_k^L \bar{V}_k^L$. Assuming ideal solutions where $\bar{V}_k^L = V_k^L$,

$$\tau - \sum_{k=1}^{n} \Gamma_k \bar{V}_k^L \simeq \tau - V^L \sum_{k=1}^{n} \Gamma_k \simeq \sum_{k=1}^{n} \Gamma_k \left(\frac{\tau}{\sum \Gamma_i} - V_k^L\right) \sim 0 \tag{13-64}$$

Thus, in this case,

$$-\left(\frac{\partial \sigma}{\partial T}\right)_L = S^\sigma - \sum_{k=1}^{n} \Gamma_k \bar{S}_k^L \quad \text{(constant composition in } L \text{ phase)} \tag{13-65}$$

Equation (13-64) is a good approximation to eliminate the "dP" term since the small residue $\sum \Gamma_i \bar{V}_i^L$ must still be divided [as from Eq. (13-63)] by a large number, $(\bar{V}_k^V - \bar{V}_k^L) \sim \Delta V_{\text{vaporization}}$, to reduce further the magnitude of the multiplier of the "dP" term. The form of Eq. (13-65) is identical to Eq. (13-55) and the discussion following the latter equation is also applicable to Eq. (13-65). That is, the equation is invariant with respect to the position or thickness of the surface layer as long as each bounding surface of the interfacial layer is in a different phase.

There is, however, a limitation to either Eq. (13-61) or (13-62). That is, the composition of all components in the liquid phase must be held constant. This restriction originates from the elimination of the $d\mu_j$ terms in Eq. (13-60) with Eq. (13-61) as simplified by forcing the terms containing $(\partial \mu_j/\partial x_k^L) dx_k^L$

to be zero. There is no thermodynamic inconsistency here since from the phase rule there are n independent variables and if $(n - 1)$ compositions in the liquid phase are fixed, then σ can be expressed as a function of temperature only. Similar statements may be made if the derivation of Eq. (13-62) were made by using a relation of the form of Eq. (13-61) but for the α phase. The final resulting equation (13-62) or (13-65) does have this composition restriction and as such, the general applicability is reduced. As in the case of many other multicomponent thermodynamic relations, other composition restrictions may be imposed and different results obtained.

13.6 Surface Tension–Composition Relationships

Euler integration of Eq. (13-14) allows one to define a Gibbs energy for the surface phase.

$$\underline{G}^\sigma = \underline{U}^\sigma - T\underline{S}^\sigma + P\underline{V}^\sigma = \sigma A + \sum_{j=1}^{n} \mu_j N_j^\sigma \tag{13-66}$$

Let us restrict our presentation to a binary system of 1 and 2. Then, if Eq. (13-66) is divided by the total moles in the surface layer, $N^\sigma = N_1^\sigma + N_2^\sigma$,

$$G^\sigma = \frac{\underline{G}^\sigma}{N^\sigma} = \sigma\Omega + \mu_1 x_1^\sigma + \mu_2 x_2^\sigma \tag{13-67}$$

where G^σ is the Gibbs energy *per mole* in the surface layer and x_1^σ, x_2^σ are the mole fractions of 1 and 2 in this layer. Ω is the area per mole and is often written

$$\Omega = x_1^\sigma \bar{\Omega}_1 + x_2^\sigma \bar{\Omega}_2 \tag{13-68}$$

$\bar{\Omega}_1$ and $\bar{\Omega}_2$ are partial molar areas for components 1 and 2. Substituting Eq. (13-68) into Eq. (13-67), we obtain

$$G^\sigma = x_1^\sigma(\mu_1 + \sigma\bar{\Omega}_1) + x_2^\sigma(\mu_2 + \sigma\bar{\Omega}_2) \tag{13-69}$$

Next, we can express the *surface* Gibbs energies for *pure* components 1 and 2 at the same temperature as

$$\begin{aligned} G_1^\sigma &= \mu_1^\circ + \sigma_1\Omega_1 \\ G_2^\sigma &= \mu_2^\circ + \sigma_2\Omega_2 \end{aligned} \tag{13-70}$$

In Eq. (13-70), μ_1° and μ_2° refer to chemical potentials of pure 1 and 2, σ_1 and σ_2 are the pure-component surface tensions, and Ω_1 and Ω_2 are the pure-component surface areas per mole of material.

With Eqs. (13-69) and (13-70), we can form a mixing function,

$$\begin{aligned} \Delta G^\sigma = G^\sigma - x_1^\sigma G_1^\sigma - x_2^\sigma G_2^\sigma &= x_1^\sigma[(\mu_1 + \sigma\bar{\Omega}_1) - (\mu_1^\circ + \sigma_1\Omega_1)] \\ &\quad + x_2^\sigma[(\mu_2 + \sigma\bar{\Omega}_2) - (\mu_2^\circ + \sigma_2\Omega_2)] \end{aligned} \tag{13-71}$$

If we then write the analogous equation for the Gibbs energy mixing function for the bulk liquid phase,

$$\begin{aligned} \Delta G &= x_1(\mu_1 - \mu_1^\circ) + x_2(\mu_2 - \mu_2^\circ) \\ &= x_1 RT \ln a_1 + x_2 RT \ln a_2 \end{aligned} \tag{13-72}$$

Comparing Eqs. (13-71) and (13-72) suggests that it might be convenient to define "surface chemical potentials" as follows:

$$\begin{aligned} \zeta_1 &\equiv \mu_1 + \sigma\bar{\Omega}_1 \\ \zeta_2 &\equiv \mu_2 + \sigma\bar{\Omega}_2 \\ \zeta_1^\circ &\equiv \mu_1^\circ + \sigma_1\Omega_1 \\ \zeta_2^\circ &\equiv \mu_2^\circ + \sigma_2\Omega_2 \end{aligned} \tag{13-73}$$

and

$$\begin{aligned} \zeta_1 - \zeta_1^\circ &= RT \ln a_1^\sigma = RT \ln \gamma_1^\sigma x_1^\sigma \\ \zeta_2 - \zeta_2^\circ &= RT \ln a_2^\sigma = RT \ln \gamma_2^\sigma x_2^\sigma \end{aligned} \tag{13-74}$$

Equation (13-74) then allows us to introduce surface activities and activity coefficients in terms of surface chemical potentials.

Using Eqs. (13-73) and (13-74) for component 1, and noting that $\mu_1 - \mu_1^\circ = RT \ln a_1 = RT \ln \gamma_1 x_1$, we get

$$\begin{aligned} \zeta_1 - \zeta_1^\circ &= (\mu_1 - \mu_1^\circ) + (\sigma\bar{\Omega}_1 - \sigma_1\Omega_1) = RT \ln \gamma_1^\sigma x_1^\sigma \\ &= RT \ln \gamma_1 x_1 + (\sigma\bar{\Omega}_1 - \sigma_1\Omega_1) \end{aligned} \tag{13-75}$$

Then

$$x_1^\sigma = \frac{x_1\gamma_1}{\gamma_1^\sigma} \exp \frac{\sigma\bar{\Omega}_1 - \sigma_1\Omega_1}{RT} \tag{13-76}$$

In Eq. (13-76), the surface composition of component 1 is related to the bulk composition (x_1) and activity coefficient (γ_1). The surface properties involved are γ_1^σ, σ, σ_1, $\bar{\Omega}_1$, and Ω_1. Writing the comparable equation for component 2 and adding,

$$x_1^\sigma + x_2^\sigma = 1 = \frac{x_1\gamma_1}{\gamma_1^\sigma} \exp \frac{\sigma\bar{\Omega}_1 - \sigma_1\Omega_1}{RT} + \frac{x_2\gamma_2}{\gamma_2^\sigma} \exp \frac{\sigma\bar{\Omega}_2 - \sigma_2\Omega_2}{RT} \tag{13-77}$$

Before discussing the use of Eq. (13-77), let us simplify it to illustrate an important point. Assume ideal solutions in both the bulk and surface layers and let $\bar{\Omega}_1 = \bar{\Omega}_2 = \Omega_1 = \Omega_2 = \Omega$; then, after rearrangement and expansion, we get

$$\sigma = x_1\sigma_1 + x_2\sigma_2 - \frac{\Omega}{2RT} x_1 x_2 (\sigma_1 - \sigma_2)^2 \tag{13-78}$$

With an estimate of Ω, the mixture surface tension may then be determined. Also, one should note that Eq. (13-78) predicts that σ is smaller than the mole fraction (bulk liquid) average of the pure-component surface tensions.

Returning to Eq. (13-76) or (13-77), these were used by Sprow and Prausnitz[8] to estimate surface compositions and mixture surface tensions for simple liquid mixtures. In this case the surface was modeled as a regular solution to obtain surface activity coefficients. Also, surface areas were correlated with liquid volumes raised to the two-thirds power.

A different approach to estimate surface compositions uses the Gibbs–

[8] F. B. Sprow and J. M. Prausnitz, *Trans. Faraday Soc.*, **62**, 1105 (1966).

Duhem equation for the surface layer [Eq. (13-50)].[9] For a binary of 1 and 2 at constant temperature and pressure,

$$A\,d\sigma + N_1^\sigma\,d\mu_1 + N_2^\sigma\,d\mu_2 = 0 \tag{13-79}$$

Dividing by N^σ, the moles in the surface layer, we obtain

$$\Omega\,d\sigma + x_1^\sigma\,d\mu_1 + x_\sigma^2\,d\mu_2 = 0 \tag{13-80}$$

where $\Omega = A/N^\sigma$ and $x_1^\sigma = N_1^\sigma/N^\sigma$. From the Gibbs–Duhem equation written for the bulk liquid, at constant temperature and pressure,

$$x_1\,d\mu_1 + x_2\,d\mu_2 = 0 \tag{13-81}$$

Eliminating $d\mu_2$ between Eqs. (13-80) and (13-81) and making use of Eq. (13-68) yields

$$x_1^\sigma = \frac{x_1 - x_2\bar{\Omega}_2(d\sigma/d\mu_1)}{1 + x_2(\bar{\Omega}_1 - \bar{\Omega}_2)(d\sigma/d\mu_1)} \tag{13-82}$$

As in Eq. (13-76), partial areas $\bar{\Omega}_1$ and $\bar{\Omega}_2$ are necessary to estimate surface compositions when employing Eq. (13-82). The derivative $(d\sigma/d\mu_1)$ could be written as

$$\frac{d\sigma}{d\mu_1} = \frac{d\sigma/dx_1}{d\mu_1/dx_1} = \frac{d\sigma/dx_1}{RT\,d\ln(\gamma_1 x_1)/dx_1} \tag{13-83}$$

and evaluated from data showing how the mixture surface tension and bulk-liquid activity coefficients vary with composition.

Example 13.4

Estimate the ethanol surface mole fraction in a 20 mole% ethanol–water solution at 298 K. Some mixture data are given below.

$T = 298$ K

Mole fraction ethanol	Surface tension (N/m × 10³)	Ethanol activity coefficient	Liquid molar volume (cm³/mol)
1.0	22.0	1.0	58.7
0.9	22.6	1.0	54.3
0.8	23.2	1.02	49.9
0.7	23.85	1.06	45.4
0.6	24.6	1.13	41.0
0.5	25.4	1.25	37.7
0.4	26.35	1.45	33.1
0.3	27.6	1.76	29.1
0.2	29.7	2.27	25.4
0.1	36.6	3.02	21.6
0.04	47.9	3.42	19.5
0	72.2	—	18.05

[9] Y.-P. Shih and S.-A. Chen, *AIChE J.*, **14**, 973 (1968); E. A. Guggenheim, *Thermodynamics*, 4th ed. (Amsterdam: North-Holland, 1959), pp. 194–201, 261–274, 296–299.

Solution

Equation (13-82) is used. To estimate the partial molar areas for ethanol and water, the relation given by Sprow and Prausnitz[10] is used:

$$\bar{\Omega}_j = \left(\frac{V_j^\sigma}{N_0}\right)^{2/3} \times N_0$$

where $\bar{V}_j^\sigma$ is the partial molar volume of component j in the surface layer. We will assume that this value is the same as in the bulk phase and, for ethanol mole fractions greater than about 0.4, the molar volume data show that $\bar{V}_{alc} = 58.7\ cm^3/mol$ and $\bar{V}_{H_2O} = 15.8\ cm^3/mol$. Thus,

$$\Omega_{alc} = \left(\frac{58.7}{6.023 \times 10^{23}}\right)^{2/3} (6.023 \times 10^{23})$$

$$= 1.28 \times 10^9\ cm^2/mol = 1.28 \times 10^5\ m^2/mol$$

$$\Omega_{H_2O} = 5.3 \times 10^8\ cm^2/mol = 5.3 \times 10^4\ m^2/mol$$

To determine $(d\sigma/d\mu_{alc})$, this derivative is written as $(1/RT)[d\sigma/d\ln(\gamma x)_{alc}]$ and σ plotted versus $\ln(\gamma x)_{alc}$. The slope when $x_{alc} = 0.2$ is $-1.6 \times 10^{-2}\ N/m^2$. Therefore,

$$\frac{d\sigma}{d\mu_{alc}} = \frac{-1.6 \times 10^2}{(8.314)(298)} = -6.4 \times 10^{-6}\ mol/m^2$$

With Eq. (13-82),

$$x_{alc}^\sigma = \frac{0.2 - (0.8)(5.3 \times 10^4)(-6.4 \times 10^{-6})}{1 + (0.8)(1.28 \times 10^5 - 5.3 \times 10^4)(-6.4 \times 10^{-6})}$$

$$= 0.76$$

It is interesting to note that if Eq. (13-76) is used to estimate x_{alc}^σ, and γ_{alc}^σ is equated to γ_{alc} in the data table at $x_{alc} \sim 0.76$, then $\gamma_{alc}^\sigma = 1.04$.

$$x_{alc}^\sigma = \frac{(0.2)(2.27)}{1.04} \exp\left[\frac{(29.7 - 22.0)(10^{-3})(1.28 \times 10^5)}{(8.314)(298)}\right]$$

$$= 0.65$$

The agreement between the two methods to estimate x_{alc}^σ is not particularly good. In Eq. (13-76), due to the exponential term, the calculated value of x_{alc}^σ is quite sensitive to the surface tensions and areas used in the computation.

13.7 Nucleation

Nucleation refers to the birth process of a new phase. It obviously has many important ramifications in science and engineering.

As is well known, if a homogeneous, particle-free, pure vapor is cooled until the pressure is equal to the vapor pressure, thermodynamics would indicate that a liquid (or solid) phase could form. The actual facts show that this does not occur. In such a case, the temperature must be decreased below the normal dew-point temperature before the second phase appears. Such undercooling is,

[10] *Op. cit.*

however, not found if the gas contains many "dust-like" particles or if the gas were "seeded" with a small quantity of the liquid (or solid) phase when the dew point was reached. Similar statements apply to the precipitation of solids from liquids or the superheating of liquids.

The reason for these phenomena lies in the fact, developed in Section 13.3, that small "nuclei" of the new phase have higher vapor pressures (or more accurately, higher chemical potentials) than those of a bulk, planar phase. These nuclei will only be stable when present in a supersaturated mother phase.

The equilibrium and stability criteria that are applicable to the formation of a new phase have already been given in Eqs. (13-9) and (13-10); that is, at equilibrium, the availability function, $\underline{\mathcal{V}}$, is either a minimum for stable equilibrium or a maximum for unstable equilibrium.

Let us assume that we have initially a β-free phase, α, which is always at T and P as dictated by isothermal and isobaric reservoirs (see Figure 13.2). We shall allow the formation of a β phase but at the same time insist that phase α remain at T and P. The moles in the α-β system are constant. We wish to determine the change in availability during this process. From the definition of $\underline{\mathcal{V}}$ in Eq. (13-8),

$$\Delta\underline{\mathcal{V}} = \underline{\mathcal{V}} - \underline{\mathcal{V}}_i \tag{13-84}$$

$$\begin{aligned}\Delta\underline{\mathcal{V}} = {} & (\underline{U}^{\alpha} + P\underline{V}^{\alpha} - T\underline{S}^{\alpha}) + (\underline{U}^{\beta} + P\underline{V}^{\beta} - T\underline{S}^{\beta}) \\ & + (\underline{U}^{\sigma} + P\underline{V}^{\sigma} - T\underline{S}^{\sigma}) - (\underline{U}_i^{\alpha} + P\underline{V}_i^{\alpha} - T\underline{S}_i^{\alpha})\end{aligned} \tag{13-85}$$

Now

$$N_i^{\alpha} = N^{\alpha} + N^{\beta} + N^{\sigma} \tag{13-86}$$

and

$$\begin{aligned} U^{\alpha} &= TS^{\alpha} - PV^{\alpha} + \mu^{\alpha} \\ U^{\beta} &= TS^{\beta} - P^{\beta}V^{\beta} + \mu^{\beta} \\ U^{\sigma} &= T\frac{\underline{S}^{\sigma}}{N^{\sigma}} - P^{\sigma}\frac{\underline{V}^{\sigma}}{N^{\sigma}} + \sigma\frac{A}{N^{\sigma}} + \mu^{\sigma} \end{aligned} \tag{13-87}$$

so that substitution of Eqs. (13-86) and (13-87) into Eq. (13-85) yields

$$\begin{aligned}\Delta\underline{\mathcal{V}} = {} & N^{\beta}[(\mu^{\beta} - \mu^{\alpha}) + (P - P^{\beta})V^{\beta}] + \sigma A \\ & + N^{\sigma}[(\mu^{\sigma} - \mu^{\alpha}) + (P - P^{\sigma})V^{\sigma}]\end{aligned} \tag{13-88}$$

Equation (13-88) is, of course, very similar to Eq. (13-15); the principal difference is that in the former we assumed that the temperatures were everywhere equal in the final system. A finite, rather than infinitesimal change was also proposed.

If we proceed one step further and assume that the final state is an equilibrium state, $\Delta\underline{\mathcal{V}}$ equals the change in availability when a homogeneous phase α changes to an α-β system with small fragments of the β phase in equilibrium with the residual α phase. Also, at this terminal state, as we have shown in Eqs. (13-19) and (13-20), there is equality of chemical potentials throughout the system, so that Eq. (13-88) simplifies to

$$\Delta\underline{\mathcal{V}} = N^{\beta}(P - P^{\beta})V^{\beta} + \sigma A \tag{13-89}$$

where we have also neglected the very small contribution due to $N^\sigma(P - P^\sigma)V^\sigma$. If phase β is a sphere, then from Eq. (13-23) with $A = 4\pi r^2$ and $N^\beta = (4/3)\pi r^3/V^\beta$,

$$\Delta \underline{\mathcal{V}} = \frac{4}{3}\pi\sigma r^2 = \frac{\frac{16}{3}\pi\sigma^3}{(P^\beta - P)^2} \tag{13-90}$$

Expressions for $(P^\beta - P)$ that were derived in Section 13.3 may be used to determine $\Delta\underline{\mathcal{V}}$ in terms of P^α and other system variables.

Relative to the question of stability, we note from Eq. (13-88) that

$$\Delta\underline{\mathcal{V}} = f(N^\beta, r)$$

since $\underline{V}^\beta$ and A depend on r. If we expand the availability in a Taylor series around the state of equilibrium, we get

$$\underline{\mathcal{V}} - \underline{\mathcal{V}}_{\text{eq}} = \mathcal{V}_r\,\delta r + \mathcal{V}_N\,\delta N^\beta + \mathcal{V}_{rr}\,\delta r^2 + 2\mathcal{V}_{rN}\,\delta r\,\delta N^\beta + \mathcal{V}_{NN}\,\delta N^{\beta^2} + \cdots \tag{13-91}$$

where we have adopted a shorthand notation for partial derivatives (e.g., $\mathcal{V}_r = (\partial\underline{\mathcal{V}}/\partial r)_{N\beta}$, $\mathcal{V}_{NN} = (\partial^2\underline{\mathcal{V}}/\partial N^{\beta^2})_r$, etc.). The first-order terms are zero at equilibrium; that is, from Eq. (13-88), neglecting the surface phase since it contributes negligibly, and noting that μ^α, P, and σ are constants, we obtain

$$\mathcal{V}_r = N^\beta\left(\frac{\partial\mu^\beta}{\partial r}\right)_{N\beta} - \frac{4}{3}\pi r^3\left(\frac{\partial P^\beta}{\partial r}\right)_{N\beta} + (P - P^\beta)(4\pi r^2) + 8\pi\sigma r \tag{13-92}$$

The first two terms cancel since

$$\left(\frac{\partial\mu^\beta}{\partial r}\right)_{N\beta} = V^\beta\left(\frac{\partial P^\beta}{\partial r}\right)_{N\beta} = \frac{\frac{4}{3}\pi r^3}{N^\beta}\left(\frac{\partial P^\beta}{\partial r}\right)_{N\beta} \tag{13-93}$$

Thus,

$$\mathcal{V}_r = (P - P^\beta)(4\pi r^2) + 8\pi\sigma r \tag{13-94}$$

and, as expected, *at equilibrium*, using Eq. (13-23), we have

$$\mathcal{V}_r = 0$$

Next, we will show that $\mathcal{V}_N$ is also zero at equilibrium. Beginning again with Eq. (13-88), we get

$$\mathcal{V}_N = (\mu^\beta - \mu^\alpha) + N^\beta\left(\frac{\partial\mu^\beta}{\partial N^\beta}\right)_r - \frac{4}{3}\pi r^3\left(\frac{\partial P^\beta}{\partial N^\beta}\right)_r \tag{13-95}$$

but

$$\left(\frac{\partial\mu^\beta}{\partial N^\beta}\right)_r = V^\beta\left(\frac{\partial P^\beta}{\partial N^\beta}\right)_r = \frac{\frac{4}{3}\pi r^3}{N^\beta}\left(\frac{\partial P^\beta}{\partial N^\beta}\right)_r \tag{13-96}$$

Therefore,

$$\mathcal{V}_N = \mu^\beta - \mu^\alpha = 0 \tag{13-97}$$

by Eq. (13-19). Next, considering the second-order terms, they may be rearranged by forming a sum of squares:

$$\underline{\mathcal{V}} - \underline{\mathcal{V}}_{\text{eq}} = \mathcal{V}_{NN}\left(\delta N^\beta + \frac{\mathcal{V}_{rN}}{\mathcal{V}_{NN}}\,\delta r\right)^2 + \frac{(\mathcal{V}_{NN}\mathcal{V}_{rr} - \mathcal{V}_{rN}^2)(\delta r)^2}{\mathcal{V}_{NN}} \tag{13-98}$$

We need to develop expressions for $\mathcal{V}_{NN}$, $\mathcal{V}_{rN}$, and $\mathcal{V}_{rr}$.

$\mathcal{V}_{NN}$: Beginning with Eq. (13-97) and differentiating with respect to N^β, with μ^α a constant,

$$\mathcal{V}_{NN} = \left(\frac{\partial \mu^\beta}{\partial N^\beta}\right)_r = \underline{V}^\beta\left(\frac{\partial P^\beta}{\partial N^\beta}\right)_r \tag{13-99}$$

To obtain a more convenient expression for $\mathcal{V}_{NN}$, since phase β is at constant T,

$$P^\beta = f(N^\beta, \underline{V}^\beta)$$

Applying Euler's theorem yields

$$0 = \underline{V}^\beta\left(\frac{\partial P^\beta}{\partial \underline{V}^\beta}\right)_{N\beta} + N^\beta\left(\frac{\partial P^\beta}{\partial N^\beta}\right)_{\underline{V}\beta} \tag{13-100}$$

Let

$$\kappa_T = -\left(\frac{\partial \ln \underline{V}}{\partial P}\right)_{T,N} \tag{13-101}$$

Then, with Eqs. (13-100), (13-101), and $\underline{V}^\beta = \frac{4}{3}\pi r^3$,

$$\mathcal{V}_{NN} = \frac{\frac{4}{3}\pi r^3}{\kappa_T N^{\beta 2}} \tag{13-102}$$

$\mathcal{V}_{rr}$: Differentiating Eq. (13-94) with respect to r, and using Eq. (13-23), we get

$$\mathcal{V}_{rr} = -8\pi\sigma - \frac{16\pi\sigma^2}{(P^\beta - P)^2}\left(\frac{\partial P^\beta}{\partial s}\right)_{N\beta} \tag{13-103}$$

But

$$\left(\frac{\partial P^\beta}{\partial r}\right)_{N\beta} = \left(\frac{\partial P^\beta}{\partial \underline{V}^\beta}\right)_{N\beta}\left(\frac{\partial \underline{V}^\beta}{\partial r}\right)_{N\beta} = -\frac{3}{\kappa_T r} \tag{13-104}$$

So

$$\mathcal{V}_{rr} = \frac{24\pi\sigma}{(P^\beta - P)\kappa_T} - 8\pi\sigma = \frac{12\pi r}{\kappa_T} - 8\pi\sigma \tag{13-105}$$

$\mathcal{V}_{rN}$: Beginning with Eq. (13-94) and differentiating with respect to N^β, we have

$$\mathcal{V}_{rN} = -4\pi r^2\left(\frac{\partial P^\beta}{\partial N^\beta}\right)_r = -\frac{4\pi r^2}{\kappa_T N^\beta} \tag{13-106}$$

where Eqs. (13-100) and (13-101) have also been used.

Clearly, $\mathcal{V}_{NN}$ is positive since $\kappa_T > 0$. Thus, the stability of our system is determined solely by the coefficient $(\mathcal{V}_{NN}\mathcal{V}_{rr} - \mathcal{V}_{rN}^2)$ in Eq. (13-98).

$$\mathcal{V}_{NN}\mathcal{V}_{rr} - \mathcal{V}_{rN}^2 = \frac{\frac{4}{3}\pi r^3}{\kappa_T N^{\beta 2}}\left(\frac{12\pi r}{\kappa_T} - 8\pi\sigma\right) - \frac{16\pi^2 r^4}{\kappa_T^2 N^{\beta 2}} = \frac{(-8\pi\sigma)(\frac{4}{3}\pi r^3)}{\kappa_T N^{\beta 2}} \tag{13-107}$$

Also,

$$\frac{\mathcal{V}_{NN}\mathcal{V}_{rr} - \mathcal{V}_{rN}^2}{\mathcal{V}_{NN}} = -8\pi\sigma < 0 \tag{13-108}$$

The interpretation of this result is that a phase β embryo, in equilibrium with a mother phase α, is in a state of *unstable* equilibrium.

In this examination, the formation of a new phase β from a mother phase α actually proceeds by increasing the availability of a system—a progression that is counter to our intuition and to usual trends in thermodynamics. Normally, systems seek a state of lowest availability. In a molecular sense, by fluctuations in density, small fragments of a new phase are formed from a mother phase α which is in a metastable condition. All these new fragments are unstable and disappear until a fragment of a critical size is formed. Then, and only then, can the fragment grow to form a bulk new phase β.

The change in availability to form an equilibrium-sized sphere of phase β [Eq. (13-90)] is often employed in nucleation theory, wherein it is assumed that the rate of nucleation is proportional to exp $(-\Delta\mathcal{U}/RT)$. An interesting study of the superheating of liquids which demonstrates the utility of the availability concept is summarized in the work of Apfel.[11]

PROBLEMS

13.1. In a distillation tray there is considerable spray of very fine liquid droplets. Since small droplets have higher vapor pressures than plane liquid–vapor interfaces, we would like to know whether or not the small spray mists have significantly different relative volatilities than those of plane interfaces. Take, for example, ethanol–water at 298 K. Using the data given below, estimate the relative volatility of alcohol-to-water for 0.1-μm-diameter drops and compare this to the volatility for plane surfaces.

x (mole % ethanol)	p_{water}(Pa)	$p_{ethanol}$(Pa)
0	3170	0.0
10	2890	2370
20	2720	3570
30	2590	4160
40	2450	4560
50	2310	4920
60	2110	5350
70	1770	5850
80	1330	6440
90	730	7110
100	0.0	7870

(a) Liquid composition, 50 mole % water.
(b) Planar vapor–liquid equilibrium at 298 K.
(c) Surface tension of 50 mole % ethanol–water solution at 298 K = 0.0254 N/m.
(d) Densities of liquid at 298 K are given in Example 13.4.
(e) Assume that the surface tension is independent of drop size.
(f) Assume that the partial molar volumes of both components in the liquid are independent of pressure.

[11] R. E. Apfel, "Vapor Cavity Formation in Liquids," *Tech. Memo.* **62** (Cambridge, Mass.: Acoustics Research Laboratory, Harvard University, Feb. 1970); see also *J. Chem. Phys.*, **54**, 62 (1971).

13.2. Estimate the composition of the vapor in small vapor bubbles in equilibrium with a superheated liquid mixture of 95 mole% ethane and 5 mole% n-butane at 270 K and 1 bar pressure. Also, what is the pressure inside the vapor bubble and equilibrium bubble radius? Assume that the liquid mixture forms an ideal solution and that the vapor in the bubble is an ideal-gas mixture. Other data are shown below.

$$T = 270 \text{ K}, \qquad P(\text{liquid}) = 1 \text{ bar}$$

	Ethane	n-Butane
$T_r = T/T_c$	0.884	0.635
P_{vp}, Pa	2.21×10^6	9.12×10^4
V(liquid) (m^3/mol)	7.38×10^{-5}	9.68×10^{-5}

The surface tension of the mixture is estimated to be

$$\sigma = (3.54x_E + 15.3x_B - 5.02x_E x_B)(10^{-3}) \text{ J/m}^2$$

where x_E is the mole fraction ethane and x_B the mole fraction n-butane.

13.3. In the expression for surface work, Eq. (13-1), the pressure–volume term is given as $P\,d\underline{V}$. If one were dealing with a curved surface, the internal pressure differs from the external pressure. Which pressure should one use in Eq. (13-1)—or does it make any difference? Why?

13.4. Derive Eq. (13-41) and state any assumptions made.

13.5. What is the general equation showing the approximate slope of a plot of ln P (equilibrium) versus $1/T$ for small droplets of liquid? For water at 293 K where $\sigma = 0.07275$ N/m, $\Delta H_v = 4.19 \times 10^4$ J/mol, $V^L = 18$ cm^3/mol; for $r = 10^{-7}$ cm, what is the slope in K? How does the heat of vaporization vary with drop size?

13.6. A Ph.D. thesis at Syracuse University carried out by Dr. Fernandez has studied the growth rates of water-ice thin platelets in a subcooled water–salt solution. It was postulated that the platelets grew by extension of scalloped edges of the crystal perpendicular to the basal plane. His physical model studied two-dimensional growth of a single scallop as shown (see Figure P13.6).

The geometric shape of a scallop is believed to be nearly parabolic, but the curvature near the tip is almost constant; thus, it is postulated that Eq. (13-41)

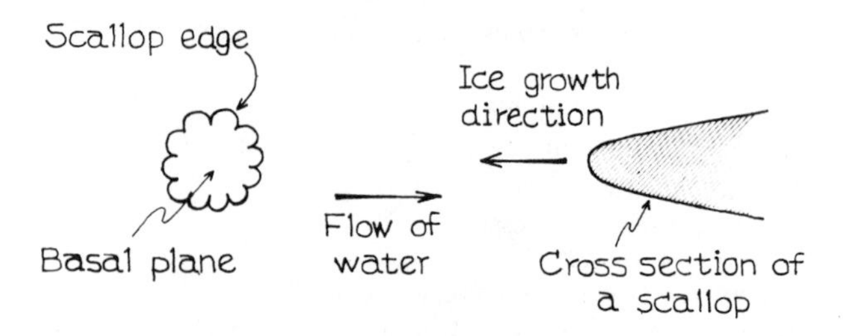

Figure P13.6

may be used to estimate the tip surface temperature. In this case, the temperature at the ice tip T_e is given by

$$T_e = T_m - \frac{2\sigma T_m V^s}{\Delta H_f r}$$

where σ is the interfacial tension between solution and pure water-ice, T_m is the freezing point of pure water, 273.2 K, V^s is the molar volume of ice, ΔH_f is the heat of fusion of ice, and r is the tip curvature. Assume no heat conduction through the ice.

(a) Derive this relationship for a salt solution containing x weight percent salt. What approximations are necessary?

Assuming that the equation given above is applicable, let the growth process be controlled by the rate of heat transfer away from the tip; that is,

$$q = h(T_e - T_\infty) \qquad \text{and} \qquad q = \frac{R\,\Delta H_f}{V^s}$$

where q is the heat flux, h the heat transfer coefficient, T_∞ the bulk temperature in the salt bath, and R the linear growth rate of the tip.

The heat transfer coefficient for flow normal to a parabolic cylinder may be approximated as

$$\frac{hr}{k} = A\left(\frac{Vr}{\nu}\right)^{1/2}$$

where r is the tip radius, k and ν the conductivity and kinematic viscosity of liquid water, V the impinging liquid velocity, and A is a constant.

(b) From these relations and the concept of a steady-state value of R during growth, determine how R depends on the experimental variable $\Delta T \equiv T_m - T_\infty$. For example, does R double if ΔT is doubled?

13.7. The surface tension of very dilute aqueous solutions of butanol has been measured at 298 K and reported by Harkins and Wampler [*J. Am. Chem. Soc.*, **53** (1931), 850]. These authors also report the activities of dilute solutions as a function of molality as follows:

Molality	Activity[a]	Surface tension (N/m × 10^3)
0.00329	0.00328	72.80
0.00658	0.00654	72.26
0.01320	0.01304	70.82
0.0264	0.02581	68.00
0.0536	0.05184	63.14
0.1050	0.09892	56.31
0.2110	0.19277	48.08
0.4330	0.37961	38.87
0.8540	0.71189	29.87

[a]Standard state, $a = m$ as $m \longrightarrow 0$.

Estimate the mole fraction butanol in the interface at a bulk molality of 0.1050. If you wish, you may assume that, on the average, one molecule of butanol occupies about 0.27 nm^2 and one molecule of water, 0.07 nm^2.

13.8. A spherical bubble is enclosed by a thin ethanol film; the bubble is 1 mm in diameter and the film is 1×10^{-4} mm thick. The bubble is in equilibrium with surrounding air, which is at 1 bar and 298 K.

Data:

For ethanol, the surface tension is believed to be insensitive to pressure variations but is related to temperature as

$$\sigma = 0.021\left(1 - \frac{T - 298}{250}\right), \qquad T \text{ in K, } \sigma \text{ in N/m}$$

The surface heat capacity is a constant:

$$C_A = T\left(\frac{\partial S}{\partial T}\right)_A = 2.72 \text{ J/g K}$$

The density of ethanol is assumed independent of temperature or pressure and equals 0.79 g/cm³.

ΔH_v, ethanol = 930 J/g at 298 K.

(a) Evaluate the surface energy of the ethanol film, $(\partial \underline{U}^\sigma/\partial A)_{P,T}$, J/m².

(b) A small tube is inserted into the bubble and air (at 298 K) blown in so that it is expanded rapidly to a diameter of 3 mm. Neglect any heat or mass transfer from the air to the liquid film and assume that the temperature of the film is uniform throughout. What is the temperature in this film after the expansion?

13.9. Consider the hypothetical system shown in Figure P13.9. A "microscopic cylinder" with piston is immersed in a constant-temperature bath at $T_0 = 293$ K.

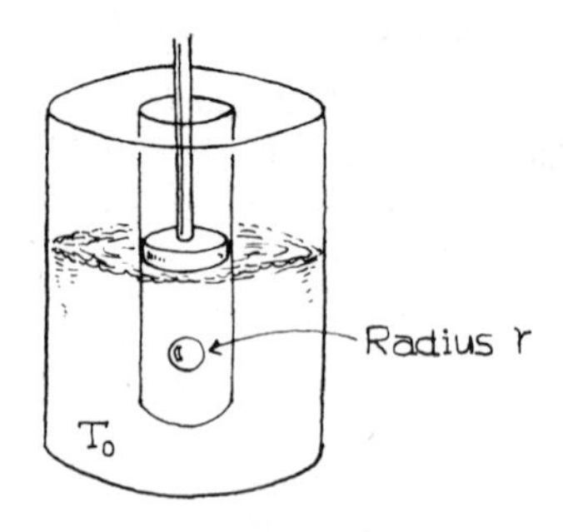

Figure P13.9

In the space enclosed by the piston is a small quantity of helium gas (10^{-16} mole), 5 mole% of which is contained within a bubble of radius r formed from a very thin film of liquid. The liquid has a surface tension of 0.040 N/m at 293 K. Assume that the surface tension is a function only of temperature and that the liquid has a negligible vapor pressure at 293 K. Helium is a perfect gas.

Calculate equilibrium values of the bubble radius, internal and external pressures, and volume fraction gas inside the bubble for two cases:

(a) The piston is held so that the total volume of the helium gas is constant at $\underline{V}_0 = 10^{-12}$ cm³.

(b) The piston is released, permitting the external pressure to go to one bar.

13.10. A fog of liquid droplets of an organic material dispersed in air has been collected in a well-insulated, rigid chamber. Initially, the fog and air are at 20°C. The droplets are allowed to coalescence to a bulk phase.

Data:

Air is an ideal gas with a heat capacity, C_p, of 29.3 J/mol K independent of temperature.

The chamber is 10^{-3} m^3 and the initial pressure is 1 bar. Initially, the fog has a surface area of 10^3 m^3/g and there is 100 g of organic liquid present.

The organic liquid is nonvolatile and may be assumed to have essentially zero values of $\alpha_p = [(1/V)(\partial V/\partial T)_P]$ and $\kappa_T = [(1/V)(\partial V/\partial P)_T]$.

The density of the liquid is 1 g/cm^3.

The heat capacity of the liquid is

$$C_A \equiv T\left(\frac{\partial S}{\partial T}\right)_A = 4.19 \text{ J/g K}$$

where the restriction is at constant area.

At 20°C, $\sigma = 0.0281$ J/m^2

$$\frac{\partial \sigma}{\partial T} = -1.26 \times 10^{-4} \text{ J/m}^2\text{ K} \qquad \text{(not a function of } T\text{)}$$

What is the final equilibrium temperature of the air–organic liquid mixture?

Summary of the Postulates

I. For closed simple systems with given internal restraints, there exist stable equilibrium states that can be characterized completely by two independently variable properties in addition to the masses of the particular chemical species initially charged.

II. In processes for which there is no net effect on the environment, all systems (simple and composite) with given internal restraints will change in such a way that they approach one and only one stable equilibrium state for each simple subsystem. In the limiting condition, the entire system is said to be at equilibrium.

III. For any states, (1) and (2), in which a closed system is at equilibrium, the change of state represented by (1) $\longrightarrow$ (2) and/or the reverse change (2) $\longrightarrow$ (1) can occur by at least one adiabatic process and the adiabatic work interaction between this system and its surroundings is determined uniquely by specifying the end states (1) and (2).

IV. If the sets of systems A, B and A, C each have no heat interaction when connected across nonadiabatic walls, there will be no heat interaction if systems B and C are also so connected.

Mathematical Relations of Functions of State B

Let B be any property, primitive or derived, of a system and let x, y, and z be independently variable properties of a single-component system. (The results can readily be generalized to $n + 2$ independent variables for an n-component system.) Since B is a function of state, a function f exists such that

$$B = f(x, y, z) \tag{B-1}$$

The function $f(x, y, z)$ is usually specified to within an arbitrary constant because derived properties are usually defined in terms of measured *differences* between two states. If the function f is known, differences in the value of B between two stable equilibrium states can be calculated as

$$\Delta B = B_2 - B_1 = f(x_2, y_2, z_2) - f(x_1, y_1, z_1) \tag{B-2}$$

In many cases, we may not know $f(x, y, z)$ explicitly, but we may have the differential form of Eq. (B-1):

$$dB = \left(\frac{\partial f}{\partial x}\right)_{y,z} dx + \left(\frac{\partial f}{\partial y}\right)_{x,z} dy + \left(\frac{\partial f}{\partial z}\right)_{x,y} dz \tag{B-3}$$

If all three partial derivatives are known, $f(x, y, z)$ can be evaluated to within an arbitrary constant by the method of indefinite integrals. Integrating first with respect to x, we obtain

$$f(x, y, z) = \int \left(\frac{\partial f}{\partial x}\right)_{y,z} dx + g(y, z) \tag{B-4}$$

where y and z are held constant in the integration and g is a function of y and z only. If we differentiate Eq. (B-4) with respect to y at constant x and z, and if we equate to the known function $(\partial f/\partial y)_{x,z}$, we get

$$\left(\frac{\partial g}{\partial y}\right)_z = \left(\frac{\partial f}{\partial y}\right)_{x,z} - \frac{\partial}{\partial y}\left[\int \left(\frac{\partial f}{\partial x}\right)_{y,z} dx\right]_{x,z} \tag{B-5}$$

Integrating with respect to y while holding x and z constant, we obtain

$$g(y, z) = \int \left(\frac{\partial f}{\partial y}\right)_{x,z} dy - \int \frac{\partial}{\partial y}\left[\int \left(\frac{\partial f}{\partial x}\right)_{y,z} dx\right]_{x,z} dy + g'(z) \tag{B-6}$$

where g' is a function of z only. Substituting Eq. (B-6) into Eq. (B-4) yields

$$\begin{aligned} f(x, y, z) = &\int \left(\frac{\partial f}{\partial x}\right)_{y,z} dx + \int \left(\frac{\partial f}{\partial y}\right)_{x,z} dy \\ &- \int \frac{\partial}{\partial y}\left[\int \left(\frac{\partial f}{\partial x}\right)_{y,z} dx\right]_{x,z} dy + g'(z) \end{aligned} \tag{B-7}$$

The function $g'(z)$ can be evaluated to within an arbitrary constant by repeating the procedure.

For functions of more than two variables, the method of indefinite integrals is somewhat laborious, and it is only worth the effort if an analytical solution of $f(x, y, z)$ is desired.

In general, we are interested in evaluating numerical differences in B, and a somewhat simpler solution can usually be obtained by integrating Eq. (B-3) over a specific path. Since B is a state function, the value of ΔB will be independent of the path chosen for integration; hence, any convenient path will suffice. One such path is to proceed from x_1, y_1, z_1 to x_2, y_1, z_1 to x_2, y_2, z_1 and then to the final state x_2, y_2, z_2. In this case it can be readily shown that

$$\Delta B = \int_{x_1}^{x_2} \left(\frac{\partial f}{\partial x}\right)_{y_1,z_1} dx + \int_{y_1}^{y_2} \left(\frac{\partial f}{\partial y}\right)_{x_2,z_1} dy + \int_{z_1}^{z_2} \left(\frac{\partial f}{\partial z}\right)_{x_2,y_2} dz \tag{B-8}$$

Finally, if we are faced with the problem of evaluating the difference in a function ϕ given a differential equation of the type

$$d\phi = M\,dx + N\,dy + Q\,dz \tag{B-9}$$

we can use any of the methods described above if it can be shown that ϕ is a state function of the variables, x, y, and z. It can be shown that the necessary and sufficient requirement for this condition is that each of the following equations be satisfied:

$$\begin{aligned} \left(\frac{\partial M}{\partial y}\right)_{x,z} &= \left(\frac{\partial N}{\partial x}\right)_{y,z} \\ \left(\frac{\partial M}{\partial z}\right)_{x,y} &= \left(\frac{\partial Q}{\partial x}\right)_{y,z} \\ \left(\frac{\partial N}{\partial z}\right)_{x,y} &= \left(\frac{\partial Q}{\partial y}\right)_{x,z} \end{aligned} \tag{B-10}$$

Alternatively, if any of equation set (B-10) is not satisfied, $d\phi$ is not an exact differential (e.g., either ϕ is not a state function or ϕ is a state function of variables other than x, y, and z). In this case $\int d\phi$ is called a *line integral* because the value of $\Delta\phi$ will depend on the specific path used for integration.

Some important thermodynamic variables are not state functions or properties (i.e., work and heat interactions). Differentials of these functions are denoted by a bar drawn through the d of the differential sign as $đ\phi$. Such functions are sometimes referred to as *Pfaffians*.

Derivation of Euler's Theorem C

Consider a function $f(a, b, x, y)$ which is homogeneous to the degree h in x and y. By definition, if the variables x and y are each multiplied by a factor k, the value of $f(a, b, kx, ky)$ will be increased by a factor of k^h. Thus, for any value of k, we have

$$f(a, b, X, Y) = k^h f(a, b, x, y) \tag{C-1}$$

where

$$X = kx \qquad \text{and} \qquad Y = ky$$

Equating the total differentials of Eq. (C-1) and treating k as a variable since Eq. (C-1) is valid for all k, we obtain

$$\begin{aligned} &\frac{\partial}{\partial a}[f(a, b, X, Y)]_{b,X,Y}\, da + \frac{\partial}{\partial b}[f(a, b, X, Y)]_{a,X,Y}\, db \\ &\quad + \frac{\partial}{\partial X}[f(a, b, X, Y)]_{a,b,Y}\, dX + \frac{\partial}{\partial Y}[f(a, b, X, Y)]_{a,b,X}\, dY \\ &= (k^h)\frac{\partial}{\partial a}[f(a, b, x, y)]_{b,x,y}\, da + (k^h)\frac{\partial}{\partial b}[f(a, b, x, y)]_{a,x,y}\, db \\ &\quad + (k^h)\frac{\partial}{\partial x}[f(a, b, x, y)]_{a,b,y}\, dx + (k^h)\frac{\partial}{\partial y}[f(a, b, x, y)]_{a,b,x}\, dy \\ &\qquad + (hk^{h-1})[f(a, b, x, y)]\, dk \end{aligned} \tag{C-2}$$

but

$$dX = k\, dx + x\, dk \qquad \text{and} \qquad dY = k\, dy + y\, dk \tag{C-3}$$

Substituting Eq. (C-3) into Eq. (C-2), and collecting terms, we obtain

$$\begin{aligned}\left\{\frac{\partial}{\partial a}[f(a, b, X, Y)] - (k^h)\frac{\partial}{\partial a}[f(a, b, x, y)]\right\} da \\ + \left\{\frac{\partial}{\partial b}[f(a, b, X, Y)] - (k^h)\frac{\partial}{\partial b}[f(a, b, x, y)]\right\} db \\ + \left\{(k)\frac{\partial}{\partial X}[f(a, b, X, Y)] - (k^h)\frac{\partial}{\partial x}[f(a, b, x, y)]\right\} dx \\ + \left\{(k)\frac{\partial}{\partial Y}[f(a, b, X, Y)] - (k^h)\frac{\partial}{\partial y}[f(a, b, x, y)]\right\} dy \\ + \left\{(x)\frac{\partial}{\partial X}[f(a, b, X, Y)] + (y)\frac{\partial}{\partial Y}[f(a, b, X, Y)] \right. \\ \left. - hk^{h-1}f(a, b, x, y)\right\} dk = 0\end{aligned} \tag{C-4}$$

Since a, b, x, y, and k are independent, Eq. (C-4) is valid only if the coefficients of da, db, dx, dy, and dk are each zero. Thus,

$$\frac{\partial}{\partial a}[f(a, b, X, Y)] = (k^h)\frac{\partial}{\partial a}[f(a, b, x, y)] \tag{C-5}$$

$$\frac{\partial}{\partial b}[f(a, b, X, Y)] = (k^h)\frac{\partial}{\partial b}[f(a, b, x, y)] \tag{C-6}$$

$$\frac{\partial}{\partial X}[f(a, b, X, Y)] = (k^{h-1})\frac{\partial}{\partial x}[f(a, b, x, y)] \tag{C-7}$$

$$\frac{\partial}{\partial Y}[f(a, b, X, Y)] = (k^{h-1})\frac{\partial}{\partial y}[f(a, b, x, y)] \tag{C-8}$$

$$(x)\frac{\partial}{\partial X}[f(a, b, X, Y)] + (y)\frac{\partial}{\partial Y}[f(a, b, X, Y)] = (hk^{h-1})[f(a, b, x, y)] \tag{C-9}$$

Substituting Eqs. (C-7) and (C-8) into (C-9), we obtain

$$(x)\frac{\partial}{\partial x}[f(a, b, x, y)] + (y)\frac{\partial}{\partial y}[f(a, b, x, y)] = h[f(a, b, x, y)] \tag{C-10}$$

Equation (C-10) is the general form of Euler's theorem. Note that it contains terms only in those variables for which f is homogeneous to degree h.

Applications of Euler's Theorem

The thermodynamic functions of interest to us are special cases of homogeneous functions. In particular, all our functions are either homogeneous to the first degree in mass (extensive) or homogeneous to the zeroth degree in mass (intensive). Thus, the arbitrary multiplier, k, will always be n (or $1/n$ as the case may be), and h will be either 1 or zero.

Energy

$$\underline{U} = f(\underline{S}, \underline{V}, N)$$

Since $\underline{U}$ is first order in mass, and since $\underline{S}$, $\underline{V}$, N are all proportional to mass, we have

$$\underline{U}(n\underline{S}, n\underline{V}, nN) = n\underline{U}(\underline{S}, \underline{V}, N)$$

Therefore,

$$x = \underline{S}$$
$$y = \underline{V}$$
$$z = N$$

so that Eq. (C-10) leads to

$$\underline{U} = \left(\frac{\partial \underline{U}}{\partial \underline{S}}\right)_{\underline{V},N} \underline{S} + \left(\frac{\partial \underline{U}}{\partial \underline{V}}\right)_{\underline{S},N} \underline{V} + \left(\frac{\partial \underline{U}}{\partial N}\right)_{\underline{S},\underline{V}} N$$

or

$$\underline{U} = T\underline{S} - P\underline{V} + \mu N$$

Enthalpy

$$\underline{H} = f(\underline{S}, P, N)$$

If we multiply the mass by n, we will increase $\underline{S}$ and N by a factor of n, but P will remain unchanged. That is,

$$\underline{H}(n\underline{S}, P, nN) = n\underline{H}(\underline{S}, P, N)$$

Thus,

$$a = P$$
$$x = \underline{S}$$
$$y = N$$

so that from Eq. (C-10),

$$\underline{H} = \left(\frac{\partial \underline{H}}{\partial \underline{S}}\right)_{P,N} \underline{S} + \left(\frac{\partial \underline{H}}{\partial N}\right)_{\underline{S},P} N$$

or

$$\underline{H} = T\underline{S} + \mu N$$

D

Derivation of the $\mathcal{M}_i$ Determinant at the Critical Point, Eq. (9-44)

We define two determinants:

$$\mathfrak{L}_i = \begin{vmatrix} y^{(i)}_{(i+1)(i+1)} & y^{(i)}_{(i+1)(i+2)} & \cdots & y^{(i)}_{(i+1)(n+1)} \\ y^{(i)}_{(i+2)(i+1)} & & \cdots & y^{(i)}_{(i+2)(n+1)} \\ \vdots & & & \vdots \\ y^{(i)}_{(n+1)(i+1)} & & \cdots & y^{(i)}_{(n+1)(n+1)} \end{vmatrix} \tag{9-19}$$

$$\mathfrak{M}_1 = \begin{vmatrix} y^{(i)}_{(i+1)(i+1)} & y^{(i)}_{(i+1)(i+2)} & \cdots & y^{(i)}_{(i+1)(n+1)} \\ y^{(i)}_{(i+2)(i+1)} & & \cdots & y^{(i)}_{(i+2)(n+1)} \\ \vdots & & & \vdots \\ y^{(i)}_{(n)(i+1)} & & \cdots & y^{(i)}_{(n)(n+1)} \\ \partial\mathfrak{L}_i/\partial x_{i+1} & & \cdots & \partial\mathfrak{L}_i/\partial x_{n+1} \end{vmatrix} \tag{9-45}$$

For a n-component system to exist in the spinoidal curve, $\mathfrak{L}_i = 0, 0 \leq i \leq n$. We wish to show that at the critical point, $\mathfrak{M}_i = 0$.

For $i = n$,

$$\mathfrak{L}_n = y^{(n)}_{(n+1)(n+1)} \tag{D-1}$$

$$\mathfrak{M}_n = \frac{\partial \mathfrak{L}_n}{\partial x_{n+1}} = y^{(n)}_{(n+1)(n+1)(n+1)} \tag{D-2}$$

From Eqs. (9-17) and (9-44) at the critical point, $\mathfrak{L}_n = \mathfrak{M}_n = 0$. Our proof will consist of showing that the ratio $\mathfrak{M}_j/\mathfrak{M}_k$ is a positive, finite number for $0 \leq j$, $k \leq n$. Thus, if $\mathfrak{M}_n = 0$, all lower values of $\mathfrak{M}_i$ are also zero.

Proof. The determinants $\mathcal{L}_0$ and $\mathfrak{M}_0$ are:

$$\mathcal{L}_0 = \begin{vmatrix} y_{11}^{(0)} & y_{12}^{(0)} & \cdots & y_{1(n+1)}^{(0)} \\ y_{21}^{(0)} & y_{22}^{(0)} & \cdots & y_{2(n+1)}^{(0)} \\ \vdots & & & \vdots \\ y_{(n+1)1}^{(0)} & y_{(n+1)2}^{(0)} & \cdots & y_{(n+1)(n+1)}^{(0)} \end{vmatrix} \tag{D-3}$$

$$\mathfrak{M}_0 = \begin{vmatrix} y_{11}^{(0)} & y_{12}^{(0)} & \cdots & y_{1(n+1)}^{(0)} \\ y_{21}^{(0)} & y_{22}^{(0)} & \cdots & y_{2(n+1)}^{(0)} \\ \vdots & & & \vdots \\ y_{(n)1}^{(0)} & y_{(n)2}^{(0)} & \cdots & y_{(n)(n+1)}^{(0)} \\ \partial\mathcal{L}_0/\partial x_1 & \partial\mathcal{L}_0/\partial x_2 & \cdots & \partial\mathcal{L}_0/\partial x_{n+1} \end{vmatrix} \tag{D-4}$$

For both $\mathcal{L}_0$ and $\mathfrak{M}_0$ we (a) factor $y_{11}^{(0)}$ from the first column and (b) for all $j = 2$ to $n + 1$, we multiply the first column by $y_{1(j)}^{(0)}$ and subtract from the jth column.

$$\mathcal{L}_0 = y_{11}^{(0)} \begin{vmatrix} 1 & 0 & \cdots & 0 \\ \dfrac{y_{21}^{(0)}}{y_{11}^{(0)}} & y_{22}^{(0)} - \dfrac{y_{12}^{(0)} y_{12}^{(0)}}{y_{11}^{(0)}} & \cdots & y_{2(n+1)}^{(0)} - \dfrac{y_{12}^{(0)} y_{1(n+1)}^{(0)}}{y_{11}^{(0)}} \\ \vdots & \vdots & & \vdots \\ \dfrac{y_{(n+1)1}^{(0)}}{y_{11}^{(0)}} & & \cdots & y_{(n+1)(n+1)}^{(0)} - \dfrac{y_{1(n+1)}^{(0)} y_{1(n+1)}^{(0)}}{y_{11}^{(0)}} \end{vmatrix} \tag{D-5}$$

$$\mathfrak{M}_0 = y_{11}^{(0)} \begin{vmatrix} 1 & 0 & \cdots & 0 \\ \dfrac{y_{21}^{(0)}}{y_{11}^{(0)}} & y_{22}^{(0)} - \dfrac{y_{12}^{(0)} y_{12}^{(0)}}{y_{11}^{(0)}} & \cdots & y_{2(n+1)}^{(0)} - \dfrac{y_{12}^{(0)} y_{1(n+1)}^{(0)}}{y_{11}^{(0)}} \\ \vdots & \vdots & & \vdots \\ \dfrac{y_{(n)1}^{(0)}}{y_{11}^{(0)}} & y_{(n)2}^{(0)} - \dfrac{y_{1n}^{(0)} y_{12}^{(0)}}{y_{11}^{(0)}} & \cdots & y_{(n)(n+1)}^{(0)} - \dfrac{y_{1(n)}^{(0)} y_{1(n+1)}^{(0)}}{y_{11}^{(0)}} \\ \dfrac{1}{y_{11}^{(0)}} \dfrac{\partial\mathcal{L}_0}{\partial x_1} & \dfrac{\partial\mathcal{L}_0}{\partial x_2} - \dfrac{y_{12}^{(0)}}{y_{11}^{(0)}} \dfrac{\partial\mathcal{L}_0}{\partial x_1} & \cdots & \dfrac{\partial\mathcal{L}_0}{\partial x_{n+1}} - \dfrac{y_{1(n+1)}^{(0)}}{y_{11}^{(0)}} \dfrac{\partial\mathcal{L}_0}{\partial x_1} \end{vmatrix} \tag{D-6}$$

Define a parameter

$${}_j\mathcal{L}_0 \equiv \frac{\partial\mathcal{L}_0}{\partial x_j} - \frac{y_{1(j)}^{(0)}}{y_{11}^{(0)}} \frac{\partial\mathcal{L}_0}{\partial x_1} \tag{D-7}$$

and a derivative operater Z_j,

$$Z_j = \frac{\partial}{\partial x_j} - \frac{y_{1(j)}^{(0)}}{y_{11}^{(0)}} \frac{\partial}{\partial x_1} \tag{D-8}$$

Then

$${}_j\mathcal{L}_0 = Z_j(\mathcal{L}_0) \tag{D-9}$$

or

$$ {}_j\mathfrak{L}_0 = Z_j\left[y_{11}^{(0)}\left(\frac{\mathfrak{L}_0}{y_{11}^{(0)}}\right)\right] \tag{D-10}$$

Expanding, we obtain

$$ {}_j\mathfrak{L}_0 = y_{11}^{(0)} Z_j\left(\frac{\mathfrak{L}_0}{y_{11}^{(0)}}\right) + \left(\frac{\mathfrak{L}_0}{y_{11}^{(0)}}\right) Z_j(y_{11}^{(0)}) \tag{D-11}$$

If we limit our consideration to states on the spinoidal curve, $\mathfrak{L}_i = 0$, and Eq. (D-11) reduces to

$$ {}_j\mathfrak{L}_0 = y_{11}^{(0)} Z_j\left(\frac{\mathfrak{L}_0}{y_{11}^{(0)}}\right) \tag{D-12}$$

The group $\mathfrak{L}_0/y_{11}^{(0)}$ may be obtained from Eq. (D-5). When a derivative operator is applied to a determinant, the result is a sum over all k of the determinant with the operator applied to each element of the kth row.

$$\frac{\mathfrak{L}_0}{y_{11}^{(0)}} = \sum_{k=2}^{n+1} \begin{vmatrix} y_{22}^{(0)} - \dfrac{y_{12}^{(0)}y_{21}^{(0)}}{y_{11}^{(0)}} & \cdots & y_{2(n+1)}^{(0)} - \dfrac{y_{12}^{(0)}y_{(n+1)1}^{(0)}}{y_{11}^{(0)}} \\ \vdots & & \vdots \\ y_{(k-1)2}^{(0)} - \dfrac{y_{12}^{(0)}y_{(k-1)1}^{(0)}}{y_{11}^{(0)}} & \cdots & y_{(k-1)(n+1)}^{(0)} - \dfrac{y_{1(k-1)}^{(0)}y_{(n+1)1}^{(0)}}{y_{11}^{(0)}} \\ Z_j\left[y_{k2}^{(0)} - \dfrac{y_{12}^{(0)}y_{k1}^{(0)}}{y_{11}^{(0)}}\right] & \cdots & Z_j\left[y_{k(n+1)}^{(0)} - \dfrac{y_{1k}^{(0)}y_{(n+1)1}^{(0)}}{y_{11}^{(0)}}\right] \\ y_{(k+1)2}^{(0)} - \dfrac{y_{12}^{(0)}y_{(k+1)1}^{(0)}}{y_{11}^{(0)}} & \cdots & y_{(k+1)(n+1)}^{(0)} - \dfrac{y_{1(k+1)}^{(0)}y_{(n+1)1}^{(0)}}{y_{11}^{(0)}} \\ \vdots & & \vdots \\ y_{(n+1)2}^{(0)} - \dfrac{y_{12}^{(0)}y_{(n+1)1}^{(0)}}{y_{11}^{(0)}} & \cdots & y_{(n+1)(n+1)}^{(0)} - \dfrac{y_{1(n+1)}^{(0)}y_{(n+1)1}^{(0)}}{y_{11}^{(0)}} \end{vmatrix} \tag{D-13}$$

Consider next the mth element in the kth row. Expanding, we obtain

$$\begin{aligned} Z_j\left[y_{km}^{(0)} - \frac{y_{1k}^{(0)}y_{m1}^{(0)}}{y_{11}^{(0)}}\right] &= \frac{\partial}{\partial x_j}\left[\;\cdot\;\right] - \frac{y_{1j}^{(0)}}{y_{11}^{(0)}}\frac{\partial}{\partial x_1}\left[\;\cdot\;\right] \\ &= y_{kmj}^{(0)} - \frac{y_{1k}^{(0)}y_{m1j}^{(0)}}{y_{11}^{(0)}} - \frac{y_{1kj}^{(0)}y_{m1}^{(0)}}{y_{11}^{(0)}} + \frac{y_{1k}^{(0)}y_{m1}^{(0)}y_{11j}^{(0)}}{(y_{11}^{(0)})^2} - \frac{y_{1j}^{(0)}y_{km1}^{(0)}}{y_{11}^{(0)}} \\ &\quad + \frac{y_{1j}^{(0)}y_{11k}^{(0)}y_{m1}^{(0)}}{(y_{11}^{(0)})^2} + \frac{y_{1j}^{(0)}y_{1k}^{(0)}y_{11m}^{(0)}}{(y_{11}^{(0)})^2} - \frac{y_{1j}^{(0)}y_{1k}^{(0)}y_{1m}^{(0)}y_{111}^{(0)}}{(y_{11}^{(0)})^3} \\ &= y_{kmj}^{(0)} - (y_{1k}^{(0)}y_{1mj}^{(0)} + y_{1m}^{(0)}y_{1kj}^{(0)} + y_{1j}^{(0)}y_{1km}^{(0)})(y_{11}^{(0)})^{-1} \\ &\quad + (y_{1k}^{(0)}y_{1m}^{(0)}y_{11j}^{(0)} + y_{11k}^{(0)}y_{1m}^{(0)}y_{ij}^{(0)} + y_{1k}^{(0)}y_{11m}^{(0)}y_{1j}^{(0)})(y_{11}^{(0)})^{-2} \\ &\qquad - (y_{1j}^{(0)}y_{1k}^{(0)}y_{1m}^{(0)}y_{111}^{(0)})(y_{111}^{(0)})^{-3} \\ &= y_{kmj}^{(1)} \qquad \text{[see Table 5.1]} \end{aligned} \tag{D-14}$$

$$ = \frac{\partial y_{km}^{(1)}}{\partial x_j} \tag{D-15}$$

Therefore, the elements in the determinant of Eq. (D-13) of the form

$$Z_j\left[y_{km}^{(0)} - \frac{y_{1k}^{(0)}y_{m1}^{(0)}}{y_{11}^{(0)}}\right]$$

may be replaced by $\partial y_{km}^{(1)}/\partial x_j$. The other elements in Eq. (D-13) can be replaced by terms in $y^{(1)}$ (see Table 5.1). Substituting, with Eq. (D-12),

$$Z_j\left(\frac{\mathfrak{L}_0}{y_{11}^{(0)}}\right) = (y_{11}^{(0)})^{-1}\,{}_j\mathfrak{L}_0$$

$$= \sum_{k=2}^{n+1} \begin{vmatrix} y_{22}^{(1)} & \cdots & y_{2(n+1)}^{(1)} \\ \vdots & & \vdots \\ y_{(k-1)2}^{(1)} & \cdots & y_{(k-1)(n+1)}^{(1)} \\ \partial y_{k2}^{(1)}/\partial x_j & \cdots & \partial y_{(k)(n+1)}^{(1)}/\partial x_j \\ y_{(k+1)2}^{(1)} & \cdots & y_{(k+1)(n+1)}^{(1)} \\ \vdots & & \vdots \\ y_{(n+1)2}^{(1)} & \cdots & y_{(n+1)(n+1)}^{(1)} \end{vmatrix} \tag{D-16}$$

$$= \frac{\partial \mathfrak{L}_1}{\partial x_j} \tag{D-17}$$

Returning to Eq. (D-6), using Table 5.1, we have

$$\mathfrak{M}_0 = y_{11}^{(0)} \begin{vmatrix} y_{22}^{(1)} & \cdots & y_{2(n+1)}^{(1)} \\ \vdots & & \vdots \\ y_{(n)2}^{(1)} & \cdots & y_{(n)(n+1)}^{(1)} \\ {}_2\mathfrak{L}_0 & \cdots & {}_{n+1}\mathfrak{L}_0 \end{vmatrix} \tag{D-18}$$

Then replacing the bottom row with Eq. (D-17)

$$\mathfrak{M}_0 = (y_{11}^{(0)})^2 \begin{vmatrix} y_{22}^{(1)} & \cdots & y_{2(n+1)}^{(1)} \\ \vdots & & \vdots \\ y_{(n)2}^{(1)} & \cdots & y_{(n)(n+1)}^{(1)} \\ \partial \mathfrak{L}_1/\partial x_1 & \cdots & \partial \mathfrak{L}_1/\partial x_{n+1} \end{vmatrix} \tag{D-19}$$

$$= (y_{11}^{(0)})^2 \mathfrak{M}_1 \tag{D-20}$$

By similar reasoning, it can be shown that

$$\mathfrak{M}_r = (y_{(r+1)(r+1)}^{(r)})^2 \mathfrak{M}_{r+1} \tag{D-21}$$

Therefore, if, at the critical state, $\mathfrak{M}_n = 0$, then $\mathfrak{M}_{n-1} = 0$, etc. The critical-state criteria are, then,

$$\mathfrak{L}_i = \mathfrak{M}_i = 0, \qquad 0 \le i \le n \tag{D-22}$$

Index